Lecture Notes in Mathematics

Edited by A. Dold and B. Eckmann
Series: Mathematisches Institut der Universität Bonn
Adviser: F. Hirzebruch

730

Functional Differential Equations and Approximation of Fixed Points

Proceedings, Bonn, July 1978

Edited by
Heinz-Otto Peitgen and Hans-Otto Walther

Springer-Verlag
Berlin Heidelberg New York 1979

Editors

Heinz-Otto Peitgen
Fachbereich Mathematik
Universität Bremen
D-2800 Bremen 33 / FRG

Hans-Otto Walther
Mathematisches Institut
Universität München
D-8000 München 2 / FRG

AMS Subject Classifications (1970): 34 J XX, 34 K XX, 65 D 15, 65 L 15

ISBN 3-540-09518-7 Springer-Verlag Berlin Heidelberg New York
ISBN 0-387-09518-7 Springer-Verlag New York Heidelberg Berlin

Library of Congress Cataloging in Publication Data
Summerschool and Conference on Functional Differential Equations and Approximation of Fixed Points, University of Bonn, 1978.
Functional differential equations and approximation of fixed points.
(Lecture notes in mathematics ; 730)
Bibliograhy: p.
Includes index.
1. Functional differential equations--Congresses. 2. Fixed point theory--Congresses.
3. Approximation theory--Congresses. I. Peitgen, Heinz-Otto, 1945-
II. Walther, Hans-Otto: III. Title. IV. Series:
Lecture notes in mathematics (Berlin) ; 730.
QA3.L28 no. 730 [QA372] 510'.8s [515'.35]
ISBN 0-387-09518-7
79-16224

Printed in Germany

Printing and binding: Beltz Offsetdruck, Hemsbach/Bergstr.
2141/3140-543210

Lecture Notes in Mathematics

Edited by A. Dold and B. Eckmann

Series: Mathematisches Institut der Universität Bonn

Adviser: F. Hirzebruch

730

Functional Differential Equations and Approximation of Fixed Points

Proceedings, Bonn, 1978

Edited by
Heinz-Otto Peitgen and Hans-Otto Walther

Springer-Verlag
Berlin Heidelberg New York

Lecture Notes in Mathematics

For information about Vols. 1–521, please contact your bookseller or Springer-Verlag.

Vol. 522: C. O. Bloom and N. D. Kazarinoff, Short Wave Radiation Problems in Inhomogeneous Media: Asymptotic Solutions. V. 104 pages. 1976.

Vol. 523: S. A. Albeverio and R. J. Høegh-Krohn, Mathematical Theory of Feynman Path Integrals. IV, 139 pages. 1976.

Vol. 524: Séminaire Pierre Lelong (Analyse) Année 1974/75. Edité par P. Lelong. V, 222 pages. 1976.

Vol. 525: Structural Stability, the Theory of Catastrophes, and Applications in the Sciences. Proceedings 1975. Edited by P. Hilton. VI, 408 pages. 1976.

Vol. 526: Probability in Banach Spaces. Proceedings 1975. Edited by A. Beck. VI, 290 pages. 1976.

Vol. 527: M. Denker, Ch. Grillenberger, and K. Sigmund, Ergodic Theory on Compact Spaces. IV, 360 pages. 1976.

Vol. 528: J. E. Humphreys, Ordinary and Modular Representations of Chevalley Groups. III, 127 pages. 1976.

Vol. 529: J. Grandell, Doubly Stochastic Poisson Processes. X, 234 pages. 1976.

Vol. 530: S. S. Gelbart, Weil's Representation and the Spectrum of the Metaplectic Group. VII, 140 pages. 1976.

Vol. 531: Y.-C. Wong, The Topology of Uniform Convergence on Order-Bounded Sets. VI, 163 pages. 1976.

Vol. 532: Théorie Ergodique. Proceedings 1973/1974. Edité par J.-P. Conze and M. S. Keane. VIII, 227 pages. 1976.

Vol. 533: F. R. Cohen, T. J. Lada, and J. P. May, The Homology of Iterated Loop Spaces. IX, 490 pages. 1976.

Vol. 534: C. Preston, Random Fields. V, 200 pages. 1976.

Vol. 535: Singularités d'Applications Differentiables. Plans-sur-Bex. 1975. Edité par O. Burlet et F. Ronga. V, 253 pages. 1976.

Vol. 536: W. M. Schmidt, Equations over Finite Fields. An Elementary Approach. IX, 267 pages. 1976.

Vol. 537: Set Theory and Hierarchy Theory. Bierutowice, Poland 1975. A Memorial Tribute to Andrzej Mostowski. Edited by W. Marek, M. Srebrny and A. Zarach. XIII, 345 pages. 1976.

Vol. 538: G. Fischer, Complex Analytic Geometry. VII, 201 pages. 1976.

Vol. 539: A. Badrikian, J. F. C. Kingman et J. Kuelbs, Ecole d'Eté de Probabilités de Saint Flour V-1975. Edité par P.-L. Hennequin. IX, 314 pages. 1976.

Vol. 540: Categorical Topology, Proceedings 1975. Edited by E. Binz and H. Herrlich. XV, 719 pages. 1976.

Vol. 541: Measure Theory, Oberwolfach 1975. Proceedings. Edited by A. Bellow and D. Kölzow. XIV, 430 pages. 1976.

Vol. 542: D. A. Edwards and H. M. Hastings, Čech and Steenrod Homotopy Theories with Applications to Geometric Topology. VII, 296 pages. 1976.

Vol. 543: Nonlinear Operators and the Calculus of Variations, Bruxelles 1975. Edited by J. P. Gossez, E. J. Lami Dozo, J. Mawhin, and L. Waelbroeck, VII, 237 pages. 1976.

Vol. 544: Robert P. Langlands, On the Functional Equations Satisfied by Eisenstein Series. VII, 337 pages. 1976.

Vol. 545: Noncommutative Ring Theory. Kent State 1975. Edited by J. H. Cozzens and F. L. Sandomierski. V, 212 pages. 1976.

Vol. 546: K. Mahler, Lectures on Transcendental Numbers. Edited and Completed by B. Diviš and W. J. Le Veque. XXI, 254 pages. 1976.

Vol. 547: A. Mukherjea and N. A. Tserpes, Measures on Topological Semigroups: Convolution Products and Random Walks. V, 197 pages. 1976.

Vol. 548: D. A. Hejhal, The Selberg Trace Formula for PSL $(2,\mathbb{R})$. Volume I. VI, 516 pages. 1976.

Vol. 549: Brauer Groups, Evanston 1975. Proceedings. Edited by D. Zelinsky. V, 187 pages. 1976.

Vol. 550: Proceedings of the Third Japan – USSR Symposium on Probability Theory. Edited by G. Maruyama and J. V. Prokhorov. VI, 722 pages. 1976.

Vol. 551: Algebraic K-Theory, Evanston 1976. Proceedings. Edited by M. R. Stein. XI, 409 pages. 1976.

Vol. 552: C. G. Gibson, K. Wirthmüller, A. A. du Plessis and E. J. N. Looijenga. Topological Stability of Smooth Mappings. V, 155 pages. 1976.

Vol. 553: M. Petrich, Categories of Algebraic Systems. Vector and Projective Spaces, Semigroups, Rings and Lattices. VIII, 217 pages. 1976.

Vol. 554: J. D. H. Smith, Mal'cev Varieties. VIII, 158 pages. 1976.

Vol. 555: M. Ishida, The Genus Fields of Algebraic Number Fields. VII, 116 pages. 1976.

Vol. 556: Approximation Theory. Bonn 1976. Proceedings. Edited by R. Schaback and K. Scherer. VII, 466 pages. 1976.

Vol. 557: W. Iberkleid and T. Petrie, Smooth S^1 Manifolds. III, 163 pages. 1976.

Vol. 558: B. Weisfeiler, On Construction and Identification of Graphs. XIV, 237 pages. 1976.

Vol. 559: J.-P. Caubet, Le Mouvement Brownien Relativiste. IX, 212 pages. 1976.

Vol. 560: Combinatorial Mathematics, IV, Proceedings 1975. Edited by L. R. A. Casse and W. D. Wallis. VII, 249 pages. 1976.

Vol. 561: Function Theoretic Methods for Partial Differential Equations. Darmstadt 1976. Proceedings. Edited by V. E. Meister, N. Weck and W. L. Wendland. XVIII, 520 pages. 1976.

Vol. 562: R. W. Goodman, Nilpotent Lie Groups: Structure and Applications to Analysis. X, 210 pages. 1976.

Vol. 563: Séminaire de Théorie du Potentiel. Paris, No. 2. Proceedings 1975–1976. Edited by F. Hirsch and G. Mokobodzki. VI, 292 pages. 1976.

Vol. 564: Ordinary and Partial Differential Equations, Dundee 1976. Proceedings. Edited by W. N. Everitt and B. D. Sleeman. XVIII, 551 pages. 1976.

Vol. 565: Turbulence and Navier Stokes Equations. Proceedings 1975. Edited by R. Temam. IX, 194 pages. 1976.

Vol. 566: Empirical Distributions and Processes. Oberwolfach 1976. Proceedings. Edited by P. Gaenssler and P. Révész. VII, 146 pages. 1976.

Vol. 567: Séminaire Bourbaki vol. 1975/76. Exposés 471–488. IV, 303 pages. 1977.

Vol. 568: R. E. Gaines and J. L. Mawhin, Coincidence Degree, and Nonlinear Differential Equations. V, 262 pages. 1977.

Vol. 569: Cohomologie Etale SGA $4^1/_2$. Séminaire de Géométrie Algébrique du Bois-Marie. Edité par P. Deligne. V, 312 pages. 1977.

Vol. 570: Differential Geometrical Methods in Mathematical Physics, Bonn 1975. Proceedings. Edited by K. Bleuler and A. Reetz. VIII, 576 pages. 1977.

Vol. 571: Constructive Theory of Functions of Several Variables, Oberwolfach 1976. Proceedings. Edited by W. Schempp and K. Zeller. VI. 290 pages. 1977

Vol. 572: Sparse Matrix Techniques, Copenhagen 1976. Edited by V. A. Barker. V, 184 pages. 1977.

Vol. 573: Group Theory, Canberra 1975. Proceedings. Edited by R. A. Bryce, J. Cossey and M. F. Newman. VII, 146 pages. 1977.

Vol. 574: J. Moldestad, Computations in Higher Types. IV, 203 pages. 1977.

Vol. 575: K-Theory and Operator Algebras, Athens, Georgia 1975. Edited by B. B. Morrel and I. M. Singer. VI, 191 pages. 1977.

Vol. 576: V. S. Varadarajan, Harmonic Analysis on Real Reductive Groups. VI, 521 pages. 1977.

Vol. 577: J. P. May, E_∞ Ring Spaces and E_∞ Ring Spectra. IV, 268 pages. 1977.

Vol. 578: Séminaire Pierre Lelong (Analyse) Année 1975/76. Edité par P. Lelong. VI, 327 pages. 1977.

Vol. 579: Combinatoire et Représentation du Groupe Symétrique, Strasbourg 1976. Proceedings 1976. Edité par D. Foata. IV, 339 pages. 1977.

continuation on page 511

Math

Sep

dedicated to Heinz Unger

on occasion of his

65 th birthday

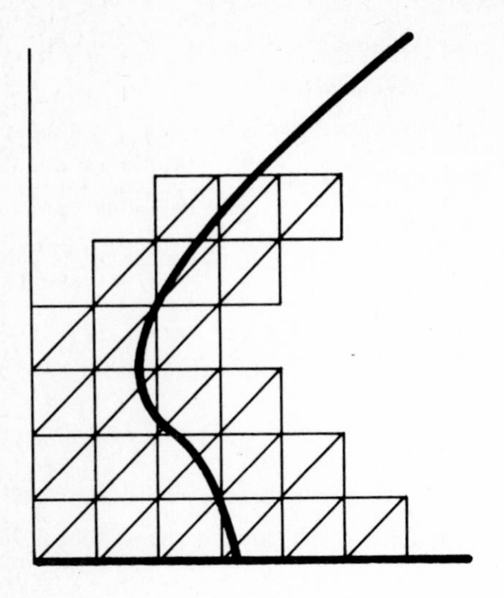

P R E F A C E

This volume comprises the proceedings of the Summerschool and Conference on Functional Differential Equations and Approximation of Fixed Points held at the University of Bonn, July 17 - 22, 1978. The conference was sponsored by *Sonderforschungsbereich 72, Deutsche Forschungsgemeinschaft,* at the *Institut für Angewandte Mathematik der Universität Bonn.*

Morning sessions were occupied by comprehensive survey talks given by

J. C. Alexander	University of Maryland
B. C. Eaves	Stanford University
J. K. Hale	Brown University
J. Mallet-Paret	Brown University
R. D. Nussbaum	Rutgers University
R. Saigal	Northwestern University

In the afternoon there were seminar meetings and invited addresses about recent progress in the two fields. Seminars covering informal discussions and research announcements were conducted by *E. Allgower, B. C. Eaves, J. K. Hale,* and *R. D. Nussbaum.*

The motivation to bring together the two topics in one conference was twofold: First, our group (N. Angelstorf, K. Georg, H. Peters, M. Prüfer, H. W. Siegberg and myself) had made some progress and was interested (cf. publications in these proceedings) in understanding functional differential equations using simplicial fixed point algorithms*. Moreover, we felt that it might be of interest for younger colleagues in this country to find an introduction into these mushrooming fields of current research in numerical and nonlinear analysis. In view of this it was a great pleasure to welcome in the meeting more than 140 participants from 16 different countries.

1978 has been the 50 th anniversary of Sperner's Lemma which was certainly at the heart of many of the results presented. The conference was delighted to have *E. Sperner* as a participant and lecturer.

I wish to express my thanks to the lecturers, to all participants who supported the meeting, to the authors having forwarded their papers and to all those who refereed for this volume.

A particular measure of gratitude is due to *E. Allgower* for his advice during the early planning of the meeting, to *N. Angelstorf, H. Peters, M. Prüfer,* and *H. W. Siegberg* for their indispensable assistance in the preparation and conduction of the conference and this volume, to *H. O. Walther* for his careful cooperation in editing these proceedings, to *S. Hildebrandt* who as chairman of Sonderforschungsbereich 72 encouraged us to keep on with the idea and format of the meeting, and finally to Mrs. *I. Kreuder* for her invaluable help and patience in all secretarial and administrative matters.

February 1979

Heinz-Otto Peitgen
Universität Bremen and
Sonderforschungsbereich 72,
Universität Bonn

---

* The symbol of the conference was designed to express a link between the two topics: it shows simplicial approximation techniques in the study of global bifurcation for functional differential equations.

PARTICIPANTS

D.	ABTS	Aachen	C.	FENSKE	Gießen
J.C.	ALEXANDER	College Park	B.	FIEDLER	Heidelberg
E.	ALLGOWER	Fort Collins	P.M.	FITZPATRICK	College Park
W.	ALT	Heidelberg	D.	FLOCKERZI	Würzburg
H.	AMANN	Kiel	W.	FORSTER	Southampton
J.	ANDRES	Bonn	G.	FOURNIER	Sherbrooke
N.	ANGELSTORF	Bonn	S.	FRYDRYCHOWICZ	Berlin
J.	APPELL	Berlin	M.	FURI	Florenz
O.	ARINO	Pau	J.	GAMST	Bremen
B.	AULBACH	Würzburg	K.	GEORG	Bonn
L.	BÄCKSTRÖM	Växjö	K.P.	HADELER	Tübingen
H.T.	BANKS	Providence	J.K.	HALE	Providence
F.S.	DE BLASI	Florenz	G.	v.HARTEN	Paderborn
R.	BORELLI	Bonn	S.	HEIMSATH	Bremen
B.	BROSOWSKI	Bonn	U.	HELMKE	Bremen
H.	BULEY	Aachen	G.	HETZER	Aachen
J.	CANAVATI	Mexico	D.	HINRICHSEN	Bremen
S.N.	CHOW	East Lansing	W.	HOFMANN	Hamburg
F.	COLONIUS	Bremen	T.	HORION	Aachen
G.	COOPERMAN	Providence	V.J.	ISTRATESCU	Frankfurt
L.	CROMME	Göttingen	J.	IZE	Mexico
N.	DANCER	Armidale	W.	JÄGER	Heidelberg
J.	DANES	Prag	H.	JEGGLE	Berlin
K.	DEIMLING	Paderborn	H.	JÜRGENS	Bremen
D.	DENNEBERG	Bremen	J.L.	KAPLAN	Boston
W.	DESCH	Graz	F.	KAPPEL	Graz
M.L.	DIVICCARO	Neapel	S.	KASPAR	Graz
E.L.	DOZO	Brüssel	R.B.	KELLOGG	College Park
B.	DUPUIS	Louvain-la-Neuve	J.S.	KIM	Seoul
J.	DYSON	Oxford	K.	KIRCHGÄSSNER	Stuttgart
B.C.	EAVES	Stanford	U.	KIRCHGRABER	Zürich
H.	ENGL	Linz	D.	KRÖNER	Aachen
G.	EISEN	Bonn	K.	KUNISCH	Graz
H.	ENGLER	Heidelberg	G.	v.d. LAAN	Amsterdam
C.	FABRY	Louvain-la-Neuve	J.M.	LASRY	Paris

A.	LETTIERI	Neapel
T.Y.	LI	East Lansing
S.	LUCKHAUS	Heidelberg
W.	MACKENS	Bochum
J.	MALLET-PARET	Providence
E.	MALUTA	Mailand
G.	MANCINI	Bologna
S.	MASSA	Mailand
W.	MANTHEY	Bremen
N.	MATZL	Graz
H.	MAURER	Münster
K.	MERTEN	Bremen
H.	MÖNCH	Paderborn
G.	MÜLLER	Aachen
H.F.	MÜNZNER	Bremen
J.A.	NOHEL	Madison
R.D.	NUSSBAUM	New Brunswick
M.E.	PARROTT	Memphis
H.O.	PEITGEN	Bremen
H.	PETERS	Bonn
M.	PRÜFER	Bremen
S.	RANKIN	Morgantown
N.	REIF	Hamburg
M.	REICHERT	Frankfurt
R.	SAIGAL	Evanston
D.	SALAMON	Bremen
D.	SAUPE	Bremen
G.	SEIFERT	Aachen
S.	SESSA	Neapel
R.	SEYDEL	München
H.W.	SIEGBERG	Bonn
G.	SIMON	Graz
H.L.	SMITH	Salt Lake City
J.	SMOLLER	Ann Arbor
D.	SOCOLESCU	Karlsruhe
R.	SOCOLESCU	Karlsruhe
M.	SORG	Stuttgart
E.	SPERNER	Sulzburg-Laufen
K.W.	SCHAAF	Heidelberg
R.	SCHAAF	Heidelberg
B.	SCHAFFRIN	Bonn
W.	SCHAPPACHER	Graz
K.	SCHERER	Bonn
J.	SCHEURLE	Stuttgart
D.	SCHMIDT	Essen
A.	SCHMIEDER	Bremen
K.	SCHMITT	Salt Lake City
E.	SCHOCK	Kaiserslautern
K.	SCHUMACHER	Tübingen
R.	SCHÖNEBERG	Aachen
C.	SCHUPP	München
J.	SPREKELS	Hamburg
U.	STAUDE	Mainz
H.	STEINLEIN	München
A.J.	TALMAN	Amsterdam
H.	THIEME	Münster
M.	TODD	Ithaca
G.	TONAR	Berlin
H.	TROGER	Wien
A.	TROMBA	Santa Cruz
H.	UNGER	Bonn
H.A.	VENOBLER	Portsmouth
A.	VIGNOLI	Cosenza
R.	VILLELA-BRESSAN	Padua
H.	VOSS	Hamburg
H.O.	WALTHER	München
P.	WALTMAN	Iowa City
H.W.	WARBRUCK	Bonn
J.	WEINBERG	Bonn
M.	WILLEM	Louvain-la-Neuve
D.	WILLIAMS	College Park
T.E.	WILLIAMSON	Upper Montclair
C.	ZANCO	Mailand

L E C T U R E S

J. C. ALEXANDER	Numerical continuation methods - continuous and discrete
W. ALT	Periodic solutions of a functional differential equation
N. ANGELSTORF	Multiple periodic solutions for functional differential equations
O. ARINO	Oscillatory solutions of a delay equation
H. T. BANKS	Approximation and control of functional differential equations
S. N. CHOW	1. Fuller index and existence of periodic solutions 2. Some results on nonlinear wave equations
G. COOPERMAN	Properties of α-condensing maps
E. N. DANCER	Bifurcation with symmetries
B. C. EAVES	Solving equations with PL-homotopies
P. M. FITZPATRICK	Global bifurcation
D. FLOCKERZI	Hopf bifurcation where the eigenvalue crosses the axis with zero speed
W. FORSTER	On approximating fixed points in $C[0,1]$
K. P. HADELER	Delay equations in biology
J. K. HALE	1. Dissipative processes 2. Phase space for infinite delays 3. Stability with respect to delays

CONTENTS

NUMERICAL CONTINUATION METHODS AND BIFURCATION

J. C. Alexander

This work was done at the Mathematisches Institut der Universität Bonn where the author was partially supported by SFB 40 and the NSF.

This report should be considered a continuation of [A_1] which itself continues the work of [A-Y]. There, it was shown that the so-called "continuous" ("differentiable" is a more accurate adjective) homotopy methods of Kellogg-Li-Yorke, et al. and the piecewise linear methods of Scarf, Eaves, Saigal, et al. are potentially applicable to a wide variety of problems. (For a survey of these techniques and a list of references, see [A-G]). Indeed, one of the themes of [A_1] and [A-Y] is that there are general topological considerations that logically precede the choice of which type of method to use or how to implement it. Isolating and studying the basic topology allows the formulation of a general framework into which a variety of problems can be fit.

In particular, problems concerning the bifurcation of zeroes of parametrized functions fit into the framework. M. Prüfer [P_1], using some topological ideas of H.-O. Peitgen about Sperner simplices, has implemented a one-parameter bifurcation problem and used it to find bifurcating periodic solutions of functional delay equations. In the one-parameter case, ordinary topological degree detects the bifurcation, and the desired solutions may be located by finding completely labelled codimension one Sperner simplices. The method of finding bifurcating solutions is different than the ones proposed in [A_1] and [A-Y]. It is more related to the "generic" proof [A-Y] of the Rabinowitz bifurcation result. Here we show that the procedure also is valid for multi-parameter bifurcation problems if the bifurcation is detected by certain bifurcation invariants lying in stable homotopy groups of spheres — exactly the invariants needed for the procedure in [A_1] and [A-Y].

One might call the arguments with these bifurcation invariants generalized degree arguments in that they are quite analogous to classical degree proofs, except that invariants more powerful than degree must be used. One of the points of [A_1] and [A-Y] is that if a problem is shown to have a solution by a generalized degree argument, then continuation methods, if they can be implemented, can be used to numerically locate such a solution. On the other hand, other standard methods of proof — in particular, variational or max-min methods — do not guarantee continuation methods can be used. An example due to Mallet-Paret was exhibited in [A_1] to illustrate this point. Here we use that example to reconstruct an example due to Böhme [B] of "bifurcation without branches" to which no continuation method is applicable.

A continuation method applied to a k-dimensional bifurcation problem leads to a k-dimensional "path" to be followed. It was briefly discussed in [A_1] and [A-Y] how one might try to implement following a higher-dimensional manifold in the differentiable case. Here we discuss some general principles that come into play in the simplicial situation.

Following bifurcation branches

The general setup is the following. Let $F: R^k \times R^n \longrightarrow R^n$ be continuous. Regard $\lambda \in R^k$ as a k-dimensional parameter and $u \in R^n$ as a variable. Suppose $F(\lambda, 0) = 0$ for $\lambda \in R^k$. Suppose for some neighborhood N of 0 in R^k that for each compact subset C of $N - \{0\}$, there exists $\varepsilon = \varepsilon(C) > 0$ such that $F(\lambda, u) \neq 0$ if $0 < |u| \leq \varepsilon$ (If the u-derivative, $D_u F(\lambda, 0)$ exists continuously on N, this condition is satisfied if $D_u F(\lambda, 0)$ is non-singular for $\lambda \in N - \{0\}$). Then the origin $(0, 0)$ is a potential bifurcation point for the zeroes of F.

One considers a (k-1)-dimensional sphere S^{k-1} around 0 in N, and using this sphere one defines an element $\gamma = \gamma_F \in \pi^s_{k-1}$, the (k-1)st stable homotopy group of spheres. If $\gamma \neq 0$, global bifurcation is guaranteed. In particular, if $k = 1$, then $\gamma \in \pi^s_0 = Z$ is the difference of the degrees around 0 of $F|(\lambda^+ \times D^n_\varepsilon)$, $F(\lambda^- \times D^n_\varepsilon)$ where $\lambda^\pm$ are in N, $\lambda^- < 0 < \lambda^+$ and D^n_ε is the disk of radius $\varepsilon = \varepsilon(\{\lambda^\pm\})$ around 0 in R^n. This is the Rabinowitz bifurcation result. In [A_1] it was shown that the proof that global bifurcation occurs amounts to a generalized degree argument and it was shown how to apply continuation methods. A "cap" $D^k \times S^{n-1}$ is attached to $S^{k-1} \times S^{n-1} = S^{k-1} \times \partial D^n_\varepsilon$; that is, a $D^k \times S^{n-1}$ is embedded in $R^k \times R^n$ so that its boundary is $S^{k-1} \times S^{n-1}$. The bifurcation result guarantees that the bifurcation set intersects the cap. Moreover, it is guaranteed that as F is continuously varied by a homotopy F_t, $0 \leq t \leq 1$, the solutions on the cap of the equations $F_t = 0$ connect those of $F_0 = 0$ and $F_1 = 0$.

Here we consider a different formulation. Modify F slightly to $\bar{F}$ so that 0 is a regular value (in either the piecewise linear or differentiable sense) of $\bar{F}$. (Alternatively, one can choose a regular value v of F near 0 and work with $F^{-1}(v)$. If we define $\bar{F}(\lambda, u) = F(\lambda, u) - v$, we reduce this version to the first). Then $\bar{F}^{-1}(0)$ is a (piecewise linear or differentiable) manifold. Moreover $\bar{F}^{-1}(0) \cap (S^{k-1} \times D^n_\epsilon) \neq \emptyset$ and can be taken to be a manifold. Using $\bar{F}^{-1}(0) \cap (S^{k-1} \times D^n_\epsilon)$ as "initial conditions", one follows $\bar{F}^{-1}(0)$ towards the bifurcation point. If all goes well, $\bar{F}^{-1}(0)$ will closely follow the bifurcation branch of F. That in fact all does go well is the content of the following two propositions. The first asserts that $\bar{F}^{-1}(0)$ approximates $F^{-1}(0)$, the second asserts $\bar{F}^{-1}(0)$ follows the bifurcation branch.

<u>Proposition.</u> Let U be an open neighborhood of $F^{-1}(0)$ in $R^k \times R^n$. Let B_r be a ball of radius r around the origin in $R^k \times R^n$. Then there exists $\delta > 0$ such that if $|\bar{F}(\lambda, t) - F(\lambda, t)| < \delta$ for $(\lambda, t) \in B_r$ then $(\bar{F}^{-1}(0) \cap B_r) \subset U$.

To state the next proposition, we need to make precise the idea of going in towards the bifurcation point from $S^{k-1} \times D^n_\epsilon$. Let V be an open annulus in N bounded by S^{k-1} and a slightly larger sphere. For each bifurcation point $p = (\lambda, 0)$ of F in B_r with $\lambda \neq 0$ let W_p be a neighborhood of p in $R^k \times R^n$. Let $\delta_0 = \min\{|F(\lambda, u)| \mid \lambda \in S^{k-1}, u \in \partial D^n_\epsilon\}$.

<u>Proposition.</u> Let U and B_r be as above and suppose $\gamma_F \neq 0$. For δ sufficiently small, then some component K of $\bar{F}^{-1}(0) - (V \times D^n_\epsilon)$

with $K \cap D^n_\epsilon \neq \emptyset$ satisfies one (or both) of the following:

a) $K \cap \partial B_r \neq \emptyset$

b) $K \cap W_p \neq \emptyset$ for some p

If the set of p is empty (so that b) is impossible), it suffices to take $\delta < \delta_0$.

Thus K must follow a bifurcation branch of F and it must either contain points of distance r away from the bifurcation point or approximate some other bifurcation point of F.

The proofs of both propositions are straightforward. To prove the first, we consider the compact set $B_r - U$ and note that $F(B_r - U)$ is bounded away from 0 in R^n. Choose δ less than this bound. To prove the second, we note that if $\delta < \delta_0$ then the bifurcation invariant $\gamma_{\bar{F}}$ is defined and $\gamma_{\bar{F}} = \gamma_F$. Thus the standard machinery applies to $\bar{F}$ (see [A_2]), and $\bar{F}^{-1}(0)$ must link $S^{k-1} \times \partial D^n_\epsilon$ in the one-point compactification of $R^k \times R^n$. Conditions a), b) are an interpretation of what linking means.

Remarks. 1. Note that these results are purely topological, and do not involve either a piecewise linear or differentiable structure. Thus either type of numerical method can be used, depending on the particular problem and the user's prejudices.

2. Technically, this is not a continuation result, in that it does not involve modifying a problem with known solution into the given problem. The two approaches to bifurcation — that of [A_1], [A-Y] and the present one — are analogous respectively to the two approaches to Brouwer fixed

point problems involving maps $f\colon D^n \longrightarrow D^n$ — the one involves homotoping f through f_t to a known map and the second involves following an interval that goes from the boundary of D^n to the fixed point set of f.

3. If $F|(S^{k-1} \times D^n_\epsilon)$ already has 0 as a regular value (for example, if $D_u F(\lambda, 0)$ is defined and non-singular for $\lambda \in S^{k-1}$), then for sufficiently small δ, the set $\bar{F}^{-1}(0) \cap (S^{k-1} \times D^n_\epsilon)$ will also be a (slightly jiggled) S^{k-1}. Thus one has nice initial conditions. Alternatively, one could require require that $\bar{F} = F$ on $S^{k-1} \times D^n_\epsilon$. Then of course $\bar{F}^{-1}(0) = S^{k-1} \times \{0\}$.

4. If $k = 1$, ordinary degree is the detecting invariant and using Sperner simplices, the degree can be "kept track of" throughout the numerical procedure. See $[P_1]$ for more details and precision. In particular, one can explicitly see that $\bar{F}^{-1}(0)$ follows the branch of $F^{-1}(0)$ that "carries the invariant". For k larger than 1, it is still true that $\bar{F}^{-1}(0)$ follows the branch that carries the invariant γ_F, but it seems to be exceedingly difficult to recover γ_F from the numerical procedure. In general, Sperner simplices are a combinatorial way of computing degree (see $[P_2]$), but there seems to be no good combinatorial way of computing invariants involved in more general generalized degree arguments.

5. Finally note that $\bar{F}^{-1}(0)$ is a k-dimensional manifold. Thus, as in $[A_1]$, [A-Y], one is faced with the problem of following a higher dimensional manifold. It is the author's opinion that it is better to follow a complete k-dimensional component, rather than, say try to follow some

curve lying within it. The problem of following k-dimensional manifold is discussed — in general terms — below.

Bifurcation without branches

We would like to advertise here an example due to Böhme [B] which shows that continuation methods —in particular the one just presented— are not applicable to all bifurcation problems. An example due to Mallet-Paret which illustrates the basic phenomenon was presented in [A_1]. A homotopy of functions $f_t: S^1 \longrightarrow R$, $0 \leq t \leq 1$, is exhibited such that f_0 has one maximum and one minimum (a Morse function with critical points of index ± 1). As t increases, the maximum and minimum coalesce, cancel each other out, and disappear. Meanwhile, a new critical point appears and becomes the maximum and minimum of f_1. Thus the critical points of the f_t do not connect the $t = 0$ and $t = 1$ levels. Moreover, on $S^1 \times I$, the union of the functions f_t form a Morse function. This example thus illustrates the fact that through a homotopy, critical points of Morse functions can cancel each other.

We use Mallet-Paret's example to reconstruct that of Böhme. Consider the following non-linear eigenvalue problem. Let $g: R^n \longrightarrow R$ be a smooth even function, such that g is $o(|u|^2)$ near zero. Let Dg denote the derivative of g, and consider the equation

$$(\ast) \qquad Dg(u) = \lambda u \quad , \quad \lambda \in R \ .$$

Note that $u = 0$ is a solution of $(\ast)$ for all λ and that $\lambda = 0$ is a eigenvalue of $Dg(0)$ of multiplicity n. Fadell and Rabinowitz [F-R] have proved that if 0 is an isolated solution of (✻) for $\lambda = 0$ then there exist non-negative integers k, l with $k+l = n$ such that for small enough $\lambda^{\pm}$, $\lambda^{-} < 0 < \lambda^{+}$, then (✻) has at least k (resp. l) pairs of non-trivial solutions for $\lambda = \lambda^{-}$ (resp. $\lambda = \lambda^{+}$). Moreover Böhme [B] and Marino [M] have proved that for ϵ sufficiently small, there exist n pairs of non-trivial solutions of (✻) with $|u| = \epsilon$. That is, in these two senses, the point $(\lambda, u) = (0, 0)$ is a bifurcation point of (✻) of multiplicity n.

However the bifurcation need not be in branches. That is, it is possible that for any neighborhood U of $(\lambda, u) = (0, 0)$, the non-trivial solutions of (✻) do not connect the boundary of U to the bifurcation point. Thus there are branches to follow, and the method of the previous section cannot, even on a basic topological level, succeed. More philosophically, the point seems to be that if a problem is proved to have a solution by a generalized degree argument, then continuation methods might be used to numerically locate that solution, but if the existence of the solution is proved by variational techniques, very possibly continuation methods are completely inapplicable and other techniques must be used.

It is easy to modify Mallet-Paret's example so that:

1) each f_t is even on S^1,

2) f_t is smooth as a function of $\theta \in S^1$ and t,

3) $|f_t(\theta)| > 0$ for all θ, t,

4) $f_t(\theta) = f_{t'}(\theta)$ for t in a neighborhood of 0 and 1,

5) the critical points of the f_t do not connect the $t = 0$ and $t = 1$ levels.

Using 4) we can extend f_t periodically so that it is defined for all real t. Let (r, θ) denote polar coordinates in R^2. Define $g: R^2 \to R$ by $g(\theta, r) = e^{-\frac{1}{r}} f_{\frac{1}{r}}(\theta)$. It is routine to check that g satisfies the required conditions. However the set of non-trivial solutions of (∗) are a pair of sequences of disjoint circles which accumulate at $(\lambda, u) = (0, 0)$.

Higher codimension pivoting

Pivoting is the local iterative step in following a simplicial submanifold. If the submanifold is one-dimensional, the basic process is completely understood. Here we discuss some aspects for submanifolds of higher dimensions.

Since we are interested only in the local situation, assume we have a triangulation τ of R^{n+k} and a map $f: R^{n+k} \longrightarrow R^n$ that is linear on each simplex of τ. We are interested in $f^{-1}(0)$. Such a map f is completely determined by its values on the vertices. Such data is called a vector labelling of the vertices. A cruder specification is integer labelling, which specifies the image of a vertex only to within some region of R^n. We restrict ourselves to integer labelling for two reasons. First the underlying structure and mechanics are simplier and it is easier to see what is happening. Second, the major application in mind is bifurcation, and Prüfer [P_1] has obtained excellent results for $k = 1$ using integer labelling.

To assign a labelling, divide R^n into the n+1 regions $V_0, \ldots, V_n$, where

$$V_i = \{(x_i, \ldots, x_n) \in R^n \mid x_j < 0 \text{ for } j<i,\ x_i \geq 0\}.$$

A vertex u of the triangulation τ is assigned label i, denoted $l(u) = i$, if $f(u) \in V_i$. A simplex σ with vertices $u_0, \ldots, u_r$ is called completely labelled if $\{l(u_i) \mid 0 \leq i \leq r\} = \{0, \ldots, n\}$. It is assumed $0 \in f(\sigma)$ if and only if σ is completely labelled.

This can be geometrically vizualized as follows. Let v_i, $i = 0, \ldots, n$, be the point in R^n with coordinates $(x_1^i, \ldots, x_n^i)$ where

$$x_j^i = \begin{cases} -1 & j = i-1 \\ +1 & j = i \\ 0 & \text{otherwise} \end{cases}$$

Let ν be the simplex spanned by the v_i. The origin is in the interior of ν. Since the integer label $l(u)$ is all the information we have about $f(u)$ for u a vertex of τ, we assume $f(u) = v_{l(u)}$ for each vertex u of τ. Then $\sigma \in f(\sigma)$ if and only if σ is completely labelled.

Geometrically then, the origin 0 is a regular value of f and $\Sigma = f^{-1}(0)$ is a k-dimensional simplicial submanifold of R^{n+k}. Clearly $\Sigma \cap \sigma \neq \emptyset$ if and only if σ is completely labelled. Note however that τ restricted to Σ is not necessarily a triangulation of Σ. The numerical process involves pivoting from some connected set of completely labelled (n+k)-dimensional simplices (which intersect the component Σ_0 of Σ, say), and eventually covering all such σ which intersect Σ_0.

In the codimension 1 case ($k = 1$), a completely labelled (n+1) simplex σ has exactly two completely labelled faces σ', σ''. (Faces are codimension 1 subsimplices or facets in the terminology of [A-G]).

The numerical process enters σ via one of them, say σ' (called a pivot around σ'), determines which other face of σ is σ'' via a function evaluation, and exists out of σ'' via a pivot around σ''.

The correct analogy for higher codimensions is: always pivot around completely labelled codimension k subsimplices.

For $k > 1$, a completely labelled (n+k) dimensional σ has more than 2 completely labelled codimension k subsimplices. However the value is independent of n.

Proposition. The number # of completely labelled codimension k subsimplices of a completely labelled (n+k) simplex satisfies

$$k+1 \leq \# \leq 2^k.$$

(It would be interesting to know the expected value of #).

To prove this we let $u_0, \ldots, u_{n+k}$ be the vertices of σ. Assume $l(u_i) = i$ for $0 \leq i \leq n$. Thus the subsimplex spanned by $u_0, \ldots, u_n$ is completely labelled. Other completely labelled codimension k subsimplices are obtained by replacing one or more u_i, $0 \leq i \leq n$, by u_j, $j > n$ with $l(u_j) = l(u_i)$. If all the $l(u_j)$, $j > n$, are the same, there are k such replacements possible, and thus k+1 completely labelled codimension k subsimplices. If the $l(u_j)$, $j > n$, are all different, there are 2^k completely labelled codimension k subsimplices. Other possibilities for the $\{l(u_j)\}$ yield intermediate values.

For $k = 2$, $n = 1$, the two possibilities $\# = 3, 4$ for a completely labelled 3-dimensional σ are illustrated, together with the piece of Σ

intersecting σ. The numbers by the vertices are the labels.

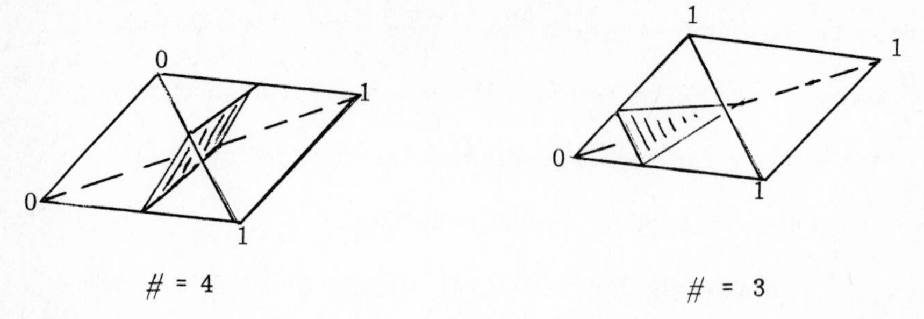

$\# = 4$ $\qquad$ $\# = 3$

The numerical procedure consists of pivoting into σ through some completely labelled σ' of codimension k, making k function evaluations to determine the other completely labelled codimension k subsimplices σ'', ... and pivoting out via all of them into all other (n+k) dimensional σ which contain one or more of the σ'', ... In this way all of a component of Σ is eventually covered.

The process of following a k-dimensional submanifold is definitely slower and more complicated than following a 1-dimensional submanifold, however possibly not for the reasons that immediately come to mind from the above discussion. The initial condition invariably consists of a manifold of dimension k-1. The process can be informally vizualized as a wave front emanating from this initial data, and enveloping the component of Σ. Thus the process should pivot outward from all points on the wave front, which is of dimension k-1. Thus locally the process is "moving" in one dimension. The pivoting process out of a simplex involves a number of new simplices, and it seems the number of computations may grow exponentially. Most of these new simplices however will already have been covered from other points on the wave front. Thus the number of computations at each step will stabilize.

On the other hand, the process has to work completely around the wave front, and this of course is time consuming. A codimension k problem is considerably more complicated than a codimension 1 problem, and one must expect to use more time. For codimension 1 problems, it has been found that using "round" simplices for the triangulation τ, ones with all measurements roughly the same, leads to a more efficient process. For higher codimension problems, to reduce the number of simplices on the wave front, it might be advantageous to use simplices that are "fat" in directions within the wave front.

Another source of computational difficulty is that the process must store the whole wave front in memory, unless it can operate on the whole wave front all at once in a parallel manner. For one thing, the process has to know where the wave front is. But there is another reason. In the codimension 1 case, it is guaranteed that the process cannot loop back to where it has already been. However for larger codimensions, it is possible for the wave front to bend around and meet itself. The process must be able to recognize when that happens. Fortunately, the wave front cannot meet the part of Σ already covered, but only the leading edge.

Clearly, there are a host of new programming techniques that must be developed to implement following higher dimensional manifolds. However, even the case $k = 2$ — for 2 parameter bifurcation — would be well worth it.

References

[A_1] J.C. Alexander, The topological foundations of an embedding method, in Continuation Methods, H.-J. Wacker, ed. Academic Press (1978), 37 - 68

[A_2] J.C. Alexander, Bifurcation of zeroes of parameterized functions, J. Func. Anal. 29 (1978), 37 - 53

[A-Y] J.C. Alexander and J.A. Yorke, Homotopy continuation Methods: numerically implementable topological procedures, Trans. Am. Math. Soc. 242, (1978), 271 - 284

[A-G] E.L. Allgower and K. Georg, Simplicial and continuation methods for approximating fixed points and solutions to systems of equations, to appear in SIAM Review

[B] R. Böhme, Die Lösung der Verzweigungsgleichungen für nichtlineare Eigenwertprobleme, Math. Z. 127 (1972), 105 - 126.

[F-R] E.R. Fadell and P.H. Rabinowitz, Bifurcation for odd potential operators and an alternative topological index, J. Func. Anal. 26 (1977), 48-67.

[M] A. Marino, La biforcazione nel caso varizionalle, in Proc. Conference del Seminario de Mathematica dell'Universita di Bari, Nov. 1972.

[P_1] M. Prüfer, Calculating global bifurcation, in Continuation Methods, H.-J- Wacker, ed. Academic Press (1978), 187 - 214

[P_2] M. Prüfer, Sperner simplices and the topological fixed point index, Sonderforschungsbereich 72 Universität Bonn, preprint No 134.

J. C. Alexander
Department of Mathematics
University of Maryland
College Park, Maryland, 20742

PERIODIC SOLUTIONS OF SOME AUTONOMOUS DIFFERENTIAL EQUATIONS WITH VARIABLE TIME DELAY

Wolfgang Alt

In many situations the time delay $\sigma_t > o$ of a functional differential equation, for example

$$(1) \quad \dot{x}(t) = -f\big(x(t), x(t-\sigma_t)\big)$$

or more generally

$$(2) \quad \dot{x}(t) = -f\Big(x(t), \int_{-r}^{o} g(x(t-\sigma_t+\theta))d\eta(\theta)\Big)$$

depends on the past of the system x, for instance by a

(3) <u>threshold condition</u>

$$\int_{t-\sigma_t}^{t} k(x(s))ds = k_o \quad ,$$

where $k : \mathbb{R} \to \mathbb{R}_+$ is continuous and positive with $k(o) = k_o$ without restriction.

In this case the time delay is an autonomous function $\sigma_t = \sigma(x_t) > o$ with derivative

$$(4) \quad d_t\sigma_t = 1 - \frac{k(x(t))}{k(x(t-\sigma_t))} < 1 \quad ,$$

where $x_t(\theta) := x(t+\theta)$ for $\theta \leq o$ denotes an element in the Banach space $C := BC^o\big((-\infty, o], \mathbb{R}\big)$ of bounded continuous functions. Indeed we have

<u>Remark 1</u> : Under the above assumptions there is a unique continuous, compact function $\sigma : C \to \mathbb{R}_+$ such that

$$\int_{-\sigma(\varphi)}^{o} k(\varphi(\theta))d\theta = k_o \quad \text{for } \varphi \in C$$

and the estimate

$$(5) \quad |\sigma(\varphi)-1| = \varepsilon\left(\|\varphi\|_{[-\sigma(\varphi),o]}\right)$$

holds, where $\varepsilon(z)$ tends to zero with $z \to o$.

In order to find periodic solutions of (1) or (2) oscillating around zero, the following wellknown method can be used, which requires two parts of investigation:

(I): Find suitable closed "invariant sets" $K \subset C$, homeomorphic to a convex (bounded) set, with $o \in K$ and the property: For each initial function $\varphi \in K$ there is a time $\tau(\varphi) \leq \infty$ such that the solution $x = x^{\varphi}$ of a certain functional differential equation

$$(*) \quad \dot{x}(t) = F(x_t) \quad \text{with} \quad x_o = \varphi$$

exists on $[o,\tau(\varphi)] \setminus \{\infty\}$ and satisfies $A\varphi := x^{\varphi}_{\tau(\varphi)} \in K$ for finite $\tau(\varphi)$, whereas for $\tau(\varphi) = \infty$ the solution x approaches zero for $t \to \infty$; with $A\varphi := 0$ in this case the so called shift operator $A : K \to K$ should turn out to be continuous and compact (or condensing).

(II): Show that the linearized equation

$$\dot{x}(t) = F'(o)\, x_t$$

is (strongly) unstable along an eigenspace in C which points "in direction of K". (For an exact formulation of the related compatibility condition see below (22).)

For equations with fixed time delay and in cases of "locally" bounded τ a general theorem on existence of periodic solutions originally was given by GRAFTON [2], see also theorem 11.2.3 in HALE [4] . Further special examples recently were handled by NUSSBAUM [6] , HADELER [3] , STECH [7] for infinite delay, WALTHER [8], CHOW, HALE and others (for a review on literature up to 1976 see [4] , 11.7), mainly using the concept of ejectivity for the operator A at zero.

In a recent paper [1] I showed that with the usual fixed point index in cones the above assumptions (I) and (II) together with a general continuity condition

(6) $|x^{\varphi}(t)| \leq M \|\varphi\|$ for $o \leq t < \tau(\varphi)$, $\|\varphi\| \leq R$, $\varphi \in K$

lead to nontrivial periodic solutions of (*). More generally I gave analogous periodicity criteria in terms of the linearization $F'(\infty)$ in theorem 1.11 [1] .

The aim of this paper is to prove that the periodicity results for a certain class of nonlinear differential equations with one time lag (prop. 3.14 in [1]) and for some equations with "distributed" time delay (prop. 3.4 in [1]) are valid also in the case of varying time delays σ_t of type (3) .

I. Definition of invariant sets

Equations (1) and (2) are of the type

(7) $\dot{x}(t) = -f\big(x(t), y(t)\big)$,

where $y(t) = G(x_{t-\sigma_t})$ is a nonlinear functional regarding values of x only in a bounded interval completely lying in the past. Thus together with the differential equation (4) for σ_t we can conclude the preliminary

Remark 2 : Let $g : \mathbb{R} \to \mathbb{R}$ and $f : \mathbb{R}^2 \to \mathbb{R}$ be locally lipschitzian, $f(\cdot, y)$ lipschitz continuous uniformly for y in bounded sets, $k : \mathbb{R} \to \mathbb{R}_+$ locally lipschitz continuous and $d\eta$ a measure of bounded variation on $[-r,o]$, $r > o$.

Then for each initial function $\varphi \in C$, which is lipschitz continuous on $[-\sigma(\varphi)-r,o]$, there exists a unique continuously differentiable solution $x : [o,\infty) \to \mathbb{R}$ of equation (2) satisfying $x_o = \varphi$ and depending only on values of φ in $[-\sigma(\varphi)-r,o]$.

Now suppose in addition to the hypotheses in remark 2 :

(8) f is weakly positive, i.e. $f(x,y) \geqq o$ resp. $\leqq o$ for x and $y \geqq o$ resp. $\leqq o$,
$f(o,y) < o$ for $y < o$,
g is positive, i.e. $x\, g(x) > o$ and $f(x,g(x)) \neq o$ } for all $x \neq o$.

(9) $f(o,\cdot)$ or g is bounded from below on $(-\infty,o]$ and

$$\nu := \sup \left\{ \frac{1}{|x|} \, | f(x,y) - f(o,y)| \; : \; \inf g < y \leqq o \neq x \right\} < \infty \; ,$$

(10) k is uniformly positive : $k \geqq \underline{k} > o$.

In case of equation (1) let these assumptions be fulfilled with $g(x) = x$ and $r = o$, while in case (2) assume without restriction $\int_{-r}^{o} d\eta = 1$.

Then the following lemma can be derived easily.

Lemma 1. Let y be a continuous function on $[t_1,t_2]$ with $\inf g \leqq y \leqq o$, then

(i) each solution x of equation (7) on $[t_1,t_2]$ with $x(t_1) < o$ changes its sign at most once, say at a time t_o with $t_1 < t_o < t_2$, such that

$$x(t)\, e^{\nu t}$$

is positive and nondecreasing for $t \in (t_o,t_2]$.

(ii) If $\partial_y f(o,o)$ exists and is positive, then there is a positive constant $b_{f,S}$, such that $-S \leqq y(t) \leqq o$ implies $\dot{x}(t) \geqq -\nu\, x(t) + b_{f,S} |y(t)|$ on $(t_o,t_2]$.

(iii) For $t \in (t_o,t_2]$ the estimates

$$|\dot{x}(t)| \leqq F \; e^{\nu(t-t_o)}$$

and

$$|x(t)| \leqq \frac{F}{\nu} \left(e^{\nu(t-t_o)} - 1 \right)$$

hold, where $F := \sup \{-f(o,y) : \inf g < y < o \}$.

Applying this lemma to

$$y(t) = x(t-\sigma_t)$$

and remarking that $(t-\sigma_t)$ strongly increases because of (4), we can conclude as usually, compare [3] for example, that the solution $x = x^{\varphi}$ with φ positive on $(-\sigma(\varphi),o]$ oscillates "slowly" in a sense specified below in lemma 2.

To argue in the same way for

$$y(t) = \int_{-r}^{o} g\big(x(t-\sigma_t+\theta)\big)\, d\eta(\theta)$$

we start with functions φ positive on $(-\sigma(\varphi),o]$ and nonpositive on $[-\sigma(\varphi)-r,-\sigma(\varphi)]$. That y changes sign only once on the intervall $[o,r]$ for instance, cannot be expected for general g and φ as above. But this turns out to be true, if $r \leqq \sigma(\varphi)$ and the <u>logarithmic concave weight function</u> hypothesis holds:

(11) $\quad d\eta(\theta) = \rho(\theta)d\theta \qquad$ with

$\rho : (o,r) \to \mathbb{R}_+$ such that $\log \rho$ is concave.

The proof of this assertion is given in [1] , lemma 3.3; it shows that condition (11) is also necessary for the sign change property formulated above. If in addition k guarantees that always $\sigma(\varphi) \geqq r$, then we derive the following lemma about "slow oscillation".

<u>Lemma 2</u>. <u>Assume the conditions</u> (8) - (10) <u>are fulfilled and</u> $\partial_y f(o,o) > o$ <u>exists; additionally in case of equation</u> (2) <u>assume that</u>

$g'(o) > o$ exists,

$d\eta$ <u>has the form</u> (11)

<u>and</u> $\sup k \leqq \frac{1}{r} k_o$.

<u>Then for initial values</u> φ <u>lying in</u>

$$K_o = \{\ \varphi \in C\ :\ \varphi \text{ lipschitz on } [-\sigma(\varphi)-r,o\,]\ ,\ o<\varphi(o) \leq R\ ,$$
$$o < \varphi(s) \leq \varphi(o)\ e^{\nu\sigma(\varphi)} \quad \text{for} \quad s\in(-\sigma(\varphi),o\,]\ ,$$
$$-\varphi(o) \leq b\varphi(s) \leq o \quad \text{for} \quad s \in [-\sigma(\varphi)-r,-\sigma(\varphi)\,]\ \}$$

<u>with some</u> $R,b > o$, <u>the solution</u> $x = x^{\varphi}$ <u>of equation</u> (1) <u>resp.</u> (2) <u>either changes sign at most once with</u> $\lim_{t\to\infty} x(t) = o$, $\tau(\varphi) = \infty$, <u>or</u> x <u>has a "second zero"</u> $t(\varphi) \geq \frac{1}{2}\kappa_R > o$ <u>and a time</u> $\tau(\varphi) < \infty$ <u>such that</u>

$$t(\varphi) = \tau(\varphi) - \sigma\left(x_{\tau(\varphi)}\right) \quad ,$$

$$o < x(\tau(\varphi)) \leq \frac{F}{\nu}\left(e^{\nu\sigma_o} - 1\right) \quad \underline{\text{with}} \quad \sigma_o := \frac{k_o}{\underline{k}} \quad ,$$

$$(12) \qquad o < x(t)\ e^{\nu t} \quad \underline{\text{nondecreasing for}} \quad t \in (t(\varphi),\ \tau(\varphi)\,]$$

$$(13) \qquad -x(\tau(\varphi)) \leq b_R\ x(t) \lneq o \quad \underline{\text{for}}\ t \in [t(\varphi)-r-\tfrac{1}{8}\,\delta_R,\ t(\varphi)]\ .$$

<u>Furthermore in all cases the following estimates for</u> $o \leq t < \tau(\varphi)$ <u>are valid</u>

$$(14) \qquad |x(t)| \leq \begin{cases} M_R \\ m_{R,b}\cdot\varphi(o) \end{cases}$$

$$(15) \qquad |\dot{x}(t)| \leq F_R \quad ,$$

<u>where</u> M_R, F_R, κ_R, b_R <u>and</u> δ_R <u>are certain positive constants depending on</u> R, f, g, ρ, r <u>and</u> σ_o <u>only, while</u> $m_{R,b}$ <u>depends on</u> b, <u>too. Moreover</u>

$$\tau(\varphi) \geq \kappa_R > \frac{2k_o}{\sup k|_{[-M_R,\,M_R]}} \quad .$$

<u>Proof</u>: Since $k_o = \int_{t-\sigma_t}^{t} k\circ x \geq \underline{k}\cdot\sigma_t$ and $\sigma_t \lneq \sigma_o$, there is in case

$$t_1(\varphi) = \sup\ \{\ s>o\ :\ x|_{[o,s]} \geq o\ \} < \infty$$

a first time $t_2 > t_1(\varphi)$ with $t_2 - \sigma_{t_2} = t_1(\varphi)$; in analogy to lemma 1 one shows that with an R-dependent constant ν_R

$$o > x(t)\ e^{\nu_R t} \quad \text{is nonincreasing for } t \in (t_1(\varphi), t_2] .$$

Repeating the same argument, using lemma 1, we get in case of

$$t(\varphi) := \sup \{ s>o : x_{|[t_1(\varphi),s]} \leq o \} < \infty$$

the existence of a time $\tau(\varphi)$ with the desired properties including assertion (12).

Now let s_1 be the unique time with $t_2 \leq s_1 < t(\varphi)$, where y first becomes negative : $s_1 = t_2$ in case (1) and $s_1 < t_2+r$ in case (2), guaranteed by condition (11). Then a careful estimation of x on $[t_1(\varphi), t(\varphi)]$ as it is done in [1] , lemma 3.10.ii, implies the existence of positive $\delta = \delta_R$ and $\varepsilon = \varepsilon_R$, such that for all $\varphi \in K_o$ the following estimates a priori hold

$$s_1 + \delta \leq t(\varphi)$$

$$o \geq x \quad \text{nondecreasing on } [s_1, t(\varphi)]$$

$$o > x(t) \geq e^{\nu(\sigma_o+r)} x(s_1) \quad \text{for } t_1(\varphi) < t \leq s_1$$

and

$$x(s_1) \leq x(t) \leq \varepsilon\, x(s_1) < o \quad \text{for } s_1 \leq t \leq s_1 + \tfrac{1}{2}\delta .$$

In <u>case of equation (1)</u> it follows for times $t \in (t(\varphi), \tau(\varphi)]$ with $t(\varphi) - \frac{1}{2}\delta \leq t - \sigma_t \leq t(\varphi) - \frac{1}{4}\delta$ the inequality

$$y(t) = x(t-\sigma_t) \leq \underline{x}_o \leq o \quad , \tag{16}$$

where $\underline{x}_r := \min \{ x(s) : t(\varphi) - r - \frac{1}{8}\delta \leq s \leq t(\varphi) \}$.

In case of equation (2) we can show as in the proof of lemma 3.10.ii [1] that for

(17) times $t \in (t(\varphi),\tau(\varphi)]$, such that $t-\sigma_t$ lies in some intervall of length $\frac{1}{4}\delta$,

the estimate

$$y(t) \leq \rho_o \cdot \sup g|[x(s_t),x(s_t+\tfrac{\delta}{8})] \leq o$$

holds, where $\rho_o > o$ and

$$s_t = \begin{cases} t-\sigma_t - \frac{1}{4}\delta \geq s_1 & , \text{ if } s_1+\delta \leq t(\varphi) \leq s_1+r+\frac{1}{4}\delta \\ \max(t-\sigma_t-r,s_1) & , \text{ if } s_1+r+\frac{1}{4}\delta \leq t(\varphi) \end{cases} .$$

But since

$$\underline{x}_r \geq \begin{cases} e^{\nu(\sigma_o+r)}\, x(s_1) & : t(\varphi)-r-\frac{1}{8}\delta \leq s_1 \\ x(t(\varphi)-r-\frac{1}{8}\delta) \geq x(s_1) & : t(\varphi)-r-\frac{1}{8}\delta \geq s_1 \end{cases}$$

it follows from $g'(o) > o$ and condition (8), that with a positive constant c_R and for t satisfying (17) we have

$$(18) \qquad y(t) \leq c_R\, \underline{x}_r \leq o .$$

Now condition (17) implies that t covers an interval in $(t(\varphi),\tau(\varphi)]$ of length at least $\chi\delta$, where

$$\chi = \frac{\underline{k}}{4 \sup k|[-M_R,M_R]} .$$

Thus in both cases the inequality for $y(t)$ in (16) resp. (18) together with lemma 1(ii) gives the desired estimate (13), with $S_R := \sup\{|g(x)| : |x| \leq M_R\}$:

$$x(\tau(\varphi)) \geq c_R\, e^{-\nu\sigma_o}\, \chi\, \delta_R\, b_{f,S_R} |\underline{x}_r| .$$

The remaining estimates (14) and (15) in lemma 2 are direct consequences of the differential equation itself.

In order to construct an invariant set $K \subset C$, homeomorphic to a convex bounded set $\tilde{K}$, we make the

Definition 1.

$$\tilde{K} := \{o\} \cup \{\psi \in C^{o,1}([-1-r-\delta_o,o], \mathbb{R}) \text{ with Lipschitz constant } \mathrm{Lip}(\psi) \leq R_1 , \ \psi(o) \leq R_o \text{ and } o < \psi(\theta) \text{ nondecreasing for } \theta \in (-1,o] , \ -\psi(o) \leq b_o\psi(\theta) \leq o \text{ for } \theta \in [-1-r-\delta_o,-1] \}$$

is a closed, bounded, convex subset of the Banach space

$$\tilde{X} = \{ \psi \in C^o([-1-r-\delta_o,o], \mathbb{R}) : \psi(-1) = o \}$$

satisfying $\tilde{K} \cap (-\tilde{K}) = \{o\}$. Defining

$$X := \{ \varphi \in C : \varphi(-\sigma(\varphi)) = o , \ \varphi \equiv \text{const. on } (-\infty, -\sigma(\varphi)-r-\delta_o] \}$$

and

$$(19) \qquad T\varphi(\theta) := \begin{cases} \varphi\big(\sigma(\varphi)\,\theta\big)\, e^{\nu\sigma(\varphi)\theta} & : -1 \leq \theta \leq o \\ \varphi\big(1-\sigma(\varphi)+\theta\big) & : -1-r-\delta_o \leq \theta \leq -1 \end{cases}$$

for $\varphi \in X$, it is evident that $T : X \to \tilde{X}$ is a homeomorphism in the C^o-topology preserving Lipschitz continuity with

$$\mathrm{Lip}(T\varphi) \leq \max \{ \mathrm{Lip}(\varphi), (\mathrm{Lip}(\varphi) + \nu R_o)\, \sigma(\varphi) \}$$

if $T\varphi \in \tilde{K}$.

Definition 2. Let

$$K := \{ T^{-1}\psi \big|_{[-\sigma(T^{-1}\psi)-r-\delta_o,o]} : \psi \in \tilde{K} \}$$

be the inverse image of $\tilde{K}$ under T , where the constants in definition 1 are chosen as

$$R_o = \frac{F}{\nu} (e^{\nu\sigma_o} - 1)$$

$$\delta_o = \frac{1}{8} \delta_{R_o} \quad , \quad b_o = b_{R_o} \qquad \text{from lemma 2} \quad ,$$

and

$$R_1 = (F_o + \nu R_o)\sigma_o$$

with

$$F_o := \sup\{|f(x,y)| : |x| \le M_{R_o} \quad , \quad \inf g < y \le \sup g|_{[o,M_{R_o}]}\} .$$

Then the properties of the solution x stated in lemma 2 can be expressed just in the statement

$$x_{\tau(\varphi)}\Big|_{[-\sigma(x_{\tau(\varphi)})-r-\delta_o,o]} \in K \quad .$$

Hence we have proved the following

<u>Proposition 1.</u> <u>Let the assumptions of lemma</u> 2 <u>be fulfilled. Then</u> K <u>is embedded in the set</u> $K_o \cup \{o\}$ <u>defined there</u> (with $R = R_o$ and $b = b_o$) <u>and the map</u>

$$\tau : K_o \to [\kappa_o, \infty] \quad \underline{\text{with}} \quad \kappa_o := \kappa_{R_o} > \frac{2k_o}{\sup\{k(x) : |x| \le M_{R_o}\}}$$

<u>defines a shift operator</u> $A : K_o \to K$, <u>compare</u> (I), <u>inducing a continuous and compact mapping</u>

$$A : K \to K \quad .$$

II. <u>Instability of the linearization at zero.</u>

Suppose f and g are differentiable at zero and let

$$\nu_o := \partial_x f(o,o) \quad \text{and} \quad \alpha_o := \partial_y f(o,o)\, g'(o) .$$

<u>Linearization of equation (2)</u> :

For $\varphi \in C$ we estimate

$$\left| \int_{-r}^{o} g\big(\varphi(-\sigma(\varphi)+\theta)\big)\, \rho(\theta)d\theta \; - \int_{-r}^{o} g\big(\varphi(-1+\theta)\big)\rho(\theta)d\theta \right|$$

$$= \omega\big(1-\sigma(\varphi)\big) \cdot O\Big(\|\varphi\|_{[-\max(1+r,\sigma(\varphi)+r),o]}\Big)$$

where (for positive ε, say)

$$\omega(\varepsilon) = \int_{\varepsilon-r}^{o} |\rho(\theta-\varepsilon)-\rho(\theta)| \; d\theta + \int_{-\varepsilon}^{o} \rho + \int_{-r}^{\varepsilon-r} \rho$$

tends to zero with $\varepsilon \to o$. Because of (5) equation (2) then is equivalent to

(20) $$\dot{x}(t) = Lx_t + \tilde{G}(x_t) \qquad \text{with}$$

$$L\varphi := -\nu_o \varphi(o) - \alpha_o \int_{-r}^{o} \varphi(-1+\theta)\, \rho(\theta) d\theta$$

and

$$g(\varphi) = |\tilde{G}(\varphi)| = o\left(\|\varphi\|_{[-\sigma(\varphi)-r-\delta_o, o]} \right) .$$

Linearization of equation (1)

Since generally the mapping $\varphi \mapsto \varphi(-\sigma(\varphi))$ is not differentiable on C , it is advantageous to rewrite the term

$$x(t-\sigma_t) = x(t-1) + d_t \left(\int_{-1}^{-\sigma_t} x(t+\theta) d\theta \right) + x(t-\sigma_t) d_t \sigma_t .$$

But with (4) we conclude

$$|d_t \sigma_t| \le \frac{1}{\underline{k}} \left(|k(x(t-\sigma_t))-k_o| + |k(x(t))-k_o| \right)$$

$$= \varepsilon\left(\|x_t\|_{[-\sigma_t, o]} \right)$$

so that equation (1) is equivalent to

(21) a neutral functional differential equation

$$d_t\left(x(t) - G(x_t) \right) = Lx_t + \tilde{G}(x_t) \qquad \text{with}$$

$$L\varphi := -\nu_o \varphi(o) - \alpha_o \varphi(-1) ,$$

$$G(\varphi) := -\alpha_o \int_{-1}^{-\sigma(\varphi)} \varphi$$

and

$$|G(\varphi)|, |\tilde{G}(\varphi)| = o\left(\|\varphi\|_{[-\sigma(\varphi)-\delta_o, o]} \right) .$$

Remark that the linearization of this neutral equation in C is a usual functional differential equation.

While in case (2) we directly can apply lemma 1.5 [1] about the strong monotonicity of the HALE-PERELLO Lyapounov function along an unstable eigenspace of L , we have to use in case (1) the related decomposition theory for neutral functional differential equations presented by HALE [4] in chapter 12.11: In the ordinary differential equation (11.7) in chapt. 12 [4] for the projection of x_t on a finite sum of generalized eigenspaces of L

$$\dot{y}(t) = B\, y(t) + B\, \Psi(o)\, G(x_t) + \Psi(o)\, \tilde{G}(x_t)$$

there arises an additional term of lower order

$$g(\varphi) := \|B\|\ |G(\varphi)| + |\tilde{G}(\varphi)| \quad .$$

With this g and with aid of condition (6) , stated in estimate (14) of lemma 2, the proof of lemma 1.5 in [1] can be carried out in the same way, so that the following proposition holds .

<u>Proposition 2</u>. <u>Assume that the linear operator</u>

$$L : C_r \to \mathbb{R} \quad \text{in (20) resp. (21)}$$

<u>with</u> $C_r := C^o([-1-r,o], \mathbb{R})$ <u>has a characteristic value</u> $\lambda \in \mathbb{C}$ <u>such that for the projection</u> π_λ <u>onto the generalized eigenspace and for the set</u> K <u>in definition</u> 2 <u>the compatibility condition</u>

$$(22) \qquad \|\pi_\lambda \varphi\| \geq \underline{\nu}\ \|\varphi\|_{[-\sigma(\varphi)-r-\delta_o,\, o]}$$

<u>is fulfilled with a</u> $\underline{\nu} > o$ <u>for all</u> $\varphi \in K$, $\|\varphi\| \leq \zeta$ <u>small</u>. <u>Then for the related nonnegative quadratic functional</u> V <u>on</u> C <u>with</u> $V(\varphi) \sim \|\pi_\lambda(\varphi)\|^2$ <u>compare</u> [4] , <u>lemma</u> 1o.1.1 , <u>there are positive constants</u>

ε_o, r_o <u>and</u> ζ_o <u>such that from</u>

$\varphi \in K$ <u>with</u> $o < \varphi(o) \leq \zeta_o$ <u>and</u> $V(\varphi) \leq r_o^2$

<u>it follows</u> $\|\varphi\|_{[-\sigma(\varphi)-r-\delta_o, o]} < \zeta_o$, <u>and the</u>

<u>solution</u> $x = x^\varphi$ <u>satisfies</u>

$$d_t V(x_t) \geq \varepsilon_o V(x_t) \quad \text{for} \quad o \leq t < \tau(\varphi) .$$

Before we conclude the final result on existence of periodic solutions in analogy to [1], theorem 1.11 , we make the

<u>Remark 3</u>:

(i) The linear equation

$$\dot{x}(t) = L x_t$$

with L in (20) resp. (21) is strongly unstable (i.e. there is a characteristic value with positive real part) iff $\nu_o < \chi(\alpha_o)$, where $\chi : [o,\infty) \to \mathbb{R}$ is an implicitely given, piecewise analytic function, depending on r and ρ in case of equation (2) , while in equation (1) $\alpha > o$ with $\nu = \chi(\alpha)$ is the smallest positive solution of

$$\nu + \alpha \cos\sqrt{\alpha^2 - \nu^2} = o .$$

Compare [1] , prop. 3.12 and the remark after 3.14.

(ii) In case of strong instability there is always a simple characteristic value λ with $\operatorname{Re}\lambda > o$ and $o < \operatorname{Im}\lambda < \pi$. For this λ the Lyapounov function $V(\varphi)$ is proportional to

$$\left| e^{is_o}\varphi(o) + \int_{-1-r}^{o} \varphi(s)\, \Phi_o(s)ds \right|^2 \qquad \text{with } |s_o| < \frac{\pi}{2}$$

and an analytic function $\Phi_o : \mathbb{R} \to \mathbb{C}$, whose real part is positive on $(-1,o]$ and nonpositive on $[-1-r,-1)$. Compare [1] , prop. 2.1.

From this we conclude, that for $\varphi \in K$ with $\|\varphi\| \leq \zeta_o$ small enough (and with $\sigma(\varphi) \leq 1$ for instance)

$$\|\pi_\lambda \varphi\| \geq C_1 \cdot V(\varphi)^{\frac{1}{2}} \geq C_2\left(\cos s_o \cdot \varphi(o) + \int_{-1}^{-\sigma(\varphi)} \varphi(s) \operatorname{Re} \Phi_o(s)ds\right)$$

$$\geq C_3 \; \varphi(o) \geq C_4 \|\varphi\|_{[-\sigma(\varphi)-r-\delta_o, o]} \;,$$

since

$$\int_{-1}^{-\sigma(\varphi)} \varphi(s) \operatorname{Re} \Phi_o(s)ds = o\left(\|\varphi\|_{[-1,o]}\right).$$

(For the other case with $\sigma(\varphi) \geq 1$ we can proceed analogically). Thus we have derived the

Lemma 3. *For the characteristic value* λ *in remark* 3.(ii) *and for the set* K *in definition* 2 *the compatibility condition* (22) *is always fulfilled*.

Then we can formulate the final result

Theorem.
Let f *and* g *be differentiable at zero, and assume that the hypotheses of lemma* 2 *are satisfied*.
If the linearization of equation (1) *resp.* (2) *with fixed time delay* $\sigma = 1$, *with the linear operator* L *in* (20) *resp.* (21), *is strongly unstable in the sense of remark* 3, *then there exists a nontrivial periodic solution of* (1) *resp.* (2) *with initial data in* K *and with period* $p \geq \kappa_o$, *estimated in proposition* 1.

Proof: From propositions 1 and 2 together with lemma 3 we conclude, that the compact operator $A : K \to K$ satisfies

$$V(A\varphi) \geq e^{\varepsilon_o \kappa_o} V(\varphi) > V(\varphi) > o$$

for all $\varphi \in K$ with $V(\varphi) \leq r_o^2$ and $o < \varphi(o) \leq \zeta_o$.

This inequality implies that the operator $\tilde{A} = T \circ A \circ T^{-1}$ on the closed convex bounded set $\tilde{K} \subset \tilde{X}$ has no fixed points in a neighborhood of o, except zero itself, which is an ejective point. Then Browder's theorem, compare NUSSBAUM [5] , Theorem 1.1, gives the existence of a function $\varphi_o \in K \setminus \{o\}$ with

$$V(\varphi_o) > r_o^2 \quad \text{or} \quad \varphi_o(o) > \zeta_o \ ,$$

such that $x = x^{\varphi_o}$ is a periodic solution with the desired estimate for the period.

<u>Remark 4</u>: Using similar invariant sets one can prove with the same methods the existence of periodic solutions for equations of type

$$\dot{x}(t) = -f\Big(x\big(t - \sigma(|x(t)|)\big)\Big)$$

with a locally lipschitz continuous and nondecreasing function σ satisfying $\sigma(o) > o$. A special example was treated by NUSSBAUM [5] , section 4.

References

1. ALT, W.: Some periodicity criteria for functional differential equations. Manuscripta math. 23, 295 - 318 (1978)

2. GRAFTON, R.B.: A periodicity theorem for autonomous functional differential equations. J. Diff. Equations 6, 87-1o9(1969)

3. HADELER, K.P.: Periodic solutions of the equation $\dot{x}(t) = -f(x(t), x(t-1))$. Applied Analysis and Math. Physics (Teubner Verlag), to appear

4. HALE, J.K.: Theory of functional differential equations. Berlin - Heidelberg - New York: Springer, Appl. Math. Sciences Vol. 3 , 1977

5. NUSSBAUM, R.D.: Periodic solutions of some nonlinear autonomous functional differential equations. Ann. Mat. pura appl. IV. Ser. 101, 263 - 306 (1974)

6. NUSSBAUM, R.D.: Differential-delay equations with two time lags. Memoirs of the Amer. Math. Soc., to appear

7. STECH, H.W.: Periodic solutions to a nonlinear Volterra integro-differential equation. (preprint) (1978)

8. WALTHER, H.-O.: Über Ejektivität und periodische Lösungen bei autonomen Funktionaldifferentialgleichungen mit verteilter Verzögerung. Habilitationsschrift. München Okt. 1977

Wolfgang Alt
Institut für Angewandte Mathematik
Im Neuenheimer Feld 294
D-6900 HEIDELBERG

Global branching and multiplicity results for periodic solutions of functional differential equations

N. Angelstorf

INTRODUCTION: In this paper we will be concerned with differential delay equations of the type

$$(E)\quad x'(t) = -f(x(t-1)),$$

where $f: \mathbb{R} \to \mathbb{R}$ is a continuous function satisfying $xf(x) \geqq 0$ for all $x \in \mathbb{R}$.

Many results have been obtained proving existence and certain stability properties of slowly oscillating periodic solutions of (E) under appropriate assumptions on f (see for example: S.N. Chow [1], J.L. Kaplan and J.A. Yorke [2,3], R.D. Nussbaum [4,5,6] and H.O. Walther [10,11]). The proofs in these papers have in common that they rely on asymptotic conditions on f, such as $\alpha < \pi/2 < \beta$ with $\alpha = \lim_{x \to 0} f(x)/x$ and $\beta = \lim_{|x| \to \infty} f(x)/x$.

We will give new conditions on f which provide the existence of slowly oscillating periodic solutions of (E) and which are independent of the behaviour of f " near 0 and ∞". These will allow to obtain multiplicity results. We will restrict our discussions to odd functions f and to periodic solutions of period 4 with special symmetry properties (see definition 1.1 below).

Furthermore we will discuss global bifurcation of periodic solutions of

$$(E_\lambda)\quad x'(t) = -\lambda f(x(t-1)).$$

Especially, we will characterize the global branches at "∞" and will show that they can be obtained in fact as a curve which has a parametric representation.

This work was supported by the Deutsche Forschungsgemeinschaft, Sonderforschungsbereich 72, Universität Bonn.

1. STATEMENT AND PROOF OF MAIN RESULTS

Let us first make precise the type of periodic solutions of (E) we are looking for.

(1.1) Definition: Let $f: \mathbb{R} \to \mathbb{R}$ be a continuous and odd function (i.e. $f(-x) = -f(x)$ for all $x \in \mathbb{R}$) that satisfies $f(x)x \geq 0$ for all $x \in \mathbb{R}$. Suppose x is a periodic solution of (E) of period 4 such that $x(-1) = 0$, $x(t) > 0$ for $-1 < t \leq 0$ and $x(t) = x(-t)$ for all $t \in \mathbb{R}$, then x will be called a periodic solution of (E) of type (H).

Note that these properties imply $x(t-1) = -x(-t-1)$ for all $t \in \mathbb{R}$.

(1.2) Example: If $f(x) = {}^{\pi}/_{2}\, x$, then $x(.) = \alpha\cos({}^{\pi}/_{2}\ .)$ is a periodic solution of (E) of type (H) for all $\alpha > 0$.

(1.3) Remarks: a) If x is a periodic solution of (E) of type (H) and we define $y(.) = x(.-1)$, then (x,y) is a periodic solution of the following system of ordinary differential equations:

$$(+)\quad \begin{cases} x'(t) = -f(y(t)) \\ y'(t) = f(x(t)) \end{cases}$$

Moreover, if (x,y) is a periodic solution of (+) of period 4, then we have that $y(t) = x(t-1)$ for all $t \in \mathbb{R}$ and with an appropriate choice of $\gamma \in [0,4)$ one observes that $x(.+\gamma)$ is a periodic solution of (E) of type (H). For a proof of this see J.L. Kaplan and J.A. Yorke [2]. We will make use of their concept in the second part of this paper.

b) A slowly oscillating periodic solution x of (E) such that $x(-1) = 0$ and $x(t) > 0$ for $-1 < t \leq 0$ need not be of type (H), even if all the hypotheses of definition 1.1 are satisfied for f. (For an example see R.D. Nussbaum [6].)

However, if x is a periodic solution of (E) of type (H) and we let $\phi := x|_{[-1,0]}$, then we have for all $t \in [-1,0]$:

$$\int_{-1}^{t} f(\phi(-s-1))\, ds = \int_{-1}^{t} f(x(-s-1))\, ds = \int_{-1}^{t} -f(x(s-1))\, ds = x(t) = \phi(t)$$

and therefore ϕ is a fixed point of the following completely continuous operator:

$$T_f : P \to P$$
$$\phi \to T_f\phi \quad \text{with } T_f\phi(t) = \int_{-1}^{t} f(\phi(-s-1))\, ds,$$

where P denotes the cone of all nonnegative functions in the Banach space $C[-1,0]$ of all continuous real-valued functions defined on the interval $[-1,0]$ with the sup-norm ($\| \ \|$) as norm.

Observe that the converse is also true, i.e. given a nonzero $\phi \in P$ such that $T_f\phi = \phi$, then there is a unique periodic solution x of (E) of type (H) such that $x_{|[-1,0]} = \phi$. For this one verifies that the unique solution $x = x(.;\phi)$ of $x'(t) = -f(x(t-1)), t \geq 0$, with $x_{|[-1,0]} = \phi$ is a peridic function which can be extended to the whole real line in such a way that it is a periodic solution of (E) of type (H).

Thus, we have the following equivalence:

Given a function f which satisfies the hypotheses of definition 1.1, then there is a periodic solution of (E) of type (H) if and only if T_f has a nonzero fixed point in P.

Furthermore, the amplitude of such a periodic solution x is the sup-norm of the corresponding fixed point of T_f, hence, no confusion should occur if we denote this amplitude by $\| x\|$. We are now ready to prove the following theorem:

(1.4) Theorem: Let $f: \mathbb{R} \to \mathbb{R}$ be a continuous and odd function satisfying $f(x)x \geq 0$ for all $x \in \mathbb{R}$. Suppose there exist constants $0 \leq c_1 < c_2 < c_3 < c_4$ such that the following inequalities hold:

(1.4.1) $f(x) \leq \dfrac{c_4 - c_3}{c_4} \, \pi/2 \; x$ for $x \in [c_3, c_4]$;

(1.4.2) $f(x) \leq \dfrac{c_4 - c_3}{c_4} \, \pi/2 \; c_3$ for $x \in [0, c_3]$;

(1.4.3) $f(x) \geq \dfrac{c_2}{c_2 - c_1} \, \pi/2 \; x$ for $x \in [c_1, c_2]$;

(1.4.4) $f(x) \geq \dfrac{c_2}{c_2 - c_1} \, \pi/2 \; c_2$ for $x \in [c_2, c_4]$;

Then there is a periodic solution x of (E) of type (H) satisfying

$$c_2 \leq \| x \| \leq c_4.$$

Proof: Using conditions (1.4.1) - (1.4.4), simple calculations show that the set

$$A := \{ \phi \in P \mid c_2\cos(\tfrac{\pi}{2}\cdot) \leq \phi \leq c_4\cos(\tfrac{\pi}{2}\cdot)\}$$

is invariant under T_f, i.e. $T_f(A) \subseteq A$. Therefore, since A is a bounded, closed and convex subset of P which does not contain zero , one can apply the Schauder fixed point theorem to obtain a nonzero fixed point $\phi \in A$ of T_f. The inequality $c_2 \leq \|\phi\| \leq c_4$ is obvious from the definition of A.

Our first result though certainly of very technical nature includes results of Nussbaum [4] for odd f and of Kaplan and Yorke [2]:

(1.5) Corollary: Let $f: \mathbb{R}\to\mathbb{R}$ be a continuous and odd function satisfying $f(x)x > 0$ for all $x \neq 0$. Suppose further $\alpha = \lim_{x\to 0} f(x)/x$ and $\beta = \lim_{x\to\infty} f(x)/x$ exist, allowing $\alpha = \infty$. Then, if $\alpha > \pi/2 > \beta$, there is a periodic solution of (E) of type (H).

Proof: Since $\beta < \pi/2$, there is an $\varepsilon > 0$ and an $x_o \in \mathbb{R}^+$ such that

$$f(x) \leq (\pi/2 - \varepsilon)x \quad \text{for all } x \geq x_o.$$

Let $x_1 := \max\{f(x) \mid x \leq x_o\}$, then one chooses

$$c_3 := \max\{x_o, x_1\},$$

and $c_4 > c_3$ such that

$$(\pi/2 - \varepsilon) \leq \frac{c_4 - c_3}{c_4}\,\pi/2 \geq 1.$$

Now since $\alpha > \pi/2$, there is an $a > 0$ such that

$$f(x) \geq \pi/2\, x \quad \text{for all } x \in [0,a].$$

Let $b := \min\{f(x) \mid a \leq x \leq c_4\}$ and choose

$$c_2 := \min\{a, 2b/\pi\}$$

and

$$c_1 := o.$$

With this choice of the constants $c_1 - c_4$ theorem 1.4 applies.

(1.6) Remark: Results for (E) which have been known so far needed conditions for $f'(0)$ and /or $f'(\infty)$ as for example in corollary 1.5. In contrast to these our conditions are somewhat conditions " between 0 and ∞ ", i.e. they are conditions on f in finite intervals. It is therefore immediate that our approach provides easily multiplicity results, simply by a suitable repetition of the arguments. It seems, however, that examples of that type are somehow artificial. Therefore, in the sequel we will restrict ourselves to monotonic increasing functions f. For these it seems that we have more adequate results.

(1.7) Theorem: Let $f:\mathbb{R}\to\mathbb{R}$ be a continuous, odd and monotonic increasing function. For $z>0$ and $0\leq x<z$ define:

$$\rho(x,z):=(1-x^2/z^2)^{-1/2}.$$

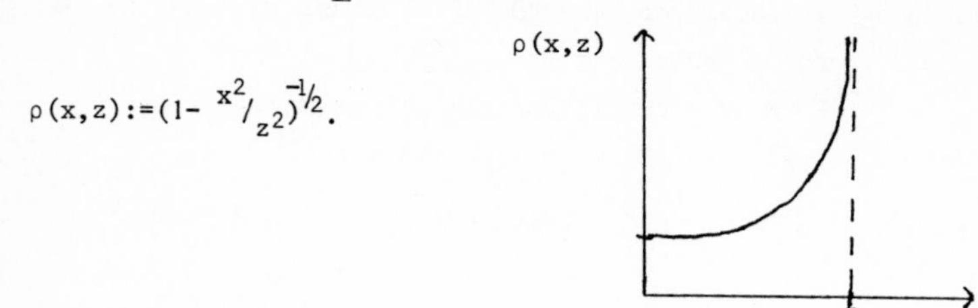

Suppose there is an $a>0$ such that

(1.7.1) $$\int_y^a (f(x)-\tfrac{\pi}{2}x)\rho(x,a)\,dx \geq 0 \qquad \text{for all } y\in[0,a)\ ,$$

and a $b\geq a$ such that

(1.7.2) $$\int_y^b (f(x)-\tfrac{\pi}{2}x)\rho(x,b)\,dx \leq 0 \qquad \text{for all } y\in[0,b)\,.$$

Then there is a periodic solution x of (E) of type (H) which satisfies

$$a\leq\|x\|\leq b\ .$$

(1.8) Remark: In particular the conditions (1.7.1) resp. (1.7.2) are satisfied provided

(1.8.1) $$\int_y^a f(x)-\tfrac{\pi}{2}x\,dx\geq 0 \qquad \text{for all } y\in[0,a)\ ,$$

resp.

$$(1.8.2) \qquad \int_y^b f(x) - \tfrac{\pi}{2}\, x \, dx \leq 0 \qquad \text{for all } y \in [0,b).$$

Proof of the theorem: Let $\phi_a := a\cos(\tfrac{\pi}{2}\,\cdot)$ and $\phi_b := b\cos(\tfrac{\pi}{2}\,\cdot)$

We will show that the following inequalities are true:

$$T_f\phi_a \geq \phi_a$$

$$T_f\phi_b \leq \phi_b \ .$$

Since T_f is monotone provided f is monotone one can have that:

$$\phi_a \leq T_f^n\phi_a \leq T_f^n\phi_b \leq \phi_b \ , \qquad \text{for all } n \in \mathbb{N}.$$

Therefore, the Picard-iterates $T_f^n\phi_a$ and $T_f^n\phi_b$ each converge and the limits are fixed points of T_f. Hence, the conclusion of the theorem follows.

Now, in order to prove that $T_f\phi_a \geq \phi_a$ we will show that

$$A(t) := T_f\phi_a(t) - \phi_a(t) \geq 0, \qquad \text{for all } t \in [-1,0].$$

This can be obtained from a change of variables and condition (1.7.1) as follows:

$$A(t) = \int_{-1}^{t} f(a\cos(\tfrac{\pi}{2}(-s-1)))\, ds - a\cos(\tfrac{\pi}{2}\, t)$$

$$= \int_{-1}^{t} f(a\cos(\tfrac{\pi}{2}(-s-1))) - a\tfrac{\pi}{2}\cos(\tfrac{\pi}{2}(-s-1))\, ds$$

$$= \int_{0}^{t+1} f(a\cos(\tfrac{\pi}{2}\, s)) - a\tfrac{\pi}{2}\cos(\tfrac{\pi}{2}\, s)\, ds$$

$$= \int_{1}^{\cos(\frac{\pi}{2}(t+1))} \tfrac{2}{\pi}\left(f(as) - a\tfrac{\pi}{2}\, s\right) \frac{d\arccos(s)}{ds}\, ds$$

$$= \frac{2}{a\pi}\left[\int_{a\cos(\frac{\pi}{2}(t+1))}^{a} (f(x) - \tfrac{\pi}{2}x)\, \rho(x,a)\, dx\right] \geqq 0.$$

The argument for $T_f\phi_b \leqq \phi_b$ is similar.

With this result one can easily formulate a general criterion for multiple periodic solutions of (E). We give an example:

(1.9) Example: Let $f(x) = \frac{\pi}{2}x - \sin(\frac{\pi}{2}x)$, then easy computations show that for each $n \in \mathbb{N}$ and for $a = 4n$ and $b = 4n+2$ the conditions (1.8.1) and (1.8.2) are satisfied. Thus for every $n \in \mathbb{N}$ we get a periodic solution x_n of (E) of type (H) satisfying

$$4n \leqq \|x_n\| \leqq 4n+2.$$

Another example of this type will be discussed in the following part.

2. GLOBAL BIFURCATION

In the sequel, we are concerned with

$$(E_\lambda)\ x'(t) = -\lambda f(x(t-1)),$$

with parameter $\lambda > 0$, where f satisfies:

(B) $f: \mathbb{R} \to \mathbb{R}$ is continuous and odd, f is differentiable at 0, $f'(0) = 1$ and $f(x)x > 0$ for all $x \in \mathbb{R}$.

Now, we are interested in families $(x(\lambda),\lambda)$ where $\lambda \in \mathbb{R}^+$ (i.e. λ is a positive real number) and $x(\lambda)$ is a non trivial periodic solution of (E_λ) of type (H) and these are equivalent to families $(\phi(\lambda),\lambda) \in P\times\mathbb{R}^+$ such that $\phi(\lambda) \neq 0$ is a fixed point of $T_{\lambda f}$.

For reasons of length, we omit the proof that the operator

$$T: P\times\mathbb{R}^+ \to P$$
$$(\phi,\lambda) \to T_{\lambda f}\phi$$

satisfies all the conditions of the global bifurcation theorem of Nussbaum [5, theorem 1.1]. Let us only state the conclusion in our case.

(2.1) Proposition: Let f satisfy property (B) and equip the set $P \times \mathbb{R}^+$ with the metric $d((\phi_1,\lambda_1);(\phi_2,\lambda_2)) := \|\phi_1-\phi_2\| + |\lambda_1-\lambda_2|$. Then there is an unbounded continuum (i.e. a closed and connected set) $S_f \subseteq P \times \mathbb{R}^+$ such that

i) $T(\phi,\lambda) = \phi$, for all $(\phi,\lambda) \in S_f$

ii) $(0,\lambda) \in S_f$ if and only if $\lambda = \pi/2$.

If for example f is the identity (i.e. f(x)=x, for all x), then

$$S_f = \{ (\phi_\alpha, \pi/2) \mid \phi_\alpha = \alpha \cos(\pi/2 \,\cdot) \text{ for all } \alpha \geqq 0 \}$$

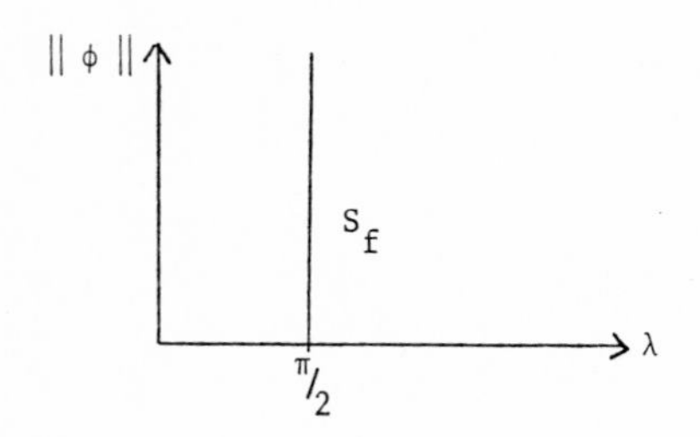

Figures like this are usually called bifurcation diagrams.

Nussbaum's result does not answer the questions

— whether S_f is a curve,

— whether all solutions of $T(\phi,\lambda)=\phi$ such that $\phi \neq 0$ are in S_f,

— how S_f " goes to infinity ".

The following theorem solves these problems.

(2.2) Theorem: a) Let $f: \mathbb{R} \to \mathbb{R}$ satisfy property (B), then

$$S_f = \{ (\phi,\lambda) \in P \times \mathbb{R}^+ \mid \phi \neq 0 \text{ and } T(\phi,\lambda) = \phi \text{ or } \phi = 0 \text{ and } \lambda = \pi/2 \}.$$

Moreover, there is a parametric representation of S_f

$$L: [0,\infty) \to P \times \mathbb{R}^+$$

such that if $L(a) = (\phi_a, \lambda_a)$, then $a = \|\phi_a\|$.

b) If in addition to property (B) we assume that f satisfies:

$$\beta := \lim_{x\to\infty} f(x)/x \quad \text{exists, allowing } \beta = \infty,$$

$$F(y) := \int_0^y f(x)\,dx \to \infty \text{ as } x \to \infty,$$

and if we let λ_a denote the second component of L(a), for $a \in \mathbb{R}^+$, then we obtain:

$$\lambda_a \to \frac{\pi}{2\beta} \quad \text{as } a\to\infty \text{ if } \beta \neq 0,$$

$$\lambda_a \to \infty \quad \text{as } a\to\infty \text{ if } \beta = 0.$$

Proof: a) Let $(\phi,\lambda) \in P\times\mathbb{R}^+$ be such that ϕ is a nonzero fixed point of $T_{\lambda f}$ and let x be the corresponding periodic solution of (E_λ) of type (H). Define $y(.) := x(./\lambda)$, then one has that

$$y'(t) = {}^1\!/_\lambda\, x'({}^t\!/_\lambda) = -f(x({}^1\!/_\lambda\,(t-\lambda))) = -f(y(t-\lambda)), \text{ for all } t \in \mathbb{R}.$$

Hence, y is a periodic solution of $y'(t) = -f(y(t-\lambda))$. Since x is of type (H), one easily verifies that if we define $z(.) := y(.-\lambda)$, then (y,z) is a periodic solution of period 4λ of

$$(+)\begin{cases} y'(t) = -f(z(t)) \\ z'(t) = f(y(t)) \end{cases}$$

such that $y(0) = \|\phi\|$ and $z(0) = 0$.

On the other hand, using the technics of Kaplan and Yorke [2] it can be shown that given any $a > 0$, then there is a unique periodic solution (y_a, z_a) of (+) satisfying $y_a(0) = a$ and $z_a(0) = 0$. Moreover, one finds out that if the period of (y_a, z_a) is 4λ, then we have that $z_a(.) = y_a(.-\lambda)$. Thus, if we define $x_a(.) := y_a(\lambda .)$, then x_a is a periodic solution of (E_λ) of type (H) which implies that $T(x_a|[-1,0], \lambda) = x_a|[-1,0]$.

Therefore, we have a one to one correspondence between the set of all (ϕ,λ)

contained in $P \times \mathbb{R}^+$ such that ϕ is a nonzero fixed point of $T_{\lambda f}$, and the set of all nontrivial periodic solutions (y,z) of (+) satisfying $y(0) > 0$ and $z(0) = 0$.

Now, the mapping

$$p: \mathbb{R}^+ \to \mathbb{R}^+$$
$$a \to \text{period of } y_a$$

is continuous and can be continuously extended at 0 by 2π (for a proof of this see Kaplan and Yorke [2]).Furthermore, if for all $a > 0$, we let ϕ_a denote the restriction of $y_a(p(a)/_4 \cdot)$ to $[-1,0]$,then the mapping $a \to \phi_a \in P$ can easily be shown to be continuous and continuously extendable at 0 by $0 \in P$. Thus,

$$L: \mathbb{R}^+ \cup \{0\} \to P \times \mathbb{R}^+$$
$$a \to (\phi_a, p(a)/_4)$$

is a parametric representation of S_f,and the first part of the theorem is now obvious.

b) The additional assumptions on f imply (for a proof of this again see Kaplan and Yorke [2]) that

$$p(a) \to \frac{2\pi}{\beta} \quad \text{as} \quad a \to \infty \quad \text{if} \quad \beta \neq 0,$$

$$p(a) \to \infty \quad \text{as} \quad a \to \infty \quad \text{if} \quad \beta = 0.$$

From this and part a) of the proof, the assertion follows.

Finally, we will discuss two examples of global bifurcation as an application of theorem 1.7 . We need the following lemma:

(2.3) Lemma: Let $f: \mathbb{R} \to \mathbb{R}$ be a continuous, odd and monotonic increasing function.

a) Suppose that there is an $a > 0$ such that

$$\text{(2.3.1)} \quad f(x) > \pi/_2\, x \, , \quad \text{for all } x \geq a \, ,$$

and

$$(2.3.2) \qquad \int_y^a f(x) - \tfrac{\pi}{2} x \quad dx > 0 \qquad \text{for all } y \in [0,a).$$

Then every fixed point ϕ of T_f satisfies:

$$\|\phi\| < \tfrac{\pi}{2} a.$$

b) Suppose that there is a $b > 0$ such that

$$(2.3.3) \qquad f(x) < \tfrac{\pi}{2} x \qquad \text{for all } x \geq b$$

and

$$(2.3.4) \qquad \int_y^b f(x) - \tfrac{\pi}{2} x \quad dx < 0 \qquad \text{for all } y \in [0,b).$$

Then every fixed point ϕ of T_f satisfies:

$$\|\phi\| < b.$$

Proof: Let us first prove part b).
Suppose there is a fixed point ϕ of T_f such that $\|\phi\| \geq b$. Set

$$c_o := \inf\{c > \|\phi\| \mid c \cos(\tfrac{\pi}{2} t) \geq \phi(t) \text{ for all } t \in [-1,0]\}$$

and

$$\phi_o(.) := c_o \cos(\tfrac{\pi}{2} .).$$

Using (2.3.3) it can easily be shown that

i) $\phi \leq \phi_o$

and

ii) $\phi(t_o) = \phi_o(t_o)$ for some $t_o \in (-1,0]$.

Now, (2.3.3) and (2.3.4) imply, parallel to arguments in the proof of (1.7) that

$$T_f\phi_o(t) < \phi_o(t) \qquad \text{for all } t \in (-1,0].$$

Hence, in particular this inequality is true for $t = t_o$. However, since f is monotonic increasing, T_f is monotone and therefore i) implies $T_f\phi \leq T_f\phi_o$.
Thus, we obtain that

$$\phi(t_o) = T_f\phi(t_o) \leq T_f\phi_o(t_o) < \phi_o(t_o) = \phi(t_o),$$

and this is a contradiction.

Proof of part a):

Suppose there exists a fixed point ϕ of T_f satisfying $\|\phi\| \geq \pi/2\, a$. Then ϕ is monotonic increasing and since $\phi'(t) = f(\phi(-t-1))$, $-1 < t < 0$, the monotonicity of f implies that ϕ' is monotonic decreasing. From this and the fact that $\|\phi\| = \phi(0) \geq \pi/2\, a$ it follows that $\phi(t) \geq \pi/2\, at \geq a \cos(\pi/2\, t)$, $-1 \leq t \leq 0$. Therefore we have that

$$a \leq c_o := \sup\{c < \|\phi\| \mid c \cos(\pi/2\, t) \leq \phi(t) \text{ for } t \in [-1,0]\}.$$

The further arguments of the proof are similar to those of part b).

(2.4) Two examples:

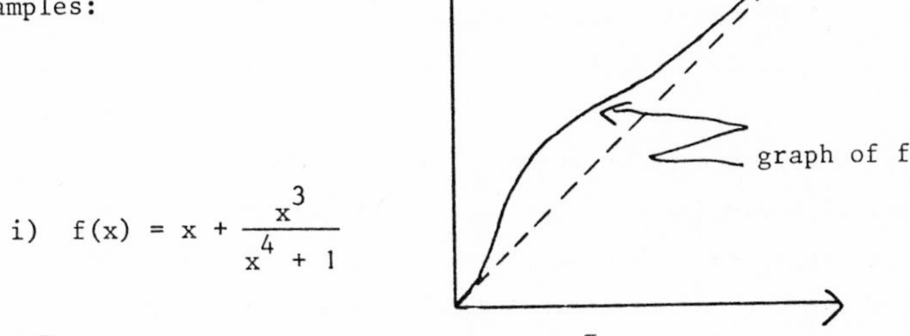

i) $f(x) = x + \dfrac{x^3}{x^4 + 1}$

For every $\lambda < \pi/2$ which is sufficiently close to $\pi/2$ we know from theorem 1.7 that $T_{\lambda f}$ has a nonzoro fixed point ϕ_λ and, moreover, that $\|\phi_\lambda\| \to \infty$ as $\lambda \to \pi/2$. On the other hand lemma 2.3 b) will imply that there is no nonzero fixed point for $T_{\lambda f}$ if $\lambda \geq \pi/2$. Furthermore, we know from theorem 2.2 that the set S_f is a curve which branches of from $(0, \pi/2) \in P \times \mathbb{R}^+$. Hence, the following figure should be a reasonable illustration of the global behaviour of S_f. In fact, this has been observed with numerical studies (cf. Prüfer [9] and Peitgen and Prüfer [7]).

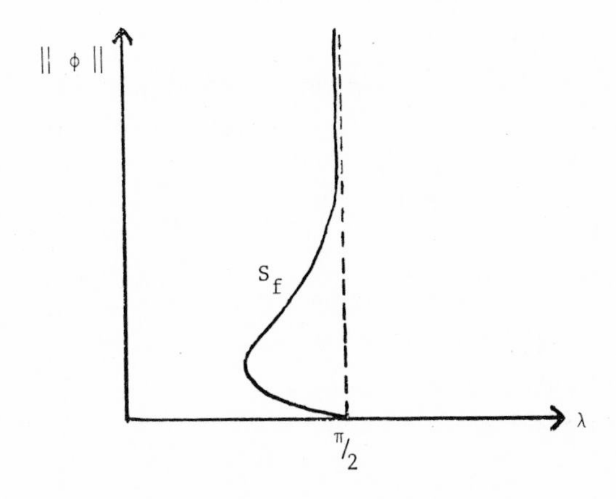

ii) $f(x) = x + c \sin^3(\pi/_2\, x)$, where $c > 0$ is assumed to be so small that f is monotonic increasing.

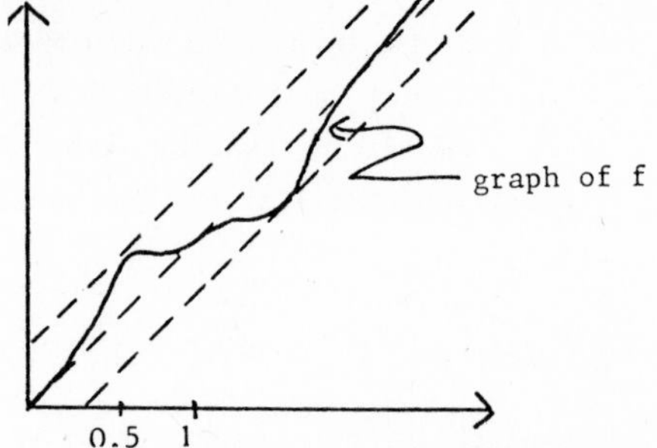

Using theorem 1.7 one can easily show that for $\lambda = \pi/_2$ and for each $n \in \mathbb{N}$ there is a fixed point ϕ_n for $T_{\lambda f}$ which satisfies $4n-2 \leqq \| \phi_n \| \leqq 4n$. If $\lambda \neq \pi/_2$, then lemma 2.3 will imply that the set of fixed points for $T_{\lambda f}$ is bounded. On the other hand, if λ is sufficiently close to $\pi/_2$, then theorem 1.7 can be applied repeatedly to obtain finitely many fixed points for $T_{\lambda f}$. However, the total number of these fixed points is increasing with $\lambda \to \pi/_2$. Thus, the following figure should be a reasonable illustration of the global behaviour of S_f, and in fact, this has been observed with numerical studies.

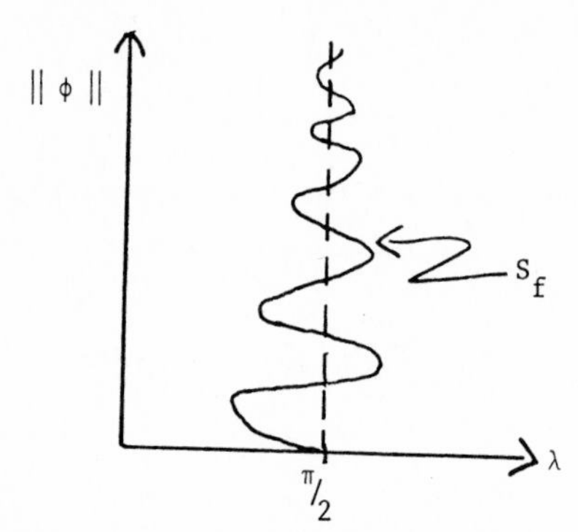

For further examples see Prüfer [9] and Peitgen and Prüfer [7].

ACKNOWLEDGMENT:

I should like to thank Prof. H.O. Peitgen and my colleagues for their advice and many helpful discussions.
I am especially grateful to M. Prüfer who placed the results of his numerical experiments at my disposal which were of great importance for my intuitive

understanding of global bifurcation. Likewise, I wish to thank H. Peters who made, in our discussions about his and my studies, many important suggestions which became the origine of many an idea concerning this paper.

REFERENCES:

1. S.N. Chow, Existence of periodic solutions of autonomous functional differential equations, J. Diff. Equations 15 (1974), 350 - 378.

2. J.L. Kaplan, J.A. Yorke, Ordinary differential equations which yield periodic solutions of differential delay equations, J. Math. Anal. Appl. 48 (1974), 317 - 324.

3. J.L. Kaplan, J.A. Yorke, On the stability of a periodic solution of a differential delay equation, SIAM J. Math. Anal. 6 (1975), 268 - 282.

4. R.D. Nussbaum, Periodic solutions of some nonlinear, autonomous functional differential equations II, J. Diff. Equations 14 (1973), 360 - 394.

5. R.D. Nussbaum, A global bifurcation theorem with applications to functional differential equations, J. Func. Anal. 19 (1975), 319 -338.

6. R.D. Nussbaum, Uniqueness and nonuniqueness for periodic solutions of $x'(t) = -g(x(t-1))$, to appear.

7. H.O. Peitgen, M. Prüfer, The Leray-Schauder continuation method is a constructive element in the numerical study of nonlinear eigenvalue and bifurcation problems, these proceedings

8. M. Prüfer, Calculating global bifurcation, Continuation methods, Ed. H.J. Wacker (1978) Ac. Press, 187 -213.

9. M. Prüfer, Simpliziale Topologie und globale Verzweigung, Thesis, Bonn (1978).

10. H.O. Walther, A theorem on the amplitudes of periodic solutions of delay equations with applications to bifurcation, J. Diff. Eq. 29 (1978) 396 - 404.

11. H.O. Walther, On instability, ω-limit sets and periodic solutions of nonlinear autonomous differential delay equations, these proceedings.

ADDRESS: N. Angelstorf, Institut für Angewandte Mathematik der Universität, 5300 Bonn, Wegelerstr.6, Fed. Rep. Germany

EXISTENCE OF OSCILLATING SOLUTIONS FOR CERTAIN DIFFERENTIAL EQUATIONS WITH DELAY

O. ARINO and P. SÉGUIER

ooo

INTRODUCTION.

Consider the equation : $x'(t) = f(x(t), x(t-\omega))$, where $\omega > 0$, $f(0,0) = 0$, $v \to f(u,v)$ is strictly increasing.

This equation has the constant 0 as a solution which is semi-ejective (see hypothesis 1.3.H_1), moreover, the application : $\varphi \to x$ is non decreasing (see 1.2.1).

The problem then set is to determine the existence of solutions oscillating near to 0.

Our main result (theorem 3.2.5) can be given the following form : suppose that

(A) | 0 is a saddle-point with an unstable manifold of a dimension not less than two ;

(B) | There is no non zero oscillating solution defined on $\mathbb{R}$ going to 0 at $\pm\infty$;

Then : there exist non damping oscillating solutions.

In part 4, we apply this result to the following equation :

$$x'(t) = \alpha x(t-\omega) + \beta g(x(t-\omega)),$$

where essentially g is C^1 ; $g(0) = g'(0) = 0$; g' is bounded ; $\alpha > 0$ such that (A) is true ; β is small enough.

A quite different example will be mentioned in a remark.

PART 1

1.1 Let ω be a strict. posit. real constant.

a) We denote by C the space of the continuous functions from $[-\omega,0]$ into $\mathbb{R}$ with the sup. norm, that is : $|\varphi| = \sup_{\theta \in [-\omega,0]} |\varphi(\theta)|$.

b) C_+ is the cone of the positive functions of C.

c) We define on C the usual order :

$$\Psi \leqslant \varphi \Longleftrightarrow \forall t \in [-\omega,0] \; , \; \Psi(t) \leqslant \varphi(t) .$$

Sometimes, we will have to make the following distinction.

1. $\Psi \lneqq \varphi$ (resp. $\gneqq$) $\Longleftrightarrow$ $\Psi \leqslant \varphi$ and $\Psi \neq \varphi$. (resp. $\Psi \geqslant \varphi$ and $\Psi \neq \varphi$).

2. $\Psi < \varphi$ (resp. >) $\Longleftrightarrow$ $\forall t$, $\psi(t) < \varphi(t)$ (resp. $\Psi(t) > \varphi(t)$).

1.2 The problem set. More notations and fundamental definitions. We consider the equations of the type $x'(t) = f(x(t),x(t-\omega))$ with $x(t) = \varphi(t)$, $t \in [-\omega,0]$ where φ is given in C. The function f should satisfy some hypotheses that will be enumerated below and which will ensure first of all the existence and the unicity of the solution of this problem. Let us recall a few classical notations :

1.2.1 Notations : for a given φ, we denote $x(.,\varphi)$ the corresponding solution. However, when there is no possible confusion, this same object is denoted either x, or x(.), or x(t), or even $x(\varphi)$ when only the dependance on φ is considered.

1.2.2 Notations : If x is a solution to the equation, for every compatible t, we denote by x_t the function defined by :

$$x_t(\theta) = x(t+\theta) \; , \; \theta \in [-\omega,0] .$$

When we wish to insist on the dependence on φ, we write : $x_t(.,\varphi)$ or $x_t(\varphi)$ if there is no possible confusion.

The objective of our work is to obtain theorems giving oscillating solutions (more particularly, oscillating near to 0, which is a stationary point in the considered equations) and specifying the behaviour of these solutions.

It is now necessary to recall some definitions and to specify the hypotheses.

1.2.3 Definition : We call a function oscillating near to 0, any function x such that : $\forall t$, $\exists t'$, $t' > t$, $x(t') = 0$.

1.2.4 Definition : We denote by P (resp. P_b), (resp. $\bar{P}_o$), the set of the functions in C such that $x(\varphi)$ is oscillating near to 0, (resp. $\varphi \in P$ and $x(\varphi)$ is bounded at the infinity), (resp. $\varphi \in P_b$ and $x(\varphi)$ goes to 0 when t goes to the infinity).

Let us say here that we will also use the notions of a "non damping" solution and of a "totally non damping" solution of which we will give later on the definitions.

1.3 Hypotheses on f :

H_o : The equation can be written in the classical way ([8]) :

$x'(t) = L(x_t) + N(x_t)$ with L linear and continuous from C into $\mathbb{R}$;

N nonlinear from C into $\mathbb{R}$ verifying : $N(0) = 0$, and there exists $\mu : \mathbb{R}_+ \to \mathbb{R}_+$, continuous and non decreasing with : $\mu(0) = 0$, such that :

$$|N(\varphi_1) - N(\varphi_2)| \leq \mu(\delta) \quad |\varphi_1 - \varphi_2| \ , \ \forall \varphi_1, \varphi_2 \in V_\delta = \{\varphi \in C : |\varphi| < \delta\}.$$

$H_{1,\alpha}$ (semi-ejectivity of the constant solution 0) $(\alpha > 0)$:

For any φ in V_α , $\varphi \neq 0$, the following properties are equivalent :

1. $\exists t_o$, $x_{t_o}(\varphi) \geq 0$ (resp. $x_{t_o}(\varphi) \leq 0$) ;

2. $\exists t'$; $\forall t \geq t'$, $x_t(\varphi) > 0$ (resp. $x_t(\varphi) < 0$).

$H_{1,\alpha,\beta}$: $H_{1,\alpha}$ with 0 replaced by $\beta > 0$ (resp. $\beta < 0$), for some β depending on α, in the inequality of 2.

H_{1,α,β^o} : $H_{1,\alpha,\beta}$ with β depending on φ.

H^*_{1,α,β^o} : H_{1,α,β^o} with $t_o = 0$.

H_1 : that will be another name of $H_{1,\infty}$.

$\underline{H_2}$ (resp. $\underline{H_2'}$) : monotony ; (resp. strict monotony) with the notations in 1.1.c.

$$\varphi \geqslant \Psi \implies x(\varphi) \geqslant x(\Psi) ,$$

$$\varphi > \Psi \implies x(\varphi) > x(\Psi) ,$$

(resp. the same as above and : if φ and Ψ are in P and satisty : $\varphi > \Psi$, then there exists a τ such that : $t > \tau \implies x_t(\varphi) > x_t(\Psi)$.

$\underline{H_3}$: each solution can be extended into a solution defined on $\mathbb{R}^+$.

1.3 <u>Remark</u> : $H_2' \implies H_1$

PART 2

We obtain here the existence of solutions oscillating near to 0 for the equations satisfying all or part of the hypotheses (H_i) in part 1.

Let us recall first of all some classical results ([8]). We suppose the hypothesis H_o holds in 2.1.

2.1.1 <u>Definition</u> : We denote U (respectively : S) the set of the data of the solutions x of $x'(t) = L(x_t)$ such that : x(t) exists in $\mathbb{R}_-$ and goes to 0 when t goes to $-\infty$ (respectively : x(t) goes to 0 when t goes to $+\infty$).

2.1.2 <u>Remarks</u> :

a) U is in fact the generalized eigenspace corresponding to the eigenvalues with a strictly positive real part of the infinitesimal generator of the semi-group associated with L, that is, corresponding to the zeros with a strictly positive real part of :

$$D(\lambda) = \det(\lambda I - \int_{-\omega}^{0} e^{\lambda\theta}\, d\eta(\theta)) = 0 \quad ([8]) .$$

b) U is finite dimensional.

2.1.3 <u>Theorem</u> [8] : <u>If $D(\lambda) = 0$ has no imaginary root, then C can be decomposed as follows : $C = U \oplus S$</u> .

2.1.4 An example : Let us consider the equation

(1) : $x'(t) = -x(t) + \sqrt{x(t-\omega)}$. It can be written (once the function x has been changed into y, $x = y + 1$) :

$$y'(t) = -y(t) + \frac{1}{2} y(t-\omega) + \left[\sqrt{y(t-\omega) + 1} - 1 - \frac{1}{2} y(t-\omega) \right]$$

$$= L(y_y) + N(y_t) \text{ , with } L(\varphi) = - \varphi(0) + \frac{1}{2} \varphi(-\omega) .$$

The characteristic equation is : $\lambda = -1 + \frac{1}{2} e^{-\lambda\omega}$, for which there are only roots with strictly negative real parts.

We can then assert (anticipating the result 2.1.5) that all the solutions of the equation (1) which are globally near to 1 tend towards 1 at the infinity. But, here, the monotony leads to a better result : all the solutions have 1 as a limit at the infinity ([19]). This result is the same as in the linear case.

2.1.5 Theorem ([8]) : Let us consider the equation $x'(t) = L(x_t) + N(x_t)$, with the hypothesis H_o. Let $D(\lambda) = \det(\lambda I - \int_{-\omega}^{o} e^{\lambda\theta} d\eta(\theta))$. If the equation $D(\lambda) = 0$ has no imaginary root and at least one root with a strictly positive real part, then :

1. $C = U \oplus S$;

2. a) There exists $\delta > 0$, $\delta' > 0$ and $\sum^+$, a Lipschitz-submanifold of C, tangent to S at 0, such that :

$$(\varphi \in C \text{ , } |\Pi_S \varphi| < \delta' \text{ , } |x_t(\varphi)| < \delta \text{ , } t \geqslant 0) \Leftrightarrow \varphi \in \sum^+ ,$$

where Π_S is the projection along U of C onto S.

3. a) (Π_S, S) is a local chart of $\sum^+$.

2. b) There exist $\xi > 0$, $\xi' > 0$ and $\sum^-$, a Lipschitz submanifold of C ; tangent to U at 0, such that :

$$(\varphi \in C \text{ , } |\Pi_U \varphi| < \xi' \text{ , } |x_t(\varphi)| < \xi \text{ , } t \leqslant 0) \Leftrightarrow \varphi \in \sum^- ,$$

where Π_U is the projection along S of C onto U.

3. b) (Π_U, U) is a local chart of $\sum^-$.

4. Moreover, there exist constants $M > 0$, $\gamma > 0$, such that :

a) $$|x_t(\varphi)| \leqslant Me^{-\gamma t}|\varphi| \quad , \quad t \geqslant 0 \quad , \quad \varphi \in \Sigma^+$$

b) $$|x_t(\varphi)| \leqslant Me^{\gamma t}|\varphi| \quad , \quad t \leqslant 0 \quad , \quad \varphi \in \Sigma^- .$$

2.1.6 Notation : We will denote by

$$\sigma_{+,-,0} = \{\lambda \; ; \; D(\lambda) = 0 \; , \; \mathrm{Re}\lambda > 0 \; , \; < 0 \; , \; = 0\} .$$

2.2 Existence theorem for solutions oscillating near to 0.

2.2.1 Theorem : with the following hypotheses : 1/ $H_{1,\alpha}$; 2/ The mapping $\varphi \to x_t(\varphi)$ is continuous, for every $t \geqslant 0$; 3/ H_3 ; then : $P \setminus \{0\} \neq \emptyset$. (*)

Proof : Let φ be a positive function in V_η, $\varphi \neq$ const., where η is any positive real with $\eta < \alpha$, α given in 1.3.$H_{1,\alpha}$ consider the family $(\varphi_\lambda)_{\lambda \in [0,1]}$ defined by : $\varphi_\lambda = \lambda\varphi - (1-\lambda)\frac{\eta}{2}$. obviously, $\varphi_\lambda \in V_\eta$.

Moreover, $\varphi_o = -\frac{\eta}{2}$, $\varphi_1 = \varphi$. Therefore, from $H_{1,\alpha}$, it follows that : there exists t', such that : $t > t' \Rightarrow x_t(\varphi_o) < 0$, $x_t(\varphi_1) > 0$.

Define the sets :

$$\Lambda^{\pm} = \{\lambda \in [0,1] \; ; \; \exists t_\lambda \; ; \; \pm\, x_t(\varphi_\lambda) > 0 \; , \; t \geqslant t_\lambda\} .$$

They are non empty open sets. Indeed, $\Lambda^+ \ni 1$, $\Lambda^- \ni 0$, and, if $\lambda \in \Lambda^+$, then, there exists t_λ such that : $x_{t_\lambda}(\varphi_\lambda) > 0$. Now, define the set

$$M = \{\mu \in [0,1] \; : \; x_{t_\lambda}(\varphi_\mu) > 0\} .$$

M is a neighbourhood of λ, as a result of the continuity of the mapping $\mu \to x_{t_\lambda}(\varphi_\mu)$; $M \subset \Lambda^+$. Hence, Λ^+ is open. The same argument works for Λ^-.
As $\Lambda^+ \cap \Lambda^- \neq \emptyset$, the connexity of $[0,1]$ implies that :

$$\Lambda^+ \cup \Lambda^- \neq [0,1].$$

Then, all the functions φ_λ, $\lambda \in [0,1] \setminus \Lambda^+ \cup \Lambda^-$ are data of solutions oscillating near to 0.

(*) an equivalent proof could be done using the sets :

$$V^{\pm} = \{\varphi \; , \; |\varphi| < \alpha \quad ; \quad \exists t_\varphi \; , \; \pm\, x_t(\varphi) > 0 \; , \; t > t_\varphi\}$$

2.2.2 (closedness of P) : with the same hypotheses as in 2.2.1 and, in addition, the hypothesis of the continuity locally uniform in t, then : P is closed in

$$V_\alpha = \{ \varphi , |\varphi| < \alpha\} .$$

Proof : Let $(\varphi_n)_{n \in \mathbb{N}}$ be a convergent sequence in P, with φ as a limit, φ and φ_n being in V_α. By continuity, we have : $x_t(\varphi_n) \to x_t(\varphi)$. Let us show that $x_t(\varphi)$ has at least one zero : now, for every $n \in \mathbb{N}$, there exists $\theta_n \in [-\omega,0]$, such that : $x_t(\theta_n,\varphi_n) = 0$ because of condition $H_{1,\alpha}$. We can find a convergent subsequence of $(\theta_n)_{n \in \mathbb{N}}$ with the limit θ. Then :

$$x_t(\theta,\varphi) = x(t+\theta,\varphi) = \lim_{n_k \to \infty} x(t+\theta_{n_k},\varphi_{n_k})$$

(by the hypothesis of uniformity) = 0.

2.2.3 Remark :

With $H_{1,\alpha,\beta}$, it follows immediately that the solutions which stay in an appropriate neighbourhood of 0 are oscillating near to 0.

With H^*_{1,α,β^o}, we can deduce, with the hypotheses and notations of the theorem 2.1.3, that there exists $\varepsilon > 0$, s.t. : $\sum^+ \cap V \subset P$.

In paragraph 3., we shall obtain additional information on the nature of the oscillations by comparing $\sum^+$ and P.

PART 3

We will suppose in the following the hypotheses H_o and H_1, and, to make it easier, we will suppose that f is C^1-which ensures in the theorem 2.1.3 that $\sum^+$ is of class C^1.

We will now specify the character of the oscillations. We give two results related to the number of the roots with strictly positive real part of the equation $D(\lambda) = 0$.

We need the following definitions and lemmas :

3.1.1 Definition : We shall say that a solution oscillating near to 0 is non damping if : $\exists \varepsilon > 0$ such that : $\forall t$, $\exists t' > t$ with : $|x(t')| > \varepsilon$.

3.1.2 Definition : We shall say that a solution oscillating near to 0 is totally non-damping if : $\exists \varepsilon > 0$ such that : $\forall t$, $\exists t' > t$, $t'' > t$ with : $x(t') > \varepsilon$ and $x(t'') < -\varepsilon$.

3.1.3 Lemma : If a bounded solution x is oscillating near to 0 and non-damping, then, with the hypothesis H_1, we can say that this solution is totally non-damping.

Proof : Let x be a bounded solution, oscillating near to 0 and non-damping. Since x is non damping, there exists $\varepsilon_o > 0$, such that : $\forall t$, $\exists t' > t$ with $|x(t')| > \varepsilon_o$, which corresponds to the definition 3.1.1. and to at least one of the two properties of the definition 3.1.2. We will show that the hypothesis H_1 gives them both at the same time. Let us assume the contrary, and, for example, that for each $\varepsilon > 0$, there exists t such that : $\forall t'' > t$, $x(t'') > -\varepsilon$.

Then, we would have : $\inf_{s \geq t} x(s) \geq -\varepsilon$, and so : $\lim_{t \to +\infty} (\inf_{s \geq t} x(s)) \geq 0$.

Let us choose a sequence $(t_n)_{n \in \mathbb{N}}$, which verifies : $t_n \to +\infty$ when $n \to +\infty$ and $x(t_n) > \varepsilon_o$, and define $\tilde{\varphi}_n = x_{t_n}$. The sequence $(\tilde{\varphi}_n)_{n \in \mathbb{N}}$ is bounded in C^1, since x is bounded ; it has a convergent subsequence $(\tilde{\varphi}_{n_k})_{k \in \mathbb{N}}$ in C which converges to $\tilde{\varphi}$.

We have : $\tilde{\varphi}(0) \geq \varepsilon_o$ and $\tilde{\varphi} \geq 0$, because :

$$\tilde{\varphi}(\theta) = \lim_{k \to \infty} \tilde{\varphi}_{n_k}(\theta) = \lim_{k \to \infty} x(t_{n_k} + \theta)$$

$$\geq \lim_{k \to \infty} (\inf_{s \geq t_{n_k} + \theta} x(s)) = \lim_{t \to \infty} (\inf_{s \geq t} x(s)) \geq 0.$$

If then we consider the solution corresponding to $\tilde{\varphi}$, we obtain by construction (see 2.2.2) an oscillating solution, but as $\tilde{\varphi} \geq 0$ and $\tilde{\varphi}(0) \geq \varepsilon_o$, according to the hypothesis H_1, we should have : $x(t) > 0$, $t > t'$ for some t', which is a contradiction.

3.1.4 Lemma : If the equation has two bounded oscillating solutions corresponding to data φ and Ψ such that : $\varphi \gneq \Psi$, then, with the hypotheses H_2' and if $\sigma_o = \emptyset$ and $\sigma_+ \neq \emptyset$ (see 2.1.6), the solutions $x(\chi)$, $\Psi \leqslant \chi \leqslant \varphi$ are non damping (and of course, oscillating near to 0).

Proof :

1/ Let us assume first of all that $x(\varphi)$ and $x(\Psi)$ go to zero at the infinity. It will be the same for the $x(\chi)$, from the monotony hypothesis. Therefore, for each $\varepsilon > 0$, there exists t_o, such that : $\forall s \geqslant 0$,

$$(1) : \quad \{x_s(\Phi) \; ; \; x_{t_o}(\Psi) < \Phi < x_{t_o}(\varphi)\} \subset V_\varepsilon \; .$$

Define $\Omega_\varepsilon = \{\Phi \; , \; x_{t_o}(\Psi) < \Phi < x_{t_o}(\varphi)\}$; from H_2', it follows that Ω_ε is a non empty open set ; and, with Ω_ε, (1) can be written :

$$\Phi \in \Omega_\varepsilon \Longrightarrow |x_s(\Phi)| < \varepsilon \; , \; s \geqslant 0 \; .$$

Take ε small enough, and it follows, with the notations of theorem 2.1.5, that : $\Omega_\varepsilon \subset \sum^+$, which is impossible since $\sum^+$ is a submanifold of a finite and $\neq 0$ codimension in C.

2/ All we need to do now is to show that, for example, if $x(\varphi)$ is non damping, so is $x(\Psi)$. For this, we merely notice the following implications :

$x(\varphi)$ non damping $\Longrightarrow$ $x(\varphi)$ totally non damping, from lemma 3.1.3.

$x(\varphi)$ totally non damping $\Longrightarrow$ $x(\chi)$ non damping, for each χ, $\chi \geqslant \varphi$ or $\chi \leqslant \varphi$, from the monotony hypothesis.

3.1.5 Remarks :

a) The hypothesis H_2' is in some way necessary, in the sense that, without H_2', the set Ω_ε could be empty. We should have to take, in that case, a set of the type $\Omega_\varepsilon' = \{\Phi : x_{t_o}(\Psi) \lneq \Phi \lneq x_{t_o}(\varphi)\}$ which can be a part of a strict submanifold.

For example, $Z = \{\Phi \; ; \; \Phi(0) = 0\}$ is a submanifold of codimension 1 dans Z contains the set Ω, $\Omega = \{\Phi : a \lneqq \Phi \lneqq b\}$, where $a : t \to t^2$, $b : t \to t$,

$t \in [0,1]$.

b) Notice that we do not need in the above lemmas the technical hypothesis H_3.

3.2 Nature of the oscillations : In what follows, we will assume that σ_o is empty.

3.2.1 Theorem : With the hypothesis H_2' and if σ_+ has a single term, which is a simple eigenvalue then : $\exists \eta' > 0$, such that : $\varphi \in P_b \cap V_{\eta'} \Longrightarrow \varphi \in \Sigma^+$. So, there exists a neighbourhood of 0 in C, such that the solutions oscillating near to 0 which are bounded and correspond to data in this neighbourhood, go to 0 at $+\infty$).

Proof : All we need to do is to show that there exists a neighbourhood of 0 in C such that : these φ in the intersection of P_b and the neighbourhood are in Σ^+.

Choose $\tilde{\varphi} < 0$ (*) in $C \setminus S$ and a mapping $f : C \to \mathbb{R}$, C^1, defining Σ^+ in a neighbourhood of 0, that is,

$$\Sigma^+ = \{\sigma \in C \; ; \; f(\sigma) = 0\} \; , \text{ with } f(0) = 0 \; , \; Df(0) \neq 0, \text{ and } \ker Df(0) = S.$$

Show first of all that any straight line starting from $\tilde{\varphi}$ which stays in a cone of the type $\tilde{\varphi} + \mathbb{R}^+ V_\varepsilon$, for some $\varepsilon > 0$, meets Σ^+. (see figure (1)) :

(1)

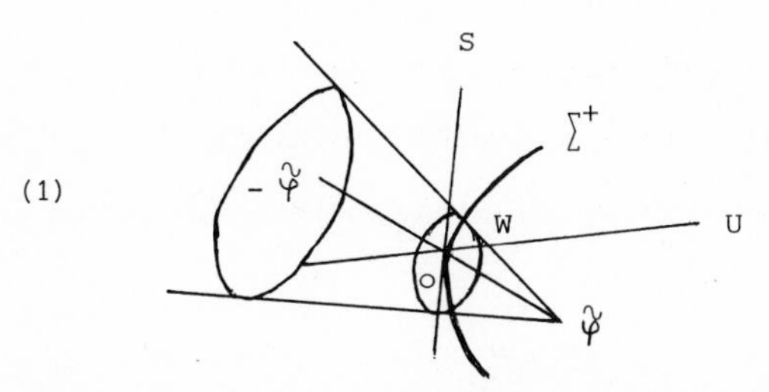

From the characterization of Σ^+ by the mapping f, it is sufficient to show that for (μ,Ψ) near to $(1,-\tilde{\varphi})$ the equation $f(\mu\Psi + \tilde{\varphi}) = 0$ has a solution. Set $g(\mu,\Psi) = f(\mu\Psi + \tilde{\varphi})$: g is C^1 ; moreover, $D_\mu g(\mu,\Psi) = Df(\mu\Psi + \tilde{\varphi}) \cdot \Psi$; in particular ; $D_\mu g(1,-\tilde{\varphi}) = -Df(0) \cdot \tilde{\varphi} \neq 0$ since $\tilde{\varphi} \notin S$. So, the implicit function theorem gives

(*) It is possible : if not, it would imply that $C^- = \{\varphi \; ; \; \varphi \in C \; , \; \varphi(t) < 0 \; , \; t \in [-\omega,0]\}$ is a part of S and since $C = \mathbb{R} C^-$, then $S = C$.

us the existence of a neighbourhood V de $-\tilde{\varphi}$, and of a function $h : V \to \mathbb{R}$, C^1, such that : $h(-\tilde{\varphi}) = 1$, and, for each Ψ in V, $g(h(\Psi),\Psi) = 0$, that is : $\tilde{\varphi} + h(\Psi)\cdot\Psi \in \sum^+$.

It is then always possible to find a neighbourhood V of 0, such that any half straight line starting from $\tilde{\varphi}$ and passing through W meets V and therefore $\sum^+$.

We can also choose W small enough in such a way that any $\varphi \in W$ verifies $\varphi > \tilde{\varphi}$. Let there be a function φ in W such that the solution $x(\varphi)$ is oscillating and bounded. The half line starting from $\tilde{\varphi}$ and passing through φ meets $\sum^+$ at a point σ. σ is such that one of the two following situations is true :

$$1. \quad \sigma \lneq \varphi \quad ; \qquad 2. \quad \varphi \lneq \sigma .$$

Applying the lemma 3.1.4 to the data σ and φ, we can see that this situation is only possible if $x(\varphi)$ and $x(\sigma)$ do not go to 0 at $+\infty$ (which is not the case of $x(\varphi)$), or if $\varphi = \sigma$.

Hence all the functions in $W \cap P_b$ are in $\sum^+$.

This theorem has the following corollary :

3.2.2 <u>Corollary</u> : <u>The mapping $T(.) : P_b \to C_b(\mathbb{R}^+,C)$ defined by : $\varphi \to x(\varphi)$ is continuous in 0 (where $C_b(\mathbb{R}^+,C)$ is given the sup norm on $\mathbb{R}^+$).</u>

<u>Proof</u> : Follows directly the theorem. Since, if φ goes to 0 in P_b, the theorem ensures that there exists $\eta' > 0$, such that : $|\varphi| < \eta' \Rightarrow \varphi \in \sum^+$; and, from the theorem 2.1.5. 4/, we have :

$$|x_t(\varphi)| \leqslant Me^{-\gamma t}|\varphi| \Rightarrow |x(\varphi)| \leqslant M|\varphi| \text{, which gives the result.}$$

Notice that if there exist bounded solutions oscillating near to 0 and non damping, their amplitude in each interval of length ω must necessarily be greater than η'. Hence, $\overline{(P_b \setminus P_o)} \cap P_o = \emptyset$.

3.2.3 <u>Theorem</u> : <u>With the hypotheses H_2', H_3, if U is at least two-dimensional and if $T(.) : P_o \to C_b(\mathbb{R}^+,C)$ is continuous in 0, then there exist solutions oscillating near to 0 and non damping.</u>

Proof : Let us assume the contrary. Then, since all the oscillating solutions go to 0 at $+\infty$, there exists a neighbourhood V such that : $V \cap P \subset \sum^+$ (T(.) is continuous in 0). Let there be $\varphi \in V$, with $\varphi > -\eta$, where η is a strictly positive given constant ; consider the family of functions $(\lambda\varphi - (1-\lambda)\eta)_{\lambda \in \mathbb{R}^+}$. According to the above hypothesis and the result of the lemma 3.1.4., there can only be a single walue of λ for which $\lambda\varphi - (1-\lambda)\eta \in P$. Set $\Lambda : \varphi \to \lambda_\varphi$, the mapping from V into $\mathbb{R}_+$, which, at each φ , associates the corresponding λ.

Admit for the moment that Λ is C^1 (in a neighbourhood of 0).

Let then $E = S_1 \oplus U$, where S_1 is a finite dimensional subspace of S, chosen so that $\mathbb{R}.1 \subset E$ (denoting by $\mathbb{R}.1$ the set of the constant functions). As S and $\sum^+$ are diffeomorphic, the restriction of Π_S^{-1} on S_1 is a diffeomorphism of S_1 onto $\sum^+ \cap E$. So, $\sum^+ \cap E$ is a submanifold of the same dimension as S_1, therefore has a codimension greater or equal to two in E from the hypothesis on U.

Consider $\Lambda|_E$, the restriction of Λ on E ; it is a mapping of class C^1 and verifying : $D(\Lambda|_E)(0) \neq 0$.

In fact, it is sufficient to show that one directional derivative is non zero. We merely determine the derivative along the constants. For that, we must, first of all, know λ_φ for such a φ. Now, if φ is a constant, $\lambda\varphi - (1-\lambda)\eta$ is a constant too, and according to the semi-ejectivity hypothesis (H_1) the only possibility for $\lambda\varphi - (1-\lambda)\eta$ being in P is that $\lambda\varphi - (1-\lambda)\eta = 0$; therefore : $\lambda_\varphi = \frac{\eta}{\eta + \varphi}$, and $D\Lambda|_E(0).1 = D\Lambda(0).1 = -\frac{1}{\eta}$ $(\neq 0)$.

Let us return to the proof of the contradiction :

as $\lambda_o = 1$, there exists a neighbourhood W of 0 in E, such that : $W \cap \Lambda|_E^{-1}(1)$ is a submanifold of codimension 1 in E, contained in $\sum^+ \cap E$ which, as we mentioned above, has a codimension greater or equal to two in E, and this gives a contradiction.

There remains to show that Λ is C^1 in a neighbourhood of 0. Let us verify first of all that it is continuous in 0.

Let ε be positive ; we show that there exists $\delta > 0$ such that :

$$|\varphi| < \delta \Rightarrow |\Lambda(\varphi) - \Lambda(0)| < \varepsilon .$$

Consider the family $\lambda\varphi - (1-\lambda)\eta$, $\lambda \in \mathbb{R}_+$:

$$|\varphi| < \delta \Rightarrow \forall\lambda > 0, - \lambda\delta - (1-\lambda)\eta \leqslant \lambda\varphi - (1-\lambda)\eta \leqslant \lambda\delta - (1-\lambda)\eta .$$

By the hypothesis H_2', we can then say :

a) $\lambda_\varphi \delta - (1-\lambda_\varphi)\eta > 0$ (otherwise, $\lambda_\varphi\varphi - (1-\lambda_\varphi)\eta$ being a negative function would not be in P).

b) $- \lambda_\varphi \delta - (1-\lambda_\varphi)\eta < 0$ (same argument).

It follows that : $\lambda_\varphi(\delta+\eta) \geqslant \eta$ and $\lambda_\varphi(\delta+\eta) \geqslant -\eta$, so : $\frac{\delta}{\delta + \eta} \leqslant \lambda_\varphi - 1 \leqslant \frac{\delta}{\eta - \delta}$, which ensures the continuity of Λ in 0. Now, we show that Λ is C^1 in a neighbourhood of 0. Σ^+ is a submanifold of codimension equal to the dimension of U. So, there exists a mapping $K : C \to \mathbb{R}^{\dim U}$, such that : K is C^1 ; $K(0) = 0$; Ker $DK(0) = S$; $DK(0)$ is surjective ; and : $\Sigma^+ = \{\varphi : K(\varphi) = 0\}$.

By the surjectivity of DK(0), each component of DK(0) is non zero.

Next, for φ quite close to 0, $\lambda_\varphi\varphi - (1-\lambda_\varphi)\eta$ is in Σ^+, therefore : $K[\Lambda(\varphi)\ \varphi - (1-\Lambda(\varphi))\ \eta] = 0$.

Consider the mapping $g_i : R \times C \to \mathbb{R}$ defined by $g_i(\lambda,\varphi) = K_i(\lambda\varphi - (1-\lambda)\eta)$ (i^{th} component of K). Then : g_i is C^1, $D_1 g_i(1,0) = DK_i(0)\cdot\eta$.
Since $1 \notin S$, $DK(0)\cdot 1 \neq 0$; so, there exists i such that $DK_i(0)\cdot 1 \neq 0$. So, Λ is nothing other than the mapping whose graph is the intersection of a neighbourhood of (1,0) with the set $\{(\lambda,\varphi) : g_i(\lambda,\varphi) = 0\}$, that is : Λ is C^1 in a neighbourhood of 0.

3.2.4 <u>Theorem</u> : <u>With the hypothesis H_3, if $\sigma_o = \emptyset$; $\sigma_+ \neq \emptyset$ and $P \setminus P_o = \emptyset$, then the following properties are equivalent</u> :

1. <u>T(.) is continuous in 0 on P_{o-}</u>;

2. There is no non trivial solution oscillating near to 0, which can be extended into a solution on the whole of $\mathbb{R}$, going to 0 at $\pm\infty$.

Proof :

1) $\Rightarrow$ 2). Suppose that $T(.)$ is continuous and there exists a function x, $x \neq 0$, defined on $\mathbb{R}$, verifying the equation, oscillating near to 0 and going to 0 at $\pm\infty$. Then, define the sequence $\varphi_n = x_{-n}$. It goes to 0 in C, since x goes to 0 at $-\infty$. But, obviously, $T(.)(\varphi_n)$ does not go to 0 in $C(\mathbb{R}^+, C)$, since $x \neq 0$. Notice that the hypothesis $P \setminus P_o = \emptyset$ is not necessary in this part.

2) $\Rightarrow$ 1). Suppose now that there is no solution oscillating near to 0, defined on $\mathbb{R}$, and going to 0 at $\pm\infty$. We will show that $T(.)$ is continuous in 0. Assume it is not so, and that there exists a sequence $(\varphi_n)_{n \in \mathbb{N}}$, $\varphi_n \in C$, $\varphi_n \to 0$ with $T(.)\varphi_n$ not going to 0, that is : there exist $\varepsilon > 0$ (which we can choose arbitrarily small) and a sequence $(t_n)_{n \in \mathbb{N}}$, $t_n > 0$, going to $+\infty$, such that : $|x(t_n, \varphi_n)| = \varepsilon$ and $|x(t, \varphi_n)| \leqslant \varepsilon$, for $t \in [-\omega, t_n]$.

Set $y(.,n)$, the function defined on $[-\frac{t_n}{2}, \infty]$ by $y(s,n) = x(t_n + s, \varphi_n)$.

There exists a subsequence of $(y(.,n))_{n \in \mathbb{N}}$ which converges in $C([-M,+M] ; \mathbb{R})$ for any $M > 0$ towards a function y defined on $\mathbb{R}$ and such that : $|y(s)| \leqslant \varepsilon$, $s \leqslant 0$; $|y(0)| = \varepsilon$; y is oscillating in $\mathbb{R}^+$.

Then, with ε small enough, $y_t \in \sum^-$, $t \leqslant 0$ and, from theorem 2.1.5 it follows that y goes to 0 at $-\infty$. As y is oscillating, by the hypotheses of the theorem, it goes to 0 at $+\infty$, which completes the contradiction.

We can now state the "main result" of this paper :

3.2.5 Theorem : With H_2', H_3, if U is at least two dimensional and there is no non zero oscillating solution defined on $\mathbb{R}$ going to 0 at $\pm\infty$, then there exist non-damping oscillating solutions.

Proof : Suppose the contrary. Then, the theorem 3.2.4 implies that T is continuous in 0 on P_o (= P). So the hypotheses of the theorem 3.2.3 are verified, which implies the existence of non-damping oscillating solutions, contradictorily.

3.2.6 Remark : In the linear case, the theorem 3.2.5 gives, but it is trivial, the points of U. If σ_o is empty, the solutions damping at $-\infty$ are oscillating and non-damping at $+\infty$, the oscillations becoming even larger and larger. This property does not hold in the nonlinear case.

PART IV

EXAMPLES :

4.1 $x'(t) = \alpha x(t-\omega).(1-x(t))$, $\alpha > 0$; $\omega > 0$; $\frac{\Pi}{2} < \alpha\omega < 2$.

Notice that : 0 and 1 are constant solutions ; moreover 1 is a separatrix, that is, any solution is, for $t \geqslant 0$, greater than 1 or less than 1, because of the term $(1-x(t))$ in the second member, and the unicity. In view of this remark, we can restrict the study of the equation to the set A of the functions φ such that : $x(\varphi) < 1$, that is $A = \{\varphi, \varphi < 1\}$.

4.1.1 Verification of the hypotheses H_i.

a) The equation can be written : $x'(t) = L(x_t) + N(x_t)$ with :

$$L(\varphi) = \alpha\varphi(-\omega) \quad , \quad N(\varphi) = -\alpha\varphi(-\omega)\varphi(0) \ ;$$

$$|N(\varphi_1) - N(\varphi_2)| \leqslant \alpha(|\varphi_1| + |\varphi_2|) \ |\varphi_1 - \varphi_2| \ .$$

b) In this example, $f(t,u,v) = \alpha.v(1-u)$; there $\frac{\partial f}{\partial v} = \alpha\cdot(1-u) \geqslant 0$, since $u < 1$; moreover, $|f(t,u_1,v) - f(t,u_2,v)| \leqslant \alpha \ |v|.|u_1 - u_2|$.

So, the hypothesis H_2 is verified ([19]).

c) The strict monotony comes from the injectivity of the mapping $\varphi \to x_t(\varphi)$ restricted to A and the fact that if $\varphi \leqslant \Psi$ with $\varphi(0) < \Psi(0)$ then $x(t,\varphi) < x(t,\Psi) \quad t > 0$.

d) We can easily show that, if $\varphi \not\leqslant 0$, then : $x(\varphi)$ is a decreasing function, going to $-\infty$ when t goes to $+\infty$; and, if $\varphi \not\geqslant 0$ then $x(\varphi)$ is an increasing function with the limit 1 at $+\infty$. So, 0 is semi-ejective.

4.1.2 About the characteristic equation : $D(\lambda) = 0$.

Here, $D(\lambda) = \lambda - \alpha e^{-\lambda\omega}$. There is always a single real root, which is strictly positive. Moreover, with the hypothesis $\frac{\Pi}{2} < \alpha\omega < 2$, we have the following results : firstly, $\sigma_o = \emptyset$. For, if we set $\lambda = a + ib$, $(ib = \alpha e^{-ib\omega} = \alpha \cos b\omega - i\alpha \sin b\omega) \Longrightarrow (\cos b\omega = 0) \Longrightarrow b\omega = \frac{\Pi}{2} + k\Pi$ $(b = \pm \alpha \Longrightarrow \alpha\omega = \pm (\frac{\Pi}{2} + k\Pi)$ which is incompatible with the above hypothesis. Secondly, σ_+ has a single term which is simple.

4.1.3 Conclusion of the example : All the conditions of the theorem 3.2.1 are satisfied. So, there exists a neighbourhood V_η of 0 in C such that all the bounded solutions oscillating near to 0, corresponding to data in V_η, go to 0 at $+\infty$. Here, we can say a little more because all the solutions oscillating near to 0 are bounded.

We do not know if there exist non-damping oscillating solutions but we can say from the above, that such a solution has an amplitude greater than η on each interval of length ω.

4.2 (E_β). $x' = \alpha x(t-\omega) + \beta g(x(t-\omega))$; $\beta . g$ is a small perturbation of the linear part ; we use on g the following hypotheses : g is C^1, $g(0) = g'(0) = 0$, g' is bounded, which leads to : $|g(r)| \leqslant c|r|$.

With $|\beta|$ small enough, $\alpha > 0$, we can verify in the same way as in the preceding example the hypotheses H_o, H_2'.

4.2.1 Application of the theorem 3.2.3. (existence of non-damping oscillating solutions) : Choose α and ω such that $\sigma_o = \emptyset$ and U is at least two dimensional.

We will now show that we can apply the theorem 3.2.5, that is, that there is no non zero oscillating solution of equation (E_β) defined on $\mathbb{R}$ and going to 0 at $\pm\infty$.

Notice that with $\sigma_o = \emptyset$ the equation $x' = \alpha x(t-\omega) + \varphi$, $\varphi \in \mathcal{S}'$ (the space of the temperate distributions) has a unique solution in $\mathcal{S}'$ which can be written :

$x = \mathcal{F}^{-1}(k(.)\mathcal{F}\varphi)$, where $\mathcal{F}$ is the Fourier transform and $k(\xi) = \dfrac{1}{i\xi - \alpha e^{-i\omega\xi}}$.

If φ is bounded, $x = \theta * \varphi$, where θ is in L^1.

Let there be x an oscillating solution defined on $\mathbb{R}$, going to 0 at $\pm\infty$ of (E_β).

We have : $x = \beta\,\theta * g(x(\cdot - \omega))$, from which it follows :

$$|x|_\infty \leqslant |\theta|_{L^1}\, |\beta| . C . |x|_\infty \text{ thus, } 1 \leqslant |\theta|_{L^1} . |\beta| . C$$

In conclusion, for $|\beta|$ small enough (e.g., $|\beta| \leqslant \dfrac{1}{|\theta|_{L^1} . C}$, (α,ω) such that $\sigma_o = \emptyset$ and U is at least two dimensional, the equation (E_β) has at least one solution oscillating near to 0 and non-damping.

4.3 Remark : Another class of examples is the following : $x'(t) = \alpha x(t-\omega) . f(x_t)$, where f is $\geqslant 0$, f = 1 in a neighbourhood of 0, α is such that U is at least two dimensional and $\sigma_o = \emptyset$. Here the method consists in studying the number of zeros on x_t which we denote by $N(x_t)$. We verify that : N is non increasing along the solutions ; $N/_{\Sigma^-} < N/_{\Sigma^+}$. It is therefore impossible to connect Σ^- to Σ^+ by a non trivial solution and, as a result, there is no non zero oscillating solution going to 0 at $\pm\infty$. We will give elsewhere a short extension of this result.

To conclude, we can say that the theorem 3.2.5 provides a relatively easy way of showing the existence of non-damping oscillating solutions. Another aspect that we have not approached in this paper and to which it could contribute concerns the study of the variety and the structure of these solutions. Notice that the verification of the conditions of the theorem leads naturally to questions, new or very little treated until now, for example : the existence of oscillating solutions going to 0 at $\pm\infty$.

We are extremely grateful to the referee for his many remarks. We should also like to thank Dr. H.O. WALTHER for his encouragement from the first hearing of the paper.

REFERENCES

[1] - N. CHAFEE : A bifurcation problem for a functional differential Equation of finitely retarded type. Journal of Mathematical Analysis and Applications 35.2. (1971), 312-349.

[2] - S.N. CHOW : Existence of periodic solutions of autonomous functional differential equations. Journal of differential equations. 15 (1974), 350-378.

[3] - S.N. CHOW - J. MALLET-PARET : Integral Averaging and bifurcation. Journal of Diff. Equat. vol. 26, N° 7. Oct. 77.

[4] - B.D. COLEMAN - G.H. RENNINGER : Periodic solutions of certain nonlinear functional equation. Istituto Lombardo. Accad. Sci. Lett. Rend. A. 109 (1975), 91-111.

[5] - R.B. GRAFTON : A periodicity theorem for autonomous functional differential equations. Archiv. for rational mechanics and analysis. 65.1. (1977), 87-95.

[6] - D. GREEN Jr. : Periodic solutions of functional differential equations with applications to epidemic models.

[7] - K.P. HADELER - J. TOMIUK : Periodic solutions of differential difference equations. Archiv. for rational mechanics and analysis. 65.1. (1977), 87-95.

[8] - J.K. HALE : Theory of functional differential equations. Applied Mathematical sciences. V3. Springer Verlag. New-York (1971-1977).

[9] - J.K. HALE - C. PERELLO : The neighbourhood of a singular point of functional differential equations. Contribution differential equations. 3. (1964).

[10] - A.F. IZE : Asymptotic behavior and stability of neutral functional differential equations. Workshop on boundary value problems for ordinary differential equations and applications. Trieste (1977).

[11] - J.L. KAPLAN - A.J. YORKE : On nonlinear differential delay equations $x'(t) = - f(x(t),x(t-\omega))$. Journal of differential equations. 23.2. (1977), 293-314.

[12] - A. LEUNG : Periodic solutions for a prey-predator differential delay equations. Journal of differential equations. 26. (1977), 391-403.

[13] - J.C. LILLO : Oscillatory solutions of the equation $y'(x) = m(x)y(x-n(x))$. Journal of differential equations. 6. (1969), 1-35.

[14] - R.D. NUSSBAUM : Periodic solutions of some nonlinear autonomous functional differential equations. Annali di Matematica. 4. 51. (1974), 263.

[15] - C. PERELLO : A note on periodic solutions of nonlinear differential equations with time lag. In Differential Equations and Dynamical Systems. 185-187. Academic Press, 1967. M.R. 36, 2896.

[16] - P. VIDAL - J.F. DENES : On phenomenia in ionized gases. IX Conférence internationale. Bucarest (1969) et thèse Toulouse (1971).

[17] - H.O. WALTHER : Existence of a non constant periodic solution of a nonlinear autonomous functional differential equation representing the growth of a single species population. Journal Mathematical Biosciences 1, (1975), 227-240.

[18] - O. ARINO - P. SEGUIER : Solutions périodiques d'équations à retard, critères de non existence (à paraître).

Oscillations autour d'un point stationnaire ; conditions suffisantes de non existence. C.R. Acad. Sc. Paris. T. 284, (1977). A-145.

[19] - P. SEGUIER : Comportements de solutions d'équations différentielles à argument retardé (thèse 3ème cycle. Pau 1975). C.R. Acad. Sc. Paris 281, (1975), A-843.

[20] - A.M. ZVERKIN - G.A. KAMENSKII - S.B. NORKIN - L.E. EL'SGOL'TZ : Differential equations with retarded arguments, Uspchi Mat. Nauk. 17 (1962). 77-164. (article de synthèse).

O. ARINO et P. SÉGUIER

Université de PAU

Département de Mathématiques

Avenue Philippon - 64000 PAU (FRANCE) -

APPROXIMATION OF DELAY SYSTEMS WITH APPLICATIONS TO CONTROL AND IDENTIFICATION

H. T. Banks*
Lefschetz Center for Dynamical Systems
Division of Applied Mathematics
Brown University
Providence, Rhode Island 02912

INTRODUCTION

In this presentation, we shall discuss approximation ideas that have proved useful in developing methods for solving optimal control and parameter identification problems involving delay systems. We shall focus on linear delay systems in our discussions although, as we shall later indicate, many of the ideas and results are valid for problems with nonlinear systems.

The approximation results we describe below are based on an abstract formulation due to Trotter and Kato dealing with approximations to semigroups of linear operators. In order to make use of these approximation theorems, it is necessary to reformulate our delay system as an abstract system in an appropriate Hilbert space. To this end, consider the delay system

$$\begin{aligned} \dot{x}(t) &= L(x_t) + f(t) \qquad 0 \le t \le t_1, \\ x(0) &= \eta, \quad x_0 = \phi, \end{aligned} \tag{1}$$

where, for ψ continuous on $[-r,0]$,

$$L(\psi) = \sum_{i=0}^{\nu} A_i \psi(-\tau_i) + \int_{-r}^{0} A(\theta)\psi(\theta)d\theta \tag{2}$$

with A_i, $A(\theta)$ given $n \times n$ matrices and $0 = \tau_0 < \tau_1 < \cdots < \tau_\nu \le r$. For $0 \le t \le t_1$, the function $\theta \to x_t(\theta)$, $-r \le \theta \le 0$, is given by $x_t(\theta) = x(t+\theta)$ whenever x is defined on $[-r,t_1]$.

With an appropriate interpretation of the operator L, one can define solutions x (i.e. solutions exist) to (1) corresponding to initial data (η,ϕ) in $Z \equiv R^n \times L_2([-r,0],R^n) \equiv R^n \times L_2^n(-r,0)$ and perturbations f in $L_2^n(0,t_1)$ as functions which satisfy the initial

*This research was supported in part by the National Science Foundation under NSF-MCS 78-18858 and in part by the Air Force Office of Scientific Research under AF-AFOSR 76-3092A.

conditions in (1), are absolutely continuous on $(0,t_1)$ and satisfy the differential equation in (1) almost everywhere on $(0,t_1)$. One can thus define a homogeneous solution semigroup of operators $S(t): Z \to Z$, $t \geq 0$, by

$$S(t)(\eta,\phi) \equiv (x(t;\eta,\phi), x_t(\eta,\phi))$$

where x is the solution of (1) with $f \equiv 0$. It is not difficult to argue that this defines a strongly continuous semigroup (C_0-semigroup) $\{S(t)\}_{t\geq 0}$ of linear operators with infinitesimal generator $\mathscr{A}$ (i.e. $S(t) \sim e^{\mathscr{A}t}$) on $\mathscr{D}(\mathscr{A}) \equiv \{(\phi(0),\phi) | \phi$ is absolutely continuous with $\dot{\phi}$ in $L_2^n(-r,0)\}$ given by

$$\mathscr{A}(\phi(0),\phi) = (L(\phi),\dot{\phi}). \qquad (3)$$

This semigroup can be used to give an abstract "variation of parameters" representation for solutions of (1) in Z. That is, suppose one defines for $(\eta,\phi) \in Z$ and $f \in L_2^n(0,t_1)$ the function $t \to z(t;\eta,\phi,f)$ by

$$z(t;\eta,\phi,f) = S(t)(\eta,\phi) + \int_0^t S(t-\sigma)(f(\sigma),0)d\sigma. \qquad (4)$$

Then one can argue (see [2]) an equivalence between (1) and (4). More precisely, one has that

$$z(t;\eta,\phi,f) = (x(t;\eta,\phi,f),\ x_t(\eta,\phi,f))$$

for $t \geq 0$ where z is given by (4) and $t \to x(t;\eta,\phi,f)$ is the solution of (1).

For $\eta = \phi(0)$ and ϕ, f sufficiently smooth, it turns out that (4) is equivalent to an ordinary differential equation (ODE) in the Hilbert space Z which may be written

$$\begin{aligned} \dot{z}(t) &= \mathscr{A}z(t) + (f(t),0) \\ z(0) &= (\phi(0),\phi). \end{aligned} \qquad (5)$$

We seek to approximate (4) (or (5)) by considering approximations in finite dimensional subspaces Z^N (where we will obtain finite dimensional ODE's). The fundamental approximation result from semigroup theory that we employ can be stated roughly as "if $\mathscr{A}^N \to \mathscr{A}$

in a proper sense, then $e^{\mathscr{A}^N t} \to e^{\mathscr{A} t}$ in some sense". One version of this result due to Trotter-Kato [13] can be stated in precise terms as follows.

<u>Theorem</u>. Suppose $\mathscr{A}^N, \mathscr{A}$ are generators of C_0-semigroups $\{S^N(t)\}$, $\{S(t)\}$ on Z satisfying:

(i) there exist M and $\beta > 0$ such that

$$|S^N(t)| \le Me^{\beta t}, \quad |S(t)| \le Me^{\beta t},$$

(ii) there exists a dense subset $\mathscr{D}$ of Z such that $\mathscr{A}^N z \to \mathscr{A} z$ as $N \to \infty$ for each $z \in \mathscr{D}$,

(iii) there exist a complex number λ_0 with real part larger than β such that $(\mathscr{A} - \lambda_0 I)\mathscr{D}$ is dense in Z.

Then $S^N(t)z \to S(t)z$ as $N \to \infty$ for every $z \in Z$, uniformly in t on compact intervals.

We wish to emphasize that use of this semigroup formulation is mainly for convenience in proving convergence of particular schemes. Just as in the case of differencing schemes for solution of partial differential equations, the semigroup formulation is not an essential aspect in the development of numerical methods. Rather, the particular choice of the subspaces Z^N and the approximating operators $\mathscr{A}^N$ is the question of utmost importance.

In both the control and identification applications discussed here, the basic system to be considered is (1) or, equivalently (4). Once we have chosen Z^N and $\mathscr{A}^N$ satisfying the Trotter-Kato hypotheses, we shall also employ (orthogonal) projections P^N of Z onto Z^N to then approximate equation (4) by

$$z^N(t) = S^N(t)P^N(\eta,\phi) + \int_0^t S^N(t-\sigma)P^N(f(\sigma),0)d\sigma. \tag{6}$$

Since in the cases of interest to us $\mathscr{A}^N$ will be bounded and $S^N(t) = e^{\mathscr{A}^N t}$, we may equivalently write this as

$$\begin{aligned} \dot{z}^N(t) &= \mathscr{A}^N z^N(t) + P^N(f(t),0) \\ z^N(0) &= P^N(\eta,\phi). \end{aligned} \tag{7}$$

The convergence of S^N to S will be sufficient to guarantee a desired convergence of solutions of (7) in both the optimal control and parameter identification problems detailed below.

OPTIMAL CONTROL PROBLEMS

The basic idea employed in control problems is a classical one which is the basis for all Ritz type methods in approximation theory. Suppose one has an optimal control problem ($\mathcal{P}$) in a Hilbert space Z. By considering a sequence of approximating problems ($\mathcal{P}_N$) on finite dimensional subspaces Z^N, one seeks to obtain a sequence of more easily solved problems whose solutions will approximate (i.e., approach in the limit as $N \to \infty$) solutions of the original problem ($\mathcal{P}$). To be more specific, let $f(t) = Bu(t)$ in (1) where B is an $n \times m$ matrix and u is to be chosen from some admissible class $\mathcal{U} \subset L_2^m(0,t_1)$ of control functions. The problem ($\mathcal{P}$) might typically consist of choosing $u \in \mathcal{U}$ so as to minimize (we do not distinguish between a vector and its transpose here and below)

$$\Phi(u) = x(t_1)Qx(t_1) + \int_0^{t_1} \{x(t)Wx(t) + u(t)Ru(t)\}dt, \qquad (8)$$

where x is the solution of (1) corresponding to u and the matrices Q, W and R are symmetric with Q and W positive semi-definite, R positive definite.

Writing solutions z^N of (7) in terms of components in R^n and L_2^n, $z^N = (x^N, y^N)$, and defining

$$\Phi^N(u) = x^N(t_1)Qx^N(t_1) + \int_0^{t_1}\{x^N(t)Wx^N(t) + u(t)Ru(t)\}dt, \qquad (9)$$

we take as our approximating problem ($\mathcal{P}_N$) that of minimizing Φ^N over $\mathcal{U}$ subject to (7).

If we denote by $\bar{u}, \bar{u}^N$ solutions to problems ($\mathcal{P}$) and ($\mathcal{P}_N$) respectively, one can often use the Trotter-Kato results to guarantee convergence of the $\bar{u}^N$ to $\bar{u}$ in a desired sense. For example, in the case of Φ, Φ^N given by (8), (9) and $\mathcal{U}$ a convex closed set, one can argue that the optimal controls $\bar{u}, \bar{u}^N$ exist and are unique, $\bar{u}^N \to \bar{u}$ in L_2 as $N \to \infty$, and furthermore $\Phi^N(\bar{u}^N) \to \Phi(\bar{u})$. Similar results can be obtained for more general payoff functions in place of (8) and (9). For details along with convergence arguments see [2], [3].

IDENTIFICATION PROBLEMS

In many practical situations, it is important to be able to fit a model described by equations such as (1) to data obtained through empirical efforts. In this case one has certain parameters, say $q \in R^k$, on which the system depends and which must be "identified" or "estimated". For example suppose the operator L in (2) depends on parameters q (typically these may be some of the matrix coefficient terms in (2) and/or even a delay so that L might have the form $L(q;\psi) = A_0(\alpha)\psi(0) + A_1(\alpha)\psi(-\tau)$, $q = (\alpha,\tau)$). One then must identify q in the system

$$\begin{aligned} \dot{x}(t) &= L(q;x_t) + f(t) \\ c(t) &= Cx(t). \end{aligned} \tag{10}$$

Here $c(t)$ represents the "observables" for the system. One perturbs the system (via f) and collects data $\xi_1,\ldots,\xi_M$ at times $t_1,\ldots t_M$ (the ξ_i are observations for $c(t_i)$). The problem is to choose $\bar{q}$ from some admissible parameter set $\mathcal{Q}$ so that it is a maximum likelihood estimator (MLE) or perhaps so that it minimizes

$$J(q) = \sum_{i=1}^{M} |c(t_i;q) - \xi_i|^2. \tag{11}$$

Just as in the control problems, one reformulates this identification problem as one in the abstract space Z involving a system such as (4) (or (5)), and approximates by a sequence of identification problems. That is, one has approximating systems (again, $z^N = (x^N,y^N)$)

$$\begin{aligned} \dot{z}^N(t) &= \mathcal{A}^N(q)z^N(t) + P^N(f(t),0) \\ c^N(t) &= Cx^N(t) \end{aligned} \tag{12}$$

with observations $\{\xi_i\}$ for $\{c^N(t_i;q)\}$. One then chooses $\bar{q}^N \in \mathcal{Q} \subset R^k$ so that it is a MLE for (12) or so that it minimizes

$$J^N(q) = \sum_{i=1}^{M} |c^N(t_i;q) - \xi_i|^2, \tag{13}$$

depending on how one has decided to seek a fit to data for the original system.

Under quite reasonable assumptions, one can use the Trotter-Kato results to guarantee existence of a vector $\bar{q}$ (loosely speaking $\lim \bar{q}^N$) so that $\mathcal{A}(\bar{q})$ is the limit of $\mathcal{A}^N(\bar{q}^N)$ in an appropriate sense and so that $\overline{q}$ is a MLE (or a minimizer for (11) if (13) is used in the approximating problems) for (10) with data $\{\xi_i\}$.

There remains the basic question of how to choose Z^N and $\mathscr{A}^N$ so that the needed Trotter-Kato hypotheses are satisfied and most importantly, so that efficient numerical schemes are generated. We shall discuss and compare two particular choices here; the first we refer to as "averaging" (AVE) approximations while the second involves spline type (SPL) approximations.

AVERAGING APPROXIMATIONS

For a given positive integer N, one partitions the interval $[-r,0]$ into N equal subintervals with a partition $\{t_j^N\}$, $t_j^N = -jr/N$, and defines the characteristic functions

$$\chi_j^N = \chi_{[t_j^N, t_{j-1}^N)}, \quad j = 2,\ldots,N, \quad \chi_1^N = \chi_{[t_1^N, t_0^N]}.$$

One then takes Z^N to be the $n(N+1)$ dimensional subspace of Z defined by

$$Z^N = \{(\eta,\psi) \mid \eta \in R^n, \ \psi = \sum_{j=1}^N v_j^N \chi_j^N, \ v_j^N \in R^n\}.$$

The projections $P^N: Z \to Z^N$ are given by

$$P^N(\eta,\phi) = (\eta, \sum_1^N \phi_j^N \chi_j^N)$$

where $\phi_j^N \equiv (N/r)\int_{t_j^N}^{t_{j-1}^N} \phi(s)ds$, the integral averages of ϕ over the partition subintervals. Finally, one defines the approximating operators $\mathscr{A}^N: Z \to Z^N$ by

$$\mathscr{A}^N(\eta,\phi) = (L^N(\eta,\phi), \sum_{j=1}^N \frac{N}{r}\{\phi_{j-1}^N - \phi_j^N\}\chi_j^N)$$

where $\phi_0^N \equiv \eta$ and

$$L^N(\eta,\phi) \equiv A_0\eta + \sum_{i=1}^{\nu} \sum_{j=1}^N A_i \phi_j^N \chi_j^N(-\tau_i) + \sum_{j=1}^N A_j^N \phi_j^N$$

with $A_j^N \equiv \int_{t_j^N}^{t_{j-1}^N} A(\theta)d\theta$. With these definitions of $Z^N, \mathscr{A}^N, P^N$ one can then argue that the desired Trotter-Kato hypotheses are satisfied and moreover that the convergence behavior necessary for use in both optimal control and identification problems is obtained (see [3],[8]).

SPLINE APPROXIMATIONS

For K^{th} order spline approximations, one defines the subspaces $Z^N \equiv \{(\psi^N(0),\psi^N) \mid \psi^N$ is a spline of order K on $[-r,0]$ with knots at $t_j^N = -jr/N$, $j = 0,1,\ldots,N\}$. Letting P^N be the orthogonal projection of Z onto the closed subspace Z^N, one can then define

$$\mathscr{A}^N \equiv P^N \mathscr{A} P^N$$

where $\mathscr{A}$ is given in (3).

For example, in the case of first order ("linear elements") splines, the subspace Z^N is of dimension $n(N+1)$. If one defines the usual "roof" functions for $j = 0,1,\ldots,N$ and $-r \le \theta \le 0$ by

$$e_j^N(\theta) = \begin{cases} \frac{N}{r}(\theta - t_{j+1}^N) & t_{j+1}^N \le \theta \le t_j^N \\ \frac{N}{r}(t_{j-1} - \theta) & t_j^N \le \theta \le t_{j-1}^N \\ 0 & \text{elsewhere}, \end{cases}$$

where $t_j^N = -jr/N$ for all j, then Z^N can be written

$$Z^N = \{(\psi^N(0),\psi^N) \mid \psi^N = \sum_{j=0}^{N} a_j^N e_j^N,\ a_j^N \in R^n\}.$$

Of course, $P^N(\eta,\phi) = \hat{\phi}^N = (\phi^N(0),\phi^N)$ where $\hat{\phi}^N$ is the solution to the problem of minimizing $|\hat{\zeta} - (\eta,\phi)|$ over $\hat{\zeta}$ in Z^N.

The conditions for the Trotter-Kato theorem can be verified for these spline based approximations and, in fact, the arguments, using fundamental estimates from spline analysis, are independent of the order of splines used (see [6]).

For both the AVE and SPL approximations, one can obtain estimates on the convergence rates of the approximations. The AVE method is essentially a first order method (error like $1/N$) while one can argue that the error in the Kth order SPL scheme behaves like $1/N^K$. In actual practice our computations have revealed that the SPL methods usually converge more rapidly than these estimates predict (see the discussions in [6]) and indeed for both control and identification problems it appears that the "first order" SPL method often offers significant advantages over the AVE method at little cost in additional complexity with respect to implementation. To partially illustrate this feature and to indicate what one might typically expect in the way of convergence for these methods, we present just one of a number of control examples to which we have applied these ideas.

EXAMPLE (CONTROLLED OSCILLATOR WITH DELAYED DAMPING)

The problem is to minimize

$$\Phi(u) = 5x_1(2)^2 + \frac{1}{2}\int_0^2 u(t)^2 dt$$

over $u \in \mathscr{U} = L_2(0,2)$ subject to

$$\begin{aligned} \dot{x}_1(t) &= x_2(t) \\ \dot{x}_2(t) &= -x_1(t) - x_2(t-1) + u(t), \qquad 0 \leq t \leq 2, \end{aligned} \tag{14}$$

and

$$\begin{aligned} x_1(\theta) &= 10 \\ x_2(\theta) &= 0, \qquad -1 \leq \theta \leq 0. \end{aligned}$$

The system (14) is, of course, the vector formulation for the system dynamics $\ddot{y}(t) + \dot{y}(t-1) + y(t) = u(t)$. One can use necessary and sufficient conditions (a maximum principle for delay system problems) to solve analytically this simple example. Upon doing so one finds (see [3],[5]) the optimal control

$$\overline{u}(t) = \begin{cases} \delta \sin(2-t) + (\delta/2)(1-t)\sin(t-1) & 0 \leq t \leq 1 \\ \delta \sin(2-t) & 1 \leq t \leq 2, \end{cases}$$

where δ is the solution of a certain algebraic equation with approximate value $\delta \approx 2.5599$. The optimal value of the payoff Φ is given by

$$\overline{\Phi} = \Phi(\overline{u}) \simeq 3.399119.$$

We used a conjugate-gradient scheme to compute the solution of each of the approximate control problems $(\mathscr{P}_N)$ described above for both the AVE and SPL schemes for several values of N. Denoting the optimal

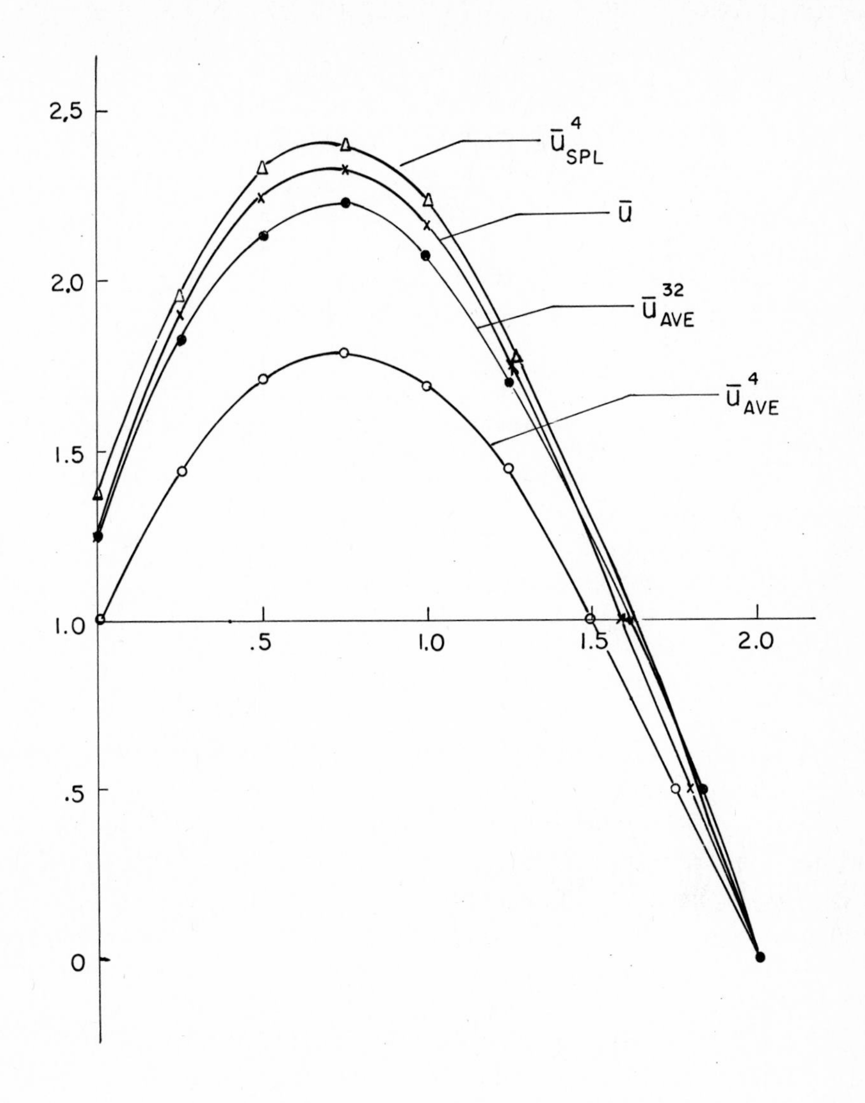

FIG. I OPTIMAL CONTROLS FOR OSCILLATOR WITH DELAYED DAMPING

values $\Phi^N(\bar{u}^N_{AVE})$ and $\Phi^N(\bar{u}^N_{SPL})$ of the payoffs by $\bar{\Phi}^N_{AVE}$ and $\bar{\Phi}^N_{SPL}$ respectively, we give a representative sample of our numerical findings in tabular form.

N	$\bar{\Phi}^N_{AVE}$	$\lvert\bar{\Phi}^N_{AVE}-\bar{\Phi}\rvert$	$\bar{\Phi}^N_{SPL}$	$\lvert\bar{\Phi}^N_{SPL}-\bar{\Phi}\rvert$
4	2.15154	1.2475	3.53549	.1363
8	2.67110	.7280	3.43454	.0354
16	3.00356	.3955	3.40857	.0094
32	3.19298	.2061	3.40193	.0028

We note that the convergence of $\bar{\Phi}^N_{AVE}$ to $\bar{\Phi}$ is like $1/N$ while that of $\bar{\Phi}^N_{SPL}$ is $1/N^2$. A graph comparing several of the optimal controls $\bar{u}^N_{SPL}, \bar{u}^N_{AVE}$ with the solution $\bar{u}$ of the original problem is given in Figure 1. One also has for this example that $|\bar{u}^N_{SPL}|_{L_2} \to |\bar{u}|_{L_2}$ like $1/N^2$ while $|\bar{u}^N_{AVE}|_{L_2} \to |\bar{u}|_{L_2}$ is like $1/N$.

We complete our presentation with brief mention of related results that have been obtained through efforts by our group at Brown University and our associates and colleagues.

The basic abstract framework employing the Trotter-Kato theorem to ensure convergence in optimization problems for delay systems was given in [2]. A detailed investigation of the AVE approximation and its use in linear system control problems along with several solved examples can be found in [3]. Additional examples of solved control problems along with use of the AVE scheme on these examples may be found in [5]. An extension of the theory of [2], [3] to treat control problems with nonlinear systems along with related nummerical findings are given in [1]. In efforts to find alternatives to the AVE scheme, Burns and Cliff discuss in [7] a scheme involving piecewise linear approximations combined with AVE type ideas. Spline approximations in the context of a general framework suitable for use in control and identification problems are developed in [6]. More recently, Kappel and Schappacher [10] have used interpolating splines for an approximation method that is applicable to nonlinear retarded equations while Kunisch [12] has developed the AVE scheme for nonlinear neutral functional differential equations. Kunisch's development is in the spirit of the earlier results of Kappel and Schappacher [9] who formulated a "local semigroup" approach to handle AVE approximations for locally Lipschitz nonlinear retarded equations. In [8], Burns and Cliff discuss the use of the AVE scheme in identification problems while results for SPL based identification methods are presented in [4].

All of the above schemes, whether for AVE or SPL, lead to the approximation of a differential equation (e.g. (5)) in the Hilbert space Z by a finite-dimensional ODE. To actually use these methods in computations, one needs a second approximation (e.g. a standard Runga-Kutta or predictor-corrector scheme) to solve the approximating ODE's. A natural question that arises is "why not go directly from the original (infinite dimensional) delay system to a difference equation?" Reber, in [14], [15], investigates this question in the spirit of the functional analytic methods of the papers cited above (factor space ideas-see [11]-are employed in place of the Trotter-Kato semigroup results). Briefly, Reber considers quite general non-autonomous linear delay equations and shows that they can be formulated as an abstract system in Z

$$Tz = \Sigma(\zeta,f) \tag{15}$$

where

$$(Tz)(t) \equiv z(t) - \int_{t_0}^{t} \mathscr{A}(\sigma)z(\sigma)d\sigma$$

$$\Sigma(\zeta,f)(t) \equiv \zeta + \int_{t_0}^{t} (f(\sigma),0)d\sigma.$$

Here ζ is the initial data (η,ϕ) and f is the perturbing function as in (1) or (5). Equation (15) is then approximated by

$$T_N z_N = \Sigma_N p_N(\zeta,f)$$

where T_N is essentially a first order difference operator and the p_N corresponds to an AVE type approximation in the state. This results in a simultaneous discretization of both "state" and "time". Reber discusses the use (advantages vs. disadvantages) of these ideas in control problems (theoretically, the analysis is slightly more complicated than that for the simple "state" discretization - i.e. ODE approximations - ideas discussed above; numerically, implementation on the computer is quite straightforward).

REFERENCES

[1] H.T. Banks, Approximation of nonlinear functional differential equation control systems, J. Opt. Theory Appl., to appear.

[2] H.T. Banks and J.A. Burns, An abstract framework for approximate solutions to optimal control problems governed by hereditary systems, in International Conf. on Differential Equations (H. Antosiewicz, ed.), Academic Press, 1975, p. 10-25

[3] H.T. Banks and J.A. Burns, Hereditary control problems: numerical methods based on averaging approximations, SIAM J. Control Opt. 16(1978), 169-208.

[4] H.T. Banks, J.A. Burns and E.M. Cliff, Spline based approximation methods for control and identification of hereditary systems, to appear.

[5] H.T. Banks, J.A. Burns, E.M. Cliff and P.R. Thrift, Numerical solutions of hereditary control problems via an approximation technique, LCDS Tech. Rep. 75-6, Brown University, Providence, 1975.

[6] H.T. Banks and F. Kappel, Spline approximations for functional differential equations, to appear.

[7] J.A. Burns and E.M. Cliff, Methods for approximating solutions to linear hereditary quadratic optimal control problems, IEEE Trans. Automatic Control 23(1978), 21-36.

[8] E.M. Cliff and J.A. Burns, Parameter identification for linear hereditary systems via an approximation technique, March, 1978; to appear in Proc. Workshop on the Linkage between Applied Mathematics and Industry.

[9] F. Kappel and W. Schappacher, Autonomous nonlinear functional differential equations and averaging approximations, J. Nonlinear Analysis: Theory, Methods and Appl. 2(1978), 391-422.

[10] F. Kappel and W. Schappacher, Nonlinear functional differential equations and abstract integral equations, Preprint No. 1, May, 1978, Universität Graz.

[11] S.G. Krein, Linear Differential Equations in Banach Space, Trans. Math. Mono., vol. 29, Amer. Math. Soc., Providence, 1971.

[12] K. Kunisch, Neutral functional differential equations and averaging approximations, J. Nonlinear Analysis: Theory, Methods and Appl., to appear.

[13] A. Pazy, Semigroups of Linear Operators and Applications to Partial Differential Equations, Math. Dept. Lecture Notes, vol. 10, Univ. Maryland, College Park, 1974.

[14] D. Reber, Approximation and Optimal Control of Linear Hereditary Systems, Ph.D. Thesis, Brown University, November, 1977.

[15] D. Reber, A finite difference technique for solving optimization problems governed by linear functional differential equations, J. Differential Equations, to appear.

A HOMOTOPY METHOD FOR LOCATING ALL ZEROS OF A SYSTEM OF POLYNOMIALS

By

Shui-Nee Chow*
John Mallet-Paret**
and
James A. Yorke***

§1. Introduction

In [2] the authors gave a constructive proof of the Brouwer fixed point theorem by means of a homotopy (continuation) method. The basic idea used in the proof is the concept of transversality; other applications were also given using this idea. In [7] and [8] actual numerical computations based on our method were shown. In particular, in [8] a nonlinear boundary value problem, which had not been solved previously by standard schemes, was numerically solved. Recently the same idea was used in [5] to find all the d complex roots (counting multiplicity) of a dth order polynomial in a single unknown. The purpose of this paper is to show that ideas and techniques from transversality theory may again be used to find numerically all the zeros of a system of n polynomial equations in n unknowns.

---

*Supported in part by the National Science Foundation under MCS 76-06739

**Supported in part by the National Science Foundation under MCS 76-07247-A03 and by the US Army Research Office under ARO-DAAG-76-G-0294.

***Supported in part by NSF Grant MCS76-24432.

The problem of finding all the zeros of such a system has in fact recently been solved by Drexler [3] using a homotopy method. However, the justification of his method relies on relatively sophisticated ideas from algebraic geometry. It is , therefore, to provide a homotopy method based on transversality theory, in the spirit of [2] and [5]. The difficulty to be overcome in extending the transversality techniques of [5] to systems in finding a class of homotopies rich enough that transversality theory is applicable, yet simple enough that the trivial problem is truly trivial -- that is, its solutions are explicitly given. Moreover, the trivial problem should have no more solutions than is necessary.

To be more precise, consider first the case $n = 1$. Let $P(z)$ be a monic polynomial of degree d (that is, the coefficient of z^d is 1) where the variable $z \in \mathbb{C}$ is complex; we wish to find all d roots of P. Assume $P(z)$ is embedded in a smooth homotopy $H(z,t)$ with, say,

$$H : \mathbb{C} \times [0,1] \to \mathbb{C}$$

$$H(z,1) = P(z)$$

$H(\cdot,t)$ is a monic polynomial of degree d for all $t \in [0,1]$.

If 0 is a regular value of H on $\mathbb{C} \times [0,1)$, then $H^{-1}(0)$ is a submanifold of real codimension two in $\mathbb{C} \times [0,1)$, hence consists only of arcs and closed curves. (In this paper we will only count real dimensions.) These curves are integral curves of the ordinary differential equation

$$(1.1) \qquad D_z H(z,t) \frac{dz}{ds} + D_t H(z,t) \frac{dt}{ds} = 0$$

where D_z and D_t denote partial derivatives, and s denotes arc length. Because $H(\cdot,t)$ is monic, there is a bound $|z| \leq K$ for points of $H^{-1}(0)$, independent of $t \in [0,1]$; therefore, with any root of $H(\cdot,0)$ serving as initial condition for (1.1), the resulting solution $(z(s),t(s))$ would eventually lead to a root of $H(\cdot,1)$ (that is, $t(s) \to 1$) or would turn to some root of $H(\cdot,0)$ (so $t(s) \to 0$). The latter case, however, is impossible: at some point $\frac{dt}{ds}$ would vanish and $D_zH(z,t)$, considered as a two-by-two real matrix, would be singular. By the Cauchy-Riemann equations, then, $D_zH(z,t) = 0$ at this point; and so 0 would not be a regular value of H, a contradiction.

In other words, $\frac{dt}{ds} > 0$ as long as $t \in [0,1)$, hence z can be regarded as a function of t. The d distinct roots of $H(\cdot,0)$ give rise to d distinct roots $z_1(t),\ldots,z_d(t)$ of $H(\cdot,t)$ as solutions of (1.1) for this range of t. As $t \to 1$, all roots of $H(\cdot,1)$ are approached. Indeed, if $H(z_0,1) = 0$ then for $0 < 1-t \ll 1$, $H(\cdot,t)$ has a root near z_0 by Rouche's Theorem; and such a root must equal $z_k(t)$ for some k.

An obvious choice for the homotopy H is

$$H(z,t) = (1-t)\prod_{h=1}^{d}(z-b_k) + tP(z)$$

where $b_i,\ldots,b_d \in \mathbb{C}$ are distinct complex constants. A transversality argument shows that for almost every choice of b_k (in the sense of Lebesgue measure) 0 is a regular value of H, as required. Hence, for such choices, all roots of P can be obtained by following the solutions of (1.1) from the initial points $(z,t) = (b_k,0)$ up to $t = 1$.

Now consider the higher dimensional problem. Let $z \in \mathbb{C}^n$ and $P(z) \in \mathbb{C}^n$ be vectors with components z_k and $P_k(z)$, $1 \leq k \leq n$, where $P_k(z)$ is a complex valued polynomial of degree d_k. We assume P is nondegenerate in the sense that its Jacobian $\det DP(z)$ is not identically zero. (Such a condition can be effectively verified, and is presumably the case if there exists an isolated root of P). We seek a homotopy from a trivial polynomial

$H(z,0) = Q(z)$ to $H(z,1) = P(z)$, where $Q_k(z)$ also has degree d_k; a simple choice of Q is

$$(1.2) \qquad Q_k(z) = z_k^{d_k} - b_k, \quad b_k \in \mathbb{C}, \quad 1 \le k \le n.$$

(The choice $Q_k(z) = \prod_{j=1}^{d_k} (z_k - b_{jk})$ as in the one-dimensional case could also be made, but we prefer to analyze (1.2) as fewer constants are introduced. In fact, many choices of $Q_k(z)$ are feasible.) The obvious choice of homotopy

$$H(z,t) = (1-t)Q(z) + tP(z),$$

however, may not work as there may not be a bound $|z| \le K$ on zeros of H for all $t \in [0,1]$; that is, some solution curves $z(t)$ may not extend to $t = 1$, but rather become unbounded as t approaches some $t_0 < 1$. This will be the case even for $n = 1$ if the coefficient of $z^d = z_1^{d_1}$ in $P(z)$ is negative.

To remedy the situation we introduce a third term $t(1-t)R(z)$, vanishing at $t = 0$ and 1, into the definition of H. The kth component

$$R_k(z) = \sum_{j=1}^{n} a_{jk} z_j^{d_k}, \quad a_{jk} \in \mathbb{C},$$

of R is homogeneous of degree d_k, and is such that for almost every a_{jk} a bound $|z| \le K(t_0)$ can be established for $t \in [0,t_0]$, whenever $t_0 < 1$. For such a homotopy the solution curves extend to $t = 1$; in fact, the following result holds.

Theorem 1.1. Let $P : \mathbb{C}^n \to \mathbb{C}^n$ be a polynomial whose Jacobian $\det DP(z)$ is not identically zero. Let $d_k \ge 1$ denote the degree of the kth component, and consider the homotopy

$$H : \mathbb{C}^n \times [0,1] \times \mathbb{C}^n \times \mathbb{C}^{n^2} \to \mathbb{C}^n$$

whose k^{th} component is

$$(1.3) \qquad H_k(z,t;b,a) = (1-t)(z_k^{d_k} - b_k) + tP_k(z) + t(1-t)\sum_{j=1}^{n} a_{jk} z_j^{d_k};$$

here $b \in \mathbb{C}^n$ and $a \in \mathbb{C}^{n^2}$ are parameters and $t \in [0,1]$ is the homotopy parameter. Then for almost every (b,a) (in the sense of either Baire category or Lebesgue measure), the zero set

$$\{(z,t) \in \mathbb{C}^n \times [0,1) \,|\, H(z,t;b,a) = 0\}$$

for $t < 1$ consists of $d = \prod_{k=1}^{n} d_k$ disjoint analytic arcs $z_1(t),\ldots,z_d(t)$ emanating from the d roots of $H(\cdot,0;b,a)$, and parameterized by t in the interval $[0,1)$. (Thus if such a curve is parameterized by arc length s, we have $\frac{dt}{ds} > 0$.) For each k, either

$$\lim_{t\to 1-} |z_k(t)| = \infty; \quad \text{or}$$

(1.4) $\qquad$ the set of limits $z_k(t_\alpha) \to z_{k0}$, $t_\alpha \to 1$, forms a connected subset of $P^{-1}(0)$.

Moroever, each isolated root z_0 of P is obtained in this manner; that is, for each such z_0

$$\lim_{t\to 1-} z_k(t) = z_0 \tag{1.5}$$

holds for some k. In particular P has at most d isolated roots. (This last statement is a classical theorem of Bézout.)

The remainder of this paper constitutes the proof of Theorem 1.1 In Section 2 transversality conditions are imposed on H. They ensure that the zero set of H does indeed consist of curves, and that these can become unbounded only as $t \to 1$. In Section 3 it is shown these curves are monotone with respect to t; and finally we observe that each isolated root of P is approached by some such curve.

We note here that in a recent paper [4] Garcia and Zangwill propose a different choice of trivial polynomial, namely

$Q_k(z) = z_k^{d_k+1} - b_k$. In this case the term involving $t(1-t)$ is not necessary. However, a drawback is that Q_k is of one degree higher than P_k, so there are $\prod_{k=1}^{n} (d_k + 1)$ curves to be followed, as opposed to $\prod_{k=1}^{n} d_k$. Many of these curves would be expected to become unbounded as $t \to 1$, rather than approach roots of P.

§2. The Transversality Conditions

We begin the proof of Theorem 1.1 by recalling the Transversality Theorem [1] as it applies to regular values. If

$$F : U \to R^p, \quad U \subseteq R^m \text{ open},$$

is a smooth map, then a point $y \in R^p$ is called a regular value of F on the set $S \subseteq U$ provided

$$\text{range } DF(x) = R^p \quad \text{for all} \quad x \in S \cap F^{-1}(y).$$

Theorem 2.1 (Transversality Theorem). Let $U \subseteq R^m$ and $V \subset R^q$ be open sets, and

$$F : U \times V \to R^p$$

be C^r smooth, where $r > \text{mas}\{0, m-p\}$. Suppose for some set $S \subseteq U$ that $y \in R^p$ is a regular value of F on $S \times V$. Then for almost every $c \in V$ (in the sense of either Baire category or Lebesgue measure), y is a regular value of $F(\cdot, c)$ on S.

Now we apply this theorem to the homotopy H (1.3) and to the function $\widetilde{H}$ given by the highest order homogeneous terms of H. Let

$$\widetilde{P}(z) = (\widetilde{P}_1(z), \ldots, \widetilde{P}_n(z))$$

where

$\widetilde{P}_k : \mathbb{C}^n \to \mathbb{C}$ is the part of P_k homogeneous of degree d_k,

and set

$$\widetilde{H} : \mathbb{C}^n \times R \times \mathbb{C}^{n^2} \to \mathbb{C}^n, \text{ given by}$$

$$\widetilde{H}_k(z,t;a) = (1-t)z_k^{d_k} + t\widetilde{P}_k(z) + t(1-t)\sum_{j=1}^{n} a_{jk} z_j^{d_k}.$$

In applying transversality theory we regard complex space as a space of two real dimensions.

Lemma 2.2. For almost every $(b,a) \in \mathbb{C}^n \times \mathbb{C}^{n^2}$, $0 \in \mathbb{C}^n$ is a regular value both of $H(\cdot,\cdot;b,a)$ on $\mathbb{C}^n \times [0,1)$, and of $\widetilde{H}(\cdot,\cdot;a)$ on $(\mathbb{C}^n - \{0\}) \times [0,1)$.

Proof: It follows immediately from the Transversality Theorem that for almost every (b,a), 0 is a regular value of H as stated, and of $\widetilde{H}$ on the smaller set $(\mathbb{C}^n - \{0\}) \times (0,1)$; and it can be seen directly that 0 is a regular value of $\widetilde{H}(\cdot,\cdot;a)$ on $(\mathbb{C}^n - \{0\}) \times \{0\}$, no matter what a is.

Henceforth, let (b,a) be fixed as in Lemma 2.2, and for simplicity write

$$H(z,t) = H(z,t;b,a),$$
$$\widetilde{H}(z,t) = \widetilde{H}(z,t;a).$$

Lemma 2.3. If $\widetilde{H}(z_0,t_0) = 0$ and $t_0 \in [0,1)$, then $z_0 = 0$.

Proof: If not, then since 0 is a regular value of $\widetilde{H}$ at (z_0,t_0), $\widetilde{H}^{-1}(0)$ forms a one-dimensional curve passing through this point. But $\widetilde{H}(\lambda z_0,t_0) = 0$ for any $\lambda \in \mathbb{C} = R^2$, since $\widetilde{H}_k$ is homogeneous in z; thus $\widetilde{H}^{-1}(0)$ contains a two-dimensional surface, a contradiction.

The above lemma allows us to conclude that the highest order terms $\tilde{H}(z,t)$ of $H(z,t)$ dominate, and thereby give rise to an a priori bound for zeros of H.

Lemma 2.4. For any $t_0 \in [0,1)$, there exists a constant $K(t_0) > 0$ such that all zeros of H in $\mathbb{C}^n \times [0,t_0]$ enjoy the estimate $|z| \leq K(t_0)$.

Proof: Suppose not; then there exists $t_0 \in [0,1)$ and a sequence $(z_\alpha, t_\alpha) \in \mathbb{C}^n \times [0,t_0]$, where $H(z_\alpha, t_\alpha) = 0$ and $|z_\alpha| \to \infty$. We may suppose $\frac{z_\alpha}{|z_\alpha|} \to w \in \mathbb{C}^n$ and $t_\alpha \to \tau \in [0,t_0]$. Since $H_k - \tilde{H}_k$ is a polynomial in z of degree less that d_k, and $\tilde{H}_k$ is homogeneous of degree d_k it follows that

$$|z_\alpha|^{-d_k}[H_k(z_\alpha, t_\alpha) - \tilde{H}(z_\alpha, t_\alpha)] \to 0$$

and so

$$\tilde{H}_k(\frac{z_\alpha}{|z_\alpha|}, t_\alpha) = |z_\alpha|^{-d_k} \tilde{H}_k(z_\alpha, t_\alpha)$$

$$= |z_\alpha|^{-d_k}[\tilde{H}_k(z_\alpha, t_\alpha) - H_k(z_\alpha, t_\alpha)] \to 0.$$

Hence $\tilde{H}(w,\tau) = 0$; but clearly $|w| = 1$, so this contradicts Lemma 2.3.

§3. Following the Curves

As immediate consequence of the choice of (b,a) by Lemma 2.2 is that each $b_k \neq 0$ and that from each of the d roots

$$z_k = b_k^{1/d_k}, \quad 1 \leq k \leq n$$

of $H(\cdot,0)$, given by various choices of the d_kth root, there emanates an arc of zeros of H. This arc may be given parametrically as $(z(s),t(s))$ for $s \geq 0$ representing arc length. If we show

$\frac{dt}{ds} \neq 0$ along each such arc whenever $t < 1$, then it follows from Lemma 2.4 that $t(s) \to 1$ as s increases to some limiting value. In other words, the arcs extend across the interval $0 \leq t < 1$ and may be described as analytic functions $z_k(t)$, $1 \leq k \leq d$, in this range. Moreover, it is clear that (1.4) holds.

In order to show $\frac{dt}{ds} \neq 0$, the following lemma is needed; it is also used in proving that each isolated root of P is reached by some arc (1.5).

<u>Lemma 3.1</u>. Let the n by n complex matrix M describe a linear transformation of the space $\mathbb{C}^n$ of complex variables $(z_1, z_2, \ldots, z_n)$ into itself. If this transformation is regarded as one on the space R^{2n} of real variables $(x_1, y_1, x_2, y_2, \ldots, x_n, y_n)$, where $z_k = x_k + iy_k$, and is represented by the $2n$ by $2n$ real matrix N, then

$$\det N = |\det M|^2 \geq 0$$

and

$$\dim_R \text{ kernel } N = 2 \dim_{\mathbb{C}} \text{ kernel } M \quad \text{is even.}$$

Here $\dim_R$ and $\dim_{\mathbb{C}}$ refer to real and complex dimension.

<u>Proof</u>: The relation between M and N is the following: if the (j, k)-entry of M is the complex number $\zeta_{jk} = \xi_{jk} + i\eta_{jk}$, and N is written in block form as an n by n array of two by two blocks, then the (j, k)-block of N is the real matrix

$$\begin{pmatrix} \xi_{jk} & -\eta_{jk} \\ \eta_{jk} & \xi_{jk} \end{pmatrix} .$$

Denote this relation by $\alpha(M) = N$. Now if M is upper triangular the result is easily seen. And for general M we may assume $A^{-1}MA$ is upper triangular for some complex nonsingular A; the result then holds because $\alpha(A^{-1}MA) = \alpha(A)^{-1}N\alpha(A)$.

Lemma 3.2. If $H(z_0, t_0) = 0$ and $t_0 \in [0,1)$, then $D_zH(z_0, t_0)$ is nonsingular. Hence $\frac{dt}{ds} \neq 0$ at that point.

Proof: Regard H as a map from $R^{2n} \times R$ to R^{2n}; since 0 is a regular value on $R^{2n} \times [0,1)$, the kernel of $D_zH(z_0, t_0)$ is at most one-dimensional. By Lemma 3.1, then, this matrix has zero kernel so is nonsingular. Equation (1.1) shows $\frac{dt}{ds} \neq 0$.

All that remains in proving Theorem 1.1 is to show that each isolated root z_0 of P is reached by some arc $z_k(t)$ as in (1.5); to do this it is sufficient to show that for t arbitrarily close to 1, $H(\cdot, t)$ has zeros arbitrarily close to z_0. We prove this by calculating the degree of the map P with respect to a small ball $|z - z_0| \leq \delta$; the hypothesis on the Jacobian of P implies this degree is positive.

The degree of a smooth map $F : U \subseteq R^m \to R^m$ with respect to a sufficiently small neighborhood V of an isolated zero $x_0 \in U$ can be calculated as follows [6]. Let $\mu : R^m \to R$ be a smooth function with compact support in a sufficiently small neighborhood $|y| \leq \epsilon$ of the origin, and such that $\int \mu(y)\,dy = 1$. Then the degree of F with respect to V is the quantity

$$\int_V \mu(F(x)) \det DF(x)\,dx$$

provided ϵ is sufficiently small. The integrand here has compact support, and the value of the integral is an integer. We may assume without loss of generality that $\mu(y) \geq 0$ for all y, and $\mu(y) > 0$ near the origin.

Therefore, by regarding P as a mapping in R^{2n}, with $z = (x,y) \in R^{2n}$, it follows from Lemma 3.1 that the degree of P with respect to a neighborhood V of z_0 equals

$$\int_V \mu(P(x,y))\,|\det DP(z)|^2\,dxdy. \tag{3.1}$$

By assumption the Jacobian of P does not vanish on any open set; hence the integral (3.1) is positive, as required. This completes the proof of Theorem 1.1.

References

[1] Abraham, R. and Robbin, J., Transversal Mappings and Flows, Benjamin, New York, 1967.

[2] Chow, S.-N., Mallet-Paret, J. and Yorke, J.A., Finding zeros of maps: homotopy methods that are constructive with probability one, Math. Comp. 32 (1978), 887-899.

[3] Drexler, F.J., A homotopy method for the calculation of all zero-dimensional polynomial ideals, Continuation Methods, 69-93, H. Wacker (ed.), Academic Press, New York, 1978.

[4] Garcia, C.B. and Zangwill, W.I., Global continuation methods for finding all solutions to polynomial systems of equations in N unknowns, preprint.

[5] Li, T.-Y. and Yorke, J.A., Finding all the roots of polynomials by homotopy method - numerical investigation, in preparation.

[6] Nirenberg, L., Topics in Nonlinear Functional Analysis, Courant Institute, New York University, 1974.

[7] Watson, L., A globally convergent algorithm for computing fixed points of C^1 maps, preprint.

[8] Watson, L., Wang, C.-Y. and Li, T.-Y., The elliptic porus slider -- a homotopy method, J. Applied Mechanics, 45 (1978), 435-436.

A VIEW OF COMPLEMENTARY PIVOT THEORY
(or Solving Equations with Homotopies)

B. Curtis Eaves

1. Introduction

Our purpose here is to give a brief, valid, and painless view of the equation solving computational method variously known as complementary pivot theory and/or fixed point methods. Our view begins in Section 2 with a bit of history and some of the successes of the method. In Section 3 the unique convergence proof of the method is elucidated with a riddle on ghosts. In Section 4 the general approach for solving equations with complementary pivot theory is encapsulated in the "Homotopy Principle". In Section 5 a simple example is used to illustrate both the convergence proof and the "Homotopy Principle". Rudiments of the general theory are stated and the "Main Theorem" is exhibited in an example in Section 6. Two representative complementary pivot algorithms are presented vis-a-vis the "Homotopy Principle" and "Main Theorem" in Section 7. Finally, in Section 8 the principal difficulty of the method is discussed and some of the studies for dealing with this difficulty are mentioned.

2. A Bit of History

If a point in time can be specified as the beginning of complementary pivot theory it is with the paper of Lemke and Howson [20]. In this paper a startling convergence proof was given for a finite algorithm for computing a Nash equilibrium of a bimatrix game. To understand their contribution, it was previously known that a Nash equilibrium existed via the Brouwer fixed point theorem and that exhaustive search offered a finite procedure for computing such an equilibrium. Furthermore, there is no theoretical proof that the Lemke-Howson algorithm has advantages over exhaustive search, and, in fact, one can construct examples where the Lemke-Howson algorithm is no better than exhaustive search. However, the point is, as a practical matter, the Lemke-Howson algorithm versus exhaustive search enables one to solve bimatrix games with characteristic size of, say, one thousand versus thirty. The situation is analogous to that of Dantzig's simplex method and linear programs.

In the paper [19] Lemke specified what is now known as "Lemke's Algorithm" and, thereby, showed that the convergence proof could be used for a much broader class of problems including quadratic programs.

The next steps came from Scarf in [26,27,28]. Using the convergence

---

This research was sponsored in part by the Army Research Office - Durham, Contract No. DAAG-29-78-G-0026 and the National Science Foundation, Grant No. MCS-77-05623.

proof of Lemke and Howson he proved for the first time that a balanced game has a nonempty core and he described algorithms for computing a Brouwer fixed point and an equilibrium point for the general competitive equilibrium model. Once again, it has not been proved that the algorithm improved upon exhaustive search, but as a practical matter, problems could be solved that could not be solved before (note, for example, that Brouwer's theorem can be proved by repeated applications of Sperner's Lemma and that only finitely many simplexes need be examined at each iteration in order to find a complete simplex).

There are now at least two hundred papers in complementary pivot theory, and many very exciting developments have occurred. The convergence proof of Lemke and Howson is now understood to be intimately related to homotopy theory, a matter which is the crux of this paper. Many classical results have been given new "complementary pivot" proofs; to mention a few: Freidenfelds [9] and a connected set theorem of Browder, Kuhn [18] and the fundamental theorem of algebra, Garcia [11] and the last theorem of Poncairé, and Meyerson and Wright [23] and the Borsak-Ulman theorem.

A great deal of effort and ingenuity has been expended in making the complementary pivot algorithms more efficient; however, we shall not discuss these matters until the last section wherein we will describe the principal weakness of the complementary pivot algorithms.

Complementary pivot theory has been used to solve a number of specific problems; for instance, the following papers are concerned with complementary pivot theory applied to the solution of differential equations: Allgower and Jeppson [1,2], Wilmuth [34], Cottle [4], Netravali and Saigal [24], and Kaneko [14]. Katzenelson's [15] algorithm, and subsequent developments thereof, for electrical network problems also fits comfortably into the framework of complementary pivot theory as discussed in this paper.

This paper is based upon Eaves and Scarf [8] and Eaves [6]. The reader might also want to consult Hirsch [12], Scarf [29] and Todd [31].

3. Convergence Proof

The following riddle and its solution illustrates the extraordinary convergence proof of Lemke and Howson.

Ghost Riddle: I eased through the front door of the allegedly haunted house. Just as a ghost appeared, the front door slammed shut behind me. He spoke, "You are now locked inside our house, but it is your fate that except for this room which has one open door, every other room with a ghost has two open doors." I thought, "Is there a room without a ghost?" □

The diagram below spills the beans.

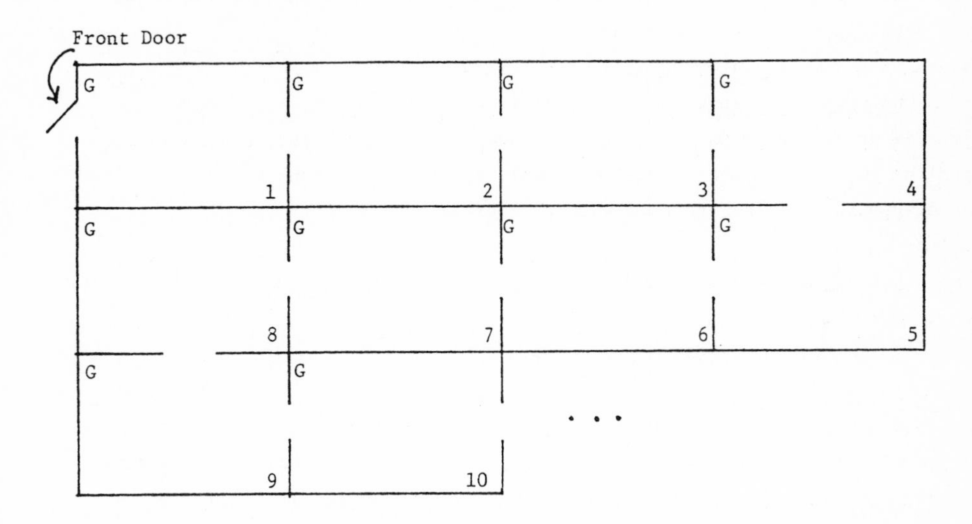

Assume we are standing in Room 1 with the front door closed. According to the riddle there is exactly one open door, so let us pass through the door into the adjoining room which we now call Room 2. In Room 2, assuming the presence of a ghost, there are two open doors available to us, one of which we just entered; so let us exit the other and enter Room 3. We continue in this fashion to Rooms 4, 5, etc. The essential property of this process is that no room is entered (i.e., numbered) more than once, which is to say, there is no cycling. A proof of this fact is available by assuming the contrary and examining the first room entered twice. Consequently, if there are only m rooms in the house, then the process must stop with m steps or less, and there is a room without a ghost. On the other hand, if the house is sufficiently haunted so as to have infinitely many rooms (George Dantzig supports this possibility) then either the process stops with a solution, that is, a room without a ghost, or it proceeds forever always entering new rooms.

In this isolated form the convergence proof appears uselessly simple. When we apply the convergence principle, the rooms will become pieces of linearity of some function and circumstances will not be quite so transparent.

4. Homotopy Principle

Now consider the continuous function $f : R^n \to R^n$ on n-dimensional Euclidean space and the system of equations $f(x) = y$. As a general procedure for solving such a system we offer the following.

Homotopy Principle: To solve a system of equations, the system is first deformed to one which is trivial and has a unique solution. Beginning with the solution to the trivial problem a route of solutions is followed as the system is deformed, perhaps with retrogressions, back to the given system. □

Let us be more specific. First we introduce a family of problems

$$F(x,\theta) = y \qquad 0 \leq \theta \leq 1$$

where F is continuous in (x,θ), $F(\cdot,0) = f$, and $F(x,1) = y$ is a trivially solved system with a unique solution. We think of θ as deforming the given system $f(x) = y$ to the system $F(x,1) = y$ with a unique solution, say x_1.

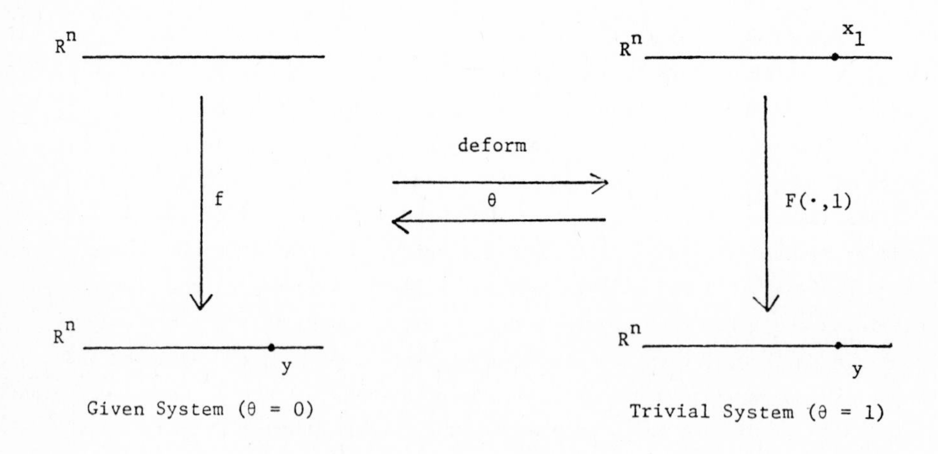

To obtain a solution to the given system $f(x) = y$ we follow the solution of $F^{-1}(y)$ beginning with $(x_1,1)$. Except for degenerate (rare) cases the component of $F^{-1}(y)$ that meets $(x_1,1)$ is a route, that is, a path.

Assuming F is piecewise linear on $R^n \times (0,1]$ the next schema illustrates the situation quite well.

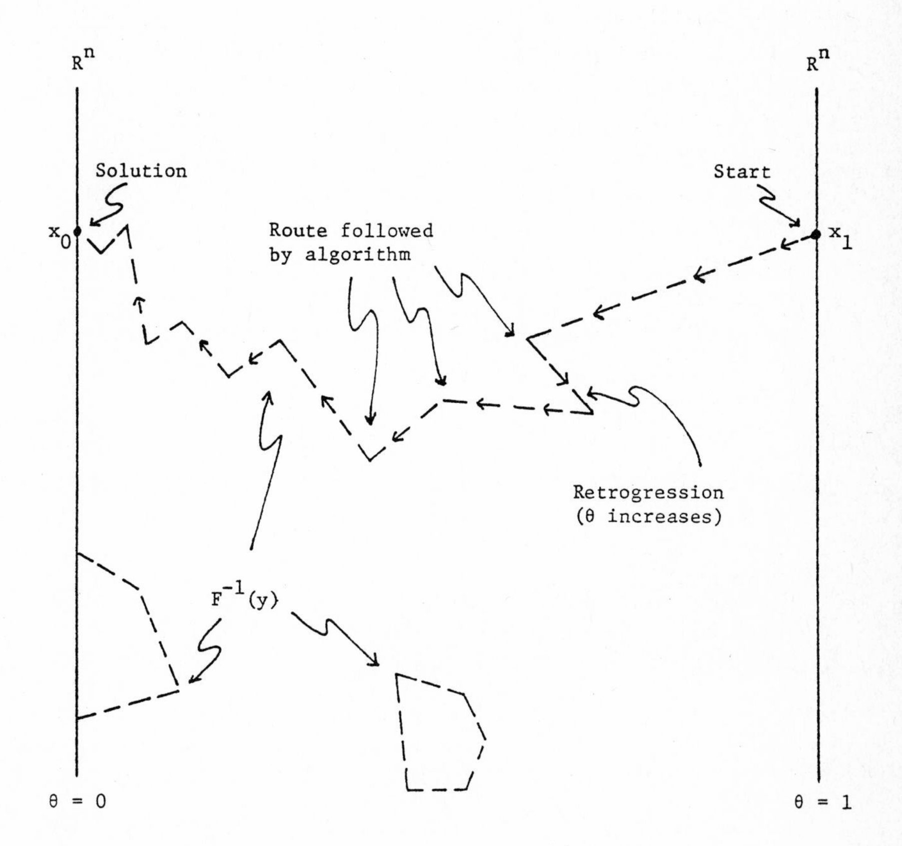

The algorithm begins with the point $(x_1,1)$ and follows the route of $F^{-1}(y)$. Under various conditions it can be shown that the route eventually leads to $R^n \times 0$ and thus yields a solution of the given problem.

This principle is illustrated in Section 5 and given theoretical credence in Section 6.

5. <u>An Example</u>

In this section we exhibit the complementary pivot convergence proof and the "Homotopy Principle" by solving a system of piecewise linear equations. This particular system of equations was chosen for its pedagogial value; later we examine merely continuous functions.

Let S be an n-simplex in R^n with extreme points $s_0, s_1, \ldots, s_n$.

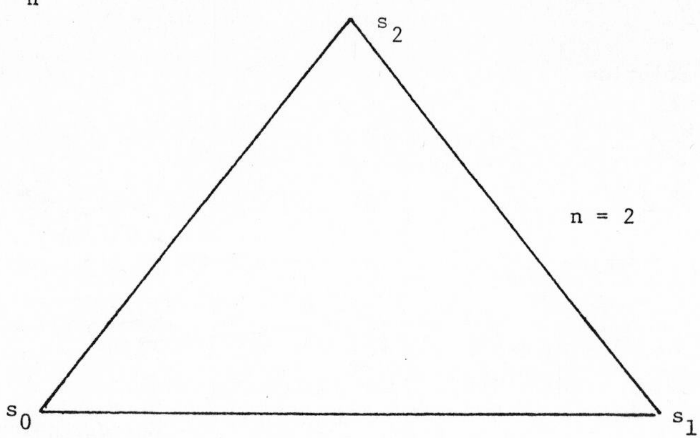

Let $\mathscr{S}$ be a collection of smaller n-simplexes which subdivides S.

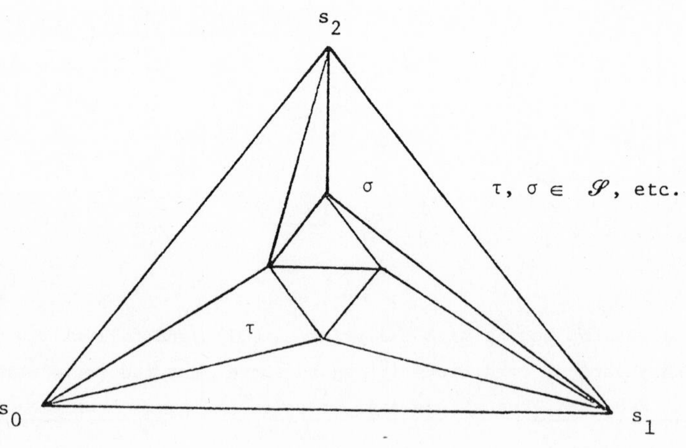

By a vertex of $\mathscr{S}$ we mean a vertex of any element of $\mathscr{S}$.

Now let $f : S \to S$ be a continuous function on the simplex with the following three properties.

a) On each element of $\mathscr{S}$, f is linear, that is, affine.

b) On the boundary of S, f is the identity, that is, $f(x) = x$ for x in ∂S.

c) Vertices of the subdivision are carried by f into extreme points of S, that is, $f(\text{vertices}) \subset \{s_0, \ldots, s_n\}$.

Let y be some interior point of S and we consider the system of equations $f(x) = y$. Toward solving this system label the vertices of $\mathscr{S}$ according to the extreme point to which it is mapped, that is, define the labeling function ℓ on the vertices by $\ell(v) = i$ if $f(v) = s_i$.

Assume that we thus obtain the following.

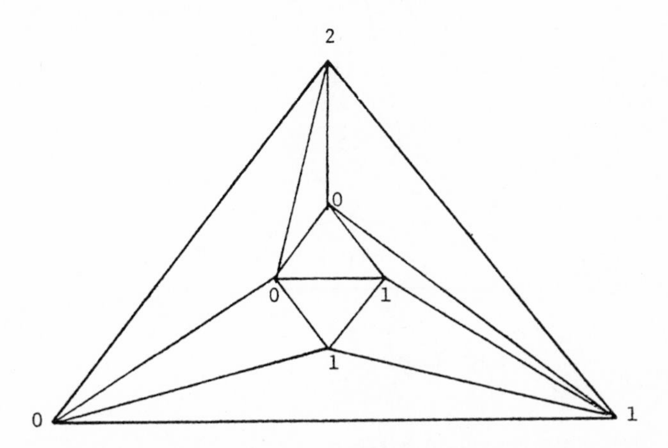

We call a simplex σ of $\mathscr{S}$ completely labeled only if all labels $0, 1, \ldots, n$ are present on its vertices. In the figure above there is exactly one such, namely the upper right one.

Let us observe that solving $f(x) = y$ is equivalent, modulo solving a system of linear equations, to finding a completely labeled simplex. If a simplex τ of $\mathscr{S}$ has only labels $\{0, \ldots, n\} \sim \{i\}$ then in view of the linearity of f on τ, f would map the entirety of τ into the face of S spanned by $\{s_0, \ldots, s_n\} \sim \{s_i\}$. Consequently, no point of τ would hit the interior point y. On the other hand, if a simplex τ of $\mathcal{S}$ is completely labeled then τ is mapped onto S, and here some point of τ hits y. So for the moment we focus on the task of finding a completely labeled simplex.

To execute the task of finding a completely labeled simplex we shall employ the "Ghost Riddle" as follows to obtain "Cohen's Algorithm", see [3]. We regard as a room a simplex of $\mathcal{S}$ and as an open door a face of a simplex of $\mathcal{S}$ with labels 0 and 1 (if $n > 2$ a face of a simplex of $\mathcal{S}$ with labels $(0, \ldots, n - 1)$). By passing through open doors the path followed is indicated below and terminates with a completely labeled simplex.

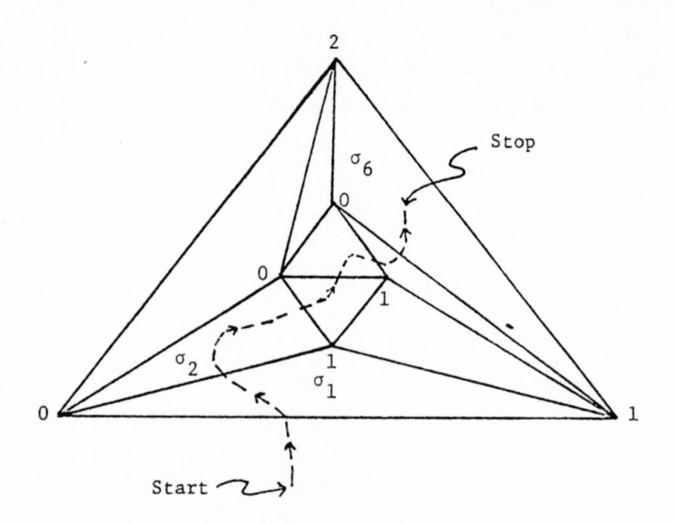

Prima facia this procedure works so smoothly that it seems rigged. But suppose that the n-simplex σ of $\mathcal{S}$ has vertices $(t_0, \ldots, t_{n-1})$ with labels $(0, \ldots, n - 1)$. If the remaining vertex t_n has label n, then one has a completely labeled simplex. But, if the remaining label is i for some $0 < i < n - 1$, then there is exactly one other face of σ, namely, that spanned by $\{t_0, \ldots, t_n\} \sim \{t_i\}$, that has labels $\{0, \ldots, n - 1\}$. Thus, if a room has at least one open door, then either it is the target (i.e., a completely labeled simplex) or it has exactly two open doors. Since only one door passes from outside the simplex S to the inside and since there are only finitely many simplexes, the procedure must terminate with a completely labeled simplex.

Our aim in the above exercise was principally to exhibit the convergence proof in action. Next we exhibit use of the "Homotopy Principle" and show that it yields the algorithm just given, "Cohen's Algorithm", for solving $f(x) = y$.

Before applying the "Homotopy Principle" to solve $f(x) = y$ let us recall an elementary fact from linear algebra. Let $L : R^{n+1} \to R^n$ be a linear map of rank n from $(n+1)$-space to n-space. If y is any point in the range R^n, then $L^{-1}(y)$ is a line. Taking matters one step further let σ be an $(n+1)$-cell, that is, a closed polyhedral convex set of dimension $n + 1$. Let $L : \sigma \to R^n$ be a linear map where $L(\sigma)$ has dimension n. Then for most values of y in R^n either $L^{-1}(y)$ is empty or is a chord of σ

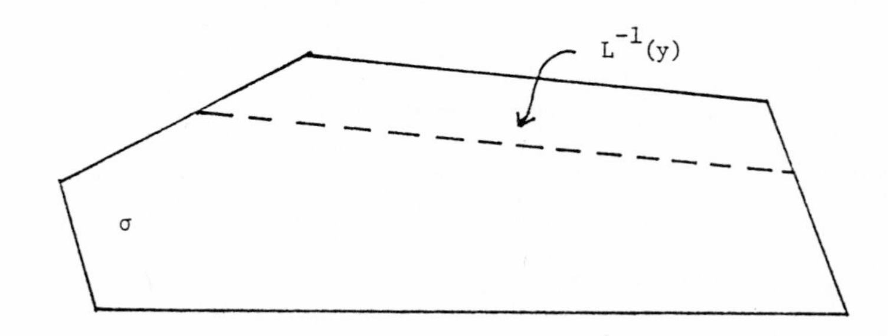

whose endpoints lie interior to n-faces of σ. There are a few values of y where $L^{-1}(y)$ meets an $(n-1)$-face of σ; these y's we call degenerate (or critical) but for convenience we shall always assume that our y is regular, that is, not degenerate. There are measures for dealing with degenerate y's but the treatment for them will not be discussed in this paper, (see Eaves [6]).

Given the system $f(x) = y$ we introduce a family of problems $F(x,0) = y$. Let x_1 be any point in the interior of the face of S spanned by s_0 and s_1 (for $n > 2$ by $s_0, s_1, \ldots, s_{n-1}$) and define the homotopy, that is, function, F by

$$F(x,\theta) = f(x) + (y-x_1)\,\theta$$

for $0 \leq \theta \leq 2$. So F carries points of the cylinder $M = S \times [0,2]$ into R^n.

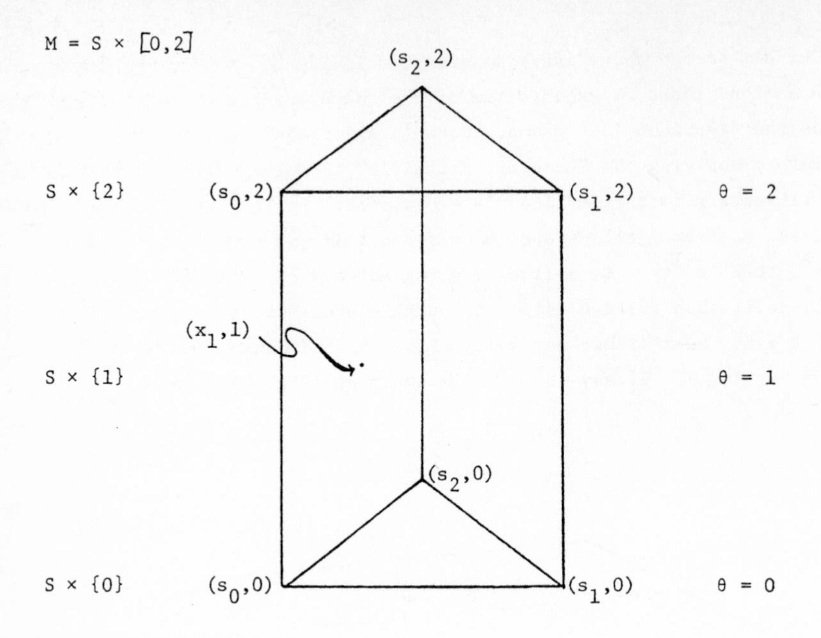

We can subdivide the cylinder $M = S \times [0,2]$ by letting $\mathcal{M}$ be the collection of cells of form $\sigma \times [0,2]$ where σ is a cell of $\mathcal{S}$. Now observe that F is piecewise linear with respect to $\mathcal{M}$, that is, F is affine on each cell of $\mathcal{M}$.

At $\theta = 0$ the system $F(x,\theta) = y$ is the system $f(x) = y$. For $\theta > 0$ the system of particular interest is

$$* \begin{cases} F(x,\theta) = y \\ (x,\theta) \in \partial M \\ \theta > 0 \end{cases}$$

The second condition requires that (x,θ) is in the boundary of M, that is, in the top, bottom or some side of M. We argue that the system $*$ has exactly one solution, namely, $(x_1,1)$.

If $\theta > 1$ then $y - \theta(y - x_1)$ is not in S and clearly $f(x) = y - \theta(y - x_1)$ can have no solution. If $0 < \theta < 1$ then $y - \theta(y - x_1)$ is interior to S and $f(x) = y - \theta(y - x_1)$ with $x \in \partial S$ can have no solution, since f is the identity on the boundary. If $\theta = 1$ our system becomes $f(x) = x_1$ with $x \in \partial M$, and again since f is the identity on the boundary, clearly $x = x_1$ is the only solution.

So there is one solution $(x_1, 1)$ away from the bottom of M and we seek a solution on the bottom of M; this is precisely our desired situation. Assume that y is a regular value, that is to say, let us assume that $F^{-1}(y)$ does not meet any $(n-1)$-faces of elements of $\mathcal{M}$. Given $(x_1, 1)$ and the cell $\sigma_1 \times [0,2]$ of $\mathcal{M}$ which contains it we have a linear map from the $(n+1)$-cell $\sigma_1 \times [0,2]$ to R^n, so we apply our result from linear algebra. $F^{-1}(y) \cap (\sigma_1 \times [0,2])$ is a chord of $\sigma_1 \times [0,2]$

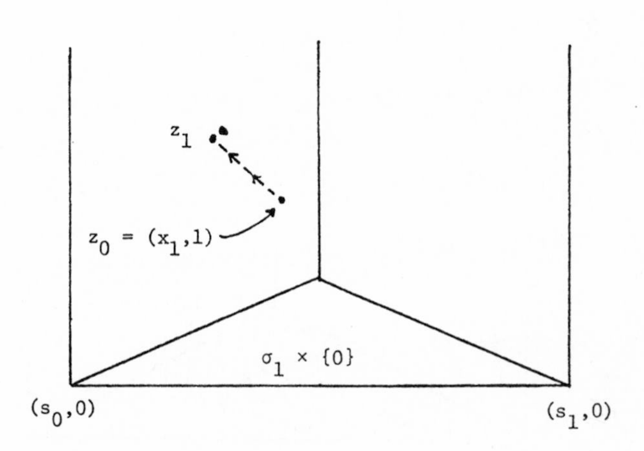

Let $z_0 = (x_1, 1)$ be one end of the chord and z_1 the other; calculating z_1 is just a matter of solving a linear system of equations. Next we go to the $(n+1)$-cell $\sigma_2 \times [0,2]$ of $\mathcal{M}$ that contains z_1 but not z_0 and repeat the procedure to get a chord $F^{-1}(y) \cap (\sigma_2 \times [0,2])$ of $\sigma_2 \times [0,2]$, etc.

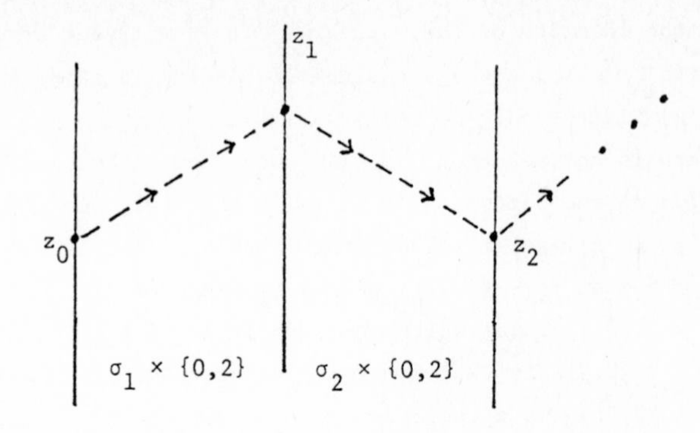

In this manner we continue to follow the route of $F^{-1}(y)$ beginning with $(x_1,1)$. Observe that this route can have no forks. Eventually the route yields a point z_k in $S \times \{0\}$ and the system $f(x) = y$ is solved.

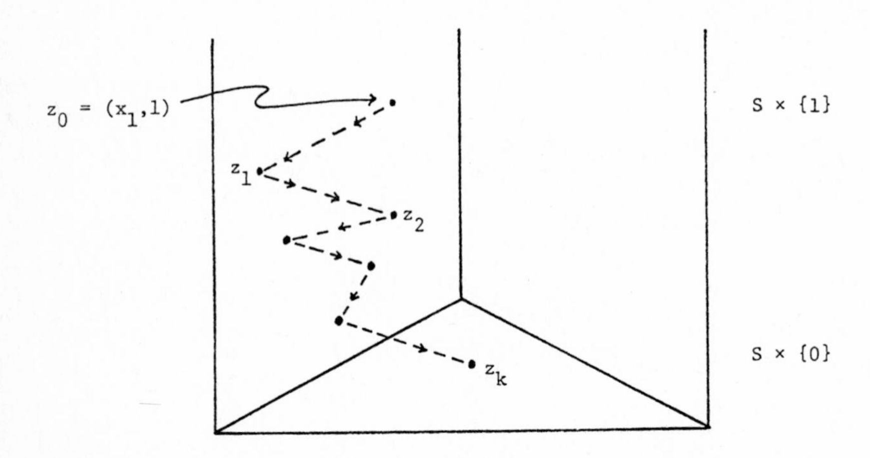

What is the relation of the route of $F^{-1}(y)$ beginning at $(x_1,1)$ and "Cohen's Algorithm" applied to the problem? Well, they are in essence identical once the smoke has cleared. If the route of $F^{-1}(y)$ beginning with $(x_1,1)$ is projected down to the base $S \times \{0\}$ of M we see that it passes through the same sequence $\sigma_1, \sigma_2, \ldots$ of rooms as "Cohen's Algorithm"! In this sense we regard "Cohen's Algorithm" for $f(x) = y$ as that yielded by the "Homotopy Principle".

6. General Theory

The general theory is to be sketched here. As the example of the previous section contains most of the ideas involved in the general theory, the conceptual step to the material presented here is small.

Cells are our building blocks. We define an m-cell to be a closed polyhedral convex set of dimension m.

Let $\mathcal{M}$ be a collection of m-cells and let M be the union of these cells. $(M, \mathcal{M})$ is defined to be a subdivided m-manifold if the following three conditions hold.

a) Given any two cells of $\mathcal{M}$ either they do not meet or they meet in a common face.

b) Any (m-1)-face of a cell of $\mathcal{M}$ lies in at most two cells of $\mathcal{M}$.

c) Given any point of M there is a neighborhood which meets only finitely many cells of $\mathcal{M}$.

Condition a) prohibits construction such as

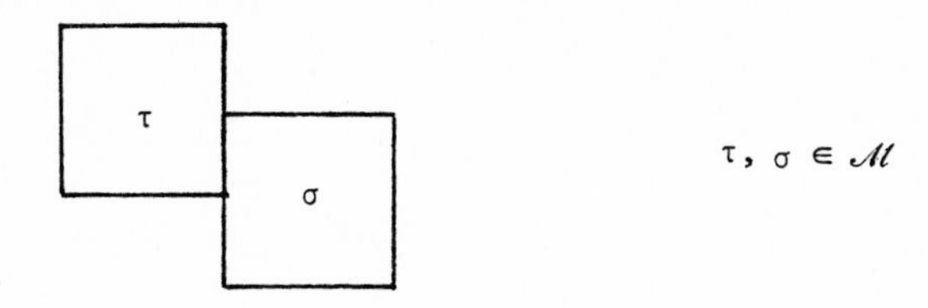

and requires constructions such as

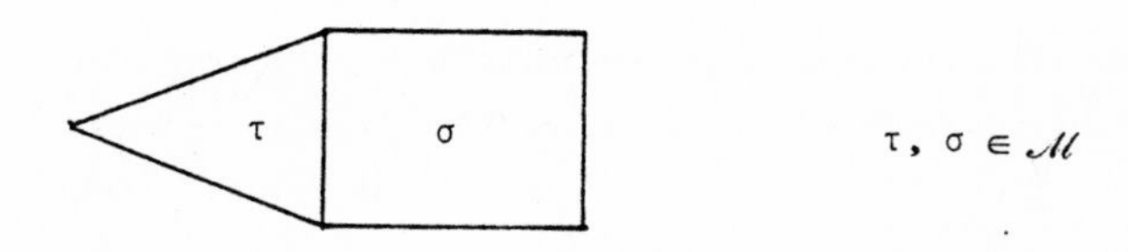

or

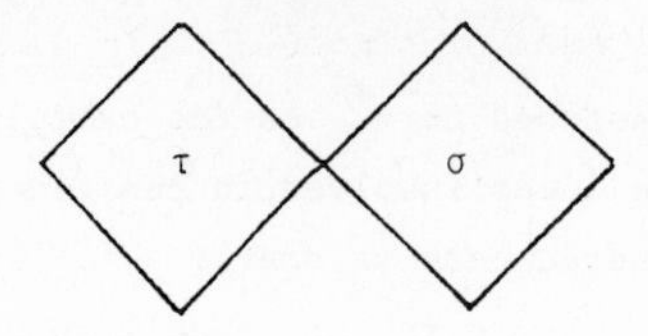

Condition b) prohibits construction such as

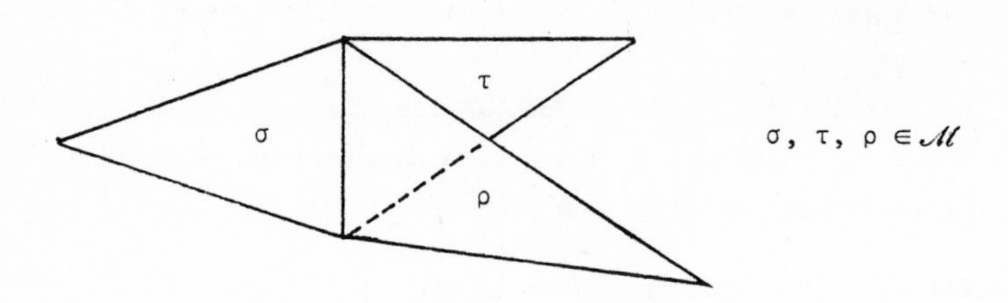

As an example of a 1-manifold we have

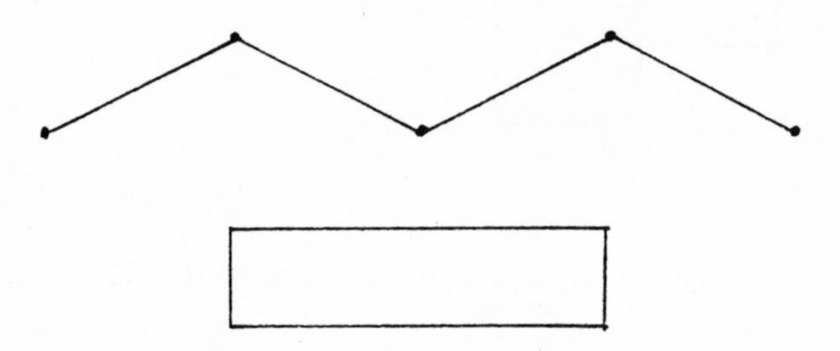

where $\mathscr{M}$ contains 8 1-cells. A 1-manifold is a disjoint collection of routes and loops; the previous example contains one route and one loop. Note that a route or loop of a 1-manifold contains no forks; a proof of this point requires essentially the "Ghost Argument" of Section 3. As examples of a subdivided 2-manifold we have $(S, \mathscr{S})$ of the previous section of the surface of cube where $\mathscr{M}$ is the set containing the top, bottom, and sides. We call M an m-manifold if for some $\mathscr{M}$, $(M, \mathscr{M})$ is a subdivided m-manifold.

By the boundary ∂M of a subdivided m-manifold we mean the union of all (m-1)-faces of cells of $\mathcal{M}$ that lie in exactly one m-cell. Thus, for example, the boundary of the surface of a cube is empty. For further examples consider

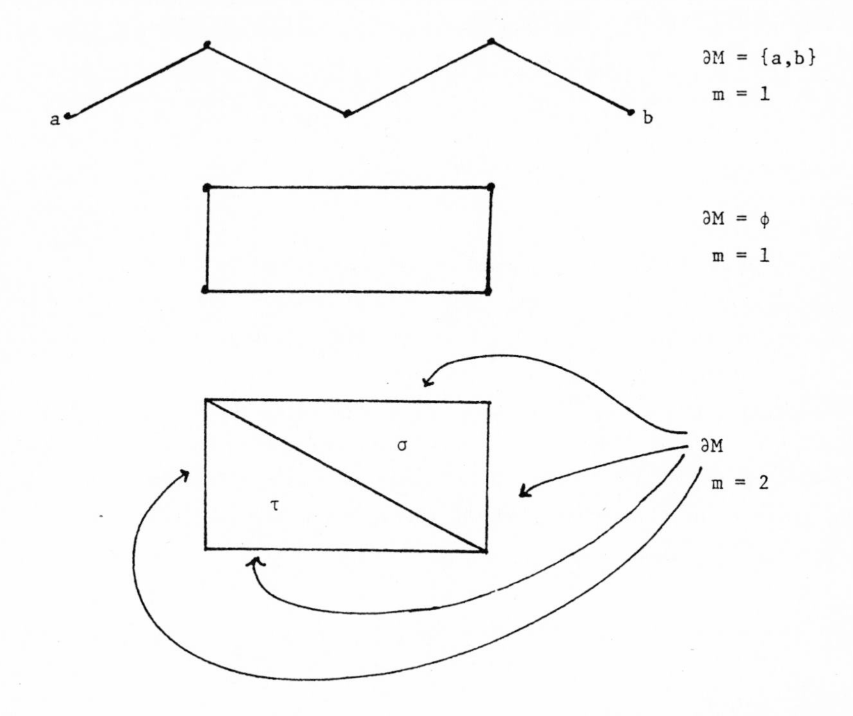

Let N be a 1-manifold and $(M,\mathcal{M})$ a subdivided m-manifold, where N is contained in M. We say that N is neat in $(M,\mathcal{M})$ if the following three conditions hold.

a) N is closed in M.

b) The boundary of N lies in the boundary of M.

c) The collections of nonempty sets of form $N \cap \sigma$ with σ in $\mathcal{M}$ forms a subdivision of N.

In the following diagram we show a 1-manifold (which is a route) N neat in a subdivided 2-manifold M.

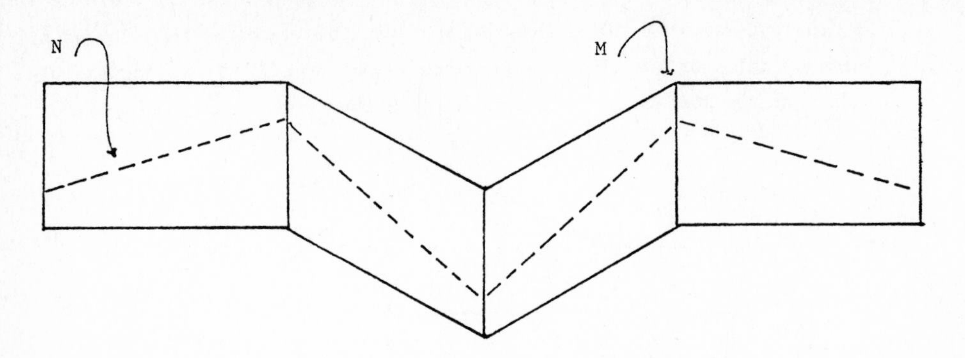

Now we can state the main theorem for regular values y.

<u>Main Theorem</u> (Regular Values): Let $(M,\mathcal{M})$ be an $(n+1)$-manifold and $F : M \to R^n$ be $\mathcal{M}$ piecewise linear. If y in $F(M)$ is regular, then $F^{-1}(y)$ is a 1-manifold neat in $(M,\mathcal{M})$. □

Recall that y is a regular value if $F^{-1}(y)$ does not meet any $(n-1)$-faces of cells of $(M,\mathcal{M})$. Note that for y regular, a loop of $F^{-1}(y)$ cannot meet the boundary of M.

For purposes of illustration of the theorem, consider the following subdivided 2-manifold $(M,\mathcal{M})$ in R^2

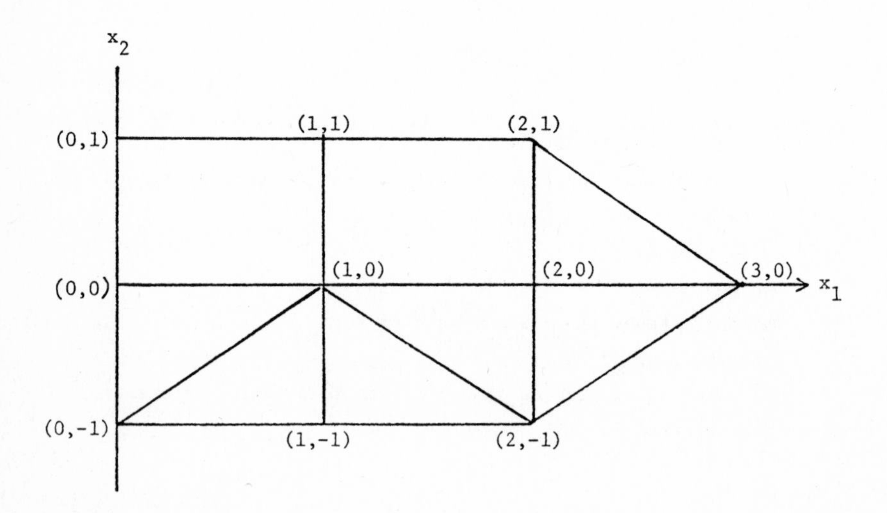

Let $F = F_1 F_2 F_3 F_4$ where the F_i's are defined in the schema below together with the requirement that the F_i's are piecewise linear with respect to the indicated subdivisions. Let v represent any vertex.

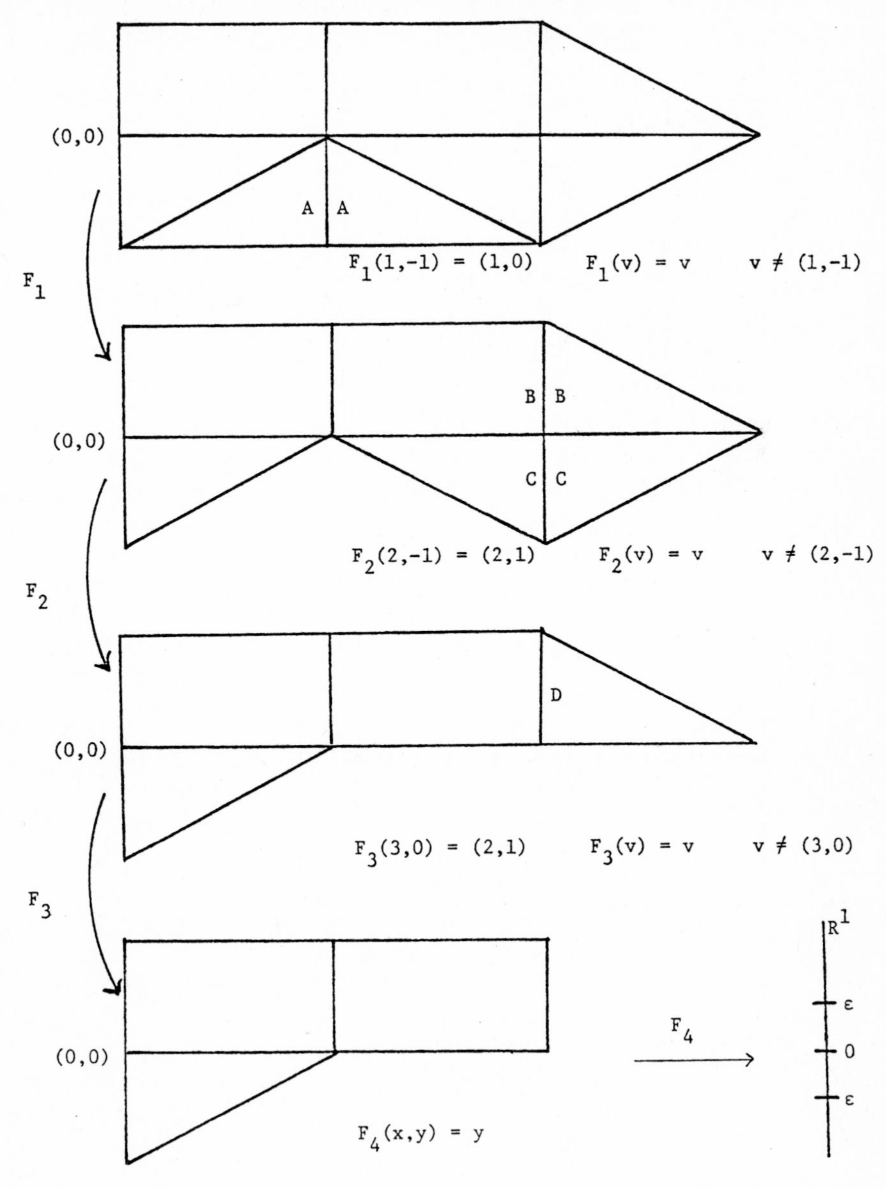

F_1 collapses the region A, F_2 flips C over to B, F_3 collapses the region D, and F_4 projects to the vertical axis. So F_i^{-1} of any set can be discerned by inspection.

By calculating $F_4^{-1}(\varepsilon)$, $F_3^{-1}[F_4^{-1}(\varepsilon)]$, $F_2^{-1}[F_3^{-1}(F_4^{-1}(\varepsilon))]$ and finally $F^{-1}(\varepsilon) = F_1^{-1}[F_2^{-1}(F_3^{-1}(F_4^{-1}(\varepsilon)))]$ we can see that $F^{-1}(\varepsilon)$ is the 1-manifold (that is, route) N which is neat in $(M,\mathcal{M})$.

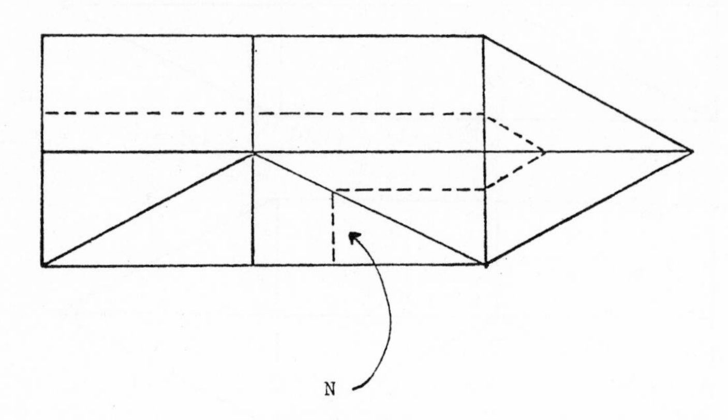

$F^{-1}(y)$ will be a neat 1-manifold in $(M,\mathcal{M})$ for all y in $(-1,0) \cup (0,1)$. The values -1, 0, and 1 are degenerate and the reader might want to investigate F^{-1} of one or all of them, especially $F^{-1}(0)$.

The "Main Theorem" justifies the notion of following a route in $F^{-1}(y)$ in the "Homotopy Principle". That is, assuming y is a regular value, then in Section 5 we were following a 1-manifold (route) neat in $(M,\mathcal{M})$ to solve $f(x) = y$.

7. Lemke's Algorithm

The linear complementarity problem, which was first stated by Cottle, is: Given an $n \times n$ matrix A and n-vector q find a z and w such that

$$Az - Iw = q$$

$$z \geq 0 \qquad w \geq 0 \qquad z \cdot w = 0 .$$

To solve such problems Lemke introduces an n-vector $d > 0$ and considers the augmented system

$$Az - Iw + d\theta = q$$

$$z \geq 0 \qquad w \geq 0 \qquad \theta \geq 0 \qquad z \cdot w = 0$$

Lemke's algorithm proceeds by generating a path of solutions to the augmented system. To explain his algorithm from the perspective of the general theory first define $f_i : R^1 \to R^n$ for $i = 1, 2, \ldots, n$ by

$$f_i(x_i) = \begin{cases} A_i x_i & \text{if} \quad x_i \geq 0 \\ I_i x_i & \text{if} \quad x_i \leq 0 \end{cases}$$

where A_i and I_i are the ith columns of A and I, the identity, respectively. Define $f : R^n \to R^n$ by

$$f(x) = \sum_1^n f_i(x_i)$$

where

$$x = (x_1, \ldots, x_n) \ .$$

The complementary problem is equivalent to $f(x) = q$ and the augmented system is equivalent to $f(x) + \theta d = q$ with $\theta \geq 0$. So, define $F(x,\theta) = f(x) + \theta d$ as the homotopy; note that F is piecewise linear with respect to the orthants of $R^n \times R^1_+$. To solve $f(x) = q$ Lemke's algorithm follows the path of $F^{-1}(q)$ beginning with θ large and $x \leq 0$. Observe that $F^{-1}(q)$, if q is regular, is a 1-manifold neat in $R^n \times R^1_+$ which is subdivided by its orthants. To show that the route of the algorithm yields a solution to $f(x) = q$ requires conditions on the matrix A, an issue that will not be treated here.

8. The Eaves-Saigal Algorithm

Let $g : S \to \overset{\circ}{S}$ be a continuous function from a simplex S in R^n to its interior $\overset{\circ}{S}$. Let us compute a fixed point $x = g(x)$ of g, or equivalently, a zero of $f(x) \triangleq g(x) - x$. The first step is to subdivide the cylinder $M = S \times (0,1]$ with $\mathcal{M}$ as indicated in the picture below.

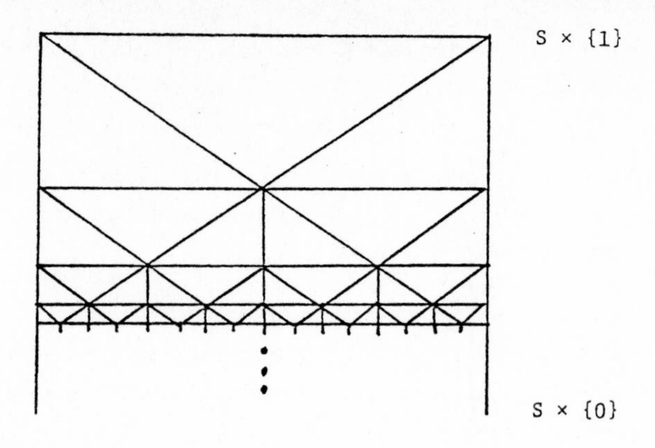

$S \times \{1\}$ should not be subdivided and the size of the simplexes of the subdivision should tend to zero uniformly as the simplexes near $S \times \{0\}$. We pause to note here that in a computer the subdivision exists only in the sense that there is a formula that can be used to generate portions of the subdivision as needed.

Let $F(x,\theta) = f(x)$ for all vertices (x,θ) of the subdivision. Define $F : M \to R^n$ by extending F to all of M in such a way that F is affine on the cells of $\mathcal{M}$; this extension is unique. Once again, in a computer F is only generated as needed. We now have the property that $F(\cdot,1)$ is linear and $F(\cdot,t)$ tends to f as t tends to 0. $F(x,1) = 0$ is a linear system and is easily shown to have a unique solution $(x_1,1)$.

Beginning with the point $(x_1,1)$ the route of $F^{-1}(0)$ is followed; $F^{-1}(0)$ is a neat 1-manifold in $(M,\mathcal{M})$ if 0 is regular. The x component of the route tends (as the subdivision gets finer and finer) to a solution (or solutions) of $f(x) = 0$, that is, fixed point (or fixed points) of g.

8. Internal Developments

The principle weakness of complementary pivot theory is simply that there are too many cells to traverse along the route of $F^{-1}(y)$. Many studies have improved the situation; let us mention those that seem to be the most important. The "restart" method of Merrill [22] permits, in effect, many cells to be skipped. In the presence of differentiability Saigal [25] has shown that

Merrill's method can be used to obtain quadratic convergence; in addition, it becomes clear from his analysis that one should formulate the system of equations to be solved so that they are as smooth as possible; the routes to be followed are then less inclined to turn radically. In the absence of special structure Todd has shown that the simplexes (and cells) should be as round as possible, that is, not long slivers.

Kojima [17], and recently Todd [31], have shown how to use special structure of the function as linearity and separability to drastically reduce the number of cells. Van der Laan and Talman [33] have revived an idea of Shapley [30] which, in effect, enables one to move through some fraction of the cells more quickly; for lack of a better term this technique is often referred to a "variable dimension method". Garcia and Gould [10] have proposed an idea with similar affect. The works of Kellogg, Li, and Yorke [16] and Hirsch and Smale [13] avoid cells altogether by using differential homotopies, see Li [21], but, then, following a differential path is more difficult than following a piecewise linear route; the tradeoff is not yet understood.

BIBLIOGRAPHY

[1] ALLGOWER, E., and M. JEPPSON, The Approximation of Solution of Nonlinear Elliptic Boundary Value Problems Having Several Solutions, to appear soon in Springer Lecture Notes, Ed: R. Ansorge and W. Törnig, Numerische, Insbesondere Approximationstheoretische Behandlung von Funktionslgleichungen, 1973.

[2] ALLGOWER, E., and M. JEPPSON, Numerical Solution of Nonlinear Boundary Value Problems with Several Solutions, to appear soon.

[3] COHEN, D.I.A., On the Sperner Lemma, J. Comb. Theory 2 (1967), 585-587.

[4] COTTLE, R.W., Complementarity and Variational Problems, TR SOL 74-6, May 1974, Department of Operations Research, Stanford University.

[5] EAVES, B.C., Homotopies for Computation of Fixed Points, Mathematical Programming 3, 1 (1972) 1-22.

[6] EAVES, B.C., A Short Course in Solving Equations with PL Homotopies, SIAM-AMS Proceedings, Vol. IX (1976) 73-143.

[7] EAVES, B.C. and R. SAIGAL, Homotopies for Computation of Fixed Points, Mathematical Programming 3, 2 (1972), 225-237.

[8] EAVES, B.C. and H. SCARF, The Solution of Systems of Piecewise Linear Equations, Mathematics of Operations Research 1, 1 (1976), 1-27.

[9] FREIDENFELDS, J., A Set of Intersection Theorem and Applications, Mathematical Programming 7, 2 (1974), 199-211.

[10] GARCIA, C. B. and F.J. GOULD, An Improved Scalar Generated Homotopy Path for Solving $f(x) = 0$, Center for Mathematical Studies in Business and Economics, Report #7633, University of Chicago, September 1976.

[11] GARCIA, B.C., A Fixed Point Theorem Including the Last Theorem of Poincaré, Mathematical Programming 8, 2 (1975), 227-239.

[12] HIRSCH, M.W., A Proof of the Nonretractability of a Cell onto its Boundary, Prodeedings of AMS, 14 (1963), 364-365.

[13] HIRSCH, M.W., and S. SMALE, On Algorithms for Solving $f(x) = 0$, Department of Mathmentics, University of California, Berkeley.

[14] KANEKO, I., A Mathematical Programming Method for the Inelastic Analysis of Reinforced Concrete Frames, TR 76-2 (1976), Department of Industrial Engineering, University of Wisconsin.

[15] KATZENELSON, J., An Algorithm for Solving Nonlinear Resistor Networks, Bell Telephone Technical Journal, 44 (1965), 1605-1620.

[16] KELLOGG, R.B., T.Y. LI, and J. YORKE, A Constructive Proof of the Brouwer Fixed Point Theorem and Computational Results, unpublished paper, University of Maryland and University of Utah (1975).

[17] KOJIMA, M., On the Homotopic Approach to Systems of Equations with Separable Mappings, B-26, Department of Information Sciences, Tokyo Institute of Technology (1975).

[18] KUHN, H.W., A New Proof of the Fundamental Theorem of Algebra, Mathematical Programming, Study 1 (1974), 148-158.

[19] LEMKE, C.E., Bimatrix Equilibrium Points and Mathematical Programming, Management Sciences, 11 (1965), 681-689.

[20] LEMKE, C.E. and J.T. HOWSON, Jr., Equilibrium Points of Bimatrix Games, SIAM Journal on Applied Mathematics 12, 2 (1964), 413-423.

[21] LI, T-Y, Path Following Approaches for Solving Nonlinear Equations: Homotopy, Continuous Newton and Projection, Department of Mathematics, Michigan State University, East Lansing, Michigan.

[22] MERRILL, O.H., Applications and Extensions of an Algorithm that Computes Fixed Points of Certain Upper Semi-Continuous Point to Set Mappings, Ph.D. Dissertation, Department of Industrial Engineering, University of Michigan (1972)

[23] MEYERSON, M.D., and O.H. WRIGHT, A New and Constructive Proof of the Borsak-Ulam Theorem, to appear in the Proceedings of the A.M.S.

[24] NETRAVALI, A.N. and R. SAIGAL, Optimum Quantizer Design Using a Fixed-Point Algorithm, The Bell Telephone Technical Journal 55, 9 (1976), 1423-1435.

[25] SAIGAL, R., On the Convergence Rate of Algorithms for Solving Equations that are Based on Complementarity Pivoting, Bell Telephone Laboratories, Holmdel, N.J.

[26] SCARF, H., The Approximation of Fixed Points of a Continuous Mapping, SIAM Journal on Applied Mathematics 15 5 (1967).

[27] SCARF, H., The Core of an N Person Game, Econometrica 35, 1 (1967), 50-69.

[28] SCARF, H., On the Computation of Equilibrium Prices, in Ten Economic Studies in the Tradition of Irving Fisher, John Wiley, New York (1967).

[29] SCARF, H., and T. HANSEN, Computation of Economic Equilibria, Yale University Press, New Haven 1973).

[30] SHAPLEY, L.S., On Balanced Games Without Side Payments, Mathematical Programming, Ed: T.C. Hu and S.M. Robinson, Academic Press, New York-London (1973), 261-290.

[31] TODD, M.J., The Computation of Fixed Points and Applications, Springer-Verlag, Berlin-Heidelberg, 1976.

[32] TODD, M.J., Exploiting Structure in Fixed Point Computation, Discussion Paper, Mathematics Research Center, University of Wisconsin, Madison, Wisconsin.

[33] VAN DER LAAN, G., and A.J.J. TALMAN, A New Algorithm for Computing Fixed Points, Free University, Amsterdam (March 1978).

[34] WILMUTH, R.J., The Computations of Fixed Points, Ph.D. Thesis (1973), Department of Operations Research, Stanford University.

ON NUMERICAL APPROXIMATION OF FIXED POINTS IN $C[0,1]$

by

W. Forster

Summary

The present paper discusses the connection between fixed points in function spaces and finite dimensional approximations. Utilizing algebraic properties of the space of continuous functions $C[0,1]$ we obtain

(i) a simple criterion for testing consistency of an approximation.

Theorem: Consistency is equivalent to $T_N(a) = a$ for all N and for all generators $\{a\} = \{a_1, a_2, \ldots, a_n\}$ of the algebra,

(ii) an understanding of stability from an algebraic point of view,

(iii) a simple proof for the equivalence theorem.

1. Introduction

The availability of algorithms for finding fixed points numerically in finite dimensional spaces makes it possible to find fixed points numerically for certain problems defined in function spaces. The first applications of fixed point algorithms to the numerical solution of ordinary and elliptic partial differential equations were published in [1], [2]. For a more recent survey see [3]. For other material on fixed point algorithms see e.g. [4].

The numerical solution of problems defined in function spaces has attracted the attention of many mathematicians. Important results have been obtained without the use of fixed point theorems or topological considerations. An early paper [5] discusses e.g. the relation between the wave equation and discrete approximations to the wave equation. For initial value problems for partial differential equations precise conditions for a discrete approximation to give the desired results are discussed in [6]. For ordinary differential equations important results connecting the original equation with discrete approximations have been obtained in [7] and further clarified in [8], [9]. Work based on set theoretical topology is reported in [10]. For a "school" with a strong topological flavour see e.g. [11].

The problem of computing fixed points in infinite dimensional spaces can under certain assumptions be reduced to the solution of a finite dimensional problem. The main fixed point theorem used in this context is Schauder's fixed point theorem [12], [13]: "If T is a completely continuous mapping which maps a closed and bounded convex set S of a Banach space into itself, then T has a fixed point in S, i.e. $Ty = y$, $y \varepsilon S$". The compactness condition in above theorem allows the infinite dimensional problem to be approximated by finite dimensional problems. Nevertheless, one has to ensure that the sequence of finite dimensional mappings converge to the infinite dimensional mapping (this is known as "consistency") and that the sequence of solutions obtained from the finite dimensional approximations converges to the solution of the infinite dimensional problem (this is known as "convergence"). The relation between the original problem and approximations of the original problem is usually expressed as a theorem (known as "equivalence theorems"). Such theorems are regarded to be key theorems governing the numerical approximation of the equations in question. Many such theorems are known, but no such theorems have been given for problems formulated in fixed point form and utilizing algebraic properties of continuous functions.

2. A novel approach to equivalence theorems

The following novel approach to equivalence theorems utilizes the Stone-Weierstrass theorem and algebraic properties of continuous functions. Together with the fixed point formulation of the problem this leads to an extremely short proof of equivalence theorems.

The solution of our original problem is assumed to be a continuous function. Continuous functions C form a commutative algebra (for details see e.g. [14]). The result of the computation is a set of discrete points. We can interpolate by linear functions and we obtain a continuous function consisting of piecewise linear functions. If we consider continuous functions consisting of piecewise polynomial functions then they form an algebra. We define

$C_N = \{f \in C : f$ is a piecewise polynomial function consisting of N pieces of polynomials$\}$.

Next we construct the following space consisting of piecewise polynomial functions

$$C_\infty^- = \bigcup_{\forall N} C_N .$$

One can easily see that C_∞^- is an algebra.

We complete the space C_∞^- by constructing Cauchy sequences of piecewise polynomial functions consisting of an increasing number of pieces of polynomials (the maximum diameter of the region over which the polynomials are taken is assumed to go to zero). We append the continuous functions obtained by this limiting process to our space of continuous functions C_∞^- and obtain a complete metric space (we use the uniform norm) C_∞ .

We now look at the approximation process. We have the following diagram.

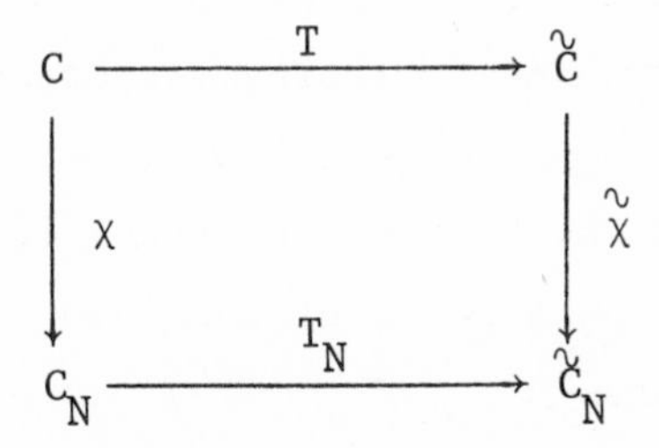

Diagram 1.

A mapping $T: C \to \tilde{C}$ (in general nonlinear) is approximated by $T_N: C_N \longrightarrow \tilde{C}_N$. The mapping T maps continuous functions C into continuous functions $\tilde{C}$. The mapping T_N maps C_N functions defined on a specified grid into C_N functions on the same grid. T_N approximates T in some sense.

C_N (respectively $\tilde{C}_N$) approximates C (respectively $\tilde{C}$). In the limiting case $N \to \infty$ we have the diagram:

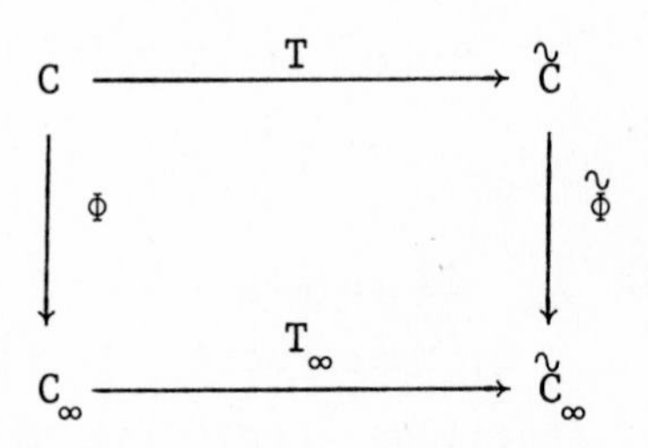

Diagram 2.

For an approximation to make sense we require $||y_N - y|| \to 0$ with $N \to \infty$, i.e. the solution of the approximated problem $y_N \varepsilon C_N$ should tend to the solution of the original problem $y \varepsilon C$. This is usually called "convergence". Another way of looking at convergence is the following. For every $y_\infty \varepsilon C_\infty$ we have to have a $y \varepsilon C$ which is mapped to y_∞ , and vice versa. We therefore have the following algebraic definition of convergence.

Definition: A numerical method is convergent if $\Phi: C \longrightarrow C_\infty$ is an isomorphism (of the algebra of continuous functions) with the same generators.

That means the spaces C and C_∞ have to be identical.

Note: The continuous functions $C[0,1]$ can be generated by $\{1, x\}$, the generators of the algebra, i.e. by using the constant function, the identity function and the three operations addition, multiplication and multiplication by scalars we can generate all polynomials. Via the Stone-Weierstrass theorem we can approximate all continuous functions.

Furthermore, we require for a reasonable approximation $||T_N y_N - Ty|| \to 0$ with $N \to \infty$ for a set of functions $y_N \in C_N$ and $y \in C$ and a class of mappings T_N and T. This property can again be formulated as an isomorphism (of the algebra of continuous functions). We split this isomorphism into an epimorphism part and into a monomorphism part.

Definition: A numerical method is consistent if $\tilde{\Phi}: \tilde{C} \longrightarrow \tilde{C}_\infty$ is an epimorphism.

Definition: A numerical method is stable if $\tilde{\Phi}: \tilde{C} \longrightarrow \tilde{C}_\infty$ is a monomorphism.

It will be shown later that above consistency definition leads to exactly the same consistency condition as follows from the consistency definition given e.g. in [8] for ordinary differential equations.

From a computational point of view it only makes sense to allow a finite error for all functions and this requires uniform boundedness for all functions involved. This is usually used for defining numerical stability. It will be shown later how above stability definition connects with the usually required uniform boundedness condition.

Note: To consider morphisms of the algebra of continuous functions it is best to look at the various substructures, i.e. group homomorphisms, ring homomorphisms (i.e. homomorphisms of the multiplicative monoid together with homomorphisms of the additive group) and homomorphisms of linear spaces.

Next we consider the simplest of all equivalence theorems, the equivalence theorem for ordinary differential equations (see e.g. [8]).

We consider one ordinary differential equation

$$dy/dt = f(t,y) \quad , \quad y(0) = y_o \quad .$$

We rewrite the problem as an equivalent fixed point problem (integral equation)

$$y(t) = y_o + \int_0^t f(s, y(s))\, ds \quad .$$

This is a fixed point equation with

$$Ty \equiv y_o + \int_0^t f(s, y(s))\, ds \quad .$$

We assume the conditions of Schauder's fixed point theorem are satisfied. The mapping T then maps the appropriate subspace of $C[0,1]$ into itself. $C[0,1]$ is the space of continuous functions defined on the interval $[0,1]$.

Theorem (St): Numerical stability is equivalent to $T_\infty(0') = \tilde{0}'$.

Proof: From the stability definition we have the requirement that $\tilde{\Phi}: \tilde{C} \longrightarrow \tilde{C}_\infty$ is a monomorphism (for the additive group) and this is equivalent to the identity $\tilde{0}$ (zero function: every element is mapped to zero) of the additive group $\tilde{C}$ to be mapped to the identity $\tilde{0}'$ (zero function) of the additive group $\tilde{C}_\infty$. By commutativity of diagram 2. this then implies $T_\infty(0') = \tilde{0}'$.
On the other hand we obviously have $T(0) = \tilde{0}$ and if $T_\infty(0') = \tilde{0}'$, then by commutativity of diagram 2. it follows that $\tilde{0}$ is mapped to $\tilde{0}'$ by $\tilde{\Phi}$, i.e. we have a monomorphism.

We have shown that numerical stability is equivalent to $T_\infty(0') = \tilde{0}'$ (stability condition).

The stability condition leads to uniform boundedness of the functions involved. Whatever the approximation to $T_\infty(0') = \tilde{0}'$, if the step size max h goes to zero, then the function has to go to the zero function 0 and this implies that all function values have to be bounded. Furthermore, this has to hold for all functions involved (uniform boundedness). This argument is very similar to one given in [8] for linear multistep methods.

Next we consider consistency.

Theorem (Co): Consistency is equivalent to $T_N(1) = 1$, $T_N(x) = x$ for all N.

Proof: We use the fact that a monoid homomorphism is totally determined by its values on the generators $\{g\}$, see e.g. [15]. In our case the algebra of continuous functions contains a multiplicative monoid generated by the function x (identity function: maps every element to itself) and we have an identity 1 (constant function: maps every element to the constant one). The requirement that $\tilde{\Phi}: \tilde{C} \longrightarrow \tilde{C}_\infty$ is an epimorphism means in our case that

$\tilde{\Phi}: \tilde{x} \longmapsto \tilde{x}'$, $\{\tilde{x}\}$ is the monoid generator in $\tilde{C}$ and
$\{\tilde{x}'\}$ is the monoid generator in $\tilde{C}_\infty$.

Furthermore, we map the identity to the identity, i.e.

$\tilde{\Phi}: \tilde{1} \longmapsto \tilde{1}'$, $\tilde{1} \varepsilon \tilde{C}$, $\tilde{1}' \varepsilon \tilde{C}_\infty$.

If the numerical method is consistent, then $T_\infty(1) = 1$ and $T_\infty(x) = x$ (by commutativity of diagram 2.). For the functions $\{1, x\}$ the map $\tilde{\chi}$ (respectively χ) in diagram 1. can be considered to be the identity map (this is an additional assumption), i.e. for each N $\tilde{\chi}$ maps the function $\tilde{1} \varepsilon \tilde{C}$ to $\tilde{1} \varepsilon \tilde{C}_N$ and $\tilde{x} \varepsilon \tilde{C}$ to $\tilde{x} \varepsilon \tilde{C}_N$. Commutativity of diagram 1. for the generators $\{1, x\}$ of the algebra then gives $T_N(1) = 1$ and $T_N(x) = x$ for all N.

On the other hand if $T_N(1) = 1$ and $T_N(x) = x$ for all N, then we have by completion that $T_\infty(1') = \tilde{1}'$ and $T_\infty(x') = \tilde{x}'$, and we can generate all functions in $\tilde{C}_\infty$. $\tilde{\Phi}$ maps the identity to the identity and the generator to the generator, i.e. we have an epimorphism.

We call $T_N(1) = 1$ and $T_N(x) = x$ consistency condition.

We can now prove the well known equivalence theorem for ordinary differential equations (see e.g. [8]).

Theorem: The necessary and sufficient condition for a numerical method to be convergent is that it is consistent and stable.

Proof: We consider diagram 2.
First we give the sufficiency part of the proof.
If we have a monomorphism (stability) and an epimorphism (consistency) on the right hand side of diagram 2., i.e. $\tilde{\Phi}$ is an isomorphism, then via the fixed point formulation of the problem (condition on T) and the stability and consistency condition (conditions on T_∞) all functions available in $\tilde{C}$ and $\tilde{C}_\infty$ are also available in C and C_∞. For each function diagram 2. commutes and Φ is an isomorphism of the algebra and we have the same generators, i.e. we have convergence.
Necessity part of the proof.
If $\tilde{\Phi}$ is only a monomorphism (stability) but not an epimorphism (consistent), or vice versa, then the fixed point formulation again makes all the functions which are available in $\tilde{C}$ available in C, but the consistency or stability condition (condition on T_∞) is not satisfied and the functions available in $\tilde{C}_\infty$ are not the same as in C_∞. It follows, that diagram 2. does not commute for all functions and Φ

is not an isomorphism, we do not have the same generators of the algebra and in general we do not have convergence.

We can summarize part of our results in form of a diagram.

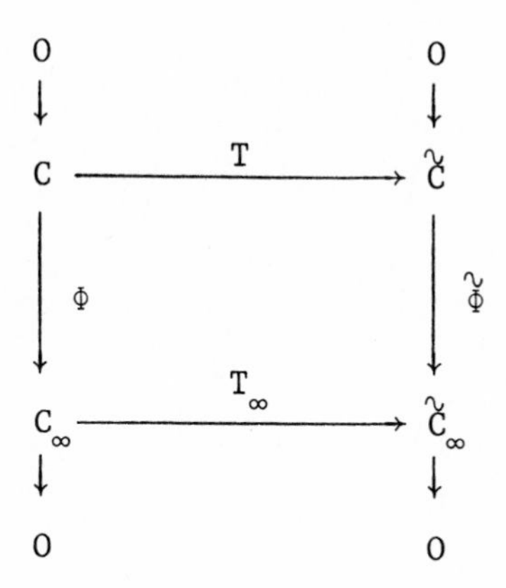

Diagram 3.

On the left hand side and on the right hand side of diagram 3. we have exact sequences, which is just another way to say that C and C_∞ are isomorphic and $\tilde{C}$ and $\tilde{C}_\infty$ are isomorphic (for group homomorphisms).

Now we consider the general case of the equivalence theorem. We assume that T maps a finite number of Cartesian products of $C[0,1]$ into itself.
The previously given definition of convergence does not change.
In the previously given definitions of stability and consistency $\tilde{C}$ and $\tilde{C}_\infty$ now stand for the appropriate Cartesian products.
There is no change in Theorem (St).

Theorem (Co) now reads:
Consistency is equivalent to $T_N(a) = a$ for all N and for all generators $\{a\} = \{a_1, a_2, \dots, a_n\}$ of the algebra.

The only change in the proof is that $\{1, x\}$, i.e. the generators of $C[0,1]$, are replaced by $\{a_1, a_2, \dots, a_n\}$, i.e. the generators of the appropriate algebra.

The equivalence theorem reads as before and the proof carries through as before.

The consistency condition given above is a condition for immediate practical use.

The stability condition given above can serve as an alternative method for establishing the boundedness condition required in Schauder's fixed point theorem.

3. Extensions

Extensions to other types of equations are obvious. E.g. for a system of two ordinary differential equations we could take the following generators:

$$\begin{bmatrix}1\\0\end{bmatrix}, \quad \begin{bmatrix}x\\0\end{bmatrix}, \quad \begin{bmatrix}0\\1\end{bmatrix}, \quad \begin{bmatrix}0\\y\end{bmatrix}.$$

For periodic problems (o.d.e's) with known period we could take periodic functions as generators, e.g. $\{\exp(inx), \exp(-inx)\}$. Methods of this kind are e.g. discussed in [16], [17]. They give no background theory, but considering the algebraic properties of continuous functions, in this case periodic functions, provides the explanation why such methods work.

A further conceivable extension can be achieved e.g. for

certain problems on $[0, \infty)$ by e.g. using as generators $\{\exp(-nx),\ x\exp(-nx)\}$,
or for certain problems on $(-\infty, \infty)$ by e.g. using as generators $\{\exp(-n^2x^2),\ x\exp(-n^2x^2)\}$.

Partial differential equations formulated in fixed point form seem to allow a similar treatment. For a partial differential equation e.g. in two variables we could have the following generators $\{1, x, y\}$.

4. Conclusions

The use of fixed point formulation for problems and the utilization of the rich structure of the algebra of continuous functions

(i) allows considerable simplifications in the proof of equivalence theorems,

(ii) leads to a practical criterion for testing consistency,

(iii) leads to an alternative method for establishing uniform boundedness, and

(iv) provides the necessary background theory for designing purpose built computational methods (via suitably chosen generators of the algebra of continuous functions).

References

1. Allgower, E.L.; Jeppson, M.M., The Approximation of Solutions of Mildly Nonlinear Elliptic Boundary Value Problems having several Solutions, Springer Lecture Notes 333, Berlin-Heidelberg-New York 1973.

2. Wilmuth, R.J., The Computation of Fixed Points, Ph.D. Thesis, Stanford University 1973.

3. Allgower, E.L., Application of a Fixed Point Search Algorithm to Nonlinear Boundary Value Problems having several Solutions, p. 87 - p. 111, in reference 4.

4. Karamardian, S. (Ed.), Fixed Points Algorithms and Applications, Academic Press, New York 1977.

5. Courant, R.; Friedrichs, K.; Lewy, H., Über die partiellen Differentialgleichungen der mathematischen Physik, Math. Ann. 100, 32 - 74 (1928).

6. Lax, P.D.; Richtmyer, R.D., Survey of the Stability of Linear Finite Difference Equations, Comm. Pure Appl. Math. 9, 267 - 293 (1956).

7. Dahlquist, G., Convergence and Stability in the Numerical Integration of Ordinary Differential Equations, Math. Scand. 4, 33 - 53 (1956).

8. Henrici, P., Discrete Variable Methods in Ordinary Differential Equations, Wiley, New York 1962.

9. Henrici, P., Error Propagation for Difference Methods, Wiley, New York 1963.

10. Stummel, F., Discrete Convergence of Mappings, p. 285 - p. 310, in: Topics in Numerical Analysis, Academic Press, London 1973.

11. Petryshyn, W.V., Nonlinear Equations Involving Noncompact Operators, p. 206 - p. 233, in: Nonlinear Functional Analysis, American Mathematical Society, Providence, R.I. 1970.

12. Schauder, J., Der Fixpunktsatz in Funktionalräumen, Studia Math. 2, 171 - 180 (1930).

13. Lusternik, L.A.; Sobolev, V.J., Elements of Functional Analysis, Wiley, New York 1961.

14. Simmons, G.F., Introduction to Topology and Modern Analysis, McGraw Hill, New York 1963.

15. Stone, H.S., Discrete Mathematical Structures and their Applications, Science Research Associates, Chicago 1973.

16. Gautschi, W., Numerical Integration of Ordinary Differential Equations Based on Trigonometric Polynomials, Numer. Math. 3, 381 - 397 (1961).

17. Lambert, J.D., Computational Methods in Ordinary Differential Equations, Wiley, New York 1973.

Faculty of Mathematical Studies,
The University, Southampton,
England.

An Application of Simplicial Algorithms to Variational Inequalities.

Kurt Georg

1. Introduction.

Recently Umberto Mosco suggested that certain types of variational and quasi-variational inequalities can be solved by simplicial (complementarity pivoting) algorithms in the sense of [6,8,13,16]. The following short note wants to indicate that this in fact can be done, and the approach given here is close in spirit to methods of solving nonlinear programming problems by such algorithms, see e.g. [7,13,15,17,18]. Since it is known [5] that variational inequalities can be approached by using KKM maps, and since KKM maps and Sperner's lemma are related to simplicial algorithms [17] , it is in fact not surprising that simplicial algorithms can solve variational inequalities.

Since we are here only interested in the relationship of such algorithms with variational inequalities, we do not intend to discuss the entirely different question of approximating such problems by some finite dimensional method. Hence we formulate the variational inequality directly in $\mathbb{R}^N$, though this may not be interesting from a theoretical point of view, i.e. for discussing existence of solutions. For a general treatment of existence (in infinite dimensional spaces) we refer to e.g. [2,3,5,11] and the references listed there.

Acknowledgement: I am indebted to Giovanni Troianello for several helpful discussions.

2. Solving nonlinear problems by a simplicial algorithm.

Following [1] let us describe a typical nonlinear problem that can be solved by a simplicial algorithm and which is general enough for our present purposes. By $\mathbb{R}^{N*}$ we denote the family of all convex compact non-empty subsets of $\mathbb{R}^N$. A set-valued map $G : \mathbb{R}^N \to \mathbb{R}^{N*}$ is called upper semi-continuous (u.s.c.) if for all open subsets $U \subset \mathbb{R}^N$ the inverse image $\{x \in \mathbb{R}^N : G(x) \subset U\}$ is open. The following theorem can be proved by using the homotopy invariance of the Brouwer degree and an "approximation" of G by a sequence of piecewise affine mappings.

(2.1) Theorem. Let $A : \mathbb{R}^N \to \mathbb{R}^N$ be an affine isomorphism and $G : \mathbb{R}^N \to \mathbb{R}^{N*}$ an u.s.c. map. Furthermore, let $U \subset \mathbb{R}^N$ be an open bounded set such that the following two conditions hold:

(i) $A^{-1}(0) \in U$,

(ii) $0 \notin \text{co}(\{A(x)\} \cup G(x))$ for $x \in \partial U$. Here "co" denotes the convex hull and "∂" the boundary.

Then there exists an $x^* \in U$ such that $0 \in G(x^*)$, i.e. G has a zero-point in U.

The above theorem can be viewed as a finite dimensional version of a combination of the fixed point theorems of Kakutani [9] and Leray-Schauder [10]. However, our main point is that not only the existence of x^* can be proved, but the zero-point x^* can also be efficiently approximated by a simplicial algorithm (the latter in fact also implies existence). Let us give here only a rough idea of how this is done, see e.g. [1] for details.

We triangulate $(0,1] \times \mathbb{R}^N$ in such a way that the mesh size on $\{t\} \times \mathbb{R}^N$ approaches zero as $t \to 0$. For numerical purposes, the triangulation J_3 described by Todd [17] is most convenient. Now, if (t,x) is a node of the triangulation, we label $\lambda(t,x) = Ax$ for $t = 1$ resp. $\lambda(t,x) \in G(x)$ for $t \in (0,1)$. In each simplex, we continue the labeling affinely so that $\lambda : (0,1] \times \mathbb{R}^N \longrightarrow \mathbb{R}^N$ becomes a continuous map, affine on each simplex of the triangulation. The family $\lambda_t(x) = \lambda(t,x)$ of piecewise affine maps can be viewed as an approximation of the u.s.c. map G , more precisely: $\lim_{t \to 0} \operatorname{dist}(\lambda(t,x) , G(x)) = 0$ uniformely for x in compact sets.

Now it is not difficult to prove that λ has no zero-point on $(0,1] \times \partial U$ if the initial mesh size on $\{1\} \times \mathbb{R}^N$ is sufficiently small. Hence, beginning in $(t(1),x(1)) = (1,A^{-1}(0)) \in \{1\} \times U$, the simplicial algorithm (Eaves-Saigal algorithm [8] in this case) produces a simple continuous piecewise linear curve

$$s \in (0,1] \longrightarrow (t(s),x(s)) \in (0,1] \times U$$

such that

(i) $\lambda(t(s),x(s)) = 0$ for $s \in (0,1]$,

(ii) $\lim_{s \to 0} t(s) = 0$,

(iii) $x(s)$ (as $s \to 0$) has at least one limit point, and each limit point belongs to U and is a zero-point of the u.s.c. map G .

3. Application to variational inequalities.

Let us now describe how this general concept can be used to approximate solutions to variational or quasi-variational inequalities. Here again the main point is not the existence of a solution (which may be well known from the literature) but the numerical approximability by a simplicial algorithm.

Let $T : \mathbb{R}^N \to \mathbb{R}^N$ be a continuous map, $f \in \mathbb{R}^N$, and let $\psi : \mathbb{R}^N \times \mathbb{R}^N \to \mathbb{R}$ be continuous and such that $\psi_x : \mathbb{R}^N \to \mathbb{R}^N$ (defined by $\psi_x(y) = \psi(x,y)$) is convex for each $x \in \mathbb{R}^N$. We consider the following quasi-variational inequality:

(3.1) Problem. Find $x^* \in \mathbb{R}^N$ such that

(i) $\psi(x^*,x^*) \leq 0$,

(ii) $\langle Tx^* - f , x^*-h \rangle \leq 0$ for all $h \in \mathbb{R}^N$ such that $\psi(x^*,h) \leq 0$.

Here $\langle .,. \rangle$ denotes the usual scalar product in $\mathbb{R}^N$, i.e. $\langle u,v \rangle = u^T v$. If ψ does not depend on the first variable, then we have a usual variational inequality, see e.g. [2] .

To obtain a solution of problem (3.1) with the aid of theorem (2.1) , we need the following two hypothesis. The first assumes that the constraints

$$K_x = \{h \in \mathbb{R}^N : \psi(x,h) \leq 0\}$$

have a common interior point v , the second is a typical coercivity condition.

(3.2) Hypothesis.

(i) Let $v \in \mathbb{R}^N$ be given such that $\psi(x,v) < 0$ for all $x \in \mathbb{R}^N$.

(ii) There exists an $R > 0$ such that $\langle Tx - f, x-v \rangle > 0$ for all $x \in \mathbb{R}^N$ such that $\|x\| \geq R$ and $\psi(x,x) \leq 0$.

We now define the affine isomorphism $A : \mathbb{R}^N \to \mathbb{R}^N$ and the u.s.c. map $G : \mathbb{R}^N \to \mathbb{R}^{N*}$ as needed in theorem (2.1) . Given the common interior point v as in (3.2)(i) , define

(3.3) $\qquad Ax = x - v$ for $x \in \mathbb{R}^N$.

For $u,x \in \mathbb{R}^N$, let $\partial\psi(u,x)$ denote the subgradient of ψ_u in x, i.e. $\partial\psi(u,x) \subset \mathbb{R}^N$ consists of those points $y \in \mathbb{R}^N$ such that

$$(3.4) \quad \psi(u,h) - \psi(u,x) \geq \langle y,h-x\rangle \quad \text{for all} \quad h \in \mathbb{R}^N .$$

For the general theory of convex analysis we refer to [14]. Here we need the crucial fact that the map

$$x \in \mathbb{R}^N \longrightarrow \partial\psi(x,x) \in \mathbb{R}^{N*}$$

is u.s.c. . Now we are in a position to define our map $G : \mathbb{R}^N \longrightarrow \mathbb{R}^{N*}$ by

$$(3.5) \qquad G(x) = \begin{cases} \{Tx-f\} & \text{for } \psi(x,x) < 0 , \\ co(\{Tx-f\}\cup\partial\psi(x,x)) & \text{for } \psi(x,x) = 0 , \\ \partial\psi(x,x) & \text{for } \psi(x,x) > 0 . \end{cases}$$

Clearly, with $\partial\psi$ also G is u.s.c. . Next, we choose a bounded open set $U \subset \mathbb{R}^N$ such that $v \in U$ and $x \in U$ for all $x \in \mathbb{R}^N$ such that $\|x\| < R$, where v and R are given by hypothesis (3.2) .

(3.6) Lemma. A, G and U, as defined above, verify the assumptions of theorem (2.1) , and hence the simplicial algorithm approximates a zero-point $x^* \in U$ of G .

Proof. Let $x \in \partial U$ and hence $\|x\| \geq R$. We consider three cases.

Case 1: $\psi(x,x) < 0$. Then $G(x) = \{Tx-f\}$, and $\langle Tx-f,Ax\rangle > 0$ by (3.2) . Thus zero cannot be a convex combination of $Tx-f$ and Ax .

Case 2: $\psi(x,x) > 0$. Then $G(x) = \partial\psi(x,x)$, and for a point $y \in G(x)$ we have $0 > \psi(x,v) - \psi(x,x) \geq \langle y,v-x\rangle$ by (3.2) and (3.4) , hence $\langle y,Ax\rangle > 0$, and zero cannot be in the convex hull of $G(x)$ and Ax .

Case 3: $\psi(x,x) = 0$. Then $G(x) = co(\{Tx-f\} \cup \partial\psi(x,x))$. As in cases 1 and 2 , it follows that $\langle Tx-f, Ax\rangle > 0$ and $\langle y, Ax\rangle > 0$ for $y \in \partial\psi(x,x)$, and hence also in this case zero cannot be in the convex hull of $G(x)$ and Ax .

Q.E.D.

(3.7) Lemma. A zero point x^* of G solves problem (3.1) .

Proof. Again we distinguish three cases.

Case 1: $\psi(x^*,x^*) > 0$. Then $G(x^*) = \partial\psi(x^*,x^*)$, the subgradient of the convex function $\psi_{x^*} : \mathbb{R}^N \longrightarrow \mathbb{R}$ at x^* contains zero, and x^* is a global minimum point for ψ_{x^*} , a contradiction to (3.2)(i) . Hence this case is impossible.

Case 2: $\psi(x^*,x^*) < 0$. Then $G(x^*) = \{Tx^*-f\}$, and $Tx^* = f$. Hence x^* solves (3.1) .

Case 3: $\psi(x^*,x^*) = 0$. Then $G(x^*) = co(\{Tx^*-f\} \cup \partial\psi(x^*,x^*))$, and there exists a $t \in [0,1]$ and a $y \in \partial\psi(x^*,x^*)$ such that $0 = t(Tx^*-f) + (1-t)y$. Since, as in case 1 , x^* cannot be a global minimal point of ψ_{x^*} , we have $t > 0$. Let us assume that x^* does not solve (3.1) . Then we can find an $h \in \mathbb{R}^N$ such that $\psi(x^*,h) \leq 0$ and $\langle Tx^*-f, x^*-h\rangle > 0$. By (3.4), we have also $\langle y, x^*-h\rangle \geq \psi(x^*,x^*) - \psi(x^*,h) \geq 0$, and hence $\langle 0, x^*-h\rangle = \langle t(Tx^*-f) + (1-t)y , x^* - h\rangle > 0$, a contradiction.

Q.E.D.

Summarizing the results of this section, we can say that, under the hypothesis (3.2), the quasi-variational inequality (3.1) can be numerically solved by using a simplicial algorithm.

4. A special case of constraints.

Let us indicate now that for a special class of constraints ψ a different and single-valued u.s.c. map G may be chosen to solve (3.1) by a simplicial algorithm.

In $\mathbb{R}^N$ we consider the usual partial ordering

$$(4.1) \qquad (\xi_1,\dots,\xi_N)^T \leq (\eta_1,\dots,\eta_N)^T \iff \xi_i \leq \eta_i \quad \text{for } i=1,\dots,N.$$

Let $C : \mathbb{R}^N \longrightarrow \mathbb{R}^N$ be a continuous map. For $x,h \in \mathbb{R}^N$, define $\psi(x,h)$ to be the maximum of the coordinates of $h-C(x)$. Then $\psi : \mathbb{R}^N \times \mathbb{R}^N \longrightarrow \mathbb{R}$ is continuous, convex for each fixed first argument , and

$$(4.2) \qquad \psi(x,h) \leq 0 \quad \iff \quad h \leq C(x) \quad .$$

Furthermore, ψ satisfies the hypothesis (3.2)(i) if and only if the map C is bounded below in the ordering of $\mathbb{R}^N$. Hence the procedure of section 3 can be applied. Note, however, that here the subgradient $\partial\psi(x,x)$ consists of the convex hull of those unit vectors which correspond to coordinates of $x-C(x)$ where the maximum $\psi(x,x)$ is attained. Hence the choice of $G(x)$ as indicated in section 3 is somewhat rigid.

In fact, in this special case (4.2), it is possible to define a single-valued u.s.c. map $G : \mathbb{R}^N \longrightarrow \mathbb{R}^{N*}$ by

$$(4.3) \qquad G(x) = \{(Tx-f) \vee (x-C(x))\} \ ,$$

where "$\vee$" indicates the supremum in the ordering of $\mathbb{R}^N$.

We mention in passing that looking for a zero-point of G can be regarded as a nonlinear complementarity problem, see e.g. [4] and the monograph [12] for such problems.

Now, it is easily seen that $x^* \in \mathbb{R}^N$ is a zero-point of the new u.s.c. map (4.3) if and only if x^* solves the quasivariational inequality (3.1), where ψ is taken with respect to (4.2). Furthermore, under the hypothesis (3.2), it is not hard to show that the new u.s.c. map G, defined in (4.3), again verifies the assumptions of theorem (2.1), where A and U are the same as in section 3.

Hence, if the constraints ψ have the special form (4.2) discussed in this section, the quasi-variational inequality (3.1) may be solved by a simplicial algorithm in a way which is different from the general procedure in section 3. Though numerical experience in confronting the two different methods is not yet available, we claim that this second method will be more efficient.

References.

1. E.Allgower and K.Georg : Simplicial and continuation methods for approximating fixed points and solutions to systems of equations , to appear in SIAM Review, 1979.

2. H.Brézis : Problèmes unilatéraux , J.Math. pures et appl. , 51 , pp. 1-168 , 1972.

3. F.Browder : Nonlinear monotone operators and convex sets in Banach spaces , Bull. Amer. Math. Soc. , 71 , pp. 780-785 , 1965.

4. R.W.Cottle : Nonlinear programs with positively bounded Jacobians , SIAM J. Appl. Math. , 14 , pp.147-158 , 1966.

5. J.Dugundji and A.Granas : KKM maps and variational inequalities , Annali della Scuola Normale Superiore di Pisa , Serie IV , Vol. V,4 , pp. 679-682 , 1978.

6. B.C.Eaves : Homotopies for computation of fixed points , Mathematical Programming , 3 , pp. 1-22 , 1972

7. B.C.Eaves : Computing stationary points , Math. Programming Study , 7 , M.L.Balinski and R.W.Cottle , eds. , pp. 1-14 , North-Holland , Amsterdam , 1978.

8. B.C.Eaves and R.Saigal : Homotopies for computation of fixed points on unbounded regions , Mathematical Programming , 3 , pp. 225-237 , 1972

9. S.Kakutani : A generalization of Brouwer's fixed point theorem , Duke Math. J. , 8 , pp. 457-459 , 1941.

10. J.Leray and J.Schauder : Topologie et équations fonctionelles , Ann.Sci. , Ecole Norm. Sup. , 51 , 3 , pp. 45-78 , 1934.

11. J.L.Lions and G.Stampacchia : Variational inequalities , Comm. Pure Appl. Math. , 20 , pp.493-519 , 1967

12. H.J.Lüthi : Komplementaritäts- und Fixpunktalgorithmen in der mathematischen Programmierung , Spieltheorie und Ökonomie , Lecture Notes in Economics and Mathematical Systems , 129 , Springer , Heidelberg , 1976.

13. O.H.Merrill : Applications and extensions of an algorithm that computes fixed points of a certain upper semi-continuous point to set mapping , Ph.D.Thesis , University of Michigan , 1972.

14. R.T.Rockafellar : Convex analysis , Princeton University Press , Princeton , 1970.

15. R.Saigal : The fixed point approach to nonlinear programming, to appear in Math. Programming Studies , 1979.

16. H.Scarf : The approximation of fixed points of a continuous mapping , SIAM J. Appl. Math. , 15 , pp. 1328-1343 , 1967.

17. M.J.Todd : The computation of fixed points and applications , Lecture Notes in Economics and Mathematical Systems , 124 , Springer , New York , 1976.

18. M.J.Todd : New fixed-point algorithms for economic equilibria and constrained optimization , Technical Report , Cornell University , Ithaca , New York , 1977.

Author's address:
Institut für Angew. Math.
Wegelerstr. 6
D - 5300 Bonn

DELAY EQUATIONS IN BIOLOGY

K.P.Hadeler
Universität Tübingen
Lehrstuhl f. Biomathematik
Auf der Morgenstelle 28

7400 Tübingen, Germany

Biological models have provided many examples of nonlinear differential equations with retarded arguments. The construction of biological models using delays has been paralleled by the mathematical investigation of nonlinear delay equations. In the following we collect some typical examples, without any claim for completeness. For proofs of results and for biological details we refer to the original papers.

1. The physical effect of a delay

From various examples we know that introducing a delay in a biological or physical system may lead to instability. In fact there seems to be a common belief that almost any system can be destabilized by incorporating delays. However, the effects of delays may be rather complicated and nothing general can be said.

As an introduction we choose a linear equation with one dependent variable. Although the example is simple and the results are well-known it may clarify the situation.

Let x describe the state of a system and I the input. At present we do not define an output. Then x(t) is the state and I(t) the input at time t. State and input are related by the delay equation

$$\delta \dot{x}(t) + ax(t) + bx(t-\tau) = I(t) \quad . \tag{1.1}$$

The time constant δ is positive, the delay τ is non-negative. With a stationary input $I(t) \equiv I$ the system has the stationary state $\bar{x} = I/(a+b)$ (except for $a + b = 0$). Since we are only interested in the deviation from the stationary state we can introduce $x - \bar{x}$ as a new state variable and obtain

$$\delta \dot{x}(t) + ax(t) + bx(t-\tau) = 0 \quad . \tag{1.2}$$

Still the system depends on four parameters. If we again switch to a new state variable kx we can use the constant k for normalization of any of the parameters a, b, δ. Also we can rescale the time variable t to obtain $\tau = 1$. Then

$$\delta k\dot{x}(t) + ka\tau\, x(t) + kb\tau\, x(t-1) = 0 \; . \tag{1.3}$$

Thus with $k = \delta^{-1}$ we have

$$\dot{x}(t) + \nu x(t) + \alpha x(t-1) = 0 \quad , \tag{1.4}$$

where

$$\nu = a\tau/\delta \; , \qquad \alpha = b\tau/\delta \; . \tag{1.5}$$

Now we try to give a physical interpretation of the constants. For $a > 0$ the term ax(t) acts as a damping or control which carries any deviation from the stationary state back to zero. For $b > 0$ the term $bx(t-\tau)$ plays a similar role, only the control acts with a delay. On the other hand, for $a < 0$ or $b < 0$ we have enhancement of perturbations instead of damping. One can imagine that to a certain extent the two terms can compensate each other.

The characteristic equation corresponding to (1.4) is

$$\lambda + \nu + \alpha e^{-\lambda} = 0 \qquad . \tag{1.6}$$

This equation was first investigated by Hayes 1950. He showed that all roots have negative real parts iff the following conditions are satisfied:

$$\nu > -1,$$
$$-\nu < \alpha < \sqrt{\zeta^2 + \nu^2} \; ,$$

where ζ is the root of the equation

$$\zeta + \nu \operatorname{tg} \zeta = 0 \tag{1.7}$$

located in

$$0 < \zeta < \pi \; .$$

The stability domain is depicted in fig. 1. (In contrast to fig. 13.1 in Bellman-Cooke 1963, chapter 13, along the boundary of the stability domain $\nu \to \infty$, $\alpha \to \infty$).

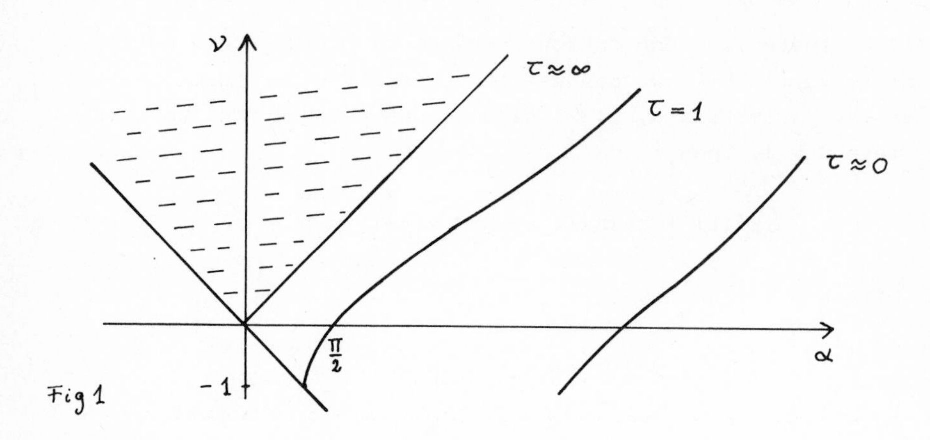

Fig 1

We make the following observations:

For $\nu + \alpha < 0$ the system is always unstable: The sum of both terms produces an enhancement.

A negative α can be always compensated by a positive ν : immediate control works.

A negative ν can only be compensated by a positive α if $\nu > -1$.

The most important feature: For any given $\nu > -1$ the system can be destabilized by increasing α : If, for the given delay 1, the delayed control acts too strong, then the system becomes unstable.

It is rather simple to investigate the dependence on τ and δ , rather τ/δ . We set $\delta = 1$. For given τ we denote by S_τ the set of all $(a,b) \in \mathbb{R}^2$ for which all zeros of the characteristic equation (1.6) have negative real parts. S_τ is the stability domain for the delay τ . We set $S_\infty = \bigcap_{\tau \geq 0} S_\tau$. Then S_0 is simply the half-plane $a + b > 0$, all S_τ are contained in S_0. The intersection S_∞ is the quarter-plane $a + b > 0$, $a > b$. One can easily see that

$$\tau_1 < \tau_2 \quad \text{implies} \quad S_{\tau_2} \subset S_{\tau_1} \quad . \tag{1.8}$$

The most interesting aspect is the following: Assume $a < b$. Then there is a critical τ_0 such that $(a,b) \notin S_\tau$ for $\tau > \tau_0$ but $(a,b) \in S_\tau$ for $\tau < \tau_0$. Thus for $a < b$ the system can be destabilized by sufficiently long delays.

For fixed $\nu \geq 0$ and α increasing from zero to infinity, condition (1.7) describes when the first characteristic root crosses the imaginary axis. Similarly one can determine those values of α for which further roots enter the right half-plane (see e.g. Hadeler and Tomiuk 1977).

2. Oscillations in populations

Several models have been designed to explain oscillations of population densities as caused by delays. First we describe Hutchinson's model. The classical Verhulst equation (Verhulst 1838) is derived from the following assumptions: At low densities the population is exponentially increasing with exponent $a > 0$. The biotop has a carrying capacity K, and at higher densities the increase is proportional to the actual density and also proportional to the remaining supplies. For densities exceeding K the increase is negative. Thus Verhulst's equation is

$$\dot{N}(t) = aN(t)(1 - \frac{N(t)}{K}) \tag{2.1}$$

Hutchinson 1948 observed that in many populations the actual increase does not so much depend on the present population density as on densities in the past. Thus he introduced a delay in the growth rate leading to

$$\dot{N}(t) = aN(t)(1 - \frac{N(t-\tau)}{K}) \quad . \tag{2.2}$$

The delay τ comprises various effects such as hatching periods, duration of pregnancy, slow replacement of food supplies etc.
If one wants to give a more detailed derivation of equation (2.2) one encounters difficulties: Suppose there is a delay between egg-laying and hatching. Let $N(t)$ be the population size at time t. Then the change in N at time t should be proportional to $N(t-\tau)$ rather then $N(t)$.

On the other hand, suppose that $N(t)$ is the number of individuals above a certain critical age. Suppose that below that age the survival of the young depends heavily on the nutritional condition of the parents, and that the latter depends on the food uptake of the parents in the past, which again is a function of the population density in the past.

Then we may consider equation (2.2) as a first step towards a more realistic model with distributed delays

$$\dot{N}(t) = aN(t)\ (1-\frac{1}{K}\int_{-T}^{0} N(t+s)\,d\sigma(s)) \quad . \qquad (2.3)$$

From the biological interpretation one expects the following behavior of solutions of equation (2.2) starting from small positive initial data: For given a and τ close to zero solutions stabilize against $N(t) = K$, for larger τ the solutions may overshoot K and start oscillating. For sufficiently large τ the solution $N(t) = K$ looses its stability (in the stability diagram Fig. 1 we have $b = 0$). One can expect that, by the non-linearity, the oscillations are shaped into a stable periodic oscillation.

This is in fact the case. A simple substitution leads to

$$\dot{x}(t) = -\alpha x(t-1)(1+x(t)) \quad . \qquad (2.4)$$

This equation was one of the first examples for existence proofs via the ejective fixed point principle. It has a periodic solution for $\alpha > \pi/2$. This solution appears stable in numerical simulations. Locally, i.e. in a neighborhood of $\pi/2$, it can be also obtained by bifurcation arguments (Nussbaum 1975). Similar results hold for

$$\dot{x}(t) = -\alpha f(x(t-1)),\ f(x)x > 0 \text{ for } x \neq 0,\ f'(0) = 1,$$

f bounded below.

Walther 1978 has shown that bifurcation may be backward, i.e. (unstable) periodic solutions may branch off the zero solution for $\alpha < \pi/2$.

Similarly equation (2.3) can be transformed into

$$\dot{x}(t) = -\alpha \int_{-1}^{0} x(t+s)\,d\sigma(s)\ (1+x(t)) \quad . \qquad (2.5)$$

Here σ is monotone, $\sigma(-1) = 0$, $\sigma(0) = 1$.

The special case of an ordinary differential equation (then σ is a step function with a jump of height 1 at $s = 0$) is not excluded and we cannot expect periodic solutions in general, not even for large α. From several investigations (Hadeler 1976, Walther 1976) it follows that destabilization of the zero solution for large α occurs only if the average in (2.5) gives greater weight to the past (i.e. the neighborhood of $s = -1$) than to the present. A rather complete description is given by Alt 1978. See also Stech 1977.

Existence of periodic solutions for equations with distributed delays has been shown by Walther 1976, 1977, Alt 1978.

3. Oscillating fly populations

In the following we describe a second model for oscillating population densities leading to conclusions rather different from Hutchinson's model.

Experiments with isolated fly populations with constant food supply have been performed by Nicholson 1954 (blow-fly Lucilia), Mourão and Tadei 1974 (fruit-fly Drosophila). A mathematical model was first proposed by Maynard Smith 1968, 1974, recently Perez, Malta and Coutinho 1978 proposed a more general model.

The idea is that a delay is introduced by the time needed for an egg to become an adult fly. The birth rate b and the death rate d are both density dependent. If N is the population density and τ is the delay then the equation reads

$$\dot{N}(t) = b(N(t-\tau))N(t-\tau) - d(N(t))N(t) \tag{3.1}$$

The birth rate and the death rate are continuously differentiable functions with the following properties: $b(0) > 0$, b is decreasing and $b(N) \to 0$ for $N \to \infty$; $0 < d(0) < b(0)$, d is increasing, and $d(N) \geq c > 0$ for $N \to \infty$.

Perez et.al. 1978 give a more detailed interpretation of the birth rate, following Maynard Smith. Assume that A is the amount of food supplied per unit time, then A/N is the amount of food per unit time per individual. Suppose B is the amount of food each individual needs for subsistence, then only the excess food $A/N - B$ will lead to egg production. In the neighborhood of the threshold value $N = A/B$ the birth rate is proportional to the excess food supply whereas in general one has to assume some saturation function $\phi(\xi)$,

$$b(N) = \phi(\frac{A}{N} - B)$$

where $\phi(\xi) = 0$ for $\xi \leq 0$, $\phi(\infty) = b(0)$, ϕ non-decreasing. These assumptions lead to a birth rate as in fig. 2.

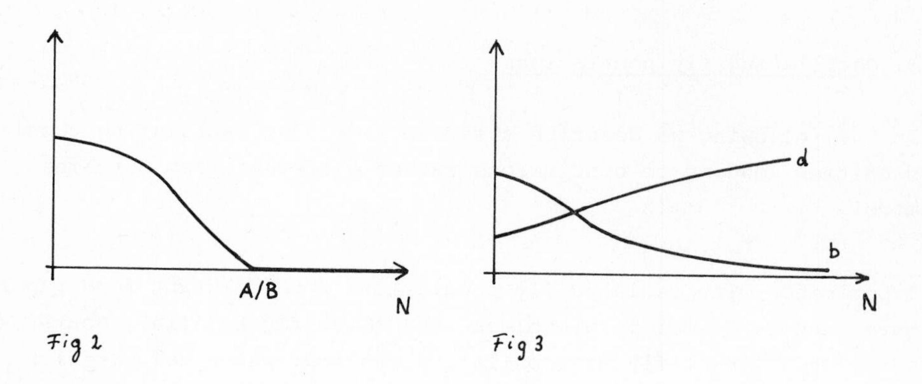

Fig 2 Fig 3

In the general case (fig. 3) we have a unique stationary solution $\bar{N}$ which is a solution to the equation

$$b(N) = d(N) \quad .$$

If we linearize at $\bar{N}$ we obtain

$$\dot{x}(t) + \nu x(t) + \alpha x(t-\tau) = 0 \quad , \tag{3.2}$$

where

$$\begin{aligned} \nu &= (d(N)N)' = d'(N)\, N + d(N) \\ \alpha &= -\,(b(N)N)' = -\,b'(N)N - b(N)\,. \end{aligned} \tag{3.3}$$

We have $\nu > 0$ since d is decreasing. However, α may be positive or negative. On the other hand we have

$$\alpha + \nu = (-b'(N) + d'(N))N > 0. \tag{3.4}$$

Thus we can say that any destabilization of the stationary state is due to α being too large rather than $\nu + \alpha$ being too small. We now ask whether the stationary state can be destabilized by sufficiently large τ.

If this is the case then $\alpha > \nu$ or

$$-(b(N)N)' > (d(N)N)' \quad \text{for } N = \bar{N} \quad .$$

Thus destabilization can occur if the decrease of the birth term at $N = \bar{N}$ is sufficiently steep or the increase of the death term is sufficiently flat. The author does not see that this condition has an obvious biological meaning.

In Hutchinson's model we have destabilization for sufficiently large delays for any choice of the other parameters, in this model destabilization occurs only for suitable choices of the other parameters.

As an example we choose a birth term and a linear death term

$$b(N) = b(1 - \frac{N}{K_o}), \qquad d(N) \equiv d \quad .$$

Then with $\tau = 0$ equation (3.1) reduces to the Verhulst equation

$$\dot{N}(t) = a(1 - \frac{N}{K})N$$

where

$$a = b - d, \qquad K = \frac{b-d}{b} K_o \, .$$

Thus the effective carrying capacity is considerably influenced by the death rate.
The stationary solution is $N(t) = K$, and

$$\nu = d \qquad\qquad \alpha = b - 2d$$

$$\alpha + \nu = b - d > 0 \qquad\qquad \nu - \alpha = 3d - b \quad .$$

Thus in this model destabilization for large τ occurs only if $b > 3d$, i.e. if the birth rate is high when compared with the death rate. This behavior is in agreement with Hutchinson's model which does not contain a proper death rate.

4. Delays in the eye of the horse-shoe crab

The horse-shoe crab Limulus has a compound eye where the units (ommatidia) are connected by "lateral" pathways which act inhibitory upon each other. We derive a model for this structure. We number the ommatidia by $j=1,\dots,n$. A light input produces a potential y_j in the sensory cell. If the potential is above a certain threshold $\bar{y}_i$ then the cell starts firing with a frequency z_j proportional to $y_j-\bar{y}_j$, thus

$$z_j = b_j \vartheta(y_j-\bar{y}_j) \quad ,$$

where $\vartheta(x) = \max(x,0)$ is the rectifier function. However, other ommatidia act inhibitory with inhibition coefficients $\beta_{jk} \geq 0$, thus

$$z_j = b_j \vartheta(y_j - \bar{y}_j - \sum \beta_{jk} z_k),\ j=1,2,\dots,n. \qquad (4.1)$$

These equations are called the Hartline-Ratliff model. The model describes a stationary input-state relation. The temporal behavior is not modeled. See Hadeler 1977 for a general review.

Starting from the Hartline-Ratliff model we can derive non-stationary equations in various ways. In order to have a differentiable state variable one first normalizes the constants by

$$y_j \ := b_j(y_j-\bar{y}_j) \quad , \quad \beta_{jk} := b_j \beta_{jk} \quad ,$$

then introduces a new state variable (reduced excitation) v_j by

$$v_j = y_j - \sum_{k=1}^{n} \beta_{jk} z_k \quad .$$

Then

$$v_j = y_j - \sum_{k=1}^{n} \beta_{jk} \vartheta(v_k), \qquad j=1,2,\dots,n \qquad (4.2)$$

is a stationary model equivalent with (4.1).

If one assumes that the system responds to a change of the input with a uniform time constant δ then

$$\delta \dot{v}_j(t) + v_j(t) = y_j(t) - \sum_{k=1}^{n} \beta_{jk} \vartheta(v_k(t)). \qquad (4.3)$$

If the right hand side would be just $y_j(t)$ then after we switch from one constant input (y_j) to another the state variable follows exponentially fast. However the nonlinear system has several stationary states corresponding to one input, the convergence problem has not been solved.

Experiments have shown that lateral inhibition acts with a delay of about 100 msec. Coleman and Renninger were the first who brought this delay into connection with certain oscillations of the spike frequency (Coleman and Renninger 1974, 1975). If we incorporate this delay τ then the equations read

$$\delta \dot{v}_j(t) + v_j(t) = y_j(t) - \sum_{k=1}^{n} \beta_{jk}\, \vartheta(v(t-\tau)) - \beta_{jj}\, \vartheta(v_j(t)). \tag{4.4}$$

The last term in equation (4.4) is the so-called self-inhibition. The effect of self-inhibition seems to be experimentally established.

Nothing can be said about periodic or oscillating solutions of the general system. Therefore one assumes a spatially homogeneous state x and obtains an equation for one dependent variable

$$\delta \dot{v}(t) + v(t) = y(t) - \tilde{\beta}\, \vartheta(v(t)) - \beta\, \vartheta(v(t-\tau)), \tag{4.5}$$

$$z(t) = \vartheta(v(t)) \quad . \tag{4.6}$$

Note that in this equation there is only one time constant δ .

Simplifying one assumes that self-inhibition is negligible. Then

$$\delta \dot{v}(t) + v(t) = y(t) - \beta\, \vartheta(v(t-\tau)) \quad . \tag{4.7}$$

For a stationary input $y(t) \equiv y$ there is a stationary solution $\bar{v} = y/(1+\beta)$. With the new variable $x = v - \bar{v}$ equation (4.7) assumes the form

$$\dot{x}(t) + \nu x(t) = -\alpha f(x(t-1)) \tag{4.8}$$

where

$$\nu = \frac{\tau}{\delta}, \quad \alpha = \frac{\tau\beta}{\delta}, \quad f(x) = \vartheta\left(\frac{y}{1+\beta} + x\right) - \frac{y}{1+\beta} \quad . \tag{4.9}$$

From general theorems (Hadeler and Tomiuk 1977) follows the existence of periodic solutions for all ß above a certain critical value which depends on τ/δ (but not on y).

If selfinhibition is present then existence of periodic solutions can be shown by application of more general existence results (see Kaplan and Yorke 1977, Hadeler 1979, an der Heiden 1979).

On the other hand Coleman and Renninger assume that selfinhibition has a different time constant. According to their model the variable z follows a law

$$z(t) = \vartheta(v(t))$$

$$\begin{aligned} v(t) &= y - \sigma(t) - \eta(t) \\ \delta_S\dot{\sigma}(t) + \sigma(t) &= K_S\,\vartheta(v(t)) \\ \delta_L\dot{\eta}(t) + \eta(t) &= K_L\,\vartheta(v(t-\tau)) \end{aligned} \tag{4.10}$$

δ_S, δ_L are the time-constants for self-inhibition and lateral inhibition, respectively.
For $\delta_S \neq \delta_L$ one has only one delay, but two different time constants. It might appear that one has to prescribe two initial functions on $[-\tau, 0]$ in order to define a solution. In fact it is sufficient to prescribe v on $[-\tau, 0]$ and $\sigma(0)$. Thus the appropriate state space is $C[-\tau, 0] \times \mathbb{R}$.
Integrating equation (4.10) from $-\infty$ to t gives

$$z(t) = \vartheta\Big(y - \int_{-\infty}^{t} e^{-(t-s)/\delta_L} z(s-\tau)\,ds - \int_{-\infty}^{t} e^{-(t-s)/\delta_S} z(s)\,ds\Big). \tag{4.11}$$

Equations (4.11) and (4.10) are equivalent. Equations (4.10) admit a stationary solution

$$\sigma = \frac{K_S y}{1+K_S+K_L}, \qquad \eta = \frac{K_L y}{1+K_S+K_L}.$$

If we linearize at the stationary solution and compute the characteristic equation we find

$$\begin{vmatrix} \lambda\delta_S + 1 + K_S & K_S \\ K_L e^{-\lambda\tau} & \lambda\delta_L + 1 + K_L e^{-\lambda\tau} \end{vmatrix} = 0$$

or

$$(\lambda\delta_S+1)(\lambda\delta_L+1) + (\lambda\delta_S+1)K_L e^{-\lambda\tau} + (\lambda\delta_L+1)K_S = 0 . \tag{4.12}$$

We establish the proper relation between the two models. Rewrite equation (4.10) as

$$\delta_L \dot{v}(t) + v(t) = y + (\frac{\delta_L}{\delta_S} - 1)\,\sigma(t) - K_L\,\vartheta(v(t-\tau)) - \frac{\delta_L}{\delta_S} K_S\,\vartheta(v(t)) \tag{4.13a}$$

$$\delta_S\,\dot{\sigma}(t) + \sigma(t) = K_S\,\vartheta(v(t)) \tag{4.13b}$$

For $\delta_S = \delta_L$ we obtain the previous model with the additional equation (4.13b). Correspondingly the characteristic equation (4.12) splits for $\delta_S = \delta_L$ as

$$(\lambda\delta+1)((\lambda\delta+1) + K_L\ e^{-\lambda\tau} + K_S) = 0$$

and the second factor is the characteristic equation of (4.5).

It is quite obvious that also equations of the form (4.10) should have non-constant periodic solutions. Up to now there is no proof.

5. The Goodwin oscillator

The Goodwin oscillator has been constructed as a non-linear differential system simulating enzyme kinetics. The enzyme x_1 enhances the formation of x_2, the product x_2 the substance x_3 a.s.o. The final product x_n, $n \geq 1$, inhibits the formation of x_1. Each formation of a product x_j is governed by a low-pass filter. Goodwin's original model is a system of ordinary differential equations. One can try to simulate slow transport of reactants by diffusion or formation of additional reactants by delays.

If one incorporates a delay in each single reaction step then the Goodwin equations assume the form

$$\begin{aligned}
\dot{x}_1(t) &= f(x_n(t-\tau_n)) - a_1x_1(t) \\
\dot{x}_2(t) &= b_2x_1(t-\tau_1) - a_2x_2(t) \\
\dot{x}_3(t) &= b_3x_2(t-\tau_2) - a_3x_3(t) \\
&\vdots \\
\dot{x}_n(t) &= b_nx_{n-1}(t-\tau_{n-1}) - a_nx_n(t)
\end{aligned} \tag{5.1}$$

with $a_j > 0$, $b_j > 0$, and $\tau_j \geq 0$. Typically the function f has the form

$$f(x) = \frac{1}{1+x^{\varrho}} \tag{5.2}$$

where $\varrho = 1,2,\ldots$.

By some suitable transformations the system can be put into a much simpler form (an der Heiden 1978). Define

$$\begin{aligned}
y_1(t) &= x_1(t-\tau_{n-1} - \ldots - \tau_1) \\
y_2(t) &= x_2(t-\tau_{n-1} - \ldots - \tau_2) \\
&\vdots \\
y_{n-1}(t) &= x_{n-1}(t-\tau_{n-1}) \\
y_n(t) &= x_n(t)
\end{aligned}$$

and

$$\tau = \sum_{j=1}^{n} \tau_j .$$

Then

$$\begin{aligned}
\dot{y}_1(t) &= f(y_n(t-\tau)) - a_1y_1(t) \\
\dot{y}_2(t) &= b_2y_1(t) - a_2y_2(t) \\
\dot{y}_3(t) &= b_3y_2(t) - a_3y_3(t) \\
&\vdots \\
\dot{y}_n(t) &= b_ny_{n-1}(t) - a_ny_n(t) .
\end{aligned} \tag{5.3}$$

Thus the system is equivalent with a system with a single delay.

The sum of the delays is the characteristic quantity. (Of course the systems (5.1) and (5.3) are equivalent only for sufficiently smooth solutions, i.e. for sufficiently large t.)

The variable y_j can be rescaled to normalize the constants b_j. Define

$$\begin{aligned} z_1(t) &= b_n \dots b_2 y_1(t) \\ z_2(t) &= b_n \dots b_3 y_2(t) \\ &\vdots \\ z_{n-1}(t) &= b_n y_{n-1}(t) \\ z_n(t) &= y_n(t) \end{aligned} \tag{5.4}$$

and

$$\tilde{f}(x) = b_n b_{n-1} \dots b_2 f(x) \quad .$$

Again, the system is equivalent with an n-th order single delay equation. Define the differential operators

$$D_j x(t) = \dot{x}(t) - a_j x(t) \quad .$$

Then

$$D_j z_j = z_{j-1}, \qquad j=2,\dots,n$$

and

$$D_1 \dots D_n z_n(t) = \tilde{f}(z_n(t-\tau)).$$

For n = 1 the system (5.4) reduces to the delay equation of the form

$$\dot{x}(t) = f(x(t-\tau)) - a_1 x(t) \quad . \tag{5.5}$$

For n = 2 one has

$$\begin{aligned} \dot{x}_1(t) &= f(x_2(t-\tau)) - a_1 x(t) \\ \dot{x}_2(t) &= x_1(t) - a_2 x_2(t) \quad . \end{aligned} \tag{5.6}$$

Again, this system is equivalent with the second order equation

$$\ddot{x}(t) + (a_1+a_2)\dot{x}(t) + a_1a_2x(t) = - f(x(t-1)) \quad . \qquad (5.7)$$

By the biological interpretation the coefficients $a_1 + a_2$ and a_1a_2 are positive.
From the results of Tyson, Webster and Hastings we know that the ordinary Goodwin oscillator admits periodic oscillations, whenever the stationary solution becomes unstable. This can happen only for n = 3. On the other hand we know that equation (5.5) has periodic solutions for appropriately chosen constants. An der Heiden 1979 has shown that equation (5.7) has a periodic solution for a typical function f ($f(x)x > 0$ for $x = 0$, f bounded below) if the zero solution becomes unstable.
Grafton 1969 has applied his method to the related generalized Liénard equation

$$\ddot{x}(t) + h(x(t))\dot{x}(t) = - g(x(t-1)) \qquad (5.8)$$

where however, as in the ordinary Lienard equation, $h(0) < 0$.

6. The sunflower equation

In a recent paper A.S. Somolinos 1978 performs a mathematical investigation of a model developed by Israelsson and Johnsson 1967 for the geotropic nutations of the sunflower. In fact the sunflower and many other higher plants perform rather complicated movements. Any deviation from the vertical position leads to compensatory unequal growth of the sides of the stem. The compensatory growth does not stop when the vertical position is achieved, and rather regular oscillations may occur. In addition the growth of the sunflower is influenced by the sun. The situation is still more complicated since in most plants, even without phototropic movements, the stem does not grow uniformly, the growth of the sides of the stem follows a rotational pattern. Some of these effects have been discovered already in the early 19th century. After plant hormones (auxins) had been discovered the most convincing explanation was that auxin is constantly and uniformly produced by the top of the stem; then, if the stem is bent, the lower (concave) side should receive more hormone than the other side and grow faster. Although the situation is probably more complex and membrane processes play a role one can accept the explanation that some substance is unevenly distributed and that there is a delay between the production of the substance and its action on growth. Following Israelsson and Johnsson (see also Johnsson, Andersen), Somolinos proposes the equation

$$\ddot{x}(t) + a\dot{x}(t) + b \sin x(t-\tau) = 0 \quad . \tag{6.1}$$

The author shows (using the ejective fixed point principle) that this equation has periodic solutions for appropriate choices of the parameters.

7. Existence of periodic solutions

In many biological situations the most interesting feature is a stable periodic oscillation. The stability of such solutions is essential, otherwise they would not be of interest from a biologist's view, because they could not be experimentally observed.
Proofs for the existence of periodic solutions have been given using various tools. The common idea is to start from an appropriately chosen set of initial data and follow the solutions until segments of the solution are again in the chosen set.

The earliest and in the long run most successful approach uses the Browder ejective fixed point principle. Such approach has been used by Jones 1962a,b, Dunkel 1968a,b, Nussbaum 1973, 1974, Hale and Chow 1978, Hadeler and Tomiuk 1977, Hadeler 1979, Walther 1975, an der Heiden 1979, Somolinos 1978 and others for various types of equations. The ejectivity is used to exclude constant periodic solutions. There are other techniques to do this. Grafton 1969, 1972, studies how trajectories leave and reenter an appropriately chosen cone. Alt 1978 develops a fixed point theorem in which the condition that a set is mapped into itself is replaced by an ejectivity type condition at infinity. Pesin 1974 applies directly Schauder's theorem constructing a convex set which excludes constant initial data. A similar approach has recently been followed by H.O.Walther. A beautiful idea is the phase plane method by Kaplan and Yorke 1977 which yields also certain stability results.
Taking all methods together, among others equations of the following types can be handled

$$\dot{x}(t) = -\alpha f(x(t-1))$$

$$\dot{x}(t) = -\alpha x(t-1)(1+\chi(t))$$

$$\dot{x}(t) = -\nu x(t) - \alpha f(x(t-1))$$

$$x(t) = - f(x(t),\chi(t-1))$$

$$x(t) = -\alpha \int_{-1}^{0} x(t+s)\,d\sigma(s)(1+\chi(t))$$

$$x(t) = - f(x(t), \int_{-1}^{0} g(x(t+s))\,d\sigma(s))$$

$$\ddot{x}(t) + a\dot{x}(t) + bx(t) = - f(x(t-1))$$

Furthermore equations have been considered, where the delay depends on the unknown function.
Another approach is to treat the transition from a stable to an unstable constant solution with appearance of periodic solutions as a Hopf bifurcation, see Hassard and Wan 1978, Kazarinoff and van den Driessche 1978. A related idea, transition from stability to ejectivity, was earlier introduced by Nussbaum 1975. Cushing 1977 uses a completely different idea, having two parameters for variation, fixing the period in advance and applying Lyapunov-Schmidt theory.
Kaplan, Sorg and Yorke 1979 exhibit a family of equations $\dot{x}(t) = - f(x(t),x(t-1))$ without periodic solutions.

8. Further applications

There are various other examples of biological models using delay equations. An der Heiden 1979 has presents quite a number of examples from physiology: The production of red blood cells has been modeled by Lasota and Wazewska . The resulting equation has been investigated by Chow 1974 . Irregular production of blood cells has been explained with a delay model by Mackey and Glass 1977, 1979 . An der Heiden 1979 has established a connection between such difference-differential equations with highly irregular behavior, discrete models and ordinary differential systems of higher order.An der Heiden 1979 discusses also the delay models for orientation of insects by Wehrhahn and Poggio 1976 finally an equation for the problem of balancing by Schürer 1948.

There is a vast literature on systems of delay equations arising from ecological models. Many of these models can be dated back to the famous

book of Volterra 1931. Recently monographs by Cushing 1978 and MacDonald 1978 have appeared.
Both monographs contain many more examples of biological models as well as existence, stability, and bifurcation results.
Particular attention has been paid to certain equations with infinite delays

$$\dot{x}_j(t) = \left[b_j + \sum_{k=1}^{n} a_{jk} x_k(t) + \sum_{k=1}^{n} \int_{-\infty}^{t} c_{jk}(t+s)\, x_k(s)\, ds \right] x_j(t),$$

$$j = 1, \ldots, n.$$

Such equations can be reduced to ordinary differential equations of higher order, if the kernels c_{jk} satisfy ordinary differential equations with constant coefficients. For this class of problems in particular the stability theory can be carried further than in the general case, see A. Wörz-Busekros 1977, Cushing 1978, MacDonald 1978.

References

Alt, W., Some periodicity criteria for functional differential equations, Manuscripta mathematica 23, 295-318 (1978).

Andersen, H., and Johnsson, A., Entrainment of geotropic oscillations in hypocotyls of Helianthus annuus - an experimental and theoretical investigation I, II. Phys. Plant. 26, 44-51, 52-61 (1972).

Andersen, H., A mathematical model for circum-nutations, Thesis, Report 2/1976, Department of Electrical Measurements, Lund Institute of Technology.

Banks, H.T., Delay systems in biological models: approximation techniques. In: V. Lakshmikantham (ed.), Nonlinear systems and applications. Proc. Conf. Arlington 1976, Academic Press, N.Y. (1977).

Bellman, R., and Cooke, K.L., Differential-difference equations, Acad. Press, N.Y. (1963).

Browder, F.E., A further generalization of the Schauder fixed-point theorem. Duke Math. J. 32, 575-578 (1965).

Chow, S.-N., Existence of periodic solutions of autonomous functional differential equations, J. Diff. Equ. 15, 350-378 (1974).

Chow, S.-N., and Hale, J., Periodic solutions of autonomous equations. J. Math. Anal. Appl. 66, 495-506 (1978).

Coleman, B.D., and Renninger, G.H., Theory of delayed lateral inhibition in the compound eye of Limulus, Proceedings Nat. Acad.Sc. 71, 2887-2891 (1974).

Coleman, B.D., and Renninger, G.H., Consequences of delayed lateral inhibition in the retina of Limulus I. Elementary theory of spatially uniform fields , J. Theor. Biol. 51, 243-265 (1975).

Coleman, B.D., and Renninger, G.H., Consequences of delayed lateral inhibition in the retina of Limulus II. Theory of spatially uniform fields, assuming the 4-point property, J. Theor. Biol. 51, 267-291 (1975).

Coleman, B.D., and Renninger, G.H., Periodic solutions of a nonlinear functional equation describing neural action, Istituto Lombardo Acad. Sci. Lett. Rend. A, 109, 91-111 (1975).

Coleman, B.D., and Renninger, G.H., Periodic solutions of certain nonlinear integral equations with a time-lag, SIAM J. Appl. Math. 31, 111-120 (1976).

Coleman, B.D., and Renninger, G.H., Theory of response of the LIMULUS retina to periodic excitation. J. Math. Biol. 3, 103-120 (1976).

Cunningham, W.J., A nonlinear differential difference equation of growth, Proc. Nat. Acad. Sci. U.S.A. 40, 709-713 (1954).

Cunningham, W.J., and Wangersky, P.I., On time lags in equations of growth. Proc. Acad. Sci. U.S.A. 42, 699-702 (1956).

Cushing, J.M., Bifurcation of periodic oscillations due to delays in single species growth models, J. Math. Biol. 6, 145-161 (1978).

Cushing, J.M., Periodic solutions of two species interaction models with lags, Math. Biosc. 31, 143-156 (1976).

Cushing, J.M., Integrodifferential equations and delay models in population dynamics, Lecture Notes in Biomath. 20, Springer-Verlag (1978).

Dunkel, G.M., Some mathematical models for population growth with lags. Univ. of Maryland, Technical Note BN - 548 (1968).

Dunkel, G.M., Single-species model for population growth depending on past history. In: Seminar on Differential Equations and Dynamical Systems, 92-99, Lecture Notes in Mathematics 60, Springer (1968).

Grafton, R.B., A periodicity theorem for autonomous functional differential equations, J. Diff. Equ. 87-109 (1969).

Grafton, R.B., Periodic solutions of certain Liénard equations with delay, J. Diff. Equ. 11, 519-527 (1972).

Hadeler, K.P., On the stability of the stationary state of a population growth model with time-lag, J. Math. Biol. 3, 197-201 (1976).

Hadeler, K.P., Some aspects of the mathematics of LIMULUS. In:Optimal estimation in approximation theory, edited by Ch.A. Micchelli and Th.J. Rivlin, Plenum Press 1977, p. 241-257.

Hadeler, K.P., and Tomiuk, J., Periodic solutions of difference-differential equations, Arch. Rat. Mech. Anal. 65, 87-95 (1977).

Hadeler, K.P., Periodic solutions of $\dot{x}(t) = - f(x(t),x(t-1))$ Math. Meth. Appl. Sciences 1, 62-69 (1979).

Hassard, B., and Wan, Y.H., Bifurcation formulae derived from center manifold theory, J. Math. Anal. Appl. 63, 297-312 (1978).

an der Heiden, U., Stability properties of neural and cellular control systems, Math. Biosciences 31, 275-283 (1976).

an der Heiden, U., Delay-induced biochemical oscillations. In: Bioph. and Bioch. Information Transfer and Recogn. Varna Conf.Proc., 533-540, Plenum Press NY 1978.

an der Heiden, U., Chaotisches Verhalten in zellulären Kontrollprozessen. In: Zelluläre Kommunikations- und Kontrollmechanismen(ed. L.Rensing, G.Roth) Universitätsverlag Bremen 1978.

an der Heiden, U., Periodic solutions of a non-linear second order differential equation with delay, J.Math.Anal. Appl. to appear

an der Heiden, U., Delays in physiological systems. In: Lecture Notes in Biomathematics (ed.R.Berger) Springer Verlag 1979.

Ísraelsson, D., and Johnsson,A., A theory for circumnutations in Helianthus annuus, Physiol. Plant. 20, 957-976 (1967).

Johnsson, A., and Israelsson, D., Application ofa theory for circumnutations to geotropic movements, Physiol. Plant. 21, 282-291 (1968).

Johnsson, A., and Israelsson,D., Phase-shift in geotropical oscillations-a theoretical and experimental study, Physiol. Plant. 22, 1226-1237 (1969).

Johnsson,A., Geotropic responses in Helianthus and the dependence on the auxin ratio.- With a refined mathematical description of the course of geotropic movements. Physiol. Plant. 24, 419-425 (1971).

Johnsson, A., Gravitational stimulations inhibit oscillatory growth movements of plants, Z. Naturforsch. 29c, 717-724 (1974).

Kaplan, J.L. and Yorke, J.A., Existence and stability of periodic solutions of $x'(t) = - f(x(t),x(t-1))$, in: Cesari Lamberto (ed.) Dynamical Systems I,II Providence (1974).

Kaplan, J.L. , Sorg,M., and Yorke, J.A., Solutions of $x'(t)=f(x(t),x(t-L))$ have limits when f is an order relation., to appear.

Kazarinoff, N.D., van den Driessche, P., and Wan, Y.-H, Hopf bifurcation and stability of periodic solutions of differential-difference and integro-differential equations, J.Inst.Math.Appl. 21,461-477 (1978).

Levin,S.A., and May,R.M., A note on difference-delay equations, Theor. Pop. Biol. 9, 178-187 (1976).

Li, T.Y., and Yorke, J.A., Period three implies chaos, Am. Math. Monthly 82, 985-992 (1975)

Mackey, M.C., A unified hypothesis for the origin of aplastic anemia and periodic haematopoiesis, Blood, to appear.

Mackey, M.C., and Glass, L., Oscillation and chaos in physiological control systems, Science 197, 287-289 (1977).

May,R.M., Conway,G.R., Hassell,M.P., and Southwood,T.R.E., Time delays, density-dependence and single-species oscillations, J.Anim.Ecol.43, 747-770 (1974).

MacDonald, N., Time delay in prey-predator models, Math.Biosc. 28, 321-330 (1976), II. Bifurcation theory Math. Biosc. 33, 227-234 (1977).

MacDonald, N., Time lag in a model of a biochemical reaction sequence with end product inhibition, J. theor.Biol. 67, 549-556 (1977).

May, R.M., Simple mathematical models with very complicated dynamics. Nature 261, 459-467 (1976).

Maynard Smith, J., Mathematical Ideas in Biology, Cambridge University Press, London 1968.

Maynard Smith, J., Models in Ecology, Cambridge University Press, London 1974.

Nussbaum, R.D., Periodic solutions of some nonlinear autonomous functional differential equations II, J. Diff. Equ. 14, 360-394 (1973).

Nussbaum, R.D., Periodic solutions of some nonlinear autonomous functional differential equations, Ann. Mat. Pura Appl. 101, 263-306 (1974).

Nussbaum,R.D., A global bifurcation theorem with applications to functional differential equations, J. Funct. Anal. 19, 319-338(1975).

Pesin, J.B., On the behavior of a strongly nonlinear differential equation with retarded argument. Differentsial'nye uravnenija 10, 1025-1036 (1974).

Perez, J.F., Malta, C.P., and Coutinho, F.A.B., Qualitative analysis of oscillations in isolated populations of flies, J. Theor. Biol. 71, 505-514 (1978).

Schürer, F., Zur Theorie des Balancierens, Math. Nachr. 1, 295-331 (1948).

Somolinos, A.S., Periodic solutions of the sunflower equation: $\ddot{x} + (a/r)\dot{x} + (b/r) \sin x (t-r) = 0$, Quart. ofAppl.Math.35, 465-477 (1978).

Stech,H.W., The effect of time lags on the stability of the equilibrium state of a population growth equation, J. Math. Biol. 5,115-120 (1978).

Walther, H.-O., On a transcendental equation in the stability analysis of a population growth model, J. Math. Biol. 3, 187-195 (1976).

Walther, H.-O., Über Ejektivität und periodische Lösungen bei autonomen Funktionaldifferentialgleichungen mit verteilter Verzögerung, Habilitationsschrift München 1977.

Walther, H.-O., On instability, ω -limit sets and periodic solutions of nonlinear autonomous delay equations, these proceedings

Wehrhahn, C., and Poggio, T., Real-time delayed tracking in flies. Nature 261, 43-44 (1976).

Wörz-Busekros, A., Global stability in ecological systems with continuous time delays, SIAM J. Appl. Math. 35, 123-134 (1978) .

Walther,H.-O., A theorem on the amplitudes of periodic solutions of differential delay equations with application to bifurcation. J.Diff. Equ. 29, 396-404 (1978).

RETARDED EQUATIONS WITH INFINITE DELAYS [+]

Jack K. Hale

Abstract: It is the purpose of these notes to describe the theory of Hale and Kato for functional differential equations based on a space of initial data which satisfy some very reasonable axioms. We also indicate some recent results of Naito showing how extensive the theory of linear systems can be developed in an abstract setting in particular, the characterization of the spectrum of the infinitesimal generator together with the decomposition theory and exponential estimates of solutions.

[+] This research was supported in part by the Air Force Office of Scientific Research under AF-AFOSR 76-3092A, in part by the National Science Foundation under NSF-MCS 78-18858, and in part by the United States Army under ARO-D-31-124-73-G130.

1. Introduction. Suppose $0 \leq r < \infty$ is given. If $x: [\sigma-r,\sigma+A) \to \mathbb{R}^n$, $A > 0$, is a given function, let $x_t: [-r,0] \to \infty$ for each $t \in [\sigma,\sigma+A)$ be defined by $x_t(\theta) = x(t+\theta)$, $-r \leq \theta \leq 0$. A retarded functional differential equation is a relation

$$\dot{x}(t) = f(t,x_t) \quad . \tag{1.1}$$

for a function $x(t) \in \mathbb{R}^n$, where $f(t,\sigma)$ is defined for $t \in \mathbb{R}$ and ϕ in some space $\mathscr{B}$ of functions from $[-r,0]$ to $\mathbb{R}^n$. The space $\mathscr{B}$ is called the state space. Let us suppose, for each $(\sigma,\phi) \in \mathbb{R} \times \mathscr{B}$, there is a well-defined solution of Equation (1.1), $x(t) = x(\sigma,\phi,f)(t)$, $t \geq \sigma - r$, $x(\sigma+\theta) = \phi(\theta)$, $\theta \in [-r,0]$, and $x_t \in \mathscr{B}$, $t \geq \sigma$. If $T(t,\sigma,f): \mathscr{B} \to \mathscr{B}$ is defined by $[T(t,\sigma,f)\phi](\theta) = x(t+\theta)$, $\theta \in [-r,0]$, then the basic problem is to discuss the properties of the function $T(t,\sigma,f)\phi$ as a function of (t,σ,ϕ,f).

The choice of the state space $\mathscr{B}$ is usually dictated by the form of the equation and the desired objectives. For $r < \infty$; that is, finite delays, most properties of $T(t,\sigma,f)\phi$ are not too sensitive to the space $\mathscr{B}$ if one defines a solution of Equation (1.1) through a point (σ,ϕ) in the natural way as a continuous extension of $\mathscr{B}$ to the right of $t = \sigma$ by Relation (1.1). After one delay interval r, $T(t,\sigma,f)\phi$ belongs to the space of continuous functions from $[-r,0] \to \mathbb{R}^n$ regardless of the space $\mathscr{B}$. Thus, it is sufficient, in most problems, to understand Equation (1.1) well in the space of continuous functions.

If the delay is infinite, that is, $r = \infty$, this is no longer the case since $x_t(\theta)$ for $\theta \leq -t$ coincides with the values of the initial function. This means that the qualitative behavior of the solution operator will depend upon the space of initial data. It thus becomes important to understand in an abstract manner those properties of the space $\mathcal{B}$ which will permit the development of the fundamental theory of existence, uniqueness, continuous dependence, continuation, etc. In addition, one needs to know some abstract properties which will imply something about global behavior of orbits; for example, when are bounded orbits precompact, when is stability in $\mathbb{R}^n$ equivalent to stability in the function space, etc.? In addition to yielding a better understanding of functional differential equations, an axiomatic development should eliminate duplication of effort, even in the case of finite delays.

The axiomatic approach for infinite delays is not new. In fact, Coleman and Mizel [4,5,6] developed a rather complete theory of fading memory spaces. Since these spaces are not appropriate for certain types of Equation (1.1) without severe restrictions on f, Hale [10] introduced some other axioms in an attempt to clarify the situation. Unfortunately, there were several omissions and confusions in [10]. On the other hand, the ideas stimulated some interest in the subject and several papers appeared showing that it was feasible to do many things in qualitative theory without knowing too much about the space $\mathcal{B}$ (see Hino [15, 16, 17], Naito [26]).

In a recent paper, Hale and Kato [13] gave a more systematic development of the subject. Independently, Schumacher [32, 33] has obtained similar results. It is the purpose of these notes to describe the status of this theory. The reader will see that further research is needed in comparing the different axioms. We also discuss recent results of Naito [28] showing how extensive the theory of linear systems can be developed in an abstract setting - in particular, the characterization of the spectrum of the infinitesimal generator together with the decomposition theory and exponential estimates of solutions.

2. Axioms for the Local Theory. In this section, we present the axioms from Hale and Kato [13] for the phase space which seem to be convenient for a local theory of functional differential equations. In a later section, we discuss the axioms of Schumacher [32].

Let I be either a fixed finite interval $[-r,0]$ or the interval $(-\infty,0]$. Let $\hat{\mathscr{B}}$ be a linear space of functions mapping I into $\mathbb{R}^n$ with elements designated by $\hat{\phi},\hat{\psi},\ldots$, where $\hat{\phi} = \hat{\psi}$ means $\hat{\phi}(t) = \hat{\psi}(t)$, $t \in I$. Suppose there is a seminorm $|\cdot|_{\hat{\mathscr{B}}}$ on $\hat{\mathscr{B}}$ and suppose $\mathscr{B} = \hat{\mathscr{B}}/|\cdot|_{\hat{\mathscr{B}}}$ is a Banach space with $|\cdot|_{\mathscr{B}}$ naturally induced by $|\cdot|_{\hat{\mathscr{B}}}$. Elements of $\mathscr{B}$ are denoted by $\hat{\phi},\hat{\psi},\ldots$ and correspond to equivalence classes of $\hat{\mathscr{B}}$. For any $\phi \in \mathscr{B}$, corresponding elements in the equivalence classes are denoted by $\hat{\phi}$ and $\phi = \psi$ in $\mathscr{B}$ means $|\hat{\phi}-\hat{\psi}|_{\hat{\mathscr{B}}} = 0$ for all $\hat{\phi} \in \phi$, $\hat{\psi} \in \psi$.

To always distinguish between elements of the equivalence class and the equivalence class itself requires cumbersome notation. Therefore, in the following, we do not use a different symbol (except where confusion may arise) to distinguish these objects. In general, the symbol "^" is omitted with the reader being expected to make the distinction by context.

The first axiom is

Axiom (α_0'): There is a constant K such that $|\hat{\phi}(0)| \leq K|\hat{\phi}|_{\hat{\mathscr{B}}}$ for every $\hat{\phi} \in \hat{\mathscr{B}}$

This axiom implies that $\hat{\phi}(0) = \hat{\psi}(0)$ for every $\hat{\phi} \in \phi$, $\hat{\psi} \in \phi$. Therefore, for every equivalence class ϕ there is associated a unique $\phi(0)$ and Axiom (α_0') can be rewritten as

Axiom (α_0): There is a constant K such that $|\phi(0)| \leq K|\phi|_{\mathscr{B}}$ for all $\phi \in \mathscr{B}$.

For any $A \geq 0$, if $x\colon I_A \to \mathbb{R}^n$ is a given function and $t \in [0,A]$, define $x_t\colon I \to \mathbb{R}^n$ by

$$x_t(\theta) = x(t+\theta), \quad \theta \in I.$$

For any $\hat{\phi} \in \hat{\mathscr{B}}$, let $F_A(\hat{\phi})$ be the set of functions $\hat{x}\colon I \cup [0,A] \to \mathbb{R}^n$ such that $\hat{x}_0 = \hat{\phi}$, $\hat{x}$ is continuous on $[0,A]$ and define

$$F_A = \bigcup \{F_A(\hat{\phi}),\ \hat{\phi} \in \hat{\mathscr{B}}\}.$$

Axiom (α_1): $x_t \in \mathscr{B}$ for all $x \in F_A$, $t \in [0,A]$.

Axiom (α_2): If $x \in F_A$, $A > 0$, then x_t is continuous in t for $t \in [0,A]$.

For the next axioms, we need two seminorms in $\mathscr{B}$, one corresponding to the restriction of ϕ to $I\backslash[-\beta,0]$ and the other to the restriction of ϕ to $[-\beta,0]$ for $-\beta \in I$. More specifically, for any $-\beta \in I$, let

$$|\phi|_\beta = \inf_{\hat{\eta}\in\hat{\mathscr{B}}} \{\inf_{\hat{\psi}\in\hat{\mathscr{B}}} \{|\hat{\psi}|_{\hat{\mathscr{B}}} : \hat{\psi}(\theta) = \hat{\eta}(\theta),\ \theta \in I\backslash[-\beta,0]\} : \eta = \phi\}$$

$$|\phi|_{(\beta)} = \inf_{\hat{\eta}\in\hat{\mathscr{B}}} \{\inf_{\hat{\psi}\in\hat{\mathscr{B}}} \{|\hat{\psi}|_{\hat{\mathscr{B}}} : \hat{\psi}(\theta) = \hat{\eta}(\theta),\ \theta \in [-\beta,0]\} : \eta = \phi\}$$

Let $\mathscr{B}^\beta = \mathscr{B}/|\cdot|_\beta$ be the Banach space generated by the seminorm $|\cdot|_\beta$ on $\mathscr{B}$. If $\{\phi\}_\beta = \{\psi \in \mathscr{B}: |\phi-\psi|_\beta = 0\}$ is a representative element of $\mathscr{B}^\beta$, then $\psi \in \{\phi\}_\beta$ if $\hat{\psi}(\theta) = \hat{\phi}(\theta)$ for $\theta \in I\backslash[-\beta,0]$. One can inprecisely think of $\mathscr{B}^\beta$ as the set of restrictions to $I\backslash[-\beta,0]$ of elements in $\mathscr{B}$.

From Axiom (α_1), for any $-\beta \in I$ and $\phi \in \mathscr{B}$, it is possible to find a $\psi \in \mathscr{B}$ such that

$$\psi(\theta) = \phi(\theta+\beta) \quad \theta \in I\backslash[-\beta,0].$$

In fact, one can take ψ to be defined as $\phi(0)$ on $[-\beta,0]$ and by the above relation on $I\backslash[-\beta,0]$. Define

$$\tau^\beta: \mathscr{B} \to \mathscr{B}^\beta$$

$$\tau^\beta\phi = \{\phi\}_\beta.$$

We can now introduce our other basic axioms

Axiom (α_3): If $\phi = \psi$ in $\mathscr{B}$, then $|\tau^\beta\phi - \tau^\beta\psi|_\beta = 0$

Axiom (α_4): $|\phi|_{\mathscr{B}} \leq |\phi|_\beta + |\phi|_{(\beta)}$ for any $-\beta \in I$ and all $\phi \in \mathscr{B}$.

Axiom (α_5): There is a continuous function $K_1(\beta)$ of $-\beta \in I$ such that $|\phi|_{(\beta)} \leq K_1(\beta)\sup_{[-\beta,0]}|\phi(\theta)|$.

Axiom (α_6): $\tau^\beta: \mathscr{B} \to \mathscr{B}^\beta$ is a bounded linear operator with norm

$$M_1(\beta) = \sup_{|\phi|_{\mathscr{B}}=1} |\tau^\beta\phi|_\beta$$

locally bounded; that is, for any $-\beta \in I$, there is a neighborhood U of β such that $\sup_{t\in U\cap(-I)} M_1(t) < \infty$.

We now give some examples of spaces which satisfy the above axioms.

Example 2.1. (Spaces $L_p(\mu) \times \mathbb{R}^n$). Suppose $g: I \to [0,\infty)$, $G: I \to [0,\infty)$ are continuous

$$g(t+s) \leq G(t)g(s), \quad t,s \in I. \tag{2.1}$$

If g is a nondecreasing function, one can satisfy this condition with $G(t) = 1$ for all $t \in I$. If $g(t) = \exp(\lambda t)$, then we can take $G(t) = \exp(\lambda t)$. For any g satisfying Relation (2.1), let

$$\mathscr{B} = \{\phi: I \to \mathbb{R}^n, \text{ measurable}, |\phi| < \infty\}$$

$$|\phi|_{\mathscr{B}} = \{|\phi(0)|^p + \int_I g(\theta)|\phi(\theta)|^p d\theta\}^{1/p}.$$

Axioms (α_0)-(α_6) are satisfied with the constants in (α_5), (α_6) chosen as

$$(2.2) \qquad K_1(\beta) = [1 + \int_{-\beta}^{0} g(\theta)d\theta]^{1/p}$$

$$(2.3) \qquad M_1(\beta) = \sup_{s\leq 0}[\frac{g(s-\beta)}{g(s)}]^{1/p}.$$

The space $\mathscr{B}$ is isomorphic to $L^p(\mu_g) \times \mathbb{R}^n$ where μ_g is the measure induced by the function g,

$$\mu_g(E) = \int_E g(\theta)d\theta, \qquad E \subset I.$$

Example 2.2. (Spaces of continuous functions C_γ). For any $\gamma \in \mathbb{R}$, let

$$\mathscr{B} = C_\gamma \overset{\text{def}}{=} \{\phi: I \to \mathbb{R}^n, \text{ continuous}, e^{\gamma\theta}\phi(\theta) \to \text{a limit as } \theta \to -\infty\}$$

$$|\phi|_{C_\gamma} = \sup_{\theta\in I} e^{\gamma\theta}|\phi(\theta)|.$$

Axioms (α_0)-(α_6) are satisfied with the constants in (α_5), (α_6) chosen as

$$(2.4) \qquad K_1(\beta) = \sup_{[-\beta,0]} e^{\gamma\theta}, \quad M_1(\beta) = e^{-\gamma\beta}.$$

3. Local theory of functional differential equations. Suppose $\mathscr{B}$ is a space of functions satisfying Axioms (α_0)-(α_6), Ω is an open set in $\mathbb{R} \times \mathscr{B}$, $f: \Omega \to \mathbb{R}^n$ is continuous and consider the retarded functional differential equation

$$(3.1) \qquad \dot{x}(t) = f(t,x_t),$$

where $x_t(\theta) = x(t+\theta)$, $\theta \in I$. For any $(\sigma,\phi) \in \Omega$, a solution $x = x(\sigma,\phi,f)$ through (σ,ϕ) is a function defined on an interval $I_\sigma \cup [\sigma,\sigma\ \alpha]$, $I_\sigma = \{\zeta \in \mathbb{R},\ \zeta = \theta + \sigma,\ \theta \in I\}$, $\alpha > 0$, such that $x_\sigma = \phi$, x satisfies (3.1) on $[\sigma,\sigma+\alpha]$.

The following results have been proved by Hale and Kato [13].

Theorem 3.1. (Existence) For any $(\sigma,\phi) \in \Omega$, there is a solution of Equation (3.1) through (σ,ϕ).

Theorem 3.2. (Uniqueness) If $f(t,\phi)$ is Lipschitzian in ϕ on Ω, then the solution through $(\sigma,\phi) \in \Omega$ is unique and there is a continuous function $K(t)$ such that

$$|x_t(\sigma,\phi) - x_t(\sigma,\psi)| \leq K(t-\sigma)|\phi-\psi| \quad , \quad t \geq \sigma.$$

Theorem 3.3. (Continuation) If x is a noncontinuable solution of (3.1) on $I_\sigma \cup [\sigma,\delta]$ and f takes closed bounded sets of Ω into bounded sets, then, for any closed bounded set W in Ω, there is a sequence $t_k \to \delta^-$ such that $(t_k,x_{t_k}) \notin W$. If, in addition, there is an $r_0 > 0$, $k > 0$, such that

$$\sup_{[-r_0,0]}|\phi(\theta)| \leq k|\phi|_{\mathcal{B}}$$

then there is a t_W such that $(t,x_t) \notin W$ for $t_W \leq t < \delta$.

<u>Theorem 3.4.</u> <u>(Continuous dependence)</u> <u>Suppose</u> $f = f_\lambda$ <u>in</u> (3.1) <u>depends continuously upon a parameter</u> λ <u>in a Banach space</u>. <u>If the solution</u> $x(\sigma,\phi,\lambda)$ <u>of</u> (3.1) <u>through</u> (σ,ϕ) <u>is unique, then</u> $x(\sigma,\phi,\lambda)$ <u>is continuous in</u> (σ,ϕ,λ). <u>If, in addition</u>, $f_\lambda(t,\phi)$ <u>is continuously differentiable in</u> (ϕ,λ), <u>then</u> $x(\sigma,\phi,\lambda)$ <u>is continuously differentiable in</u> (ϕ,λ).

To indicate the proof of the above theorems, let

$$(3.2)\qquad \begin{aligned} &S(t): \mathscr{B} \to \mathscr{B}, \qquad t \geq 0, \\ &(S(t)\phi)(\theta) = \begin{cases} \phi(0), & t+\theta \geq 0 \\ \phi(t+\theta), & t+\theta < 0 \end{cases} \end{aligned}$$

The family of operators $S(t)$, $t \geq 0$ is a strongly continuous semigroup of operators on $\mathscr{B}$ by Axioms (α_1), (α_2). In the literature, the function $(S(t)\phi)(0)$ is sometimes called the <u>static continuation of</u> ϕ.

Axioms (α_0)-(α_4) imply that x is a solution of Equation (3.1) through (σ,ϕ) on $[\sigma,\sigma+A]$ if and only if

$$(3.3)\qquad \begin{aligned} x(t+\sigma) &= (S(t)\phi)(0) + y(t) \\ y_0 &= 0 \end{aligned}$$

$$(3.4)\qquad y(t) = \int_0^t f(\tau+\sigma, S(\tau)\phi + y_\tau)d\tau, \qquad t \in [0,A]$$

To obtain existence of a solution, one first defines

$$A(\delta,\eta) = \{\zeta : I \cup [0,\delta] \to \mathbb{R}^n, \text{ continuous}, \ \zeta(t) = 0 \ \text{ for } \ t \in I,$$
$$|\zeta(t)| \leq \eta \quad \text{for} \quad t \in [0,\delta]\}$$

$$(T\xi)(t) = \begin{cases} 0 & t \in I \\ \int_0^t f(\sigma+\tau, S(\tau)\phi + \zeta_\tau)\,d\tau, & t \geq 0 \end{cases}$$

Using the axioms on $\mathcal{B}$, one shows in a fairly standard way that there are $\delta > 0$, $\eta > 0$ such that T is a completely continuous map of $A(\delta,\eta)$ into itself. The Schauder theorem completes the proof of existence.

Uniqueness Theorem 3.2 is a simple application of Gronwall's inequality. The continuation Theorem 3.3 can be proved easily by contradiction.

It is not known if the conclusion in the continuation Theorem 3.2 can be strengthened to say the following: <u>for any bounded set</u> W <u>in</u> Ω <u>there is a</u> t_W <u>such that</u> $(t,x_t) \notin W$ for $t_W \leq t < \delta$. In the case of finite delay and the space of continuous functions, this result is true (see, Hale [12]). In [13], it is shown that this stronger conclusion is true if the following hypotheses is satisfied:

(*) there is an $r > 0$ and a constant K such that

$$\sup_{\theta \in [-r,0]} |\phi(\theta)| \leq K|\phi|$$

For fading memory spaces Coleman and Mizel [4,5,6] have proven results similar to the above. For the general case of of $L_2(\mu) \times \mathbb{R}^n$, Lima [25] has done the same. For similar results in $L_2(\mu) \times V$ where V is a Banach space and f is Lipschitzian, see Coleman and Owen [7]. For fading memory spaces and integral equations see Leitman and Mizel [24].

In recent years, there has been interest in studying functional differential equations in spaces like $L^p \times \mathbb{R}^n$ when the right-hand side of Equation (3.1) is not a function on L^p; for example, it may be a differential difference equation (see, for example, Delfour and Mitter [9], Banks and Burns [1], Reber [31]). In this case, one must specify axioms through the Integral Equation (3.4). Since this introduces minor technical changes which are primarily notational, such modifications will not be discussed.

It should be apparent that our axioms do not impose any differentiability hypotheses on elements of the state space (see, for example Axiom (α_4)). On the other hand, it is sometimes necessary to consider a Banach space $\mathscr{B}_1$ as a state space where $\mathscr{B}_1$ does impose conditions on the derivatives. For example, in some control problems (see Banks and Jacobs [2]), one may have $\mathscr{B}_1$ as the Sobolev space $W^{2,1}$ and the control function in L^2. In this case, one considers $W^{2,1}$ as a subspace of a Banach space (for example, $L^2 \times \mathbb{R}^n$ or the space of continuous functions) and supposes the functional differential equation (3.1) is defined on $\mathscr{B}$ and has "nice" properties. One can then use the Equation (3.1) to discover relevant properties when the initial data is restricted to $W^{2,1}$. For specific cases, this procedure has been used effectively. It would be interesting to know something about the more general case of subspaces $\mathscr{B}_1$ continuously (or compactly) imbedded in a space $\mathscr{B}$ satisfying the above axioms.

In the previous paragraph, we mentioned Equation (3.1) defined on $\mathscr{B}$ and considered its restriction to a subspace $\mathscr{B}_1$ in $\mathscr{B}$ where the norm in $\mathscr{B}_1$ could impose conditions on the derivatives of its elements. Why not consider the Equation (3.1) on $\mathscr{B}_1$ itself without any reference to $\mathscr{B}$? If this were done, then a continuous function $f(t,\phi)$ on $\mathbb{R} \times \mathscr{B}_1$ could involve both ϕ and $\dot{\phi}$; that is, Equation (3.1) involves the derivative of $x(\tau)$ for $\tau < t$. Such equations are called equations of neutral type and are much more complicated. The axiomatic theory for such systems should be developed.

4. Other axioms for the local theory. In this section, we discuss the axioms of Schumacher [32] and the relationship to the ones in Section 2 of Hale and Kato [13]. In the theory of retarded functional differential equations, the translation operator $S(t)$ defined in Equation (3.2) by

$$(4.1) \qquad (S(t)\phi)(\theta) = \begin{cases} \phi(0), & t+\theta \geq 0 \\ \phi(t+\theta), & t+\theta < 0 \end{cases}$$

for $t \geq 0$, ϕ a function from I to R^n has played an important role for a number of years (see, for example, Coleman and Mizel [4], Hale and Lopes [14], Hale [12, 13], Stech [34], Lima [25], Hale and Kato [13], Kappel and Schappacher [23]). This operator represents the solution operator for the trivial differential equation

$$(4.2) \qquad \begin{aligned} \dot{x}(t) &= 0, \quad t \geq 0 \\ x_0 &= \phi. \end{aligned}$$

Equation (3.1) can then be considered as a perturbation of Equation (4.2) at least as far as the local theory is concerned. This remark is expressed very clearly in Equations (3.3), (3.4) for a solution of Equation (3.1) through (σ,ϕ). It is rather surprising that this operator also has implications about the global behavior of solutions of Equation (3.1) as we will see in Section 5.

Motivated by Equations (3.3), (3.4), Schumacher [32] has developed an axiomatic theory which imposes conditions on the translation operator $S(t)$ and the solutions of the nonhomogeneous differential equation

$$\begin{aligned} \dot{y}(t) &= g(t), \quad t \geq 0 \\ y_0 &= 0. \end{aligned} \tag{4.3}$$

The solution of Equation (4.3) corresponds to Equation (3.4) with $g(t) = f(t+\sigma, S(t)\phi + y_t)$.

Schumacher [32] has proved Theorems 3.1-3.4 under the following hypotheses on the state space $\mathscr{B}$.

<u>Axiom (β_0):</u> <u>There is a constant</u> K <u>such that</u> $|\phi(0)| \leq K|\phi|$

<u>Axiom (β_1):</u> <u>For each</u> $\phi \in \mathscr{B}$, $t \geq 0$, <u>the function</u> $S(t)\phi \in \mathscr{B}$.

<u>Axiom (β_2):</u> <u>If</u> $|\hat{\phi}-\hat{\psi}|_{\hat{\mathscr{B}}} = 0$, <u>then</u> $|S(t)\hat{\phi} - S(t)\hat{\psi}|_{\hat{\mathscr{B}}} = 0$.

<u>Axiom (β_3):</u> $\{S(t), t \geq 0\}$ <u>is a strongly continuous semigroup of bounded linear operators on</u> $\mathscr{B}$.

Axiom (β_4): *For each* $-\tau \in I$, $\tau > 0$, *the set* $\mathscr{B}(\tau)$ *of continuous functions from* I *to* $\mathbb{R}^n$ *with support in* $[-\tau, 0]$ *belongs to* $\mathscr{B}$.

Axiom (β_5): *If* $\mathscr{B}(\tau)$ *is endowed with the maximum norm, the inclusion map taking* $\mathscr{B}(\tau)$ *into* $\mathscr{B}$ *is continuous*.

These axioms are very similar to Axioms (α_0)-(α_6) and some are directly comparable. Axioms (α_0) and (β_0) are the same. Axiom (α_1) is equivalent to Axioms (β_1), (β_4). Note that Axioms (α_4), (α_5), (α_6) imply that

$$|S(t)\hat{\phi} - S(t)\hat{\psi}|_{\hat{\mathscr{B}}} \leq K_1(t)|\hat{\phi}(0) - \hat{\psi}(0)| + M_1(t)|\hat{\phi} - \hat{\psi}|_{\hat{\mathscr{B}}}.$$

Therefore, Axioms (α_4), (α_5), (α_6) imply Axiom (β_2). Axioms (α_4), (α_5) imply (β_5).

The other axioms are similar, but do not seem to be comparable without some additional hypotheses on the structure of the space $\mathscr{B}$.

As remarked earlier, the axioms of Schumacher are motivated by the operator $S(t)$ and the Equation (4.3). Axioms (β_1)-(β_3) deal with $S(t)$. Axioms (β_4), (β_5) imply the solutions of Equation (4.3) are well behaved in $\mathscr{B}$.

5. *Axioms for global theory*. Our original objective in trying to axiomitize the space $\mathscr{B}$ for a global theory of Equation (3.1) is the following. We want bounded orbits to be precompact, stability in $\mathbb{R}^n$ to be equivalent to stability in the function space $\mathscr{B}$ and to relate the concepts of asymptotic stability and uniform asymptotic stability for autonomous and periodic systems. All of these concepts are related to properties of the solution

operator for Equation (3.1).

Let us suppose the axioms in either Section 2 or 4 are satisfied. Suppose $f: \mathbb{R} \times \mathscr{B} \to \mathbb{R}^n$ is continuous, if x is a solution through (σ,ϕ) of the equation

$$\dot{x}(t) = f(t,x_t) \tag{5.1}$$

defined on a maximal interval of existence $[\sigma,\sigma+\gamma_\phi)$, then we can define

$$T(t,\sigma)\phi = x_t(\sigma,\phi), \qquad t \in [\sigma,\sigma+\gamma_\phi).$$

For ease in exposition, let us suppose that f satisfies enough conditions to ensure that $T(t,\sigma)\phi$ is continuous in all variables and $\gamma_\phi = \infty$ for all ϕ. Then

$$T(t,\sigma): \mathscr{B} \to \mathscr{B}, \qquad t \geq \sigma.$$

<u>Definition 5.1</u>. If B is a bounded set in a Banach space X, the <u>Kuratowskii measure of noncompactness of</u> B, $\alpha(B)$, is defined as

$$\alpha(B) = \inf\{d > 0 : B \text{ can be covered by a finite number of sets of diameter } d\}.$$

<u>Definition 5.2</u>. If X is a Banach space and $T: X \to X$ is a continuous mapping, then we say T is a <u>conditional α-contraction</u> if there is a $k \in [0,1)$ such that $\alpha(TB) \leq k\alpha(B)$ for all bounded sets $B \subseteq X$ such that TB is bounded. If T is a conditional

α-contraction and takes bounded sets into bounded sets, then T is said to be an <u>α-contraction</u>. For a linear continuous operator T,

$$\alpha(T) \overset{\text{def}}{=} \inf\{k\colon \alpha(TB) \leq k\alpha(B) \text{ for all bounded } B \subseteq X\}.$$

<u>Definition 5.3</u>. If X is a Banach space and $U(t,\sigma)\colon X \to X$ is defined for $t \geq \sigma$, then $U(t,\sigma)$ is said to be <u>conditionally completely continuous</u> if $U(t,\sigma)\phi$ is continuous in t,σ,ϕ and for any bounded set $B \subseteq X$, there is a compact set $B^* \subseteq X$ such that $U(\tau,\sigma)\phi \in B$ for $\tau \in [\sigma,t]$ implies $U(t,\sigma) \in B^*$. If $U(t,\sigma)$ is conditionally completely continuous and for any bounded set $B \subseteq X$ and any compact set $J \subseteq [\sigma,\infty)$, there is a bounded set $B_0 \subseteq X$ such that $U(t,\sigma)B \subseteq B_0$ for $t \in J$, then $U(t,\sigma)$ is completely continuous in the usual sense.

The following result is very easy to prove, but at the same time, very important (for a proof, see [13]).

<u>Theorem 5.1</u>. <u>If the axioms in either Section 2 or 4 are satisfied</u>, f <u>is completely continuous</u>, $S(t)$ <u>is defined in Equation (3.2) and</u> $U(t,\sigma)$ <u>is defined by</u>

$$T(t,\sigma)\phi = S(t-\sigma)\phi + U(t,\sigma)\phi$$

<u>then</u> $U(t,\sigma)$ <u>is conditional completely continuous</u>. <u>Furthermore, for any bounded set</u> $B \subseteq X$ <u>for which</u> $T(s,\sigma)B$ <u>is uniformly bounded for</u> $\sigma \leq s \leq t$, <u>then</u>

(5.3) $$\alpha(T(t,\sigma)B) = \alpha(S(t-\sigma)B) \leq \alpha(S(t-\sigma))\alpha(B).$$

To estimate $\alpha(S(t))$, we need the following lemma whose proof is an easy consequence of the fact that any mapping from one finite dimensional space to another is completely continuous. To state the result, we need

Axiom (γ_1): *All constant functions belong to* $\mathscr{B}$.

Lemma 5.1. *If Axiom* (γ_1) *is satisfied, then*

(5.4)
$$\alpha(S(t) = \alpha(S_0(t)) \leq |S_0(t)|$$
$$S_0(t) = S|\mathscr{B}_0, \quad \mathscr{B}_0 = \{\phi \in \mathscr{B}: \phi(0) = 0\}.$$

The norm of the operator $S_0(t)$ in Lemma 5.1 can be estimated by

Lemma 5.2. *For any* $\phi \in \mathscr{B}$,

$$|S_0(t)\phi| = |\tau^t\phi|_t$$

where τ^t *and* $|\cdot|_t$ *are defined in Section 2.*

Theorem 5.1 has several important implications. First, we need some definitions.

Definition 5.4. The solution $x = 0$ of Equation (5.1) is *stable in* $\mathscr{B}$ if, for any $\sigma \geq 0$, $\varepsilon > 0$ there is a $\delta > 0$ such that

$|\phi|_{\mathscr{B}} < \delta$ implies $|x_t(\sigma,\phi)|_{\mathscr{B}} < \varepsilon$ for $t \geq \sigma$, where $x(\sigma,\phi)$ is the solution of Equation (5.1) through (σ,ϕ). If, in addition to the stability, for any $\sigma \geq 0$, there is a constant $\delta_0 > 0$ such that $|\phi|_{\mathscr{B}} < \delta_0$ implies $|x_t(\sigma,\phi)|_{\mathscr{B}} \to 0$ as $t \to \infty$, then $x = 0$ is said to be asymptotically stable in $\mathscr{B}$. The solution $u = 0$ is said to be <u>uniformly stable</u> if δ is independent of σ. It is said to be <u>uniformly asymptotically stable</u> if δ, δ_0 above can be chosen independent of σ and, for any $\varepsilon > 0$ there is a $T > 0$ such that $|\phi|_{\mathscr{B}} < \delta_0$ implies $|x_t(\sigma,\phi)| < \varepsilon$ for $t \geq \sigma + T$ for all $\sigma \geq 0$.

<u>Definition 5.5</u>. In Definition 5.4, if the $\mathscr{B}$-norm $|x_t(\sigma,\phi)|$ is replaced by the $\mathbb{R}^n$-norm $|x(t,\sigma,\phi)|$, then we say the stability is in $\mathbb{R}^n$.

The first result is a consequence of Theorem 5.1 and general properties of α-contraction mappings proved by Cooperman [8], Izé and dos Reis [21].

<u>Theorem 5.2</u>. <u>Suppose the Axioms in either Section 2 or 4 are satisfied. Consider Equation (5.1) where</u> $f(t,\phi)$ <u>is either independent of</u> t <u>(autonomous systems) or periodic in</u> t <u>(periodic systems) and</u> f <u>is completely continuous. If</u>

<u>Axiom (γ_2)</u>: <u>There is a</u> $t_0 > 0$ <u>such that</u> $\alpha(S(t_0)) < 1$

<u>then the zero solution of Equation (5.1) is asymptotically stable if and only if it is uniformly asymptotically stable</u>.

The Axiom (γ_2) implies $T(t_0+\sigma,\sigma)$ is an α-contraction. Cooperman [8] has very general results for α-contractions concerning the existence of a maximal compact invariant set as well as the stability and global attractivity properties of this set.

The next result relates stability in $\mathbb{R}^n$ to stability in $\mathscr{B}$. The proof is elementary and may be found in Hale and Kato [13].

Theorem 5.3. *Suppose Axioms* (α_0)-(α_5), *Axiom* (γ_1) *and*

(γ_3)
$$|S_0(t)| \leq M_1, \quad t \geq 0.$$

and f *in Equation* (5.1) *is completely continuous. Then (uniform) stability in* $\mathbb{R}^n$ *is equivalent to (uniform) stability in* $\mathscr{B}$. *If, in addition,*

(γ_4)
$$|S_0(t)\phi| \to 0 \quad \text{as} \quad t \to \infty \quad \text{for each} \quad \phi \in \mathscr{B}$$

then asymptotic stability in $\mathbb{R}^n$ *is equivalent to asymptotic stability in* $\mathscr{B}$. *If Axiom* (γ_2) *is satisfied, then uniform asymptotic stability in* $\mathbb{R}^n$ *is equivalent to uniform asymptotic stability in* $\mathscr{B}$.

Axiom (γ_4) corresponds in the literature to saying that the space $\mathscr{B}$ has the property of fading memory. Axiom (γ_2) is much stronger than Axiom (γ_4). The conclusions are also stronger. With Axiom (γ_2), one can obtain results for Equation (5.1) which have some uniformity with respect to initial data ϕ whereas this is generally not possible with Axiom (γ_2).

Our next result concerns precompactness of bounded orbits.

Theorem 5.4. *If Axioms* (α_0)-(α_6), *Axiom* (γ_4) *and*

Axiom (γ_5): *If* $\{\hat{\phi}^k\} \subseteq \hat{\mathscr{B}}$ *is uniformly bounded and converges to* $\hat{\phi}$ *uniformly on compact sets in* I, *then* $\hat{\phi} \in \hat{\mathscr{B}}$ *and* $|\phi-\phi^k|_{\mathscr{B}} \to 0$ *as* $k \to \infty$,

are satisfied, $x(\sigma,\phi)$ *is the solution of Equation* (5.1) *through* (σ,ϕ) *and* f *is completely continuous, then the set* $\{x_t(\sigma,\phi),\ t \geq 0\}$ *being bounded in* $\mathscr{B}$ *implies it is precompact in* $\mathscr{B}$.

The same conclusion is valid if Axioms (α_0)-(α_6), Axiom (γ_1) and

Axiom (γ_6): *There is a* $t_0 > 0$ *such that* $|S_0(t_0)| < 1$

are satisfied.

Hino [17] proved the first part of Theorem 5.4. It should be noted that Axiom (γ_5) implies all bounded continuous functions belong to $\mathscr{B}$ (see [13]). Hino [17] also has discussed stability and uniform asymptotic stability for the zero solution of Equation (5.1) under Axiom (γ_5) and the method of Liapunov functions. For the uniform asymptotic stability, he also imposed another hypothesis [17, p. 55] which is equivalent to Axiom (γ_6). Some results on integral manifolds have been given by Naito [26].

The statement of Theorem 5.4 using Axiom (γ_6) may be found in Hale and Kato [13]. With this hypothesis, they also prove much more general results on the compactness of orbits with initial data from a compact set [13, Theorem 3.1].

These compactness results have implications in the theory of the asymptotic behavior of the solutions of almost periodic systems or, more generally, systems with a vector field which has a compact hull. Similar results had been previously obtained by Hino [15]. For a more general presentation using skew-product flows, see Palmer [30].

Schumacher [33] has recently given an interesting axiomatic treatment of dynamical systems with memory. Under certain conditions, he has given necessary and sufficient conditions for an orbit to be precompact. Related results have been previously obtained by Kappel [22].

One of the crucial axioms of Schumacher [33] is that the norm in $\mathscr{B}$ has a monotonicity property with respect to the pointwise norm in $\mathbb{R}^n$. More specifically, if $\phi, \psi \in \mathscr{B}$, $|\phi(\theta)| \leq |\psi(\theta)|$ for $\theta \in I$, then $|\phi|_{\mathscr{B}} \leq |\psi|_{\mathscr{B}}$. This type of axiom was previously used by Hino [15] in the study of asymptotic behavior of almost periodic systems.

Results on the invariance of ω-limit sets in autonomous equations are also contained in Hale and Kato [13], Schumacher [33]. We state only one result from [13].

Theorem 5.5. Suppose $\mathscr{B}$ *satisfies Axioms* (α_0)-(α_6), (γ_1), (γ_6). *If* f *is independent of* t *and completely continuous, then the* ω-*limit set of any solution* x *of Equation* (5.1) *bounded for* $t \geq 0$ *is nonempty, compact and connected. If, in addition,*

(γ_7) *If* $\{\phi^k\}$ *converges to* ϕ *uniformly on compact subsets of* I *and if* $\{\phi^k\}$ *is Cauchy in* $\mathscr{B}$, *then* $\phi \in \mathscr{B}$ *and* $\phi^k \to \phi$ *in* $\mathscr{B}$

then the ω-limit set is invariant.

For the examples considered before, it is easy to see when all axioms above are satisfied. For the space $L^p(\mu_g) \times \mathbb{R}^n$, $\mu_g(E) = \int_E g$, $E \subseteq I$, Axiom (γ_1) implies $\int_{-\infty}^0 g < \infty$,

$$(5.2) \qquad |S_0(t)| = \sup_{s \leq 0} \left[\frac{g(s-t)}{g(s)}\right]^{1/p}$$

If g is nondecreasing then Axiom (γ_3) is satisfied. The fading memory Axiom (γ_4) is satisfied since $\int_{-\infty}^0 g < \infty$. If there is a $t_0 > 0$ such that $|S_0(t_0)| < 1$, then Axiom (γ_6) is satisfied. This is much stronger than Axiom (γ_4) and requires that $g(\theta)$ decrease exponentially as $\theta \to -\infty$.

For the space C_γ, Axiom (γ_1) implies $\gamma \geq 0$ and

$$(5.3) \qquad |S_0(t)| = e^{-\gamma t}, \quad t \geq 0.$$

For this space, Axioms (γ_4) and (γ_6) are equivalent and are satisfied only if $\gamma > 0$. There is no fading memory unless $\gamma > 0$.

6. Linear autonomous equations. For linear autonomous equations, an extensive theory can be developed without imposing many hypotheses on the space-especially the axiom about $|S_0(t_0)| < 1$. On the other hand, Axiom (γ_7) seems to play a more important role than in the previous section. Consequently, in this section we discuss a theory of linear systems using only the axioms that seem necessary and follow closely the paper of Naito [28] but in less generality. We restate the hypotheses explicitly in the form that is needed.

Axiom (δ_0): *There is a constant* K *such that* $|\phi(0)| \leq K|\phi|$ *for all* $\phi \in \mathscr{B}$.

Axiom (δ_1): $x_t \in \mathscr{B}$ *for all* $x \in F_A$, $t \in [0,A]$, x_t *is continuous in* t.

Axiom (δ_2): *For any* $x \in F_\infty$, $t \in [0,\infty)$,

$$|x_t|_{\mathscr{B}} \leq K(t)\sup_{0\leq s\leq t}|x(s)| + M(t)|x_0|$$

where K,M *are continuous*, $M(t+s) \leq M(t)M(s)$.

Axiom (δ_3): *If* $\{\phi^k\}$ *converges to* ϕ *uniformly on compact sets of* I *and if* $\{\phi^k\}$ *is a Cauchy sequence in* $\mathscr{B}$, *then* $\phi \in \mathscr{B}$ *and* $\phi^k \to \phi$ *in* $\mathscr{B}$.

Axioms (α_0)-(α_6) and (γ_7) imply Axioms (β_0)-(β_3).

Suppose $L: \mathscr{B} \to \mathbb{R}^n$ is a continuous linear operator and consider the autonomous linear equation

$$\dot{x}(t) = Lx_t. \tag{6.1}$$

Since this equation is autonomous, we take the initial time to be zero. The results in [13] or [28] imply that the solution operator $T(t)$, $t \geq 0$, defined by $T(t)\phi = x_t(\phi)$ for $\phi \in \mathscr{B}$, is a strongly continuous semigroup of bounded linear operators on $\mathscr{B}$. Let A be the infinitesimal generator of $T(t)$.

The specific form of the infinitesimal generator is not known. However, it is surprising how much of the general theory of linear systems is independent of this fact. In the general theory of semigroups of transformations acting on an arbitrary Banach space, one could not possibly do very much without having specific information about the infinitesimal generator. The semigroup generated by a retarded functional differential equation is very special and $T(t)\phi$ for $\phi \in \mathscr{B}$, $t \geq 0$ is the "restriction" of some function defined from $I \cup [0,t]$ to $\mathbb{R}^n$. In fact, there is a function $x(t,\phi)$ such that

$$(T(t)\phi)(\theta) = x(t+\theta,\phi) \quad \text{for} \quad t \geq 0, \ \theta \in I.$$

This is the basic reason that a comprehensive theory can be developed without knowing too much about the infinitesimal generator.

In this section, we state some results on the spectrum $\sigma(A)$ of A, the point spectrum $P_\sigma(A)$ of A and the resolvent set $\rho(A)$ of A. The first observation concerns the point spectrum (see [13]). A proof is indicated to illustrate the remarks in the previous paragraph.

Theorem 6.1. *If A is the infinitesimal generator of $T(t)$, then $P_\sigma(A)$ is the set of λ for which there exists a $b \neq 0$, $b \in \mathbb{C}^n$, such that $e^{\lambda \cdot} b \in \mathscr{B}$ and*

$$\text{(6.2)} \qquad \det \Delta(\lambda) = 0, \qquad \Delta(\lambda) = \lambda I - L(e^{\lambda \cdot} I).$$

Proof: If $\lambda \in P_\sigma(A)$, then there exists a $\phi \neq 0$, $\phi \in D(A)$, the domain of A, such that $A\phi = \lambda\phi$. Thus,

$$\frac{d}{dt} T(t)\phi = T(t)A\phi = \lambda T(t)\phi .$$

Since $T(t)$ is a strongly continuous semigroup, this implies $T(t)\phi = e^{\lambda t}\phi$, $t \geq 0$ (in particular, $e^{\lambda t} \in P_\sigma(T(t))$). Since $(T(t)\phi)(\theta) = x(t+\theta,\phi) = T(t+\theta)\phi(0)$ for all $t + \theta \geq 0$, $\theta \in I$, this implies

$$e^{\lambda t}\phi(\theta) = e^{\lambda(t+\theta)}\phi(0), \qquad t + \theta \geq 0, \quad \theta \in I.$$

Thus, $\phi(\theta) = e^{\lambda\theta}\phi(0)$ for all $\theta \in I$. Since $\phi \neq 0$, this implies $\phi(0) \neq 0$. Also, since the function $e^{\lambda(t+\theta)}\phi(0) = x_t(\phi)(\theta)$ is a solution of Equation (6.1), it follows that $\Delta(\lambda)\phi(0) = 0$. Since $\phi(0) \neq 0$, we must have $\det \Delta(\lambda) = 0$.

The converse is much easier and will not be proved.

If $S(t)$ is the semigroup defined in Equation (3.2) by the differential equation $\dot{x}(t) = 0$ in $\mathscr{B}$ and

$$(6.3) \qquad T(t)\phi = S(t)\phi + U(t)\phi, \qquad \phi \in \mathscr{B},$$

then the Representation Theorem 5.1 says that $U(t)$ is completely continuous and

$$(6.4) \qquad \alpha(T(t)) = \alpha(S(t)) = \alpha(S_0(t)) \leq |S_0(t)|$$

where α is the Kuratowskii measure of noncompactness, $S_0(t) = S(t)\,|\,\mathscr{B}_0$, $\mathscr{B}_0 = \{\phi \in \mathscr{B}: \phi(0) = 0\}$.

For any bounded linear operator T, let $r_e(T)$ be the smallest closed disk in the complex plane with center zero which contains the essential spectrum of T. It is shown in [29] that $r_e(T) = \lim_{n\to\infty} \alpha(T^n)^{1/n}$. Since $\alpha(T(t)) = \alpha(S_0(t))$ it also follows from [29] that $r_e(T(t)) = r_e(S_0(t))$. To estimate $\alpha(S_0(t))$, observe that

$$(S_0(t+\tau)) = \alpha(S_0(t)S_0(\tau)) \leq \alpha(S_0(t))\alpha(S_0(\tau))$$

for all $t, \tau \geq 0$ and so there is a $\beta \in [-\infty,+\infty)$ such that

$$\beta = \lim_{t\to\infty} \frac{\log \alpha(S_0(t))}{t} = \inf_{t>0} \frac{\log \alpha(S_0(t))}{t} . \tag{6.5}$$

Thus,

$$r_e(T(t)) = r_e(S_0(t)) = e^{\beta t}, \qquad t > 0. \tag{6.6}$$

If μ is in $\sigma(T(t))$ and $|\lambda| > \exp(\beta t)$, then μ is a normal eigenvalue of $T(t)$ and $\mu = \exp(\lambda t)$ for some $\lambda \in P_\sigma(A)$. Thus, one obtains (see [28])

Theorem 6.2. The spectral radius $r_\sigma(T(t))$ *of* $T(t)$ *is given by*

$$r_\sigma(T(t)) = e^{t\alpha}, \qquad t \geq 0$$

where

$$\alpha = \alpha_L = \max\{\beta, \sup\{\text{Re } \lambda : \lambda \in P_\sigma(A)\}\}.$$

Also, for any $\varepsilon > 0$, *there is a* $c(\varepsilon) > 0$ *such that*

$$|T(t)| \leq c(\varepsilon)e^{(\alpha+\varepsilon)t}, \quad t \geq 0$$

Theorem 4.2 for more special spaces $\mathscr{B}$ was stated and proved in essentially the same way in Hale [13] (see, also Barbu and Grossman [36]).

A more difficult result from [28] is the following one.

Theorem 6.3. *Any point* λ *such that* $\operatorname{Re}\lambda > \beta$ *is a normal point of* A; *that is,* λ *does not lie in the essential spectrum of* A.

The proof of this result is very complicated and uses the following nontrivial facts. If $\operatorname{Re}\lambda > \beta$, then $e^{\lambda\cdot}b \in \mathscr{B}$ for every n-vector b and is analytic in λ. Also, if B is the infinitesimal generator of $S(t)$, A is the infinitesimal generator of $T(t), R(\lambda;B), R(\lambda;A)$ are the corresponding resolvent operators, then

$$R(\lambda;A)\phi = e^{\lambda\cdot}\Delta(\lambda)^{-1}L(R(\lambda;B)\phi) + R(\lambda;B)\phi, \qquad \phi \in \mathscr{B}$$

for all $\lambda \in D = \{\lambda\colon \operatorname{Re}\lambda > \beta, \det\Delta(\lambda) \neq 0$ and $\lambda \neq 0\}$.

With this result, for any $\lambda \in \sigma(A)$, $\operatorname{Re}\lambda > \beta$, the space may be decomposed as

$$\mathscr{B} = \mathfrak{N}(A-\lambda I)^k \oplus \mathscr{R}(A-\lambda I)^k$$

for some integer k, where $\mathfrak{N}(A-\lambda I)^k$ is of finite dimension, invariant under A and $\mathscr{R}(A-\lambda I)^k$ is a closed subspace of $\mathscr{B}$. The subspace $\mathfrak{N}(A-\lambda I)^k$ can be explicitly computed (see [13]). This result implies that the subspaces $P \overset{\text{def}}{=} \mathfrak{N}(A-\lambda I)^k$, $Q = \mathfrak{N}(A-\lambda I)^k$ are invariant under $T(t)$ and the spectrum of $T(t)|Q$ is the spectrum of $T(t)\setminus\{e^{\lambda t}\}$. The importance of this fact in finite delays is well known (see Hale [12]).

A more explicit representation of this decomposition is needed, but is unavailable at the present time. More specific information is needed about A and its adjoint A^*. A step in the right direction is the following theorem of Naito [28] generalizing a corresponding theorem of Stech [34].

Theorem 6.4. The domain $\mathscr{D}(A^*)$ of A^* is independent of L,

$$\mathscr{D}(A^*) = \mathscr{D}(B^*)$$

where B is the infinitesimal generator of $S(t)$.

It remains to give the variation of constants formula. Let

$$\mu = \lim_{t\to\infty} \frac{\log M(t)}{t} = \inf_{t>0} \frac{\log M(t)}{t}$$

$c > \max(\alpha_L, \mu)$, where α_L is defined in Theorem 6.2, and define

$$X(t) = \begin{cases} \lim_{T\to\infty} \frac{1}{2\pi i} \int_{c-iT}^{c+iT} e^{\lambda t}\Delta(\lambda)^{-1} d\lambda , & t > 0 \\ I , & t = 0 . \end{cases}$$

The matrix $X(t)$ is continuous and is called the <u>fundamental matrix</u> for Equation (6.1). By a nontrivial application of the Laplace transform, Naito [28] has proved the following variation of contants formula.

<u>Theorem 6.5</u>. <u>If</u> f <u>is a continuous function from</u> $[0,\infty)$ <u>to</u> $\mathbb{R}^n$, <u>then the solution</u> $x(t,\phi,f)$, $x_0(\cdot,\phi,f) = \phi$, <u>of the equation</u>

$$\dot{x}(t) = Lx_t + f(t)$$

<u>is given by</u>

$$(6.7) \qquad \begin{aligned} x(t,\phi,f) &= x(t,\phi,0) + \int_0^t X(t-\tau)f(\tau)d\tau , \quad t \geq 0 \\ x(t,\phi,0) &= \phi(0) + \int_0^t X(t-\tau)L(S(\tau)\phi)d\tau , \quad t \geq 0 \end{aligned}$$

<u>where</u> $S(t)$ <u>is the semigroup generated by the equation</u> $\dot{x}(t) = 0$ <u>in</u> $\mathcal{B}$.

One more important result is needed for linear nonautonomous systems. For any $\theta \in I$, one can write Equation (6.7) as

$$[x_t(\phi,f)](\theta) = [T(t)\phi](\theta) + \int_0^t X_{t-\tau}(\theta)f(\tau)d\tau$$

since $X(t) = 0$ for $t < 0$. This equation may be written in a symbolic way as

$$x_t(\phi,f) = T(t)\phi + \int_0^t X_{t-\tau}f(\tau)d\tau. \tag{6.8}$$

In the case of finite delays and the space $\mathcal{B}$ being continuous functions, the following important result is known. If $\mathcal{B} = P \oplus Q$ where P corresponds to the generalized eigenspace of some set Λ of elements of the point spectrum of $T(t)$, then $T(t)P \subseteq P$, $T(t)Q \subseteq Q$. Also, the Equation (4.8) can be written as

$$\begin{aligned} x_t^P(\phi,f) &= T(t)\phi^P + \int_0^t X_{t-\tau}^P f(\tau)d\tau \\ x_t^Q(\phi,f) &= T(t)\phi^Q + \int_0^t X_{t-\tau}^Q f(\tau)d\tau \end{aligned} \tag{6.9}$$

where $\phi = \phi^P + \phi^Q$, $\phi^P \in P$, $\phi^Q \in Q$, etc. The importance of Formula (6.9) in the qualitative theory has been known for a long time (see [12]).

Does Formula (6.9) hold for the general space $\mathcal{B}$ satisfying Axioms (δ_0)-(δ_3)? If $\mathcal{B} = L^p(\mu_g) \times \mathbb{R}^n$, it is known that it does and the formulas have actually been used to prove the Hopf bifurcation theorem for nonlinear autonomous equations (see Lima [25]) under the additional hypothesis that $|S_0(t)| \to 0$ as $t \to \infty$. In [25], the specific form of the infinitesimal generator was used.

In Formula (6.6), $r_e(T(t)) = r_e(S_0(t)) \leq |S_0(t)|$. If $|S_0(t)| \to 0$ as $t \to \infty$, this implies there is a $t_0 > 0$ such that

$r_e(T(t_0)) < 1$ and $T(t_0)$ is an α-contraction. In this case, with Formula (6.9), one should be able to obtain all the usual results on behavior near an equilibrium point as in [12]; for example, the saddle point property, Hopf bifurcation, nonlinear oscillations in equations with small parameters, asymptotic behavior, etc. For the space $L^p(\mu_g) \times \mathbb{R}^n$ this means

$$|S_0(t)| = \sup_{s \leq 0} \left[\frac{g(s-t)}{g(s)}\right]^{1/p} \to 0 \quad \text{as} \quad t \to \infty.$$

This is not necessarily satisfied if g satisfies only the property of fading memory; that is, $S_0(t)\phi \to 0$ as $t \to \infty$ for all $\phi \in \mathscr{B}_0$. For the space C_γ,

$$|S_0(t)| = e^{-\gamma t} \to 0 \quad \text{as} \quad t \to \infty$$

if $\gamma > 0$.

7. Summary of the axioms. In this section we list a set of axioms which are sufficient to imply that every result mentioned in the previous pages is valid. Recall that

$$|\phi|_\beta = \inf_{\hat{\eta} \in \hat{\mathscr{B}}} \{ \inf_{\hat{\psi} \in \hat{\mathscr{B}}} \{ |\hat{\psi}|_{\hat{\mathscr{B}}} : \{ \hat{\psi}(\theta) = \hat{\eta}(\theta),\ \theta \in I \backslash [-\beta, 0] \} : \eta = \phi \}$$

$$|\phi|_{(\beta)} = \inf_{\hat{\eta} \in \hat{\mathscr{B}}} \{ \inf_{\hat{\psi} \in \hat{\mathscr{B}}} \{ |\hat{\psi}|_{\hat{\mathscr{B}}} : \{ \hat{\psi}(\theta) = \hat{\eta}(\theta),\ \theta \in [-\beta, 0] \} : \eta = \phi \}.$$

Also, define

$$S(t): \mathscr{B} \to \mathscr{B}$$

$$[S(t)\phi](\theta) = \begin{cases} \phi(0) & t + \theta \geq 0 \\ \phi(t+\theta), & t + \theta < 0 \end{cases}$$

$$\mathscr{B}_0 = \{\phi \in \mathscr{B}: \phi(0) = 0\}$$

$$S_0(t) = S(t) \mid \mathscr{B}_0$$

Our axioms are:

(ε_0) There is a constant K such that $|\phi(0)| \leq K|\phi|$

(ε_1) If $\phi = 0$ in $\mathscr{B}$, then $|\tau^\beta\phi|_\beta = 0, \tau^\beta\phi = \{\psi \in \mathscr{B}: |\phi-\psi|_\beta = 0\}$.

(ε_2) $|\phi|_{\mathscr{B}} \leq |\phi|_\beta + |\phi|_{(-\beta)}$.

(ε_3) There is a constant K_1 such that $|\phi|_{(\beta)} \leq K_1 \sup_{[-\beta,0]} |\phi(\theta)|$.

(ε_4) All constant functions belong to $\mathscr{B}$.

(ε_5) For each $-\tau \in I$, the set of continuous functions from I to $\mathbb{R}^n$ with support in $[-\tau, 0]$ belongs to $\mathscr{B}$.

(ε_6) $S(t): \mathscr{B} \to \mathscr{B}$, $t \geq 0$, is a strongly continuous semigroup of bounded linear operators.

(ε_7) There is a $t_0 > 0$ such that $|S_0(t_0)| < 1$.

(ε_8) If $|\hat{\phi}^k| \subseteq \hat{\mathscr{B}}$ converges to $\hat{\phi}$ uniformly on compact sets in I and if $\{\phi^k\}$ is a Cauchy sequence in $\mathscr{B}$, then $\phi \in \mathscr{B}$ and $\phi^k \to \phi$ in $\mathscr{B}$.

It can be verified that Axioms (ε_0)-(ε_6) imply Axioms (α_0)-(α_6) and Axioms (β_0)-(β_5). Also, Axioms (α_0)-(α_8) imply Axioms (δ_0)-(δ_3) and Axioms (γ_1)-(γ_4), (γ_6), (γ_7).

Some of the results on compactness of orbits could be proved under less restrictive hypotheses that Axioms (ε_7), (ε_8). More specifically, one could replace these by the following:

(ε_9) $S_0(t)\phi \to 0$ <u>as</u> $t \to \infty$ <u>for every</u> $\phi \in \mathscr{B}$.

(ε_{10}) <u>If</u> $\{\hat{\phi}^k\} \subseteq \mathscr{B}$ <u>is uniformly bounded and converges to</u> <u>uniformly on compact subsets of</u> I, <u>then</u> $\hat{\phi} \in \hat{\mathscr{B}}$ and $\hat{\phi}$ $|\phi-\phi^k|_{\mathscr{B}} \to 0$ <u>as</u> $k \to \infty$.

Axioms (ε_9), (ε_{10}) have also been sufficient to discuss the existence of almost periodic solutions of functional differential equations (see Hino [19,20]). For the existence of almost periodic solutions based on the existence of solutions with certain stability properties, Hino [18] used Axioms (ε_9), (ε_{10}) as well as the monotonicity property

(ε_{11}) <u>If</u> $\hat{\phi}, \hat{\psi} \in \hat{\mathscr{B}}$, $|\hat{\phi}(\theta)| \le |\hat{\psi}(\theta)|$, $\theta \in I$, <u>then</u> $|\phi|_{\mathscr{B}} \le |\psi|_{\mathscr{B}}$.

As remarked earlier Schumacher [33] has found some other interesting implications of Axiom (ε_{11}) in the theory of dynamical systems with memory.

In the space $L^p(\mu_g) \times \mathbb{R}^n$, $p \ge 1$, $\mu_g(E) = \int_E g$ with g nondecreasing, $\int_I g < \infty$, Axioms (ε_0)-(ε_6), (ε_8)-(ε_{11}) are satisfied. If, in addition, $\sup_{s\le 0}[g(s-t)/g(s)] \to 0$ as $t \to \infty$, then Axiom (ε_7) is satisfied. In the space

$$C_\gamma = \{\phi\colon I \to \mathbb{R}^n \text{ continuous, } e^{\gamma\theta}\phi(\theta) \to \text{limit as } \theta \to -r\}$$

all hypotheses are satisfied for $\gamma > 0$.

REFERENCES

[1] Banks, H.T. and J. Burns, An abstract framework for approximate solutions to optimal control problems governed by hereditary systems. Proc. Int. Conf. Diff. Eqns. Univ. Southern California, 1974, Academic Press, 1975.

[2] Banks, H.T. and M. Jacobs, An attainable sets approach to optimal control of functional differential equations with function space side conditions. J. Differential Equations, 13(1973), 127-149.

[3] Barbu, V. and S. Grossman, Asymptotic behavior of linear integral differential systems. Trans. Am. Math. Soc., 173(1972), 277-289.

[4] Coleman, B.D. and V.J. Mizel, Norms and semigroups in the theory of fading memory. Arch. Rat. Mech. Ana., 23(1966), 87-123.

[5] Coleman, B.D. and V.J. Mizel, On the general theory of fading memory. Arch. Rat. Mech. Ana., 29(1968), 18-31.

[6] Coleman, B.D. and V.J. Mizel, On the stability of solutions of functional differential equations. Arch. Rat. Mech. Ana., 30(1968), 173-196.

[7] Coleman, B.D. and D. Owen, On the initial value problem for a class of functional differential equations. Arch. Rat. Mech. Ana., 55(1974), 275-299.

[8] Cooperman, G., α-condensing maps and dissipative systems. Ph.D. Thesis, Brown University, Providence, Rhode Island, 1978.

[9] Delfour, M.C. and S.K. Mitter, Hereditary differential systems with constant delays, I. General case. J. Differential Equations, 12(1972), 213-255.

[10] Hale, J.K., Dynamical systems and stability. J. Math. Ana. Appl., 26(1969), 39-59.

[11] Hale, J.K., Functional differential equations with infinite delay. J. Math. Ana. Appl., 48(1974), 276-283.

[12] Hale, J.K., Theory of Functional Differential Equations. Appl. Math. Sci., Vol. 3, Springer-Verlag, 1977.

[13] Hale, J.K. and J. Kato, Phase space for retarded equations with infinite delay. Funk. Ekv., 21(1978), 11-41.

[14] Hale, J.K. and O. Lopes, Fixed point theorems and dissipative processes. J. Differential Equations, 13(1973), 391-402.

[15] Hino, Y., Asymptotic behavior of solutions of some functional differential equations. Tôhoku Math. J., 22(1970), 98-108.

[16] Hino, Y., Continuous dependence for functional differential equations. Tôhoku Math. J., 23(1971), 565-571.

[17] Hino, Y., On the stability of the solution of some functional differential equations. Funk. Ekv., 14(1971), 47-60.

[18] Hino, Y., Stability and existence of almost periodic solutions of some functional differential equations. Tôhoku Math. J. 28(1976), 389-409.

[19] Hino, Y., Favard's separation theorem in functional differential equations with infinite retardations. Tôhoku Math. J., 30(1978), 1-12.

[20] Hino, Y., Almost periodic solutions of functional differential equations with infinite retardation. Preprint.

[21] Izé, A. and J.G. dos Reis, Contributions to stability of neutral functional differential equations . J. Differential Equations, 29(1978), 58-65.

[22] Kappel, F., The invariance of limit sets for autonomous functional differential equations. SIAM J. Appl. Math., 19(1970), 408-419.

[23] Kappel, F. and N. Schappacher , Nonlinear functional differential equations and abstract integral equations. Proc. Royal Soc. Edinburgh, Ser. A. To appear.

[24] Leitman, M.J. and V.J. Mizel, On fading memory spaces and hereditary integral equations. Arch. Rat. Mech. Ana., 55(1974), 18-51.

[25] Lima, P., Hopf bifurcation in equations with infinite delays. Ph.D. Thesis, Brown University, Providence, Rhode Island, 1977.

[26] Naito, T., Integral manifolds for linear functional differential equations on some Banach space. Funk. Ekv., 13(1970), 199-213.

[27] Naito, T., On autonomous linear functional differential equations with infinite retardations. J. Differential Eqns., 2(1976), 297-315.

[28] Naito, T., On linear autonomous retarded equations with an abstract space for infinite delay. J. Differential Equations. To appear.

[29] Nussbaum, R., The radius of the essential spectrum. Duke Math. J., 37(1970), 473-488.

[30] Palmer, J.W., Liapunov stability theory for nonautonomous functional differential equations. Ph.D. Thesis, Brown University, Providence, Rhode Island, 1978.

[31] Reber, D., Approximation and optimal control of linear hereditary systems. Ph.D. Thesis, Brown University, Providence, Rhode Island, 1978.

[32] Schumacher, K., Existence and continuous dependence for functional differential equations with unbounded delay. Arch. Rat. Mech. Ana., 67(1978), 315-335.

[33] Schumacher, K., Dynamical systems with memory on history spaces with monotonic seminorms. Preprints.

[34] Stech, H., Contribution to the theory of functional differential equations with infinite delay. J. Differential Equations, 27(1978), 421-443.

Author:

Jack K. Hale
Lefschetz Center for Dynamical Systems
Division of Applied Mathematics
Brown University
Providence, Rhode Island 02912

A DEGREE CONTINUATION THEOREM FOR A CLASS OF COMPACTLY PERTURBED DIFFERENTIABLE FREDHOLM MAPS OF INDEX O

Georg Hetzer

In [4-6] continuation theorems for operator equations

$$Lx = Nx,$$

where L is a linear Fredholm operator of index 0, and N is completely continuous or set-contractive, are used, to prove the existence of a periodic solution for some classes of nonautonomous neutral functional differential equations. In order to apply such an assertion to the "quasilinear" problem

$$(1) \quad \frac{d}{dt}D(t,u_t) = f(t,u_t),$$

where $D: \mathbb{R} \times C([-r,0],\mathbb{R}^n) \rightarrow \mathbb{R}^n$ and $f: \mathbb{R} \times C([-r,0],\mathbb{R}^n) \rightarrow \mathbb{R}^n$ are continuous ($r \in \mathbb{R}^+$), and ω-periodic concerning the first argument ($\omega \in \mathbb{R}\setminus\{0\}$), very restrictive assumptions in regard to D are necessary.

Here we describe a more adequate approach with respect to a special case of (1), which contains for example differential difference equations of the form

$$(2) \quad \sum_{0 \leq i \leq n} g_i(u(t-r_i))u'(t-r_i) = f(t,u(t-r_o),\ldots,u(t-r_n)),$$

where $0 = r_o < r_1 < \ldots < r_n$, $g_i: \mathbb{R} \rightarrow \mathbb{R}$ is twice continuously differentiable for $0 \leq i \leq n$, and $f: \mathbb{R}^{n+2} \rightarrow \mathbb{R}$ is continuous, and ω-periodic in the first argument. Applications of the here stated abstract results will appear elsewhere.

The typical abstract setting, suggested by Eq. (2), can be subsumed under the following hypotheses:

(H1) X, Y Banach spaces, $X \hookrightarrow Y$ completely continuous,

$\Omega \subseteq X$ open, bounded.

(H2) $F: cl_X(\Omega) \longrightarrow Y$ continuous, proper, bounded, $F|\Omega$ twice continuously Frechet-differentiable, $A: X \longrightarrow Y$ linear Fredholm operator of index 0, $\lambda A + F'(u)$ Fredholm operator for $\lambda \in \mathbb{R}^+$ and $u \in \Omega$.

(H3) $N: cl_Y(\Omega) \longrightarrow Y$ continuous.

Then, if $y \in Y$, we ask for $x \in cl_X(\Omega)$ with

(3) $Fx = Nx + y$.

The connection between (H1) - (H3) and Eq. (2) can be illustrated by the following simple

Example 1.

Set $n = 1$ in (2), $g_o = 1$, $r_1 = 1$, $g_1(u) = \exp(u^m)$ for some $m \in \mathbb{N}$, and $\omega = \nu_1/\nu_2$ with $\nu_1, \nu_2 \in \mathbb{N}$ and ν_1 odd. Further define:

$X = C^1_\omega(\mathbb{R},\mathbb{R})$, $Y = C_\omega(\mathbb{R},\mathbb{R})$,

$Fu(t) := u'(t) + \exp(u(t - 1)^m)u'(t - 1)$ for $u \in X$

$Nu(t) := f(t,u(t),u(t - 1))$ for $u \in X$

$Au(t) := u'(t)$ for $u \in X$,

then one can show, that (H1) - (H3) are satisfied for each $\Omega \subseteq X$, but that the map $u \longmapsto \exp(u(t - 1)) \cdot u'(t - 1)$ is not condensing on open balls with center 0.

We want to introduce a coincidence degree for Eq. (3), and need:

Lemma 1.

Assume that E, Z are Banach spaces and K is a compact subset of E. Denote by $C^\infty(K,Z)$ the set of continuous functions from K into Z, which are restrictions of C^{∞}-functions with domain containing K. Then $C^{\infty}(K,Z)$ is dense in $C(K,Z)$ with respect to the topology of uniform convergence.

Proof. Apply [1; chap. 10 § 4.-4 prop. 5] on $H := C^{\infty}(K,\mathbb{R})$.

Let now $y \in Y$ and $Fx \neq Nx + y$ for $x \in \partial_X(\Omega)$. Since N is a continuous map from the compact subset $cl_Y(\Omega)$ of Y into Y, F - N is proper as a map from $cl_X(\Omega)$ into Y, hence $\varepsilon := \frac{1}{2} \inf\{\|Fx - Nx - y\| \mid x \in \partial_X(\Omega)\} > 0$. Choose $\hat{N} \in C^{\infty}(cl_Y(\Omega),Y)$ according to Lemma 1, such that $\|Nx - \hat{N}x\| \leq \varepsilon$ for $x \in cl_Y(\Omega)$, then $\hat{N}$ is a C^{∞}-map from Ω into Y and $Fx \neq \hat{N}x + y$ for $x \in \partial_X(\Omega)$. Moreover let $P: X \longrightarrow X$ be a continuous projector on Ker(A), and $J: Ker(A) \longrightarrow Y$ be injective and linear with $Im(J) \cap Im(A) = \{0\}$, then $L := A + JP$ is a linear homeomorphism from X into Y. In order to apply the degree concept for C^2-Fredholm maps, stated in [2], on $G := FL^{-1} - \hat{N}L^{-1}$, we remark that G is clearly proper and, restricted to Ω, a C^2-function. Therefore it remains to show that

(i) $G|L(\Omega)$ is a Fredholm operator of index 0, and

(ii) $\lambda I - G'(v)$ is a Fredholm operator for each $v \in L(\Omega)$

and $\lambda \leq 0$. Now, if $u \in \Omega$, we know $ind(F'(u)) = 0$ because of the Fredholm property of $\lambda A + F'(u)$ for $\lambda \geq 0$, $ind(F'(u)) =$ $= ind(\lambda A + F'(u))$ for $\lambda \in \mathbb{R}^+$, and

$ind(\lambda A + F'(u)) = ind(A + \lambda^{-1}F'(u)) = ind(A) = 0$ for $\lambda \in \mathbb{R}^+$ sufficient large. Observing $ind(G'(v)) = ind(F'(L^{-1}v))$ for $v \in L(\Omega)$, we obtain (i). Obviously (ii) is satisfied, if $\lambda L - F'(u)$ is a Fredholm operator for $\lambda \leq 0$ and $u \in \Omega$, which directly follows from the Fredholm property of $\lambda A + F'(u)$ for $\lambda \geq 0$ and $u \in \Omega$. Since

$\deg(G,L(\Omega),y)$ is clearly independent of the choice of $\hat{N}$, we can define a coincidence degree in the sense of J. Mawhin ([3]) by

$$D_L([F,N],\Omega,y) := \deg(G,L(\Omega),y),$$

which possesses the solution property, the additivity and the homotopy invariance. We only state:

Lemma 2. (homotopy invariance)

Let (H1) be satisfied, $\Phi : cl_X(\Omega) \times [0,1] \to Y$ be continuous, proper, and bounded, $A: X \to Y$ be a linear Fredholm operator of index 0, and $\Psi : cl_Y(\Omega) \times [0,1] \to Y$ be continuous. Further suppose that $\Phi | \Omega \times (0,1) \in C^2(\Omega \times (0,1),Y)$, that $\Phi(\cdot,t)|\Omega \in C^2(\Omega,Y)$ for $t \in [0,1]$, and that $\lambda A + [\Phi(\cdot,t)]'(u)$ is a Fredholm operator for $\lambda \in \mathbb{R}^+$ and $u \in \Omega$. Finally let L be defined like above. Then we have for each $y \in Y$, satisfying $\Phi(x,t) \neq \Psi(x,t) + y$ for $x \in \partial_X(\Omega)$:
$D_L([\Phi(\cdot,0),\Psi(\cdot,0)],\Omega,y) = D_L([\Phi(\cdot,1),\Psi(\cdot,1)],\Omega,y)$.

Now we can prove:

Theorem 1.

Let (H1) - (H3) be satisfied, $0 \in \Omega$, which is convex, and $P,Q: Y \to Y$ be continuous projectors with $Im(P) = Ker(A)$, $Ker(Q) = Im(A)$ and $\Omega = P(\Omega) + (I - P)(\Omega)$. Suppose $Im(F) \subseteq Im(A)$, $0 \in F^{-1}(\{0\}) \subseteq Ker(A)$ and $Ker(F'(0)) \subseteq Ker(A)$. Moreover assume

(4) $F(u - (1 - t)Pu) \neq tNu$ for $(u,t) \in \partial_X(\Omega) \times (0,1)$

(5) $QNu \neq 0$ for $u \in \partial_X(\Omega) \cap Ker(A)$

(6) $\deg_B(QN/cl_X(\Omega) \cap Ker(A), \Omega \cap Ker(A), 0) \neq 0$,

where $\deg_B$ denotes the Brouwer degree.

Then there exists a $u \in cl_X(\Omega)$ with $Fu = Nu$.

Proof. Without loss of generality suppose $Fu \neq Nu$ for $u \in \partial_X(\Omega)$. Define Φ and Ψ by

$$\Phi(u,t) := F(u - (1 - t)Pu) \quad \text{for } (u,t) \in cl_X(\Omega) \times [0,1]$$

$$\Psi(u,t) := t(I - Q)Nu + QNu \quad \text{for } (u,t) \in cl_Y(\Omega) \times [0,1].$$

Further let $J: Ker(A) \rightarrow Im(Q)$ be injective and linear, and $L := A + JP$, then it is easy to see that Φ, Ψ, A and L satisfy the assumptions of Lemma 2. In particular $\Phi(u,t) = \Psi(u,t)$ for some $(u,t) \in cl_X(\Omega) \times [0,1)$ yields $F(u - (1 - t)Pu) = t(I - Q)Nu$ and $QNu = 0$, hence because of (4), (5) and $F^{-1}(\{0\}) \cap Ker(P) = \{0\}$ it follows $u \in \Omega$. Thus we have from Lemma 2:

$$D_L([F,N],\Omega,0) = D_L([F(I - P),QN],\Omega,0).$$

Since $0 \in \Omega$, Ω convex and $\Omega = P(\Omega) + (I - P)(\Omega)$ by assumption, the map $(u,t) \mapsto QN(Pu + t(I - P)u)$ is defined on $cl_Y(\Omega) \times [0,1]$ and continuous. Moreover it follows from

$F(I - P)u = QN(Pu + t(I - P)u)$ for some $(u,t) \in cl_X(\Omega) \times [0,1]$

that $F(I - P)u = 0$ and $QN(Pu + t(I - P)u) = 0$, hence $u \in Ker(A)$ because of $F^{-1}(\{0\}) \subseteq Ker(A)$ and $QNu = 0$, which yields $u \notin \partial_X(\Omega)$ according to (5). Therefore we can again apply Lemma 2 and obtain

$$D_L([F(I - P), QN],\Omega,0) = D_L([F(I - P), QNP],\Omega,0).$$

We calculate the right hand side of the equation. Set

$$\delta := \min\{\inf\{\|F(I - P)u - QNu\| \mid x \in \partial_X(\Omega)\},$$
$$\inf\{\|QNx\| \mid x \in \partial_X(\Omega) \cap Ker(A)\}\}/2$$

and choose $\vartheta \in C^\infty(cl_Y(P(\Omega)), Im(Q))$ according to Lemma 1 with $\|\vartheta u - QNu\| \leq \delta$ for $u \in cl_Y(P(\Omega))$, then

$$D_L([F(I - P),QNP],\Omega,0) = deg(F(I - P)L^{-1} - \vartheta PL^{-1},L(\Omega),0).$$

Since $L^{-1}(Im(A)) \subseteq Ker(P)$ and $L^{-1}(Im(Q)) \subseteq Ker(A)$, the right hand side is equal to

$\deg(FL^{-1}|L(cl_X(\Omega) \cap Ker(P)), L(\Omega \cap Ker(P)),0)\cdot$

$\cdot\deg(\vartheta L^{-1}|L(cl_X(\Omega) \cap Ker(A)),L(\Omega \cap Ker(A)),0)$.

Clearly the second factor is nonzero because of (6) and the choice of ϑ. Moreover $F'(0)$ is a linear homeomorphism from $Ker(P) \cap X$ onto $Im(A)$ because of $Im(F) \subseteq Im(A)$, which implies $Im(F'(0)) \subseteq Im(A)$, $Ker(F'(0)) \subseteq Ker(A)$, and since $F'(0)$ is a Fredholm operator of index 0. Further $F^{-1}(\{0\}) \cap Ker(P) = \{0\}$ by assumption, which ensures that the first factor is also nonzero. Thus $D_L([F(I - P),QNP],\Omega,0) \neq 0$, which yields $D_L([F,N],\Omega,0) \neq 0$, and therefore there exists a $u \in \Omega$ with $Fu = Nu$.

Remarks.

(1) The operator F, induced by Eq. (2), has a range, which is contained in the subspace $\{v|v \in C_\omega(\mathbb{R},\mathbb{R}), \int_0^\omega v(t)dt = 0\}$.

(2) The operators F, A of Example 1 also satisfy $Im(F) \subseteq Im(A)$, $F^{-1}(\{0\}) \subseteq Ker(A)$ and $Ker(F'(0)) \subseteq Ker(A)$.

(3) The hypothesis $\Omega = P(\Omega) + (I - P)(\Omega)$ is not very restrictive in applications since one usually realizes (4), (5) by verifying "a priori" bounds for operators, defined on the whole space.

We state a second continuation theorem, which can be proved by the above coincidence degree in the same way like the corresponding results of [3].

Theorem 2.

Let (H1) and (H3) be fulfilled and $A: X \longrightarrow Y$ be a linear Fredholm operator of index 0. Suppose that $G: cl_X(\Omega) \longrightarrow Y$ is continuous and bounded, $A + \mu G$ is proper for $\mu \in (0,1]$, $G|\Omega \in C^2(\Omega,Y)$, and $\mu A + G'(u)$ is a Fredholm operator for $\mu \geq 1$ and $u \in \Omega$. Moreover assume that $Im(G) \subseteq Im(A)$, that $Q: Y \longrightarrow Y$ is a continuous projector with $Ker(Q) = Im(A)$, and that we have:

(7) $Au + tGu \neq tNu$ for $(u,t) \in \partial_X(\Omega) \times (0,1)$

(8) $QNu \neq 0$ for $u \in \partial_X(\Omega) \cap Ker(A)$

(9) $deg_B(QN|cl_X(\Omega) \cap Ker(A), \Omega \cap Ker(A), 0) \neq 0$.

Then there exists a $u \in cl_X(\Omega)$ with $Au + Gu = Nu$.

Appendix.

Here we indicate that Example 1 falls into the abstract setting, given by (H1) - (H3). For simplicity we treat the case $m = 1$. If we equip $Y = C_\omega(\mathbb{R},\mathbb{R})$ with the maximum norm $\| \ \|_\infty$ and $X = C^1_\omega(\mathbb{R},\mathbb{R})$ with the norm $\| \ \|_{\infty,1}$, given by $\|u\|_{\infty,1} := \max\{\|u\|_\infty, \|u'\|_\infty\}$ for $u \in X$, and use the notations, introduced in Example 1, we immediately see that (H1) and (H3) are satisfied by X, Y and $N|cl_Y(\Omega)$ for any open bounded subset Ω of X. Moreover it is well-known that F is a C^∞-function on X, which is bounded on bounded subsets of X, and that A is a linear Fredholm operator of index 0. One easily calculates for $u,v \in X$ and $t \in \mathbb{R}$:

(10) $(F'u)(v)(t) = v'(t) + \exp(u(t-1))v'(t-1)$

$+ \exp(u(t-1))u'(t-1)v(t-1)$

We set for $u,v \in X$ and $t \in \mathbb{R}$:

$T_u v(t) := v'(t) + \exp(u(t-1))v'(t-1)$,

and because of the complete continuity of $F'u - T_u$ it suffices to show that T_u is a Fredholm operator of index 0. Since T_u is continuous, this follows from

(11) $Ker(T_u) = \{v | v \in X, \ v \text{ constant}\}$

(12) $Im(T_u) = \{w | w \in Y, \int_0^\omega [1 + \exp(\sum_{1 \leq i \leq v_1} u(t-i))]^{-1}$.

$$\cdot \sum_{0 \le i \le \nu_1 - 1} [(-1)^i w(t - i)\exp(\sum_{1 \le j \le i} u(t - j))]dt = 0 \}$$

We derive (11) and (12):

Obviously $Ker(T_u)$ contains all constant functions on $\mathbb{R}$.

Now let $v \in Ker(T_u)$, then we have for $t \in \mathbb{R}$:

$$v'(t) = -\exp(u(t - 1))v'(t - 1).$$

Hence we obtain by induction for $k \in \mathbb{N}$:

$$v'(t) = (-1)^k \exp(\sum_{1 \le i \le k} u(t - i))v'(t - k).$$

In particular this yields for $k = \nu_1$ because of $v(t) = v(t - \nu_1)$:

$$v'(t) = -\exp(\sum_{1 \le i \le \nu_1} u(t - i))v'(t),$$

hence $v' = 0$, thus (11).

Let $w \in Im(T_u)$, then there exists a $v \in X$ with

$$(13) \quad v'(t) + \exp(u(t - 1))v'(t - 1) = w(t)$$

for $t \in \mathbb{R}$ and $v(0) = 0$. Again by induction one obtains

$$v'(t) = -\exp(\sum_{1 \le i \le \nu_1} u(t - i))v'(t - \nu_1) +$$

$$+ \sum_{0 \le i \le \nu_1 - 1} [(-1)^i w(t - i)\exp(\sum_{1 \le j \le i} u(t - j))],$$

which yields because of $v'(t) = v'(t - \nu_1)$ and $v(0) = 0$.

$$v(t) = \int_0^t [1 + \exp(\sum_{1 \le i \le \nu_1} u(s - i))]^{-1}.$$

$$\cdot \sum_{0 \le i \le \nu_1 - 1} [(-1)^i w(s - i)\exp(\sum_{1 \le j \le i} u(s - j))]ds.$$

Since $v(\omega) = 0$, we have shown "$\subseteq$" in (12). On the other hand, if for some $w \in Y$ the integral in (12) is zero, the above function v is a solution of (13) and ω-periodic, hence $w \in Im(T_u)$. So (12) is derived.

In order that $\lambda A + F'(u)$ is a Fredholm operator for $u \in X$ and $\lambda \geq 0$, it suffices to establish this assertion for $\lambda A + T_u$. Now $\lambda A + T_u$ is equal to $(\lambda + 1)T_{u+\log(\lambda+1)}-1$, which is a Fredholm operator according to the above considerations.

Finally we derive the properness of F on closed, bounded subsets of X. Let $B \subseteq X$ be a closed, bounded subset and $K \subseteq Y$ be a compact subset. Since $F|B$ is continuous, the set $D := (F|B)^{-1}(K)$ is a closed subset of B, hence of X. Therefore we have only to show that each sequence in D has a convergent subsequence. Let $(u_n)_{n\in\mathbb{N}} \in D^{\mathbb{N}}$, then we can suppose by going, if necessary, to a subsequence that $(u_n)_{n\in\mathbb{N}}$ and $(Fu_n)_{n\in\mathbb{N}}$ are $\|\ \|_\infty$-convergent. Now, setting $w_n := Fu_n$, we obtain analogously like in the proof of (12) that

$$u'_n(t) = \left[1 + \exp\left(\sum_{1\leq i\leq \nu_1} u_n(t-i)\right)\right]^{-1} \cdot \sum_{1\leq i\leq \nu_1 - 1} \left[(-1)^i w_n(t-i)\exp\left(\sum_{1\leq j\leq i} u_n(t-j)\right)\right].$$

Hence $(u'_n)_{n\in\mathbb{N}}$ is $\|\ \|_\infty$-convergent, thus $(u_n)_{n\in\mathbb{N}}$ $\|\ \|_{\infty,1}$-convergent.

In order to treat the general case $m \in \mathbb{N}$, one only has to substitute $u(t-1)$ by $u(t-1)^m$ in the above proof, since F' is given by

$$(F'u)(v)(t) = v'(t) + \exp(u(t-1)^m)v'(t-1) + \\ + mu(t-1)^{m-1}\exp(u(t-1)^m)u'(t-1)v(t-1)$$

for $u,v \in X$ and $t \in \mathbb{R}$, and the last term generates a completely continuous, linear operator from X into Y for each $u \in X$.

REFERENCES

1. N. Bourbaki, Éléments de Mathématique, Topologie générale, chap. 5-10, Hermann, Paris, 1974.

2. G. Eisenack and C. Fenske, Fixpunkttheorie, Bibliographisches Institut, Mannheim-Wien-Zürich, 1978.

3. R. E. Gaines and J. L. Mawhin, Coincidence Degree and Nonlinear Differential Equations, Springer-Verlag, Berlin-Heidelberg-New York, 1977.

4. J. K. Hale and J. L. Mawhin, Coincidence degree and periodic solutions of neutral equations, J. Differential Equations 15(1974), 295-307.

5. G. Hetzer, Some applications of the coincidence degree for set-contractions to functional differential equations of neutral type, Comm. Math. Univ. Carolinae 17(1975), 121-138.

6. G. Hetzer, Alternativ- und Verzweigungsprobleme bei Koinzidenzgleichungen mit Anwendungen auf Randwertprobleme bei neutralen Funktionaldifferential- und elliptischen Differentialgleichungen, Habilitationsschrift, Aachen, 1978.

Author's address: G. Hetzer

Lehrstuhl C für Mathematik
Technische Hochschule Aachen

Templergraben 55

D 5100 Aachen

Federal Republic of Germany

CHAOTIC BEHAVIOR OF MULTIDIMENSIONAL DIFFERENCE EQUATIONS

by

JAMES L. KAPLAN[1,3] and JAMES A. YORKE[2,3]

In the first half of this paper we present results on chaotic behavior in multidimensional spaces. In the second half we discuss the problem of determining the dimension of an attractor. We believe it is possible to predict the dimension of the invariant attractor, at least generically, from a knowledge of the Lyapunov numbers of the map. Computer generated illustrations are given in evidence.

In a series of papers A. N. Sharkovskii investigated chaotic behavior of one-dimensional maps. The following theorem is in this spirit. See [1] for a proof.

Theorem 1. (Period three implies chaos). Consider the first order difference equation

$$x_{k+1} = f(x_k) , \qquad k = 0,1,2,... \tag{1}$$

where $x_k \in R$, $f:R \to R$ is continuous. Let J be an interval and let $f:J \to J$ be continuous. If there exists a point of period 3 for (1) then

(i) There is an uncountable set $S \subset J$ (containing no periodic points) satisfying:

(a) For every $p,q \in S$ with $p \neq q$, $\limsup_{n\to\infty} |f^n(p) - f^n(q)| > 0$,

(b) For every $p \in S$ and periodic point $q \in J$ $\limsup_{n\to\infty} |f^n(p) - f^n(q)| > 0$,

(c) There exists an uncountable set $S_0 \subset S$ such that $\liminf_{n\to\infty} |f^n(p) - f^n(q)| = 0$ for all $p,q \in S_0$.

[1] Department of Mathematics, Boston University, Boston, MA 02215
[2] Institute for Physical Science and Technology and Department of Mathematics, University of Maryland, College Park, MD 20742
[3] Research supported by NSF Grant MCS76-24432

Although "chaotic" behavior was originally observed in the study of a hydrodynamical system [2] this phenomenon has drawn considerable attention from mathematical ecologists. There are many situations in which population growth is a discrete process, and where the appropriate models are difference equations relating the population in generation $\tau + 1$, $N_{\tau+1}$, to that in generation τ, N_τ. Biological examples are provided by many temperate zone arthropod populations, with one short lived adult generation each year.

For example, among the many density dependent forms which have been proposed as discrete analogs of the logistic equation are the quadratically nonlinear equation:

$$N_{k+1} = N_k \left[1 + r\left(1 - {}^{N_k}/_{K}\right)\right] \tag{2}$$

and the equation

$$N_{k+1} = N_k \exp\left[r\left(1 - {}^{N_k}/_{K}\right)\right]. \tag{3}$$

(For a complete discussion of equations (2) and (3), as well as a cataloguing of other related models, see [3] and [4].)

It may be instructive to examine in greater detail the behavior of solutions of equation (2). We first make the preliminary change of variables:

$$x_k = {}^{1}/_{K} \left({}^{a-1}/_{a}\right) N_k, \qquad a = 1 + r$$

which reduces (2) to:

$$x_{k+1} = a x_k (1 - x_k) \tag{4}$$

We restrict ourselves to the parameter values $0 \le a \le 4$, since for such values $f(x) = ax(1 - x)\colon [0,1] \to [0,1]$. Equation (4) is the most frequently studied example of a chaotic scalar difference equation. (See [1], for example).

For $1 < a \le 4$, f has the nontrivial fixed point $z = {}^{a-1}/_{a}$. When $1 < a < 3$, $-1 < f'(z) < 0$ and we have a locally stable fixed point. As a increases beyond 3, the behavior is determined by examining $f^2(x) = f(f(x))$.

We easily compute:

$$\frac{d}{dx} f'(x)\Big|_{x=z} = f'(f(z))f'(z) = [f'(z)]^2$$

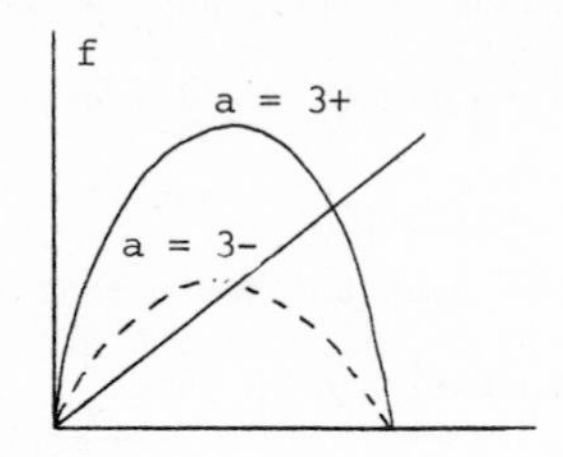

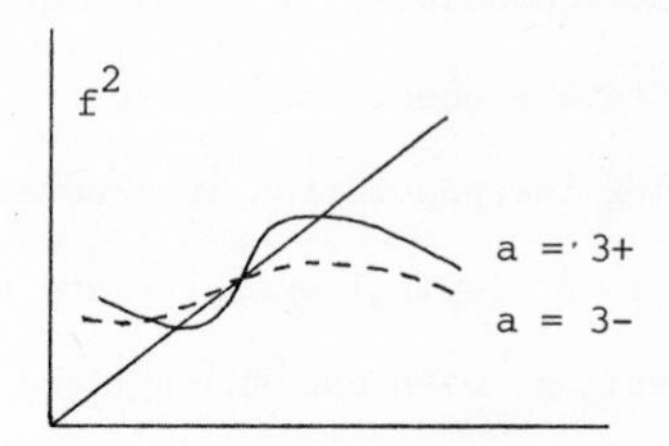

As the fixed point z remains stable with $a \to 3-$ the map f^2 has a slope less than, but approaching, +1. It follows that its graph can only cross the 45° line once in a neighborhood of z. But as a increases past 3, the slope of f^2 exceeds 1, so that the graph of f^2 must cross the 45° line three times in the neighborhood of z. Exactly at the parameter value where the fixed point becomes unstable ($a = 3$) we find a new (and locally stable) solution of period 2. Thus there is a bifurcation of the fixed point into two locally stable points of period 2, between which the population oscillates.

As a increases these two points (2-cycle) in turn become unstable and bifurcate to give four locally stable fixed points of period 4. In this way there arises, by successive bifurcations, an infinite collection of stable cycles, of period 2^n.

This sequence of stable cycles of period 2^n converges on a limiting value of a, say a_c. ($a_c \simeq 3.570$). For the difference equation (4), it can be shown by direct computation that there exists a point of period 3 for $a > 3.83$. Thus there will be chaos in this regime.

Similarly, if we consider the map f^2, this will have a point of period 3 for $a > 3.67$. Thus f^2 (and hence f) will be chaotic in this region. By considering the 3 cycles of f^n as $n \to \infty$ we find that as n increases the value of a for which Theorem 1 implies chaos decreases monotonically and approaches a_c.

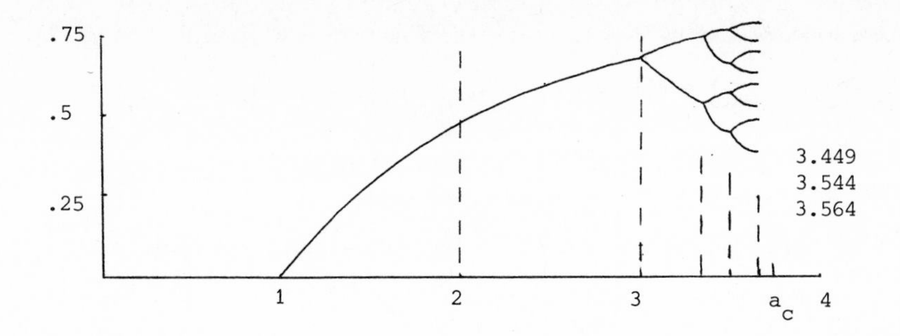

Difference schemes describing population growth are by no means restricted to single species models. There are prey-predator systems and competitive systems where generations do not overlap.

In an interesting paper, Beddington, Free and Lawton [5] investigate the particular host parasite system.

$$(5) \qquad H_{k+1} = H_k \exp[r(1 - {}^{H_k}/_K) - aP_k],$$

$$P_{k+1} = \alpha H_k [1 - \exp(-aP_k)] \qquad k = 0, 1, 2, \ldots..$$

where $r, a, \alpha, K \varepsilon R^+$. Numerical studies show regions of stable points, of stable limit cycles and of chaos. The onset of chaotic behavior takes place at lower r values than for the corresponding single species model. This paper is illustrated by fascinating photographs of oscilloscope trajectories.

In another paper, Guckenheimer, Oster and Ipaktchi [6] discuss another discrete model (Leslie model). Here a single population has been divided into two age classes x and y:

$$(6) \qquad \begin{pmatrix} x \\ y \end{pmatrix}_{k+1} = \begin{pmatrix} m_1 & m_2 \\ s & o \end{pmatrix} \begin{pmatrix} x \\ y \end{pmatrix}_k ;$$

s is the fraction of individuals in the first age class x that survive to the second age class, $m_1(\cdot)$ and $m_2(\cdot)$ are the per capita birthrates for the two age classes. The density dependence of the birthrates is described by:

$$m_i(x,y) = b_i e^{-a(x+y)} \qquad i = 1, 2.$$

This difference scheme may be described by the map $f:R^2 \to R^2$ given by:

(7) $$(x_{k+1}, y_{k+1}) = f(x_k, y_k) = [(b_1 y_k + b_2 y_k) e^{-a(x_k + y_k)}, sx_k] \quad i = 1, 2.$$

The coeffecients b_1 and b_2 give the maximum per capita reproductive rates for each age class. As in system (5) numerical investigation of the model shows that there are stable points, stable cycles, stable "invariant curves" or chaotic orbits, depending on the values of the various parameters. The authors offer a heuristic explanation of this behavior.

It is significant to note that the behavior of these multidimensional systems cannot be described or explained in terms of Theorem 1. The proof of Theorem 1 depends, in an essential way, upon a fixed point theorem which is only valid for real valued mappings. In fact, it is possible to give simple examples of difference equations on R^2 for which stable 3-cycles exist. Thus an alternate analysis is required for higher dimensional systems like (5) or (6). Numerous cases of such systems have been investigated by Mira and Gutowski in a long series of papers.

This problem was treated by F.R. Marotto, a former student of the first author, in [7]. Consider:

(8) $$x_k = f(x_k) \qquad f:R^n \to R^n, \text{ f continuous}.$$

In order to illustrate his ideas let us consider the following analog of equation (4).

$$x_{k+1} = (ax_k + bx_{k-1}) \ (1 - ax_k - bx_{k-1}).$$

If we set $y_k = x_{k+1}$ this may be written in the form of (8) with n = 2:

(9) (a) $x_{k+1} = (ax_k + by_k) \ (1 - ax_k - by_k)$

(b) $y_{k+1} = x_k$.

Since we are interested in only the positive solutions of (9) we restrict our parameters to the region R of the (a,b) parameter space:

$$R = \{(a,b): a \geq o,\ b \geq o,\ a+b \leq 4\}$$

In this case:

$$D = \{(x,y) : 0 \leq x,y \leq 0,25\}$$

is invariant. The dynamics of this difference scheme will be strongly dependent upon the eigenvalues of the Jacobian of the right hand side of (9), evaluated at the fixed points. There are two such fixed points: $x_k = y_k = 0$ (trivial fixed point and for $a+b \geq 1$, $x_k = y_k = \frac{a+b-1}{a+b}$.

A simple calculation shows that:

$$Df(x,y) = \begin{bmatrix} a - 2a(ax+by) & b - 2b(ax+by) \\ 1 & 0 \end{bmatrix}.$$

The eigenvalues of (9) at $z = \left(\frac{a+b-1}{a+b}, \frac{a+b-1}{a+b}\right)$ satisfy

$$\lambda^2 + A\lambda + B = 0$$

where

$$A = \frac{a(a+b-2)}{a+b}, \quad B = \frac{b(a+b-2)}{a+b}.$$

It is easily checked that for a + b near 1, the point z is stable. However, moving away from this line there are two ways in which z can become unstable:

(i) both eigenvalues are real and one of them exceeds one in absolute value (R_3), or

(ii) eigenvalues are complex conjugates of one another, and both exceed one in norm. (R_4)

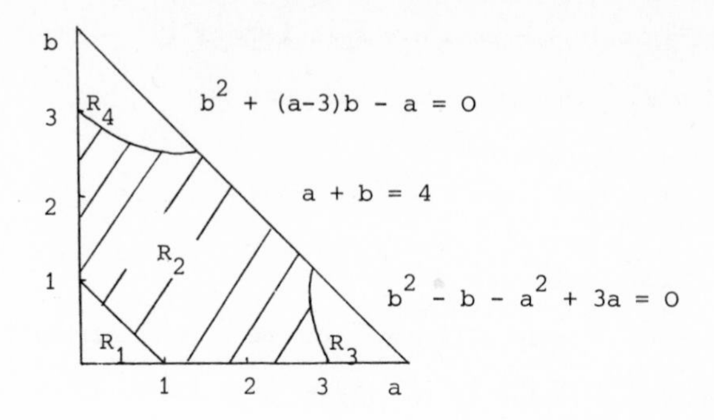

<u>Definition</u>: Let $f:R^n \to R^n$ be continuously differentiable in some ball $B_r(z)$ about the fixed point z. We will say that z is a <u>repelling fixed point</u> of f if all the eigenvalues of $Df(x)$ exceed 1 in absolute value for all $x \in B_r(z)$.

The repelling fixed point z will be called a <u>snap-back repeller</u> if there exists $x_0 \in B_r(z)$, $x_0 \neq z$, and an integer M such that:

$$f^M(x_0) = z$$

and

$$|Df^M(x_0)| \neq 0$$

Marotto proved the following result [7].

<u>Theorem 2</u>. Snap-back repellers imply chaos in R^n. More precisely, suppose z is a snap-back repeller for f . Then

(i) there is an integer N such that for every $k \geq N$ there exists a periodic point in $B_r(z)$ having period k,

(ii) there is an uncountable set $S \subset B_r(z)$ containing no periodic points satisfying:

(a) For every $p,q \in S$ with $p \neq q$

$$\limsup_{n\to\infty} | f^n(p) - f^n(q)| > 0,$$

(b) For every $p \in S$ and periodic point $q \in J$,

$$\limsup_{n\to\infty} | f^n(p) - f^n(q)| > 0,$$

(c) There is an uncountable $S_o \subset S$ such that for all $p,q \in S_o$,

$$\liminf_{n\to\infty} |f^n(p) - f^n(q)| = 0.$$

A sketch of the proof of Theorem 2 can also be found in [8].

It should be noted that if $f:R \to R$ is continuously differentiable, then the existence of a snap-back repeller for f is equivalent to the existence of a point of period 3, so that Theorem 2 is a true generalization of Theorem 1.

The arguments used to establish Theorem 2 are reminiscent of Smale's famous "horseshoe" example [9]. Smale showed that for a conditionally stable fixed point of a diffeomorphism the assumption of a transverse homoclinic orbit implies the existence of an infinite number of periodic points of different periods. In our case, the snap-back repeller plays the role of the homoclinic orbit, while the condition $|Df^M(x_o)| \neq 0$ can be viewed as a transversality condition.

Theorem 2 provides sufficient conditions for establishing chaotic behavior in multidimensional difference schemes such as (5), (6) and (9). Note, however, that Theorem 2 may be applied to (9) only in the region R_4, where the equation possesses a snap-back repeller. In order to analyze behavior in R_3 we must employ alternate methods. These are based on the observation that, for small b, equation 9(a) can be viewed as a perturbation of

$$x_k = ax_k(1 - ax_k)$$

for which chaotic behavior has already been demonstrated.

Consider

(10) $$x_{k+1} = f(x_k, bx_{k-1}),$$

$f:R^2 \to R$, f continuously differentiable. If we rewrite (10) in system form,

(a) $x_{k+1} = f(x_k, by_k)$

(b) $y_{k+1} = x_k$

then the fixed point (x_o, x_0) will be said to be neutrally stable if at least one eigenvalue of the associated linearized mapping at (x_o, x_o) has norm equal to 1.

The following Lemma 3 and Theorem 4 are modifications of a private communication from F. Marotto to the first author.

Lemma 3 Suppose (10) has a fixed point x_o which is not neutrally stable when $b = 0$. Then

(i) there exists $a, c \varepsilon \{R, \infty\}$ with $a < o$, $c > o$ such that (10) has a fixed point $x(b)$ with the same stability as x_o, for all $b \varepsilon (a,c)$,

(ii) $x(b)$ is a uniquely defined continuous function of b for $b \varepsilon (a,c)$ with $x(o) = x_o$, and

(iii) if $-\infty < a$ (or $c < \infty$) then as $b \to a$ (as $b \to c$) either $x(b)$ is unbounded or $x(b)$ becomes neutrally stable.

Sketch of proof. Define $g(x,b) = x - f(x,bx)$.

Then

$$g(x_o, o) = x_o - f(x_o, o) = 0$$

while

$$g_1(x,b) = 1 - f_1(x,bx) - bf_2(x,bx)$$

so that

$$g_1(x_o, o) = 1 - f_1(x_o, o) \neq 0$$

since x_o is not neutrally stable. It follows from the Implicit Function Theorem that there exists a unique continuous function $x(b)$ such that

$$g(x(b), b) = 0$$

for b sufficiently small. Thus $x(b) = f(x(b), b\, x(b)$.

Standard continuation arguments as well as continuous dependence of eigenvalues of a matrix upon its entries imply the remaining conclusion.

Theorem 4. Suppose 11(a) has a snap-back repeller when $b = 0$. Then (10) has a structurally stable homoclinic orbit for all $|b|<\varepsilon$ for some $\varepsilon>0$. In particular, for $|b|<\varepsilon$ (10) is chaotic.

Sketch of proof. Consider the reduced system,

(12)
$$\tilde{x} = f(x,o)$$
$$\tilde{y} = x$$

obtained from (11) by setting $b = 0$. Consider the curve C described by

$$C = \{(f(t,o),\ t) : t\varepsilon R\}.$$

The curve C is invariant under (12) because for any $(x,y)\varepsilon\ C$,

$$(\tilde{x},\tilde{y}) = (f(x,o),\ x)\varepsilon C.$$

Note that the curve C is simply the graph $x = f(y,o)$.

Now if $x_{k+1} = f(x_k,o)$ has a snap-back repeller then by definition there exists a point z satisfying $z = f(z,o)$, $|f_1(z,o)| > 1$. Note that (z,z) is a fixed point of (12). The eigenvalues of the Jacobian matrix of (12) at (z,z) satisfy:

$$\det \begin{bmatrix} f_1(z,o) - \lambda & o \\ 1 & -\lambda \end{bmatrix} = (f_1(z,o) - \lambda)(-\lambda) = o.$$

Thus, $\lambda_1 = o$ is associated with the stable manifold S_o and $|\lambda_2| = |f_1(z,o)| > 1$ is associated with the unstable manifold U_o. In fact, in a local neighborhood of (z,z) these manifolds can be identified precisely.

S_o. Since S_o is composed of those points which approach (z,z) under iteration of (12) we see that locally

$$S_o = \{(x,y) : x = z\}.$$

In fact, all points in S_o are mapped onto (z,z) precisely.

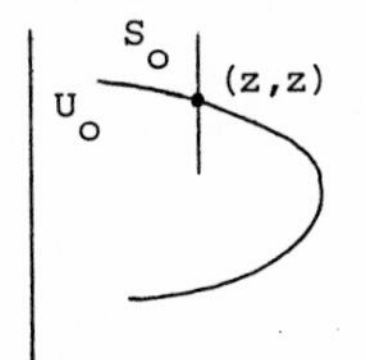

U_o. The curve C is U_o. As noted above C is an invariant manifold in the x,y plane. Moreover the points of C are mapped away from $(z,z)\varepsilon C$ under (12) (because (z,z) is a repelling fixed point).

Now since z is a snap-back repeller there must exist a homoclinic point at the intersection of S_o and U_o (i.e., the orbit in R^2 around the snap-back repeller must be a homoclinic orbit.).

By Lemma 3, for $|b| < \varepsilon$, there exists $(x(b), y(b)) = z_b$ a solution of (11) such that $(x(o), y(o)) = (z,z)$, z_b depends continuously upon b, and z_b is a fixed point of (11) with the same stability as (z,z). The structural stability of the homoclinic orbit of (12) now implies the existence of a homoclinic orbit of (11) for b sufficiently small.

Having proved Theorem 4 we can now demonstrate chaotic behavior in a manner similar to Smale's horseshoe arguments.

Remark: The small perturbation arguments outlined above can be adapted to several other situations. For instance, we could just as easily have considered the equation,

$$x_{k+1} = f(x_k, b_1 x_{k-1}, b_2 x_{k-2}, \ldots, b_m x_{k-m}).$$

An interesting case is the system described by $f,g:R^2 \to R$,

$$(13) \qquad \begin{aligned} x_{k+1} &= f(x_k, by_k) \\ y_{k+1} &= g(cx_k, y_k). \end{aligned}$$

Consider the uncoupled system

$$(14) \qquad \begin{aligned} &(a) \quad x_{k+1} = f(x_k, o) \\ &(b) \quad y_{k+1} = g(o, y_k). \end{aligned}$$

If (14)a has a snap-back repeller and 14(b) has an unstable fixed point then (13) has a snap-back repeller for $|b|$, $|c|$ sufficiently small. If 14(a) has a snap-back repeller and 14(b) has a stable fixed point then (13) has a structurally stable homoclinic orbit for $|b|$, $|c|$ sufficiently small. These ideas can be applied to the (competitive) system

$$(15) \qquad x_{n+1} = \lambda x_n [1 + a(x_n + \alpha y_n)]^{-b}$$

$$y_{n+1} = \lambda' y_n [1 + a'(y_n + \beta x_n)]^{-b'}.$$

For $\alpha = \beta = o$, each of the possibilities enumerated above can be obtained for selected parameter values, and thus we can conclude the chaotic behavior of the full system (15), which has been observed numerically by Hassell and Comins [10].

In our examination of system (9) we found that there were two chaotic regions, R_3 and R_4. Our analysis of the system for parameters chosen from R_3 indicated that the chaotic set is one dimensional (in fact, it is a perturbation of the curve $x = ay(1-ay)$) while in the region R_4 it <u>appears</u> that the chaotic set is two dimensional. In general, given a difference scheme $x_{k+1}=f(x_k)$, $f:R^n \to R^n$, is there any way to decide, a priori, what the dimension of the chaotic set will be?

In order to formulate a conjecture on this question it is first necessary for us to formulate a notion of dimension. Intuitively, the dimension of a space indicates the amount of information required to specify a location. In practice, positions are given approximately to within ε. For a space X, let $N(\varepsilon)$ be the minimum number of points that can be chosen so that the ε balls centered at these points cover the space. The dimension of X tells us how fast $N(\varepsilon)$ grows as ε shrinks to o. If the set is n dimensional we expect $N(\varepsilon) \approx \varepsilon^{-n}$. Define

$$\dim X = \lim_{\varepsilon \to o} - \frac{\ln N(\varepsilon)}{\ln \varepsilon}$$

whenever the limit exists.

This is metric space dimension (Hausdorff dimension). It is easy to give examples where dim X is not an integer (see work of Mandelbrot). This fact is an interesting sidelight which should not be distracting.

At least two other concepts of dimension are commonly used. One is linear dimension (vector space dimension) of a vector space, the maximal number of linearly independent vectors. This is not applicable to our work since the chaotic attracting set will not be a vector space. The second notion is that of topological dimension. The topological dimension is n if there is a cover of open sets such that for every refinement there are points that are in at least n+1 open sets of the refinement. One of the most striking aspects of this definition is how far removed it is from any idea of measurement. For instance, if $F:X \to X$ where X,Y are compact metric spaces, then top dimension is not well behaved since we can have $\dim F(X) > \dim X$: you can map an interval continuously onto a square.

For Lipschitz mappings the metric space dimension is better behaved. It can be shown to satisfy

$$\dim F(X) \le \dim X$$

$$\dim (X_1 + X_2) = \dim X_1 + \dim X_2 \;.$$

Remark: For Lipschitz maps we might define an integer valued dimension as follows. If there is a Lipschitz map $F:X \to R^n$ such that F(X) contains a ball in R^n we say X is at least n dimensional, and the dimension of X would then be the largest such number. We conjecture that for spaces whose Hausdorf dimension is an integer, these two definitions are equal.

The following is an example of a set in R^2 with metric space dimension 1.

Example: A space with Hausdorff dimension 1. Let Y = the Cantor set obtained by removing open middle halves from the interval [0,1], let X = Y x Y.

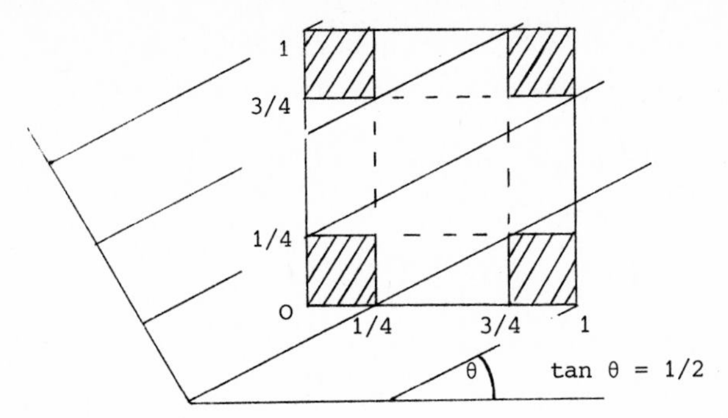

Then X has Hausdorff dimension 1. Also there is a linear projection of $R^2 \to R^1$ which takes X onto an interval.

In order to formulate our conjecture, we examined numerically a class of difference equations of the form

(16) $$x_k = f(x_{k-1})$$

where f: $B \to B^o$, B compact, $f(B) \subset \text{Int}B$.

In particular, we examined the family of equations

(17) $$\begin{aligned} x_{k+1} &= 2x_k \bmod 1 \\ y_{k+1} &= \lambda_1 y_k + p(x_k) \end{aligned}$$

and

(18) $$\begin{aligned} x_{k+1} &= 2x_k \bmod 1 \\ y_{k+1} &= \lambda_1 y_k + p(x_k) \\ z_{k+1} &= \lambda_2 z_k + q(x_k) \end{aligned}$$

where p,q are periodic with period 1. For (17)

$$B = \{(x,y): o \le x \le 1, \frac{1}{1-\lambda_1} P_{min} \le y \le \frac{1}{1-\lambda_1} P_{max}\}$$

while for (18)

$$B = \{(x,y,z): 0 \le x \le 1, \frac{1}{1-\lambda_1} P_{min} \le y \le \frac{1}{1-\lambda_1} P_{max} \frac{1}{1-\lambda_2} q_{min} \le z \le \frac{1}{1-\lambda_2} q_{max}\}$$

These systems were chosen because of the ease with which it is possible to compute

$$\text{Det Df} = 2\lambda_1$$

for equation (17), and

$$\text{Det Df} = 2\lambda_1\lambda_2$$

for equation (18).

Notation: Let A be an n×n matrix with eigenvalues $\lambda_1, \lambda_2, \ldots, \lambda_n$, where $|\lambda_1| \ge |\lambda_2| \ge \ldots \ge |\lambda_n|$. Write

$$\text{Det}_j A = |\lambda_1\lambda_2 \ldots \lambda_j| .$$

We say a map f has Lyapunov numbers if

$$\delta_j = \lim_{k\to\infty}\{\text{Det}_j (Df^k)\}^{1/k}$$

exists, and the Lyapunov numbers are $L_1 = \delta_1$, $L_j = \delta_j/\delta_{j-1}$ for $j = 2,3,\ldots$. Note that A has Lyapunov numbers $L_i = |\lambda_i|$.

For system (18) observe that the Lyapunov numbers are 2, λ_1 and λ_2.

For system (17) we found that if $2\lambda_1 > 1$ then the chaotic set appears 2 dimensional, if $2\lambda_1 \le 1$ then the chaotic set appears 1 dimensional. (See figures.)

For system (18) we found that if $2\lambda_1\lambda_2 > 1$ then the chaotic set appears 3 dimensional, if $2\lambda_1 > 1$ while $2\lambda_1\lambda_2 \le 1$ then the chaotic set is 2 dimensional. This relationship holds for "almost" every choice of functions p and q. For certain special choices, however, this relationship broke down.

For example, if

$$\lambda_1 = \lambda_2 \quad \text{and} \quad p = q$$

then $y = z$ and (18) degenerates into a 2 dimensional system, because of symmetry, regardless of the magnitude of λ_1 and λ_2.

Similarly, if $p(x) = \cos 8\pi x$ while $q(x) = \cos 4\pi x$ and $\lambda_1 = \lambda_2$ the system degenerated because $x_n = 2x_{n-1} \bmod 1$. These examples lead us to make the following conjecture.

<u>Conjecture</u>: Let $f\colon B \to B^0$, B compact and consider

$$x_k = f(x_{k-1}). \tag{16}$$

Then for nearly every (in the generic sense) f satisfying

$$\liminf_{m\to\infty} \{\mathrm{Det}_j (Df^m)\}^{1/m} > 1 \tag{19}$$

the attracting limit set $\bigcap_{m=1}^{\infty} f^m(B)$ has metric dimension $\geq j$.

Thus far we have been able to develop a heuristic argument for the validity of this conjecture, but to date we have been unable to provide a rigorous proof.

The above conjecture concerns the integer part of the dimension. More specifically, assume f has Lyapunov numbers. Assume we can choose j such that $\delta_j > 1 > \delta_{j+1}$. Then we conjecture that generically the metric dimension = j + fraction, where the fraction is given by

$$\frac{\log \delta_j}{|\log L_{j+1}|}$$

This latter number can be shown to be less than 1.

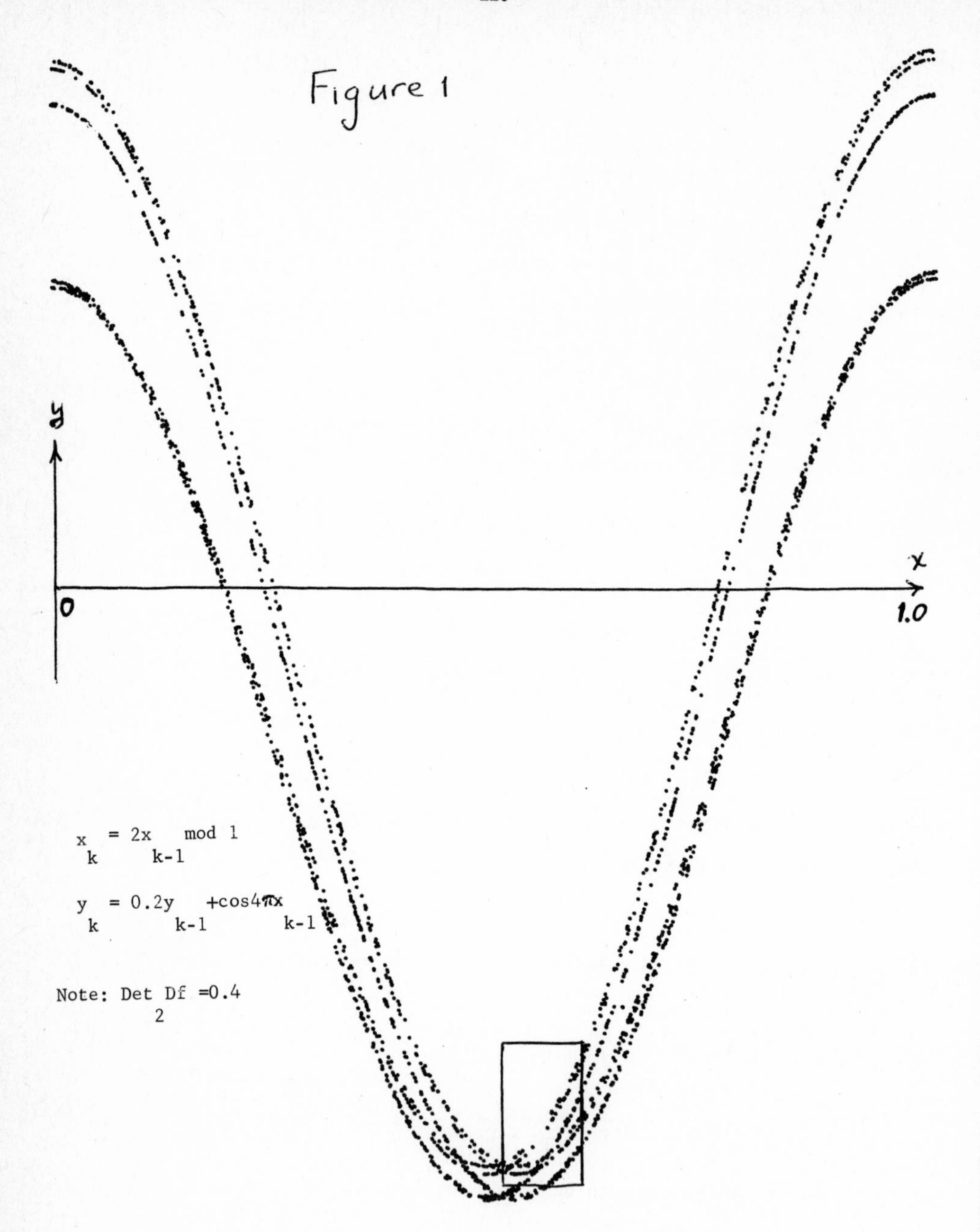
Figure 1
y
x
0
1.0
x = 2x mod 1
k k-1
y = 0.2y +cos4πx
k k-1 k-1
Note: Det Df =0.4
2

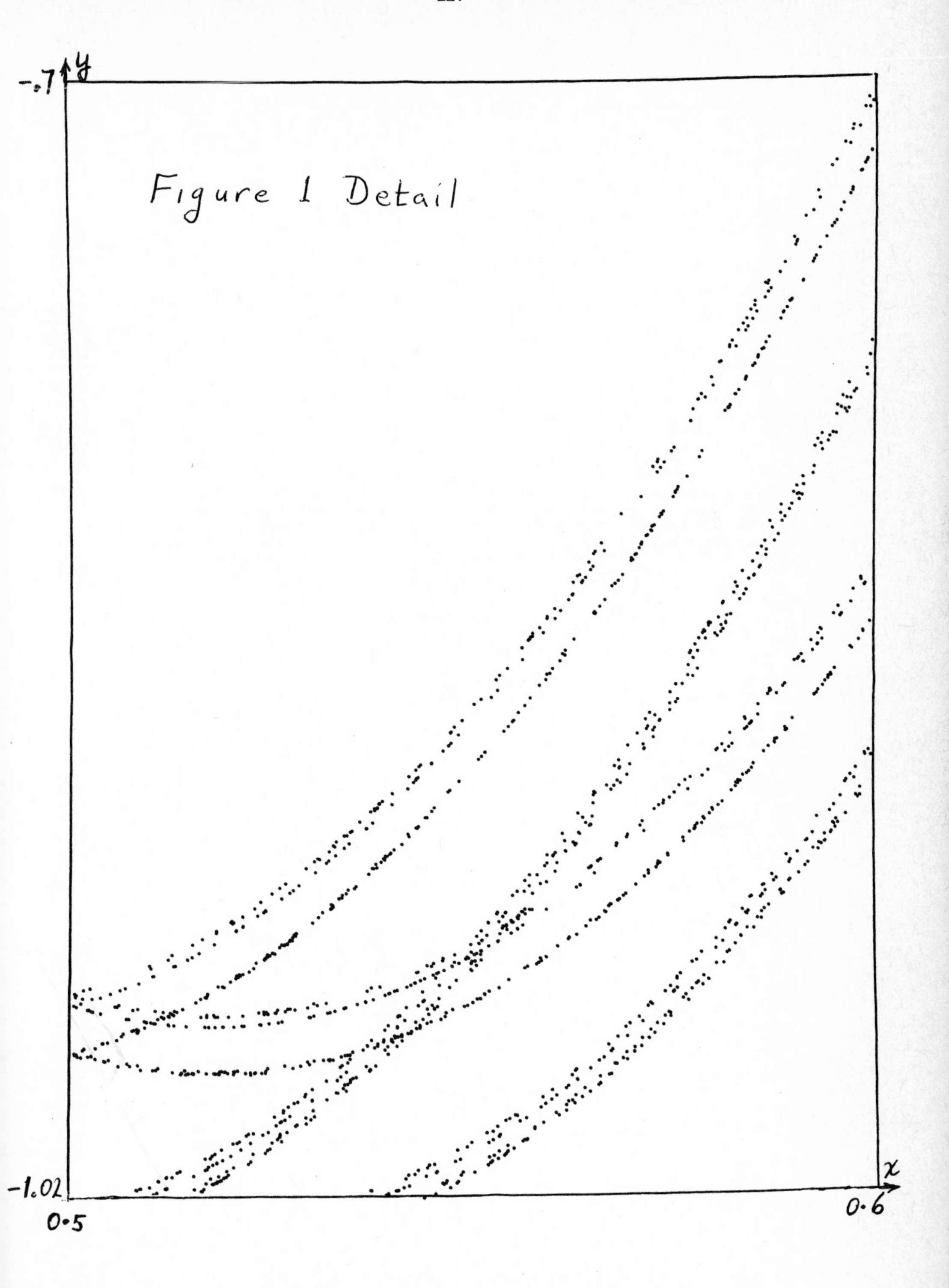
Figure 1 Detail
-.7
y
-1.02
0.5
0.6
x

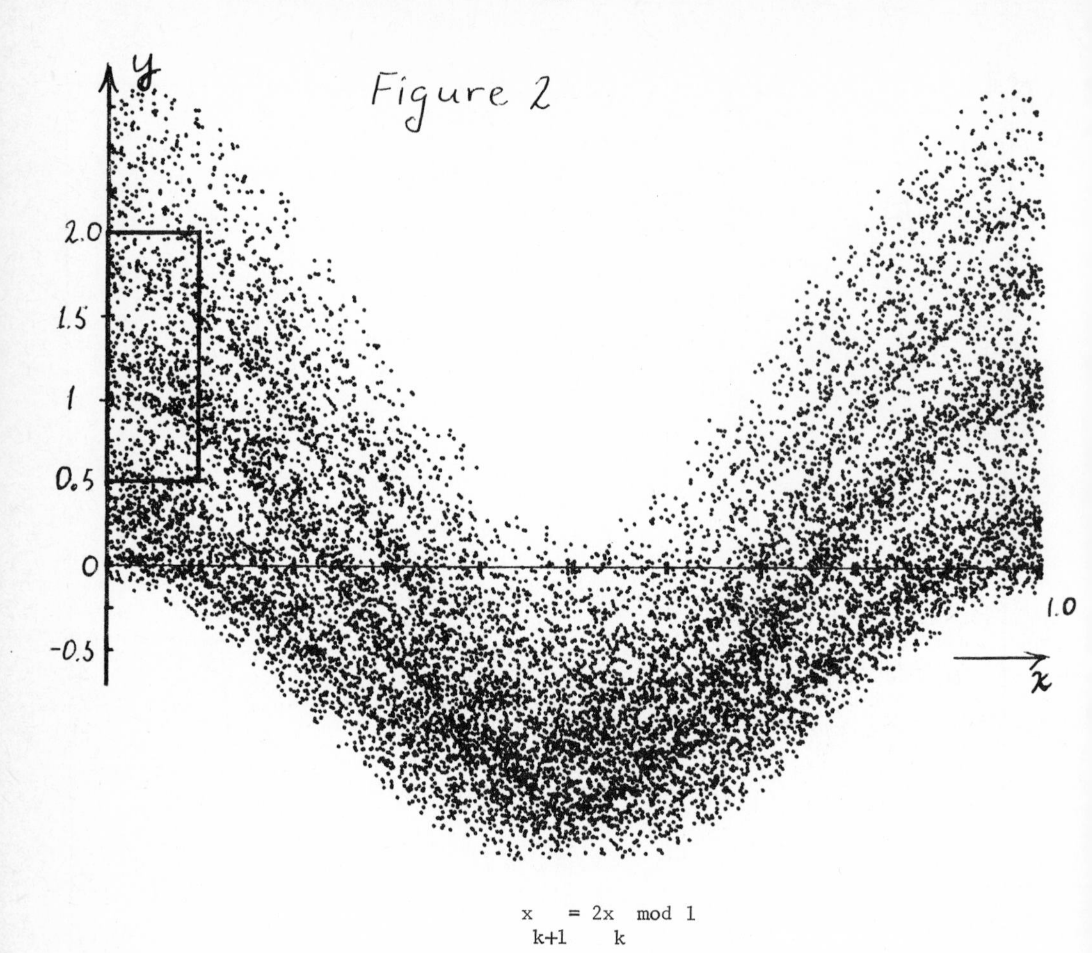

$$x_{k+1} = 2x_k \bmod 1$$

12000 points plotted

$$y_{k+1} = 2/3\, y_k + \cos(4\pi x_k)$$

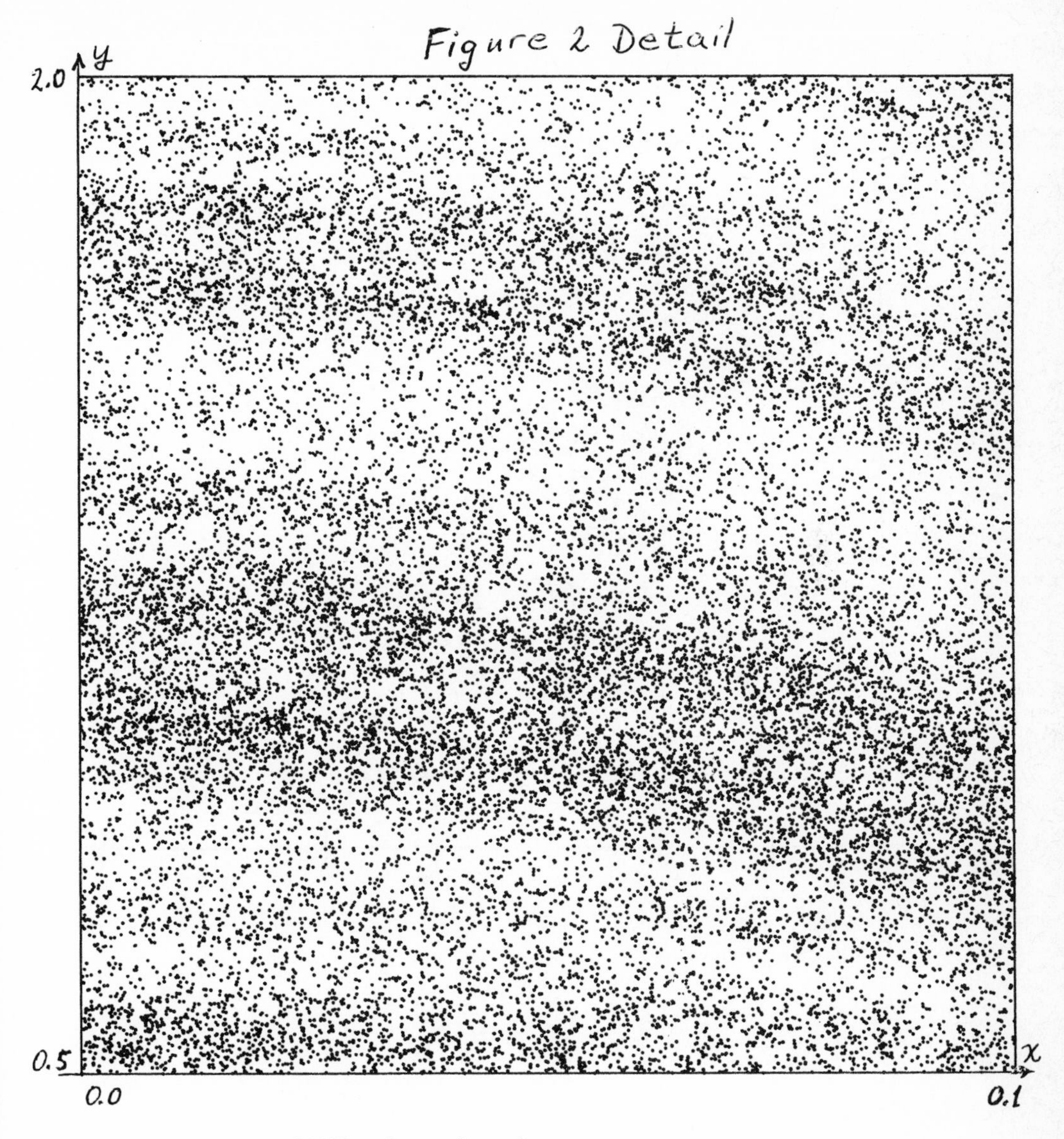

20000 points plotted

Figure 3

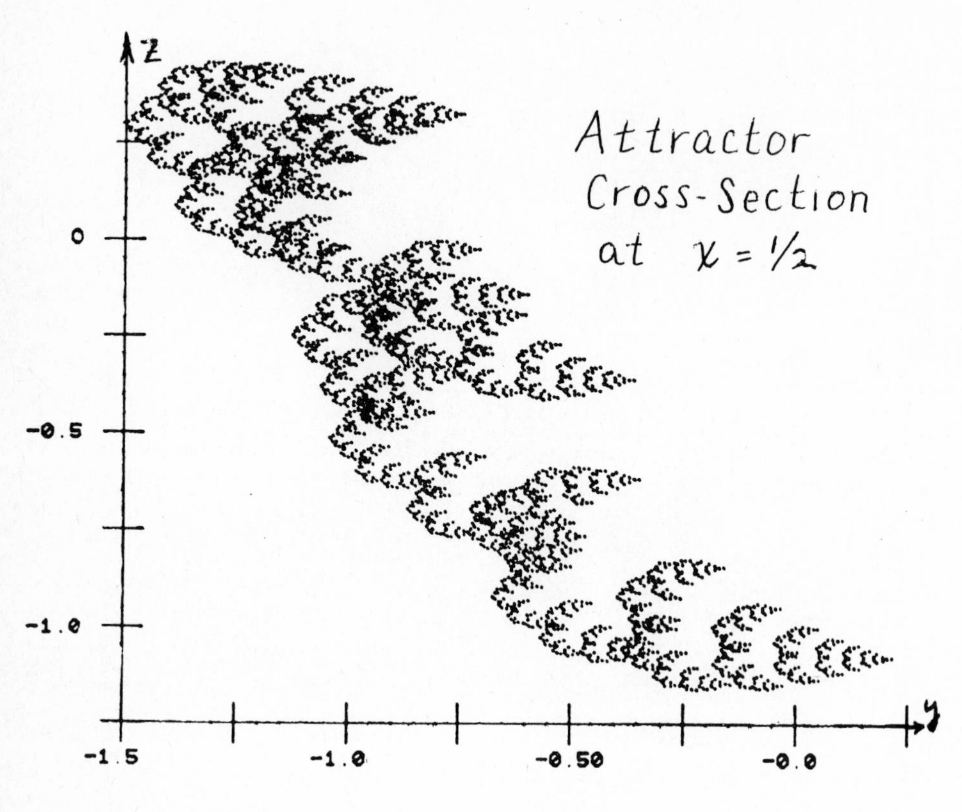

16385 points plotted

$\text{Det } Df_3 = 8/9$

$$x_{k+1} = 2x_k \mod 1$$

$$y_{k+1} = 2/3\, y_k + \cos(4\pi x_k)$$

$$z_{k+1} = 2/3\, z_k + \sin(4\pi x_k)$$

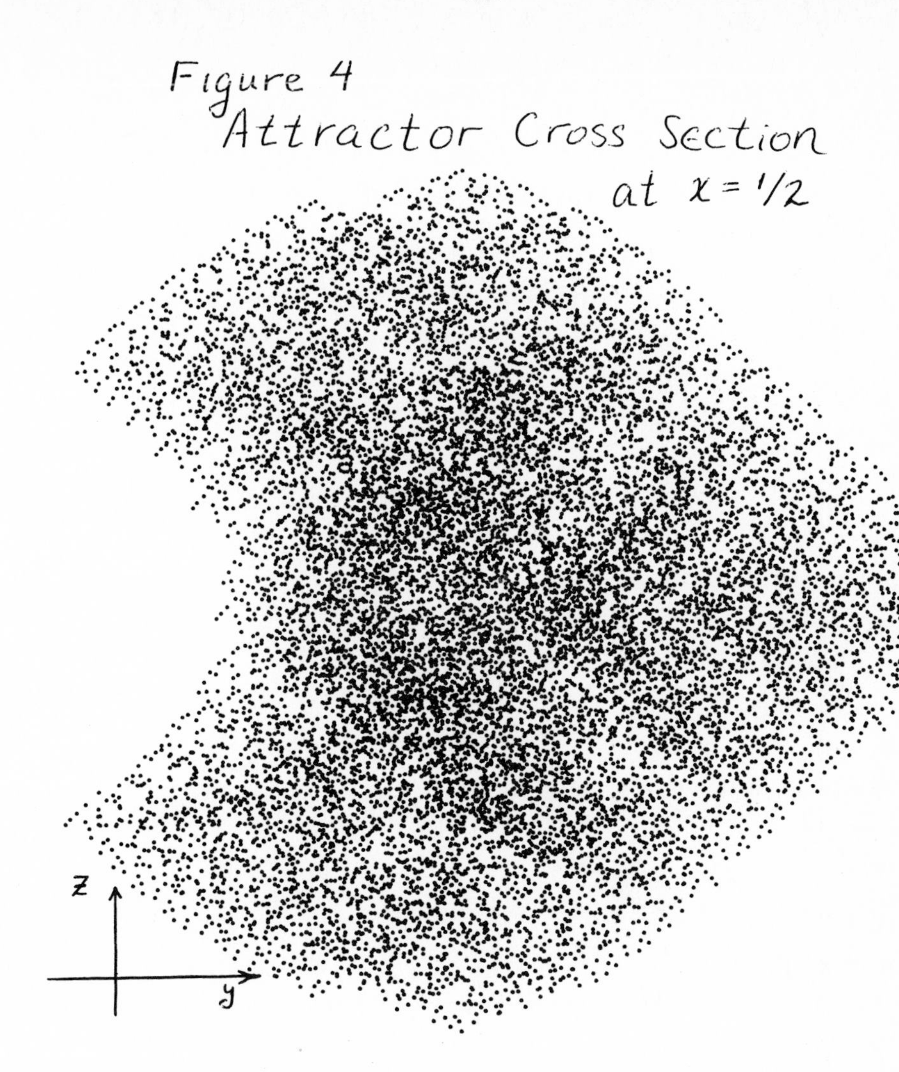

Note: The expansion coefficient here is 3.0 and Det f_3 = 4/3

14600 points plotted

$$x_{k+1} = 3x_k \mod 1$$

$$y_{k+1} = 2/3\, y_k + \cos(4\pi x_k)$$

$$z_{k+1} = 2/3\, z_k + \sin(4\pi x_k)$$

Figure 5

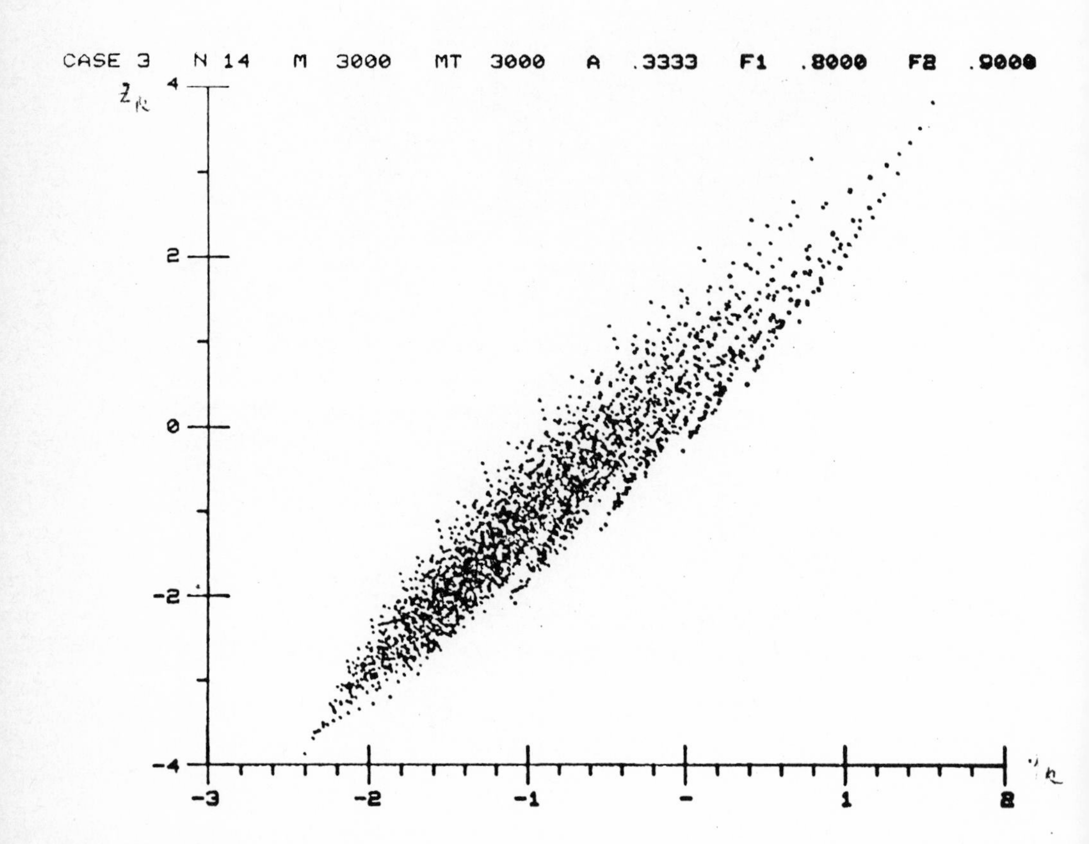

$$x_k = 2x_{k-1} \mod 1$$
$$y_k = 0.8y_{k-1} + \cos 4\pi x_{k-1}$$
$$z_k = 0.9z_{k-1} + \cos 4\pi x_{k-1}$$

REFERENCES

[1] Li, T.Y. and J.A. Yorke, Period Three Implies Chaos, American Math Monthly 82 (1975), pp. 985-992.

[2] Lorenz, E.N., Deterministic Nonperiodic Flow, J. Atmos. Sci. 20 (1963), pp. 130-141.

[3] May, R.M., Biological Populations with Nonoverlapping Generations: stable points, stable cycles, and chaos. Science, 186 (1974), pp. 645-647.

[4] May, R.M. and G. Oster, Bifurcations and Dynamic Complexity in Simple Ecological Models, to appear.

[5] Beddington, J.R., Free, C.A., and Lawton, J.H., Dynamic Complexity in Predator Prey Models Framed in Difference Equations, Nature 255.(1975), pp. 58-60.

[6] Guckenheimer, J., Oster, G. and Ipaktchi, A., The Dynamics of Density Dependent Population Models, to appear.

[7] Marotto, F.R., Snap-Back Repellers Imply Chaos in R^n, J. Math. Anal. Appl. 63, No. 1 (1978), pp. 199-223.

[8] Kaplan, J.L. and Marotto, F.R., Chaotic Behavior in Dynamical Systems, Proceedings Int. Conf. Non-linear Systems and Appl. Ed V. Lakshmikantham, Academic Press (1977) pp. 199-210.

[9] Smale, S. Differentiable Dynamical Systems, Bull. Amer. Math. Soc. 73 (1967), pp. 747-817.

[10] Hassell, M.P. and Comins, H.N., Discrete Time Models for Two Species Competition, Theor. Pop. Biol. 9 (1976), pp. 202-221.

NUMERICAL SOLUTION OF A GENERALIZED
EIGENVALUE PROBLEM FOR EVEN MAPPINGS

by James L. Kaplan[1,3] and James A. Yorke[2,3]

Newton's method has probably long been the most frequently used method for solving systems of nonlinear equations but it is most useful when an approximate solution is available. Then it will generally provide a sequence of approximations which converge rapidly to a solution. Newton's method can also be tried by picking an "approximate" solution at random, but the approximate solution is likely to lie outside the domain of convergence. In comparison, numerous topological proofs proclaim the existence of solutions in situations in which local procedures, like Newton's method or the method of steepest descent, are of less value since topological methods by their nature have not provided us with good approximations of the solutions.

In the past few years a number of surprising advances have been made in finding numerical procedures for solving such nonlinear systems of equations. Two closely related methods have become available. The simplicial methods developed by H. Scarf, B. C. Eaves, R. Saigal and others [1-7] was employed initially for finding the Brouwer fixed point. Their method is in essence based on using a simplicial approximation of the map as is used in the Sperner Lemma proof of the Brouwer Fixed Point Theorem. Kellogg, Li and Yorke developed an alternative approach [8,9] for the numerical solution of

---

1 Department of Mathematics, Boston University, Boston, MA 02215.

2 Institute for Physical Science and Technology and Department of Mathematics, University of Maryland, College Park, MD 20742.

3 Research supported by NSF Grant MCS76-24432.

the Brouwer Fixed Point Theorem. By a twist of a nonconstructive proof of M. Hirsch [10], they obtained a constructive proof. Given a smooth map F of a ball in R^n into itself they chose a point p on the boundary of the ball. For almost any choice of p there is a smooth curve Γ_p which leads to a fixed point.

Every point $q \in \Gamma_p$ is such that q lies on the line segment between p and F(q). The curve can be followed efficiently to the extent that on the Univac 1108 at the University of Maryland, they found they could solve a class of problems with dimension n = 20 in an average of 3.3 seconds. See also [11].

The simplicial method also follows a curve, a polygonal path, and the methods are closely related. Both methods seem to work well on problems they have been tested with. We feel that the numerical techniques for following smooth curves are far more thoroughly developed; the curve Γ_p is the solution of an ordinary differential equation and this fact is the basis of methods for following the curve. For highly nonlinear problems in high dimensions, this may conceivably give a competitive advantage to the smooth techniques.

The Brouwer Fixed Point Theorem itself is not used extensively in applications, though similar degree-theoretic results are. Chow, Mallet-Paret, and Yorke [12] showed that by using elementary homotopy arguments, numerical methods became available for many problems. (B. C. Eaves uses a similar homotopy approach.) Their idea is to take any existence proof based on degree theory and to convert it using elementary homotopies into a constructive, computer implementable method for finding solutions. (Problems with dimension exceeding 100 are difficult and require large amounts of computer time.) They employ a transversality-type generalization of Sard's Theorem. There is no standard procedure yet for converting degree theory proofs but there have been far more successes than failures. Chow, Mallet-Paret and Yorke showed how to carry out this homotopy-based transformation for the following situations. We assume all maps concerned are smooth.

1. The Brouwer Fixed Point Theorem.

2. "Any continuous vector field on an even dimensional sphere has a zero."

3. Finding roots of polynomials. (It is well known that the fundamental theorem of algebra can be proved using degree theory.) -- Li and Yorke are currently doing numerical experiments using polynomials cited in the literature as requiring particular care. So far our method works beautifully.

4. Solving two-point boundary value problems for second order ordinary differential equations in R^n.

5. Smale and Hirsch [13,14] have shown that "continuous Newton" methods can be employed for certain proper maps in R^n. In [12] it was shown how situations could be reformulated in terms of the homotopy approach.

Existence proofs can be given in the above situations using elementary Brouwer degrees or elementary homology proofs. Alexander and Yorke [15] extended the methods to the following situations.

6. The Borsuk-Ulam Theorem: "A continuous (smooth) odd map $f: S^n \to R^n$ has a zero, i.e. a solution of $f(x) = 0$."

7. The bifurcation results of Rabinowitz [16] in which a connected set of zeroes bifurcates from the trivial solution as a parameter is varied.

8. A scheme was suggested for finding periodic solutions under the Hopf bifurcation hypotheses used by Alexander and Yorke [15].

It is shown in [15] that the method can be expected to work quite generally and even in some situations where elementary degree theory is not applicable. In particular the homotopy method works in situation 8 above even though the method requires framed bordism groups.

In this paper we will illustrate how the homotopy continuation method can be adapted to a problem for which a constructive approach has not previously been available. In [17], Kaplan and Yorke considered the quadratic differential system

$$\frac{dx_i}{dt} = \sum_{j,k=1}^{n} A^i_{jk} x_j x_k \qquad i = 1,2,\ldots,n \tag{1}$$

where A^i_{jk} are real constants. The right hand side of (1) can be used to define a binary "multiplication" operation in R^n as follows. Let $x,y \in R^n$. Denote the ith component of the product $x*y$ by $(x*y)^i$ and define

$$(x * y)^i = \sum_{j,k=1}^{n} A^i_{jk} x_j x_k \;. \tag{2}$$

It is easily verified that under ordinary vector addition and the multiplication $*$ defined in (2), R^n is a nonassociative, real finite dimensional algebra. In [17] the following results were proved.

Theorem 1. Either

(a) there exists a nonzero $x_0 \in R^n$ such that $x_0 * x_0 = 0$ (x_0 is nilpotent), or

(b) there exists a nonzero $x_0 \in R^n$ such that $x_0 * x_0 = x_0$ (x_0 is idempotent).

Corollary. If all solutions of (1) are bounded then there exists an invariant line on which (1) reduces to $\frac{dx_i}{dt} = 0$.

These results were obtained as a consequence of the next theorem, which was not stated explicitly in [17], but was implied in the degree theoretic proof of Theorem 1.

Theorem 2. Let f be a continuous map $f: S^{n-1} \to R^n$. Assume f is even (i.e., $f(x) = f(-x)$). Then there exists a generalized eigenvector; that is, a point $x \in S^{n-1}$ and some $\lambda \in R$ (λ possibly 0) such that $f(x) = \lambda x$.

Our constructive approach to Theorem 2 will be derived as a consequence of a variation of Sard's Theorem, whose proof may be found in [18]. First we introduce some terminology.

Definition. Let $U \subset R^n$ be open and $f: R^n \to R^p$ be smooth. We say $y \in R^p$ is a regular value for f if

$$\text{Range } Df(x) = R^p \quad \text{for all } x \in f^{-1}(y)$$

where $Df(x)$ denotes the $p \times n$ matrix of partial derivatives of $f(x)$.

Theorem 3. (Generalized Sard's Theorem). Let $V \subset R^q$, $U \subset R^m$ be open and let $\phi: V \times U \to R^p$ be C^r, $r > \max\{0, m-p\}$. If $0 \in R^p$ is a regular value of ϕ, then for almost every $c \in V$, 0 is a regular value of $\phi_c(\cdot) = \phi(c, \cdot)$.

Next, define

$$G(x,t,c) = (1 - t)c + tf(x)$$

where $c \in R^n$ and f satisfies the hypotheses of Theorem 2. Let

$$\phi(\lambda,x,t,c) = G(x,t,c) - \lambda x \; . \tag{3}$$

Then, for fixed (t,c), $\phi_{(t,c)}(\lambda,x) = \phi(\lambda,x,t,c)\colon R \times S^{n-1} \to R^n$. Let Γ_a be the component of $\phi_c^{-1}(0) \cap (0,1) \times R \times S^{n-1}$ whose closure contains $(0,|c|,c/|c|)$.

Theorem 4 (Constructive Version of Theorem 2). *For the mapping* ϕ *defined in* (3) *we have*

(a) $0 \in R^n$ *is a regular value of* ϕ,

(b) *for almost every* c, Γ_a *is a smooth curve in* $(0,1) \times R \times S^{n-1}$ *joining a generalized eigenvector of* f.

Proof. Let $(\bar{\lambda},\bar{x},\bar{t},\bar{c}) \in (R \times S^{n-1} \times R \times R^n)$ and suppose $\phi(\bar{\lambda},\bar{x},\bar{t},\bar{c}) = 0$. Then $D_c\phi(\bar{\lambda},\bar{x},\bar{t},\bar{c}) = (1 - t)I$, I = identity matrix, and for $t \neq 1$,

$$\text{Range } D\phi(\bar{\lambda},\bar{x},\bar{t},\bar{c}) \supset \text{Range } D_c\phi(\bar{\lambda},\bar{x},\bar{t},\bar{c}) = R^n.$$

This proves (a). By Theorem 3, for almost every $c \in R^n$, 0 is a regular value of ϕ_c, and thus $\phi_c^{-1}(0)$ is a smooth curve in $(0,1) \times R \times S^{n-1}$. Hence each component is a smooth curve and is either diffeomorphic to a circle or an open interval.

Now $\phi_c(0,|c|,c/|c|) = 0$ and $D_{(\lambda,x)}\phi_c(0,|c|,c/|c|)$ is nonsingular, so that by the Implicit Function Theorem we can solve for $|t| \ll 1$ in a neighborhood of $(0,|c|,c/|c|)$ and write

$$\phi_c(t,\lambda(t),x(t)) = 0, \quad \text{where } \lambda(0) = |c|,\ x(0) = c/|c|.$$

This implies that $\Gamma_a \subset (0,1) \times R \times S^{n-1}$ is not diffeomorphic to a circle.

It remains for us to show

(a) λ is bounded on Γ_c,

(b) Γ_a doesn't "double back" to $t = 0$.

To establish (a), simply note that for a curve in $R \times S^{n-1}$,

$$|\lambda| = |\lambda x| = |G(x,t,c)| \leq \sup_{(t,x)} |G(x,t,c)| .$$

To prove (b), suppose that Γ_a does double back to $t = 0$. Then, from (3), we can see that the two solutions of $\phi(\lambda,x,0,c) = 0$ must be

$$(\lambda,x) = (|c|,c/|c|) \quad \text{and} \quad (-\lambda,-x) = (-|c|,-c/|c|) .$$

In addition, for any $t \neq 0$, if $\phi(\lambda,x,t,c) = 0$ then we must also have $\phi(-\lambda,-x,t,c) = 0$, since $f(x) = f(-x)$. Thus, if Γ_a contains (t,λ,x) it must contain the antipodal point $(t,-\lambda,-x)$. Parameterize Γ_a by arc length and follow the curve from $(|c|,c/|c|)$ and $(-|c|,-c/|c|)$ simultaneously. Since Γ_a is assumed to double back to $t = 0$ there must exist t_0, $0 < t_0 < 1$, such that the midpoint of the curve Γ_a occurs at t_0. At such a point the two branches of Γ_a originating at $(|c|,c/|c|)$ and $(-|c|,-c/|c|)$ must intersect. That is, we must have

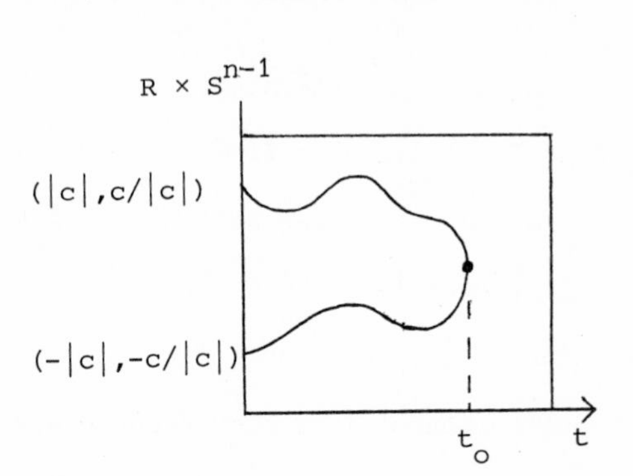

$$(t_0,\lambda(t_0),x(t_0)) = (t_0,-\lambda(t_0),-x(t_0)) .$$

But this can only occur if

$$\lambda(t_0) = -\lambda(t_0) = 0 \quad \text{and} \quad x(t_0) = -x(t_0) = 0 \quad .$$

But this contradicts the construction of x as lying on S^{n-1}, and it follows that Γ_a has its other end at some point $(1,\lambda(1),x(1))$.

For a discussion of how one can follow the curve Γ_a, see Section §7 in [12].

The homotopy method necessarily varies from problem to problem and it is useful to determine what these variations are. It should be emphasized how little change is necessary from the proof of the Brouwer fixed point theorem to our main proof here, even though the results are very different in nature. The trick of showing the antipodal solutions can not meet is also used for the Borsuk-Ulam Theorem [15]. R. Saigal has explained to us in a private communication how the proof could be handled using piecewise linear simplicial decompositions.

References

[1] H. Scarf, The approximation of fixed points of a continuous mapping, SIAM J. Appl. Math. 15 (1967), 1328-1343.

[2] B. C. Eaves, An odd theorem, Proc. Amer. Math. Soc. 26 (1970), 509-513.

[3] H. W. Kuhn, Simplicial approximation of fixed points, Proc. Nat. Acad. Sci. 61 (1968), 1238-1242.

[4] B. C. Eaves, Homotopies for the computation of fixed points, Math. Programming 3 (1972), 1-22.

[5] B. C. Eaves and R. Saigal, Homotopies for the computation of fixed points on unbounded regions, Math. Programming 3 (1972), 225-237.

[6] R. T. Willmuth, The computation of fixed points, Ph.D. Thesis, Dept. Operation Research, Stanford University, 1973.

[7] B. C. Eaves and H. Scarf, The solution of systems of piecewise linear equations, Math. of Oper. Res. 1 (1976), 1-27.

[8] R. B. Kellogg, T. Y. Li and J. A. Yorke, A method of continuation for calculating a Brouwer fixed point. Computing Fixed points with Applications, S. Karamadian, ed., (Meeting Proceedings: Clemson, June, 1974), Academic Press, New York, 1977, 133-147.

[9] R. B. Kellogg, T. Y. Li, and J. A. Yorke, A constructive proof of the Brouwer Fixed Point Theorem and computational results, SIAM J. Num. Anal. 13 (1976), 473-483.

[10] M. Hirsch, A proof of nonretractability of a cell onto its boundary. Proc. Amer. Math. Soc. 14 (1963), 364-5.

[11] L. Watson, Finding fixed points of C^2 maps by using homotopy methods, Computation and Applied Math., to appear.

[12] S. N. Chow, J. Mallet-Paret and J. A. Yorke, Finding zeroes of maps: homotopy methods that are constructive with probability one, to appear in J. Math. Computation.

[13] S. Smale, A convergent process of price adjustment and global Newton methods, J. Math. Econ. 3 (1976), 1-14.

[14] M. Hirsch and S. Smale, personal communication.

[15] J. Alexander and J. A. Yorke, Homotopy continuation method: numerically implementable topological procedure, Transactions Amer. Math. Soc., to appear.

[16] P. Rabinowitz, Some global results for nonlinear eigenvalue problems, J. Functional Anal. 7 (1971), 487-513.

[17] J. L. Kaplan and J. A. Yorke, Nonassociative, real algebras and quadratic differential equations, to appear in J. Nonlinear Analysis.

[18] R. Abraham and J. Robbin, Transversal mappings and flows, Benjamin, N.Y., 1967.

POSITIVE SOLUTIONS OF FUNCTIONAL DIFFERENTIAL EQUATIONS

K. Kunisch and W. Schappacher

1. Introduction.

Let $(H,|.|_H)$ be a real Hilbert lattice, i.e. H is a Hilbert space and a vector lattice such that $|x| \leq |y|$ implies $|x|_H \leq |y|_H$ for $x,y \in H$. For elementary properties of vector lattices we refer to Yosida [9]. We use the notation $x^+ = x \vee o$, $x^- = -(x \wedge o)$ and $|x| = x \vee (-x)$. Elements $x \in H$ satisfying $x = x^+$ are called non-negative and the cone of all such elements is denoted by H^+. For $r > 0$ let $C = C(-r,0;H)$ denote the Banach space of continuous functions from $[-r,0]$ into H endowed with its usual structure as a Banach lattice and norm $||.||$. For a continuous function $x:[-r,a]\to H$, $a > 0$, we write $x_t \in C$ for $x_t(\tau) = x(t+\tau), \tau \in [-r,0]$, $t \in [0,a]$. We consider the following functional differential equation

$$\frac{d}{dt} x(t) = f(t,x_t), \qquad t \geq s \geq 0$$
$$x_s = \varphi, \ \varphi \in C \tag{FDE}$$

where $D(f(t,.)) \subset C$ for each $t \geq 0$.

Contrary to earlier papers on applying semigroup methods to (FDE) in recent papers (e.g.[1],[8]) one does not first construct the solution operator and then calculate its infinitesimal generator but one starts by defining an operator $A(t)$ by

$$D(A(t)) = \{\varphi | \varphi \in C, \dot{\varphi} \in C, \varphi \in D(f(t,.)), \dot{\varphi}(0) = f(t,\varphi)\}$$
$$A(t)\varphi = \dot{\varphi}, \text{ for } \varphi \in D(A(t)).$$

It is known that under certain conditions on f, $A(t)$ is exactly the infinitesimal generator of the solution operator.

Definition 1.1. A family $\{V(t,s); 0 \leq s \leq t\}$ of (nonlinear) continuous operators on C is called an evolution operator if

(i) $V(s,s) = id$ for all $s \geq 0$

(ii) $V(t,s) = V(t,r)V(r,s)$ for all $0 \leq s \leq r \leq t$.

If moreover, there exist constants M, ω such that

(iii) $\|V(t,s)x - V(t,s)y\| \leq M e^{\omega t} \|x - y\|$

for all $x, y \in C$ and $0 \leq s \leq t$, then $V(t,s)$ is said to be of class $G(M,\omega)$.

There are known sets of conditions on f such that $A(t)$ generates an evolution operator $U(t,s)$ of class (P); i.e. $U(t,s)$ is an evolution operator and

$$U(t,s)x = \lim_{n\to\infty} \prod_{i=0}^{n-1} (id - \frac{(t-s)}{n} A(s + \frac{i(t-s)}{n}))^{-1} x, \qquad (1.1)$$

for $x \in C$ and $0 \leq s \leq t$.

As an example we state:

Assume that for $\varphi \in C$, $f(t,\varphi) = f_1(t,\varphi(0)) + f_2(t,\varphi)$ where

(B1) for each $t \in [0,\infty)$, there exists an $a(t) \in R$ such that $f_1(t,.) - a(t)id$ is dissipative, $R(id - \lambda f_1(t,.)) = H$ for $0 < \lambda < \frac{1}{\max(0,a(t))}$ and $\overline{D(f_1(t,.))}$ is independent of t,

(B2) $f_2(t,.)$ is lipschitz continuous.

If (B1), (B2) and a suitable condition on the t-dependence of f hold, then $A(t)$ generates an evolution operator of class (P). (see [1]).

In the remaining part of the paper we concentrate on the operator $U(t,s)$ in (1.1), whose relationship to solutions of (FDE) is well understood in certain cases. Again we describe the result from [1]: Under the above assumptions on f_1 and f_2 the function

$$x(s,\varphi)(t) = \begin{cases} \varphi(t-s) & \text{for } s-r \leq t \leq s \\ (U(t,s)\varphi)(0) & \text{for } t \geq s \end{cases}$$

is a solution of (FDE) if $\varphi \in \{\varphi \in C | \varphi \text{ is lipschitz continuous and } \varphi(0) \in \hat{D}(f_1(t,.))\}$; here $\hat{D}(f_1(t,.))$ denotes the generalized domain of f_1.

2. Positive Semigroups.

Let Y be an arbitrary Banach lattice with norm $||.||$; an evolution operator $V(t,s)$ on Y is called positive if $V(t,s)Y^+ \subset Y^+$, $0 \leq s \leq t$.

In order to characterize the infinitesimal generator of positive linear semigroups Sato and Hasegawa introduced various functionals, which are generalisations of Phillips and Lumer's semi-inner product [6].

In this paper we will use Sato's definition:

Let τ be the right hand Gateaux derivative of the norm:

$$\tau(x,y) = \lim_{\varepsilon \to 0^+} \frac{1}{\varepsilon}(||x+\varepsilon y|| - ||x||)$$

We define $\sigma : Y^+ \times Y \to R$ by

$$\sigma(x,y) = \inf \lim_{a \to \infty} \tau(x,(y+z)\vee(-ax)),$$

where the infimum is taken over all $z \in Y$ satisfying $x \wedge |z| = o$.

Lemma 2.1. [6] σ has the following properties:

(i) $-||y^-|| \leq \sigma(x,y) \leq ||y^+||$

(ii) $\sigma(x,ay) = a\sigma(x,y)$ for all $a \geq 0$

(iii) $\sigma(x,ax+y) = a||x|| + \sigma(x,y)$ for all real a

(iv) if $y \leq z$ then $\sigma(x,y) \leq \sigma(x,z)$

(v) if $x \wedge |y| = o$ then $\sigma(x,z) = \sigma(x,z+y)$

(vi) $\sigma(x,y+z) \leq \sigma(x,y) + \sigma(x,z)$

(vii) $\sigma(x,o) = \sigma(o,y) = 0$

(viii) $-\sigma(x,-y) \leq \sigma(x,y)$

(ix) $|\sigma(x,y) - \sigma(x,z)| \leq ||y-z||$

The following lemma clarifies the connection between σ and a certain class of semi-inner products.

Lemma 2.2. [7] Let $[.,.] \to R$, be a functional $Y \times Y \to R$ such that for real a,b and $x,y,z \in Y$

(i) $[ax + by, z] = a[x,z] + b[y,z]$

(ii) $|[x,y]| \leq ||x||\ ||y||$ $\qquad [x,x] = ||x||^2$

(iii) $[x,x^+] = ||x^+||^2$

(iv) $[x,y] \geq 0$ for $x \geq o$, $y \geq o$.

Then $||x||\ \sigma(x,y) = \max\ [y,x]$ and $-||x||\ \sigma(x,-y) = \min[y,x]$; where the maximum resp. minimum is taken over all functionals that satisfy (i) - (iv).

In the space C, τ and σ can be calculated explicitly. If $(.,.)$ denotes the inner product in H, we get

Lemma 2.3. In C, the functionals τ and σ are given by

$$\tau_C(x,y) = \max_{s \in X(x)} \frac{(x(s),y(s))}{||x||}, \text{ for } x,y \in C,\ x \neq o \tag{2.1}$$

where $X(x) = \{s|\ |x(s)|_H = ||x||\}$, and

$$\sigma_C(x,y) = \tau_C(x,y) \quad \text{for} \quad x \in C^+,\ x \neq o,\ y \in C. \tag{2.2}$$

Proof: Let $x \neq o$ and $s \in X(x)$; i.e. $||x||_C = |x(s)|_H$. Then

$$\frac{1}{\varepsilon}(||x + \varepsilon y|| - ||x||) = \frac{1}{\varepsilon}\frac{(||x + \varepsilon y||^2 - ||x||^2)}{||x+\varepsilon y|| + ||x||}$$

$$\geq \frac{1}{\varepsilon}\frac{((x + \varepsilon y)(s),(x + \varepsilon y)(s)) - (x(s),x(s))}{||x + \varepsilon y|| + ||x||}$$

$$= \frac{1}{\varepsilon}\frac{(2\varepsilon(x(s),y(s)) + \varepsilon^2(y(s),y(s)))}{||x + \varepsilon y|| + ||x||}$$

This implies that $\tau_C(x,y) \geq \frac{(x(s),y(s))}{||x||}$ for $s \in X(x)$. To show the converse, pick $\varepsilon_n \to 0^+$ and let $s_n \in X(x+\varepsilon_n y)$. Since $[-r,0]$ is compact, there exists a subsequence of s_n, again denoted by s_n, such that $s_n \to \bar{s}$. Obviously, $\bar{s} \in X(x)$ and we find

$$\varepsilon_n^{-1}(||x + \varepsilon_n y|| - ||x||) \leq \varepsilon_n^{-1}\left(\frac{((x(s_n) + \varepsilon_n y(s_n), x(s_n) + \varepsilon_n y(s_n))}{||x + \varepsilon_n y|| + ||x||} - \right.$$

$$- \frac{(x(s_n),x(s_n))}{||x+\varepsilon_n y|| + ||x||})$$

$$= \varepsilon_n^{-1} \frac{(2(x(s_n),y(s_n)) + \varepsilon_n^2 |y(s_n)|_H}{||x + \varepsilon_n y|| + ||x||}$$

This implies

$$\tau_C(x,y) \leq \frac{(x(\bar{s}),y(\bar{s}))}{||x||} \leq \max_{s \in X(x)} \frac{(x(s),y(s))}{||x||}$$

To see that (2.2) is true, we use the definition of σ, (2.1), Lemma 2.1 (v) and the claim is proved.

Remark: Note that all our results remain true if H is a uniformly convex Banach lattice since $\tau_C(x,y) = \max_{s \in X(x)} \tau_H(x(s),y(s)) \, ||x||^{-1}$ and (2.2) holds.

Proposition 2.1. Suppose B(t) is the infinitesimal generator of a positive evolution operator V(t,s) on Y. Then for each t

$\sigma(x, -B(t)y) \leq 0$, for $y \in D(B(t))$, $x,y \in Y^+$ and $\sigma(x,y) = 0$ (H1)

holds.

Proof: By Lemma 2.1 (i), (ii) and (vi) we find

$\sigma(x, \frac{1}{h}(y - V(t+h,t)y) \leq \sigma(x,\frac{1}{h}y) + \sigma(x,\frac{1}{h}(-V(t+h,t)y) \leq 0.$

Lemma 2.1 (ix) then implies the result.

In the space C we get:

Theorem 2.1. Let B(t) generate an evolution operator V(t,s) of class (P) on C. Then V(t,s) is positive iff for each $t \geq 0$

$\sigma_C(x,B(t)y) \geq 0$, for all $y \in D(B(t))$ with $x,y \in C^+$ and $\sigma_C(x,y)=0$ (H2)

holds.

Proof: The assertion of the theorem is proved with ideas similar to those used in [7].

The necessity of (H2) follows from Proposition 2.1 and Lemma 2.1 (viii).

To show the 'if' part, let $J_\lambda(t) = (id - \lambda B(t))^{-1}$; $t \geq 0$, which exists for sufficiently small λ, say $\lambda < \lambda'$. It suffices to proof that $y \geq o$ implies $J_\lambda(t)y \geq o$ for these λ. By iteration the result will follow. The goal is to construct functions $y \in C$ that fulfill (H2) and for which $J_{\lambda_o}(t)y(s_1) = o$ and $B(t)J_{\lambda_o}(t)y(s_1) \geq o$ for some $s_1 \in [-r,0]$ and $\lambda_o < \lambda'$ which will lead to a contradiction.
We assume $y(s) > o$ on $[-r,0]$ in the ordering of H; if this is not true we approximate y by function $y+\varepsilon$, $\varepsilon > 0$ being a constant function. Suppose now that $J_\lambda(t)y(s) < o$ for $\lambda < \lambda'$ and $s \in [-r,0]$. Since $J_\lambda(t)u \to u$ for each u as $\lambda \to 0$, and since $\inf_{s\in[-r,0]} y(s) > o$, the infimum over all λ such that $J_\lambda(t)y(s) < o$ for some $s \in [-r,0]$ exists and will be denoted by λ_o, where $\lambda_o \neq 0$. Choose λ_n and s_n so that λ_n decreases to λ_o and $J_{\lambda_n}(t)y(s_n) < o$.
Since $[-r,0]$ is compact we can always find a subsequence of s_n, again denoted by s_n, converging to s_o, for which $J_{\lambda_o}(t)y(s_n)$ converges to $J_{\lambda_o}(t)y(s_o)$. Applying the nonlinear resolvent equation one finds that J_λ is continuous with respect to λ and therefore $J_{\lambda_n}(t)y(s_n)$ converges to $J_{\lambda_o}(t)y(s_o)$. For s_o we have $J_{\lambda_o}(t)y(s_o) = o$.
Now define $x(s) = ||J_{\lambda_o}(t)y|| - J_{\lambda_o}(t)y(s)$. Using the explicit representation of σ_C we find $\sigma_C(x, J_{\lambda_o}(t)y) = o$. (H2) now implies $0 \leq \sigma_C(x, B(t)J_{\lambda_o}(t)y) = \max\,(x(s), B(t)J_{\lambda_o}(t)y(s))$, where the maximum is taken over $\{s|\ |x(s)|_H = ||x||\} = \{s|J_{\lambda_o}(t)y(s) = o\}$.
Therefore there exists an s_1 such that $J_{\lambda_o}(t)y(s_1) = o$ and $(x(s_1), B(t)J_{\lambda_o}(t)y(s_1)) \geq 0$.
Hence $y(s_1) = (id - \lambda_o B(t))J_{\lambda_o}(t)y(s_1) \leq o$, which contradicts $y(s) > o$ on $[-r,0]$. Here we used the fact that $a,b \in H^+$ implies $(a,b) \geq 0$.

3. Application to Functional Differential Equations.

In this section we apply the results of section 2 to the operator $A(t)$ defined in section 1.

Theorem 3.1. Let $A(t)$ generate an evolution operator $U(t,s)$ of class (P). $U(t,s)$ is positive iff for each t

$$(x(0), f(t,y)) \geq 0, \text{ for all } x \in M,\ y \in D(A(t)) \text{ with } y \geq o \text{ and } \sigma_C(x,y) = 0 \qquad (3.1)$$

where $M = \{x \mid x \geq o,\ |x(0)|_H = ||x||\}$.

Proof: By Theorem 2.1 and the representation of σ_C, we only have to show that (H2) holds iff (3.1) is true.

To this end, we have to investigate $(x(s), \dot{y}(s))$ for $s \in X(x)$.

So, let $x, y \in C$ be as in (3.1), and

$X(x) = \{\theta \in [-r,0] \mid |x(\theta)|_H = ||x||\}$.

(α) If $-r \in X(x)$, then $(x(-r), \dot{y}(-r)) = \lim_{h \to 0^+} \frac{1}{h} (x(-r), y(h-r) - y(-r))$

$= \lim_{h \to 0^+} \frac{1}{h} (x(-r), y(h-r)) \geq 0$, since $\sigma_C(x,y) = 0$.

(β) If $s \in (-r,0) \cap X(x)$, we find $(x(s), \dot{y}(s)) = 0$ by a similar argument.

(γ) If finally $0 \in X(x)$, then $(x(0), \dot{y}(0)) = (X(0), f(t,y)) \geq 0$ just as required by (3.1).

Remark 3.1. Let $f \in C$, with $f \geq o$. A condition similar to (3.1) guarantees that $g \leq f$ implies $U(t,s)g \leq f$.

Remark 3.2. In case $U(t,s)$ is only a local evolution operator of class (P) it is obvious that Theorem 3.1 remains true locally.

Example 3.1. Let us first treat the very simple equation

$$\dot{x}(t) = \sum_{i=0}^{n} A_i x(t - r_i) + \int_{-r}^{0} B(s) x(t+s)\, ds, \qquad t \geq 0$$

$x_0 = \varphi \in C(-r,0;R^n)$

where A_i are $n\times n$ matrices, $B(.)$ is an $n\times n$ matrix valued integrable function and $0 = r_0 < r_1 <\ldots< r_n < r$.

It is easy to see that (3.1) holds iff every element of A_i, $i=1,..,n$ is nonnegative, $(a_0)_{i,j}$, $i \neq j$, the i,j-element of A_0 for $i \neq j$ is nonnegative and all elements of B are nonnegative almost everywhere on $[-r,0]$.

Example 3.2. Suppose that in equation

$$\dot{x}(t) = f(t,x_t) = f_1(t,x(t)) + f_2(t,x_t)$$

f is such that the corresponding A(t) generates an evolution operator of class (P). Let $f_1(t,x(t))$ be the infinitesimal generator of a positive evolution operator (e.g. the infinitesimal generator of a translation semigroup, the Δ operator etc.). Then (3.1) will be fulfilled for f iff (3.1) holds for f_2. This is a simple consequence of Proposition (2.1) and Lemma 1.1 (vi).

Example 3.3. To treat the equation

$$\dot{x}(t) = \int_0^t g(s,x(s))\,ds, \quad x(0) = \eta \quad , \; g:R\times R^n \to R^n,$$

we choose for arbitrary $T > 0$ the space $C(-T,0;R^n)$; this would not be a reasonable state space for dealing questions different from the one considered here. It is easily seen that $g(s,z) \geq 0$, for almost all s, and $z \in R^n_+$ is equivalent to (3.1).

References.

[1] Dyson J. and R. Villella Bressan, Functional differential equations and nonlinear evolution operators, *Proc. Royal Soc. Edinburgh* 75 A (1975/76), 223-234

[2] Hale J.K., Theory of Functional Differential Equations, Applied Math. Sciences 3, Springer 1977

[3] Konishi Y., Nonlinear semigroups in Banach lattices, *Proc. Japan Acad.* 47 (1971), 24-28

[4] Konishi Y., Some examples of nonlinear semigroups in Banach lattices, *J. Fac. Sciences Tokyo* 18 (1972), 537-543

[5] Kunisch K. and W. Schappacher, Order preserving evolution operators of functional differential equations to appear in *Boll. U.M.I.*

[6] Sato K.I., On the generators of nonnegative contraction semigroups in Banach lattices, *J.Math. Soc. Japan* 20 (1968) 423-435

[7] Sato K.I., On dispersive operators in Banach lattices, *Pacific J. Math.* 33 (1970), 429-443

[8] Webb G.F., Autonomous nonlinear functional differential equations and nonlinear semigroups, *J. Math. Anal. Appl.* 46 (1974), 1-12

[9] Yosida K., Functional Analysis , 5th ed., Springer (1978)

Karl Kunisch
Institut für Mathematik 2
Technische Universität
Kopernikusgasse 24
A-8010 Graz

Wilhelm Schappacher
Mathematisches Institut
Universität Graz
Elisabethstraße 11
A-8010 Graz

A Restart Algorithm without an Artificial Level for Computing Fixed Points on Unbounded Regions.

G. van der Laan and A.J.J. Talman.

1. Introduction.

In this paper we generalize the algorithm introduced by Van der Laan and Talman [3] for computing a fixed point of an upper semi-continuous point-to-set mapping to unbounded regions. This algorithm is a restart method not using an extra dimension with artificial labelled points like the Sandwich method of Merrill [4] and Kuhn and MacKinnon [2]. The special feature of the algorithm is that it does not start in a certain grid with a subsimplex, but with one point only, chosen arbitrary or by prior information. From this point, a zero dimensional facet, the algorithm generates by alternating replacement and pivot steps a path of adjacent facets and terminates with a subsimplex, which provides a good approximation of the fixed point and can be used as the starting point for a new application of the algorithm in a finer grid. The factor of incrementation may be of any size and even different at each stage. So it may depend on the accuracy obtained in the last stage.

In secton 2 we give some preliminaries. In section 3 we present the algorithm including the convergence proof and a geometrical interpretation. Computational experience is given in section 4 and conclusions are drawn in section 5.

2. Preliminaries.

Let I_m be the set of integers $\{1,2,\ldots,m\}$ and let T be any subset of I_{n+1} with t is the number of elements. Define the n x (n+1) matrix Q by

$$Q = \begin{bmatrix} 1 & 0 & \cdots & 0 & -1 \\ 0 & 1 & \cdots & 0 & -1 \\ \vdots & \vdots & \ddots & \vdots & \vdots \\ 0 & 0 & \cdots & 1 & -1 \end{bmatrix}.$$

The ith column of Q will be denoted by q(i). This matrix Q induces the K-triangulation of the real space (Todd [10]). For a given T let γ^T be a permutation of T and $v^o, v^1, \ldots, v^t$ points of R^n such that $v^j = v^{j-1} + q(\gamma_j)$, j=1,...,t. Then the convex hull of the points $v^0, v^1, \ldots, v^t$ is called a t-dimensional facet of the K-triangulation. A subsimplex is an n-dimensional facet. Two t-dimensinal facets or a t-dimensional and a (t-1)-dimensional facet (t ≥ 1) will be called adjacent if they have t vertices in common.

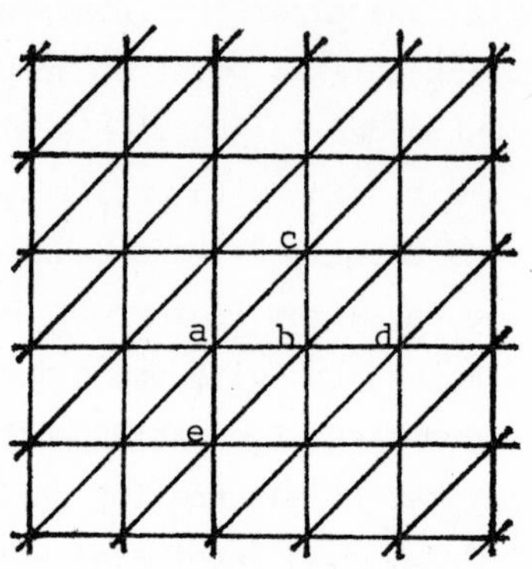

figure 2.1.

For example, see fig 2.1.; the one-dimensional facet with vertices a and b is adjacent to the zero-dimensional facets a and b, to the one dimensional facets {b,c}, {b,d} etc. and to the subsimplices {a,b,c} and {a,b,e}. Let Φ be an upper-semicontinuous mapping from R^n to R^n. The points of R^n receive a label induced by the mapping Φ. In the case of integer labelling the labels are determined by the following rule

$$\ell(x) = i \text{ if } f_i(x) - x_i \geq f_k(x) - x_k, \text{ all } k \in I_n, \text{ and } f_i(x) - x_i \geq 0;$$
$$= n+1 \text{ if } f_k(x) - x_k < 0, \text{ all } k \in I_n$$

where f(x) is some element of Φ(x).
A facet will be called completely labelled if all its vertices have a different label. Note that a completely labelled subsimplex is a good approximation of a fixed point Φ (see Saigal [5]).
In the case of vector labelling a point x in R^n receives the vector $\ell(x) \in R^{n+1}$ defined by

$$\ell_i(x) = f_i(x) - x_i + c_i, \quad i = 1,\ldots,n \quad \text{and} \quad \ell_{n+1}(x) = c_{n+1}$$

where f(x) is again some element of Φ(x) and c is an arbitrary positive vector in R^{n+1}.

A t-dimensional facet of the K-triangulation with vertices $v^0,v^1,..,v^t$ $(t \le n-1)$ induced by a certain T, γ^T and v^0, will be called completely labelled if the set of linear equations

$$\sum_{i=0}^{t} \lambda_i \ell(v^i) + \sum_{j=1}^{n-t} \mu_j e(\pi_j) = c$$

has a non-negative solution for λ_i and μ_j, where e(j) is the jth unit vector in R^{n+1} and $(\pi_1,...,\pi_{n-t})$ is some permutation of the (n-t) elements of I_{n+1} not in T.

A subsimplex $v^0,v^1,...,v^n$ will be called completely labelled if

$$\sum_{i=0}^{n} \lambda_i \ell(v^i) = c$$

has a non-negative solution λ_i^*. Observe that $\Sigma\lambda_i^* = 1$ and that $\sum_{i=0}^{n} \lambda_i^* v^i$ is a good approximation of a fixed point of Φ (see Saigal [5]).

3. The algorithm.

We give only the description of the algorithm for vector labelling. For the application with integer labelling we refer to Van der Laan and Talman [3]. Let D be a real positive number and s a point in R^n. By K(D,s) we denote the set of points x such that x=y/D+s, where y is a vertex of the K-triangulation. Observe this is a one-to-one correspondence. In the sequel a point of K(D,s) will be represented by its correspondence in K. Starting in a point v^0 and with a gridsize m, the algorithm searches via a path of adjacent facets a completely labelled simplex of $K(m,v^0)$, i.e. the algorithm starts with the zero vector of K. Assume now, that for a certain subset T of I_{n+1}, for a permutation γ^T of T and a point w^0 in R^n such that $w^0 = \sum_{j \in T} R_j q(j)$ for for some (n+1)-dimensional nonnegative integer vector R (i.e. $R_j=0$, $j \in I_{n+1}/T$), the convex hull of the points $w^0,w^1,...,w^{t-1}$, where $w^j = w^{j-1} + q(\gamma_j)$, j = 1,..,t-1, is a completely labelled (t-1)-dimensional facet.

Observe that this assumption is fullfilled for $w^0 = v^0$, $T = \{i_0\}$, $\gamma^T = i_0$ and R = (0,..,0) where i_0 is the index with the greatest ratio of $\ell_i(w^0)/c_i$, i = 1,..,n+1, or equivalently, i_0 is the index of the unit vector which is eliminated by pivoting the vector $\ell(w^0)$ in the set of linear equations Ay = c with A = I. The completely labelled (t-1)-dimensional facet $w^0,w^1,..,w^{t-1}$ is now extended to a t-dimensional facet by adding the new vertex $w^t = w^{t-1} + q(\gamma_t)$.

Then its label $\ell(w^t)$ is computed and a pivot step is made with this vector in the set of linear equations $Ay = c$, where $\ell(w^i)$ and $e(j)$, $j \in I_{n+1}/T$ form the columns of A. Without loss of generality we assume that $\ell(w^j)$ is eleminated for some j, $j \leq t-1$. Then we adapt the vector w^0, the permutation γ^T and the vector R according to tabel 3.1 with $s = j$, i.e. w_j is replaced by a new vector, say $\bar{w}$, whose label is computed, and so on. By alternating pivot and replacement steps the algorithm generates by this way a path of adjacent t-dimensional facets.

Table 3.1. s is the index of the vector which must be replaced.

	w^0 becomes	γ^T becomes	R becomes
s=0	$w^0 + q(\gamma_1)$	$(\gamma_2,\ldots,\gamma_t,\gamma_1)$	$R + e(\gamma_1)$
$1 \leq s \leq t-1$	w^0	$(\gamma_1,\ldots,\gamma_{j-1},\gamma_{j+1},\gamma_j,\gamma_{j+2},\ldots,\gamma_t)$	R
s=t	$w^0 - q(\gamma_t)$	$(\gamma_t,\gamma_1,\ldots,\gamma_{t-1})$	$R - e(\gamma_t)$

We shall prove that only two cases can occur:

a) the unit vector $e(k)$ has to be eleminated by a pivot step, for some k not in T.

b) R_h becomes negative by a replacementstep, for some $h \in T$.

In case a) the above mentioned assumption is fullfilled for $T = T \cup \{k\}$, $\gamma^T = (\tilde{\gamma}^T, k)$, $w^0 = \tilde{w}^0$ and $R = \tilde{R}$, where $\tilde{\gamma}^T$, $\tilde{w}^0$ and $\tilde{R}$ are the last found in table 3.1. The algorithm continues with the new vector $w^{t+1} = \tilde{w}^t + q(k)$ by computing its label etc.

In case b) table 3.1 was applied for $s = t$; hence $h = \tilde{\gamma}_t$. The last vertex $\tilde{w}^t$ is now dropped, $R_{\tilde{\gamma}_t}$, which became minus one, is set equal to zero, T becomes $T \setminus \{\tilde{\gamma}_t\}$ and γ^T becomes $(\tilde{\gamma}_1,\ldots,\tilde{\gamma}_{t-1})$. The algorithm continues by pivoting the $\tilde{\gamma}_t$ th unit column in the set of linear equations $\tilde{A}y = c$, where $\tilde{A}$ is the current basismatrix.

So, to approximate a fixed point (given a gridsize m) we start the algorithm in a point v^0, we make a pivot step with the vector $\ell(v^0)$ in the set of linear equations $Ay = c$ with $A = I$, and we set $T = \{i_0\}$, $\gamma^T = i_0$ and $R = (0,0,\ldots,0)$, where i_0 is defined as above. Below it is proved that the algorithm either terminates with a completely labelled subsimplex by applying the above described method or it generates a path of facets for which some components of w^0 goes to infinity permitting us to draw some new conclusions regarding the mapping. Before proving the termination of the algorithm we give a geometrical interpretation.

For a given gridsize m and a starting point v_0 the points of $K(m,v^0)$ are divided in subregions $A(T)$, $T \subsetneq I_{n+1}$, defined by

$$A(T) = \{x \in K(v^0,m) \mid x = v^0 + \frac{1}{m} \sum_{j \in T} R_j\, q(j),\ R_j \text{ nonnegative and integer}\}$$

Any element of $K(v^0,m)$ is an element of only one subregion $A(T)$ such that $R_j > 0$, all $j \in T$. Such a point is called a proper point of $A(T)$, e.g. b is a proper point of $A(1,2)$, with $R_1 = 1$ and $R_2 = 2$, whereas c is a proper point of $A(1)$ with $R_1 = 3$ (see figure 3.1).

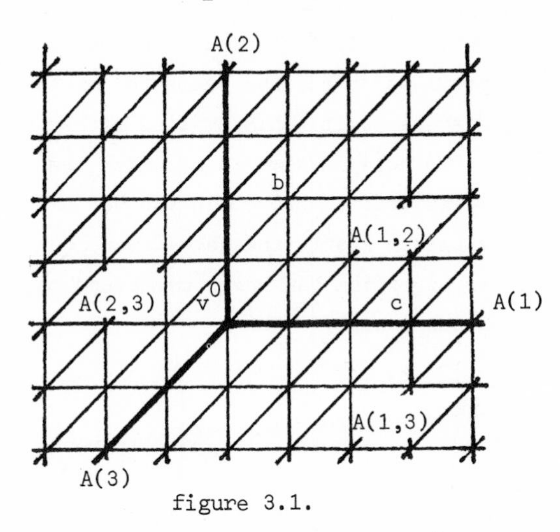

figure 3.1.

When the algorithm generates a set T, γ^T, w^0 and R then the points $w^0,..,w^t$, where $w^j = w^{j-1} + q(\gamma_j)$, $j = 1,..,t$, are all elements of $A(T)$, since $w^0 = \sum_{j \in T} R_j q(j)$. Observe that w^t is always a proper element of $A(T)$ and that if w^s is proper in $A(T)$ for some s, $0 \le s \le t-1$, w^i is also proper in $A(T)$, for all $i > s$.

If now a replacement step should imply that the jth component of R becomes negative, then a change would be made from a t dimensional facet in $A(T)$ to an adjacent t-dimensional facet in the subregion $A(T \cup \{k\} \setminus \{j\})$, for some $k \notin T$. It is easy to see that this can happen if and only if w_t, which is the candidate to be removed, is the only proper point of $A(T)$. Instead of removing, w^t is deleted and the algorithm continues with an (t-1)-dimensional facet with vertices in $A(T \setminus \{j\})$ and a pivot step is made with the unit vector $e(j)$.

Except in the above mentioned case that the algorithm generates a path of facets, which goes to infinity, we will prove now that the algorithm finds a completely labelled subsimplex within a finite number of steps. The unicity of the replacement step between two t-dimensional facets has been proved by Scarf [7]. Van der Laan and Talman [3] proved the unicity of the change from a t-dimensional facet to a (t-1)- or a (t+1)-dimensional facet.

This together implies that a t-dimensional facet $w^0,\dots,w^t$ can be reached from only two adjacent facets. They can be (t-1), t or (t+1)-dimensional. Moreover, the pivot steps are unique. Consequently, if a facet has already provided a feasible basis for the algorithm, so must the two positions adjacent to it. This implies either the algorithm cycles for ever, or if there is some facet which can never be revisited, no facet can ever be revisited. As proved in Van der Laan and Talman [3] the algorithm has a unique start (the zero-dimensional facet v^0). So it follows, that the algorithm can never cycle and that it finds a completely labelled subsimplex.

4. Computational results.

To test the algorithm we solved the six problems mentioned in table 4.1.. The economy problems were converted to problems in R^n by setting the price of the last good equal to one. Concerning the labelling rule, for the pure exchange models a price vector was labelled with the corresponding demand vector, while c_i was set equal to the total initial resources of commodity i, $i = 1,\dots,n+1$. In the fourth problem the labelling rule described in section 5.2 of Scarf [8] was used, in problem five the labelling rule

$\ell_i(x) = \frac{\partial f(x)}{\partial x_i} + 1$, $i = 1,\dots,n$ and in the last problem the standard labelling rule with $c = \iota$.

When, for gridsize m and starting point v^o, a completely labelled subsimplex $w^o,w^1,\dots,w^n$ was found, the approximated fixed point $w^* = \Sigma_i \lambda_i^* w^i$, with $\lambda_0^*,\dots,\lambda_n^*$ the last found solution of $Ay = c$, was chosen as the starting point for a next application of the algorithm. The new gridsize was then set equal to 2m for the last two problems and equal to $\max \{2m, \min_{i=1,..,n} (1/|w_i^*-v_i^o|)\}$ for the economy problems.

Problem	Dimension	Starting point in the first stage	Initial gridsize	Final gridsize	Achieved accuracy	Number of function evaluations
1	4	1,1,1,1	1.2	164	10^{-4}	47
				101,916	4.10^{-11}	69
2	7	1,1,...,1	1.125	142	6.10^{-5}	72
				11,261	2.10^{-8}	80
3	9	1,1,...,1	1.1	341	5.10^{-5}	102
				5,935	10^{-8}	123
4	5	1,1,1,1,1	1.16667	74.6	5.10^{-3}	176
5	4	-3,-1,-3,-1	0.5	128	5.10^{-4}	107
6	20	0,0,...,0	0.5	4	2.10^{-5}	114
				16	10^{-9}	158

Table 4.1. Problems 1,2,3: The pure exchange economy problems of Scarf [7].
Problem 4 : The production economy problem of Scarf [8].
Problem 5 : Unconstrained optimization problem of Colville [1].
Problem 6 : Problem 2 of Saigal [6].

The computational results are shown in table 4.1.. Only for problem 4 there was a difference between the number of pivot steps (184) and the number of function evaluations, i.e. only for problem 4 pivot steps were made with a unit vector.

The accuracy of the final solutions for the exchange economy models was the supnorm of the excess demand vector. For the production model the accuracy was the maximum absolute value of the difference between the final price vector and that reported by Scarf [8] and for the last two problems $\max_i |\ell_i(x) - 1|$.

5. Conclusions.

A comparison with the results for the above mentioned problems as reported by Wilmuth [12], Saigal [6] and Todd [11] shows that our algorithm takes significantly fewer iterations than the other restart algorithms as well as the homotopy method. The computation time of a evaluation is the same as for other algorithms, whereas the computation time of a pivot and replacement step is of the same order.

Todd [11] improved the restart algorithms by twisting the labels of the artificial level to achieve a better agreement between the labels of the points on the artificial level and the natural level. So, by twisting, the number of pivot steps with artificial labels can be decreased. In our algorithm twisting is of no importance since the algorithm operates only on the natural level.

Moreover Todd improves in his paper the restart algorithms by taking quasi-Newton steps. Of course this can also be implemented in our algorithm. In the exchange economy problems, the computational results show that the factor if incrementation increases very fast which is caused by the indifferentiability of the functions. However, in the production model, the factor of incrementation was always two, which could be expected because of the non-differentiability of the mapping. Further study is needed to decide how the information obtained from the last approximation can be used to determine the next factor of incrementation (see Saigal [6]).

We remark that for all problems the number of iterations was very small in the first levels. In problem 6 the algorithm took 37 iterations for the first stage and 28 for the second. In the approximated fixed point w^* then, $|f_i(w^*) - w_i^*|$ was in the range $(6.10^{-6}, 5.10^{-2})$. The reason of this feature is that the 21st unit vector was first removed from the feasible basis $Ay = c$ with $A = I$, implying that the second point found by the algorithm was the vector $-\iota$ by applying the last column of Q on the starting point (1,1,...,1). Observe, this is not possible in the other restart algorithms for they use a vector for the link between the natural and artificial level.

Finally, we remark that we present the algorithm only for the K-triangulation. Other triangulations (see Todd [9] and [10]) can be easily implemented. We will make a comparison between the several triangulations in a subsequent paper.

References.

[1] Colville, A.R.,"A Comparitive Study on Nonlinear Programming Codes", I.B.M.New York Scientific Center Report, no. 320-2949, June 1968.

[2] Kuhn, H.W. and MacKinnon, J.G., "Sandwich method for Finding Fixed Points", *Journal of Optimization Theory and Applications,* Vol. 17, pp. 189-204, 1975.

[3] Laan, G. van der, and Talman, A.J.J., "A New Algorithm for Computing Fixed Points", Submitted to *Mathematical Programming.*

[4] Merrill, O.H.,"Applications and Extensions of an Algorithm that Computes Fixed Points of Certain Upper Semi-continuous Point to Set Mappings", Ph.D. Dissertation, University of Michigan, 1972.

[5] Saigal, R., "Investigations into the Efficiency of the Fixed Point Algorithms", *Fixed Points: Algorithms and Applications,* S. Karamardian, Ed., Academic Press, New York, pp. 203-223, 1977.

[6] ———————, "On the Convergence of Algorithms for Solving Equations that are Based on Methods of Complementary Pivoting", *Mathematics of Operations Research,* Vol. 2, pp. 108-124, 1977.

[7] Scarf, H., "The Approximation of Fixed Points of a Continuous Mapping", *SIAM Journal on Applied Mathematics,* Vol. 15, pp. 1328-1343, 1967.

[8] ———————, "The Computation of Economic Equilibria" (with the collaboration of T. Hansen), Yale University Press, New Haven, Connecticut, 1973.

[9] Todd, M.J., "On Triangulations for Computing Fixed Points", *Mathematical Programming,* Vol. 10, pp. 322-346, 1976.

[10] ————, "Union Jack Triangulations", *Fixed Points: Algorithms and Applications,* S. Karamardian, Ed., Academic Press, New York, pp. 315-336, 1977.

[11] ————, "Improving the Convergence of Fixed-Point Algorithms" *Mathematical Programming Study,* Vol. 7, pp. 151-169, 1978.

[12] Wilmuth, R.J., " A Computational Comparison of Fixed Point Algorithms Which Use Complementary Pivoting", *Fixed Points: Algorithms and Applications,* S. Karamardian, Ed., Academic Press, New York, pp. 249-280, 1977.

Department of Actuarial Sciences and Econometrics,
Free University, Amsterdam, The Netherlands.

PATH FOLLOWING APPROACHES FOR SOLVING NONLINEAR EQUATIONS: HOMOTOPY, CONTINUOUS NEWTON AND PROJECTION

by

Tien-Yien Li[1,3]

and

James A. Yorke[2,4]

§1. Introduction

Many mathematical problems have the form (or may be reformulated to)

$$F(x) = 0 \tag{1}$$

where $F : R^n \to R^n$ is continuous and has partial derivatives of higher order. Write $DF(x)$ for the $n \times n$ matrix of partial derivatives. The Newton method for solving this equation is to choose an initial point x^1 and iteratively define

$$x^{n+1} - x^n = - DF(x^n)^{-1} F(x^n). \tag{2}$$

This method works well if x^1 is a good approximation to a solution x^0 of (1) and $DF(x^0)$ is invertible, and one often obtains convergence to a solution even when there appears to be no reason to expect it. A variant of this approach is to introduce a constant σ, and define

$$x^{n+1} = x^n - \sigma(DF(x))^{-1}F(x^n). \tag{3}$$

---

1. Department of Mathematics, Michigan State University, East Lansing, Michigan, 48824.

2. Institute for Physical Science and Technology, College Park, Maryland, 20742

3. Research was partially supported by the National Science Foundation under grant MCS-78-02420.

4. Research was partially supported by the National Science Foundation under grant MCS-76-24432.

Convergence is improved in various problems. Generally speaking, the algorithm (2) is highly localized in the sense that a 'good' initial point x^1 is often difficult, if it is not impossible, to get in higher dimension problems. In recent years, a number of ideas have been developed to search for global algorithms which are implementable by computer to solve (1). See, for example, Kellogg, Li, and Yorke [5], Smale [8], Hirsch and Smale [4], and Chow, Mallet-Paret, and Yorke [1]. The purpose of this note is to relate all these global approaches and cast into a bigger framework, which we call the projection method. We will introduce this concept in the next section and follow with examples which include all the approaches mentioned above. We emphasize here only the smooth path-following approaches for solving (1), while similar piecewise linear approaches studied by Scarf [7], Eaves [2], Eaves and Saigal [3], Kojima [6], Todd [9], will not be discussed.

§2. Projection

Let F be a C^2 function map an open subset M in R^n to R^n. Let S be an $n-1$ dimensional manifold. The projection method is a method of solving equation (1) by defining a C^2 mapping P from an open subset D of M to S, such that for any regular value y in S there is a smooth curve in $P^{-1}(y)$ leads to the solution x_0 of $F(x)$ where P is not defined. (We say y is a regular value of P if $DP(x)$ is of rank $n-1$, or full rank, for all $x \in P^{-1}(y)$.) By Sard's Theorem, for almost every y in S (in the sense of $n-1$ dimensional measure) y is regular. It follows that, by repeated application of Implicit Function Theorem, that for a regular value y, $P^{-1}(y)$ consists of smooth 1-manifold, which is either homeomorphic to the circle or an open interval on the real line.

§3. Examples

Example 1: Computing the Brouwer fixed point.

Let M be an open convex ball in R^n, and ∂M be its boundary. Let $G : \bar{B} \to \bar{B}$ be C^2. Let $C = \{x | G(x) = x\}$ be the set of the fixed point. For $x \in M = B \backslash C$ we define $P(x)$ to be the point of intersection of ∂B with the ray drawn from $F(x)$ through x. P projects M onto ∂B. In [5] we proved that for almost every $x^0 \in \partial B$ the connected component of the set $P^{-1}(x^0)$ which contains x^0 is a curve $x(t)$, $0 < t < T \leq \infty$ and for each $\epsilon > 0$ there is a $t(\epsilon) < T$ such that for $t(\epsilon) < t < T$, $\text{dist}(x(t),C) < \epsilon$. This result has been implemented successfully on the computer [5] and thus is one way of making Brouwer Fixed Point Theorem constructive.

It is interesting to see the limiting behavior of the curve $x(t)$. Since $x(t)$ satisfies the relation $P(x(t)) = x_0$, we may write, for any projection equation

$$DP(x)\frac{dx}{dt} = 0. \tag{4}$$

On the other hand, $P(x)$ lies on the ray from $G(x)$ through x, we may write

$$P(x) = (1-\mu(x))x + \mu(x)G(x)$$

where $\mu(x) : B\backslash C \to R^1$ is chosen so that $\mu(x) < 0$ and $P(x) \in \partial B$. Then, the action $DP(x)$ on any vector v is given by

$$DP(x)v = [(1-\mu(x))I + \mu(x)G'(x)]v + (\mu'(x),v)(G(x)-x).$$

Set $A = (1-\mu(x))I + \mu(x)G'(x)$. Then, if A is nonsingular, $DP(x)$ has rank $n-1$ and $v(x) = (A(x))^{-1}(G(x)-x)$ is a null vector of $DP(x)$. This leads to

$$\frac{dx}{dt} = -(DG(x)-I+\mu(x)^{-1}I)^{-1}(G(x)-x). \tag{5}$$

When $x \to C$, $\mu(x) \to -\infty$ and (5) becomes

(6) $$\frac{dx}{dt} = -(DG(x)-I)(G(x)-x).$$

By letting $F(x) = G(x) - x$, we write (6) as

(7) $$\frac{dx}{dt} = -(DF)^{-1}F(x).$$

Example 2: Continuous Newton Method.

Let $F : R^n \to R^n$ be C^2 and $C = \{x|F(x) = 0\}$. Define $P : R^n\backslash D \to S^{n-1}$ by

$$P(x) = \frac{F(x)}{\|F(x)\|}$$

where $\|\cdot\|$ denotes the Euclidean norm and S^{n-1} is the unit sphere. For $x_0 \in R^n\backslash C$, let $\Gamma = \Gamma(x_0)$ be $\{x : P(x) = P(x_0)\}$ and let Γ_0 be the connected component $x(s)$ of Γ which contains x_0. Consider

$$P(x(s)) = P(x_0) \quad \text{and} \quad DP(x)x' = 0$$

where x and x' denotes $x(s)$ and $\frac{d}{ds}x(s)$. For P defined as above, we obtain

$$0 = \frac{d}{ds}P(x(s)) = \frac{(\frac{d}{ds}F)\|F\| - F\frac{d}{ds}\|F\|}{\|F\|^2}$$

i.e. (8) $$(\frac{d}{ds}F)\|F\| - F\frac{d}{ds}\|F\| = 0.$$

Now,

$$\frac{d}{ds}\|F\| = \frac{d}{ds}\sqrt{\langle F,F\rangle}$$
$$= \frac{1}{2}[\langle F,F\rangle]^{-\frac{1}{2}}\frac{d}{ds}\langle F,F\rangle = \|F\|^{-1}\langle F,\frac{d}{ds}F\rangle$$

where $\langle\,,\,\rangle$ denotes the inner product in R^n.

Substituting into (8), we obtain

$$\|F\| \frac{d}{ds} F = F\|F\|^{-1} < F, \frac{d}{ds} F >$$

and

$$(DF)x' = \sigma(x)F(x)$$

where $\sigma(x) = < F, (DF)x' >/\|F\|^2$. If we make the questionable assumption that DF is invertable along the path we have

(9) $$x' = \sigma(x)(DF(x))^{-1}F(x).$$

The similarity between (3) and (9) results in the equation (9) to be called 'Continuous Newton Method' for solving nonlinear equations. If σ in (3) is choosen to be small the sequence x_n will approximate the solution $x(t)$ of (9). Smale [8] has pointed out circumstances under which the path $x(s)$ is smooth and leads to a solution x_0 of (1), yet there DF is not invertible. As one follows the path, $\|F(x(s))\|$ is not monotomic and at a local minimum and maximum DF is not invertible. The power of Smale's technique is that DP will generally have full rank (is rank $n - 1$) even at the points where $DF(v)$ is not invertible. The "correct" equation is (4).

Example 3: A Homotopy Method which can be Turned into a Projection Method.

Suppose $F(x) = Ax + b(x)$ where A is an invertible matrix and $b(x)$ is C^2 and bounded. We may find zeros by choosing x_0 and homotoping $A(x - x_0)$ to $F(x)$, i.e. define the homotopy

(10) $$H(x,t) = (1-t)A(x-x_0) + t\,F(x)$$

By defining $P : R^n x(0,1) \to R^n$ by essentially solving for x_0, that is

$$(11) \qquad P(x,t) = x + \frac{t}{1-t} A^{-1}F(x),$$

we see that for any regular value x_0 of P the curve $(x(s),t(s))$ starting from $x_0,0$ and satisfying

$$P(x(s),t(s)) = x_0$$

is exactly the curve satisfying $H(x(s),t(s)) = 0$. Hence, when $t(s) \to 1$, $x(s) \to C = \{x|F(x) = 0\}$. Some other homotopies can not be readily placed in the projection.

§4. Continuous Newton and Homotopy

The path found in Smale's continuous Newton in Example 2 above is also the path from a homotopy and a path from a homotopy is also a path from a continuous Newton.

First we formulate the continuous Newton as a homotopy (as in [1]. Let $F : R^n \to R^n$. Choose x_0 such that $DF(x_0)$ is non-singular. Write

$$(12) \qquad H(x,t) = F(x) - (1-t)F(x_0) = 0 \quad \text{where} \quad 0 < t < 1.$$

If $(x(s),t(s))$ is a curve satisfying (12), then

$$P(x(s)) = \frac{F(x(s))}{\|F(x(s))\|} = \frac{(1-t)F(x_0)}{\|(1-t)F(x_0)\|} = \frac{F(x_0)}{\|F(x_0)\|} \in S^{n-1}$$

which is a constant. In this context, if we choose x_0, the problem

$$F(x) - F(x_0) = 0$$

is trivial in that at least one solution is known. If we homotopy this equation to "$F(x) = 0$", then we obtain

$$(1-t)(F(x) - F(x_0)) + t\,F(x) = 0$$

which is equivalent to (12). Either both homotopy and continuous Newton Method will produce a solution or neither will. Further conditions are needed to guarantee the method will "work".

Next, we take any homotopy path and reformulate it to be a continuous Newton Method path.

Consider $H = R^n \times R \to R^n$ and the general homotopy equation

$$H(x,t) = 0 \quad \text{where} \quad H(x_0,0) = 0 \tag{13}$$

Write $\tilde{F}(x,t) = (H(t,x), 1-t) : R^{n+1} \to R^{n+1}$. A solution path $(x(s),t(s))$ satisfying $H(x,t) = 0$ also satisfies

$$\tilde{P}(x,t) = \frac{\tilde{F}(x,t)}{\|\tilde{F}(x,t)\|} = \frac{(1-t,H(x,t))}{\sqrt{(1-t)^2 + H^2}} \equiv (1,0) \in S^n \subset R \times R^n \tag{14}$$

which is a constant. Notice that if a continuous Newton problem is turned into a homotopy and the latter is turned back into a continuous Newton problem, the dimension $(n+1)$ of the space at the end is one greater than the dimension (n) of the space at the beginning. If we let

$$\tilde{P}(x,t) = \frac{\tilde{F}(x,t)}{\|\tilde{F}(x,t)\|},$$

then $\tilde{P}(x,t) = (1,0)$ implies $\frac{H(x,t)}{1-t} = 0$. We have proceeded formally here. Usually homotopies in [1] are allowed to depend on a parameter which is chosen at random. After the parameter is chosen, the resulting path can be seen to also be a continuous Newton path.

While any continuous Newton Problem may be reformulated as a Homotopy Problem, projection methods are more general and do not generally appear to be translatable directly into homotopies.

References

1. S. N. Chow, J. Mallet-Paret and J. A. Yorke, Finding zeros of maps: Homotopy methods that are constructive with probability one, J. of Math. Comp., in press.

2. B. C. Eaves, Homotopies for the computation of fixed points. Math. Programming 3 (1972), 1-22.

3. B. C. Eaves and R. Saigal, Homotopies for the computation of fixed points on unbounded regions, Math. Programming, 3 (1972), 225-237.

4. M. Hirsch and S. Smale, personal communication.

5. R. B. Kellogg, T. Y. Li and J. A. Yorke, A constructive proof of the Brouwer Fixed Point Theorem and computational results, SIAM J. Num. Anal., 13 (1976), 473-483.

6. Kojima, M., Computational methods for solving the nonlinear complementarity problem, Keio Engineering Reports, 27 1 (1974), 1-41.

7. H. Scarf, The approximation of fixed points of a continuous mapping, SIAM J. Appl. Math. 15 (1967), 1328-1343.

8. S. Smale, A convergent process of price adjustment and global Newton methods, J. Math. Econ. 3 (1976), 1-14.

9. N. J. Todd, Union Jack Triangulations, TR220, Department of Operations Research, Cornell, University.

A NONLINEAR SINGULARLY PERTURBED VOLTERRA FUNCTIONAL DIFFERENTIAL EQUATION

John A. Nohel*

University of Wisconsin
Madison, Wisconsin, USA

Abstract. We study the Cauchy problem for the nonlinear Volterra f.d.e. with infinite "memory"

$$(V)\quad \begin{cases} -\mu y'(t) = \int_{-\infty}^{t} a(t-s)F(y(t), y(s))\,ds & (t > 0) \\ y(t) = g(t) & (-\infty < t \leq 0), \end{cases}$$

where $\mu > 0$ is a small parameter, $' = d/dt$, a is a given real kernel, and F, g are given real functions; (V) models the elongation ratio of a homogeneous filament of a molten polyethelene which is stretched on the time interval $(-\infty, 0]$, then released to undergo elastic recovery for $t > 0$. Under physically reasonable assumptions concerning a, F, g we present results on qualitative behaviour of solutions of (V) and of the corresponding reduced equation when $\mu = 0$, as well as the relation between them as $\mu \to 0^+$, both for t near zero and for large t. These results, obtained jointly with A.S. Lodge and J.B. Mc Leod, show that in general the filament of polyethelene never recovers its original length, and that the affect of the Newtonian term $-\mu y'(t)$ in (V) is highly significant during the early part of the recovery, but not in the ultimate recovery. We also present an implicit finite difference scheme, recently developed by O. Nevanlinna, to obtain a discretization of (V) which exhibits most of the same qualitative properties as the continuous model.

---

* Sponsored by the United States Army under Grant Number DAAG 29-776-0004 and Contract Number DAAG 29-75-C-0024.

1. Introduction. Molten plastics exhibit large elastic recovery. In some experiments conducted by Meissner [4] and refined by A.S. Lodge a homogeneous filament of a certain polyethelene when elongated at a rate of 1 cm/sec/cm from an initial length of 1 cm to 55 cm and then released, reached a final length of 5 cm during recovery. Such rheological phenomena in materials with "memory" are of interest in the processing of plastics and rubber.

The purpose of this lecture, based primarily on recent joint work with A.S. Lodge and J.B. Mc Leod [3], is to analyse the following mathematical model which includes the above process in polymer rhelogy. We refer to [3, Appendix A] for a derivation of the equation of motion, (1.1) below, arrived at by combining a certain constitutive law with the stress equations of motion, and by neglecting certain inertial and body forces, as well as by assuming that the volume of the filament remains constant. We consider the Cauchy problem

$$(1.1)\qquad \begin{cases} -\mu y'(t) = \int_{-\infty}^{t} a(t-s)\, F(\, y(t), y(s)\,)\, ds & (t > 0\,;\ ' = d/dt\,) \\ y(t) = g(t) & -\infty < t \le 0, \end{cases}$$

where $\mu > 0$ is a small parameter, a, F, g are given real functions. In the above mentioned class of experiments μ is a parameter related to the viscosity, and

$$(1.2)\qquad a(t) = \sum_{k=1}^{m} a_k \exp(-t/\tau_k) \qquad (a_k > 0,\ \tau_k > 0,\ k = 1, \dots, m),$$

$$(1.3)\qquad F(y, z) = (y^3/z^2) - z,$$

$$(1.4)\qquad g(t) = \begin{cases} 1 & \text{if } -\infty < t \le -t_0 \quad (t_0 > 0) \\ \exp \kappa(t + t_0) & \text{if } -t_0 < t \le 0 \end{cases}$$

The unknown function y measures the ratio of the extended length to the original length of a homogeneous filament which is stretched according to the law $y = g(t)$ on $-\infty < t \le 0$ and then released. Equation (1.1) describes the process of recovery. Also of interest is the reduced problem

$$(1.5)\qquad 0 = \int_{-\infty}^{t} a(t-s)\, F(\, y(t), y(s)\,)\, ds \qquad (t > 0),$$

where $y(t) = g(t)$ on $(-\infty, 0)$, and the relationship between solutions of

(1.1) and (1.5) as $\mu \to 0^+$; for small $\mu > 0$ (1.1) is regarded as a singular perturbation of (1.5).

From the viewpoint of Volterra integrodifferential equations with positive, decreasing, convex kernels a, the theory is well understood if the nonlinearities consist of functions of one variable (see [2], [6] for an early result, and recent results respectively, where the latter gives references to other literature). A part of the novelty of the present problem is that the nonlinearity is a function of two variables. In addition, the singular perturbation nature of (1.1), (1.5), the infinite memory, and the fact that the logarithmic convexity of a plays a key role in the analysis, constitute interesting new features. An interesting extension of most of the results of [3] to nonconvolution kernels $a = a(t, s)$ has recently been obtained by G.S. Jordan [1].

In Section 2 we present the important results for (1.1), (1.5) in a setting sufficiently general to include the physical problem (1.2), (1.3), (1.4); sketches of some proofs are given in Section 3. Section 4 is devoted to the discretization of (1.1), (1.5) and is based on recent ongoing unpublished work by O. Nevanlinna [7].

2. Summary of Results. We make the following assumptions concerning the functions a, F, g throughout.

$$(H_a)\quad \begin{cases} a \in C^1[0,\infty);\ a(t) > 0,\quad a'(t) < 0 \quad (0 \le t < \infty); \\ a \in L^1(0,\infty); \\ \log a(t)\ \text{convex, i.e.}\ a'(t)/a(t)\ \text{nondecreasing},\quad 0 \le t < \infty; \end{cases}$$

$$(H_F)\quad \begin{cases} F : \mathbb{R}^+ \times \mathbb{R}^+ \to \mathbb{R};\quad F \in C^1(\mathbb{R}^+ \times \mathbb{R}^+); \\ F(x,x) = 0 \quad \text{for every}\ x > 0;\ F_1(y,z) > 0,\quad F_2(y,z) < 0 \\ \text{for}\ y, z \in \mathbb{R}^+, \end{cases}$$

where the subscripts denote partial differentiation with respect to the first and second variable;

$$(H_g)\quad \begin{cases} g : (-\infty, 0] \to \mathbb{R}^+;\ g(0) > 1,\quad g(-\infty) = 1; \\ g \in C(-\infty, 0] \quad \text{and}\quad g \quad \text{nondecreasing}. \end{cases}$$

It is readily verified that the specific functions (1.2) - (1.4) satisfy (H_a), (H_F), (H_g) respectively.

The first result concerns global existence, uniqueness, and certain useful properties of solutions of (1.1).

Theorem 2.1. Let the assumptions (H_a), (H_F), (H_g) be satisfied. Then for each $\mu > 0$ the Cauchy problem (1.1) has a unique solution $\phi(t,\mu)$ on $[0,\infty)$ having the following properties :

(2.1) $\phi'(t,\mu) < 0$, and $1 < \phi(t,\mu) \le g(0)$ $(0 \le t < \infty)$;

(2.2) if g_1, g_2 satisfy (H_g) with $g_1(t) \ge g_2(t)$ $(-\infty < t \le 0)$, then the corresponding solutions ϕ_1, ϕ_2 of (1.1) satisfy $\phi_1(t,\mu) \ge \phi_2(t,\mu)$ $(0 \le t < \infty)$;

(2.3) if $\mu_1 > \mu_2$ with g fixed in (1.1), then the corresponding solutions satisfy $\phi(t,\mu_1) > \phi(t,\mu_2)$ $(0 < t < \infty)$;

(2.4) there exist constants $K_1 > 0$ (independent of μ), $\mu_0 > 0$, $\tilde{K} = \tilde{K}(\mu_0) > 0$ such that the solution ϕ satisfies the estimate

$$0 < -\phi'(t,\mu) \le \frac{\tilde{K}}{\mu} \exp(-K_1 t/\mu) + \tilde{K}\int_t^\infty a(s)ds \quad (0 \le t < \infty;\ 0 < \mu \le \mu_0)$$

The elementary proof of Theorem 2.1 is sketched in Section 3; for details we refer the reader to [3, Theorems 1, 2].

As an easy consequence of properties (2.1), (2.3), (2.4) we have

Corollary 2.2. For each fixed $\mu > 0$

(2.5) $\alpha(\mu) = \lim_{t\to\infty} \phi(t,\mu)$ exists, and $\alpha(\mu) \ge 1$; moreover, if $\mu_1 > \mu_2$, then $\alpha(\mu_1) \ge \alpha(\mu_2) \ge 1$;

if, in addition,

$$(2.6) \qquad \int_0^\infty t\,a(t)\,dt < \infty,$$

then there exists constants $\mu_0 > 0$, $K = K(\mu_0) > 0$ such that

$$(2.7) \qquad 0 < \phi(t,\mu) - \alpha(\mu) \le K\int_t^\infty (s-t)\,a(s)\,ds \qquad (0 \le t < \infty;\ 0 < \mu \le \mu_0).$$

The last conclusion follows from integration of (2.4).

Remark 2.3. If $a(t) = A\exp(-t/\tau)$, $A > 0$, $\tau > 0$, which satisfies

(H_a) with $a'(t)/a(t) \equiv -1/\tau$, conclusions (2.1), (2.5) can be strengthened to

(2.1') $$\phi'(t,\mu) < 0 \quad \text{and} \quad 1 < y_0 < \phi(t,\mu) \le g(0) \qquad (0 \le t < \infty),$$

(2.5') $$\alpha(\mu) \ge y_0 > 1$$

respectively when y_0 is uniquely defined by the equation

$$\int_{-\infty}^{0} e^{s/\tau} F(y_0, g(s))ds = 0 .$$

The monotonicity property (2.3) and the estimate (2.4), together with a compactness argument establish the global existence of a solution of the reduced problem (1.5), once one observes that any (continuous) solution ϕ_0 of (1.5) for $t \ge 0$ must have the property that $\phi_0(0) = \phi_0(0^+)$ satisfies

(2.8) $$0 = \int_{-\infty}^{0} a(-s) F(\phi_0(0), g(s)) ds ;$$

the integral exists by $a \in L^1(0,\infty)$, (H_F), (H_g). Moreover, by (H_F) the integral in (2.8) is a continuous, strictly increasing function of $\phi_0(0)$; the integral is negative if $\phi_0(0) = 1$, and positive if $\phi_0(0) = g(0)$. Therefore, there is a unique number, $1 < y_0 < g(0)$, such that the integral vanishes when $\phi_0(0) = y_0$. The uniqueness assertion in Theorem 2.4 below is proved separately by an elementary argument, and the remaining properties of the solution ϕ_0 of (1.5) are established by elementary arguments analogous to the corresponding properties in Theorem 2.1 (for details see [3]).

Theorem 2.4. <u>Let</u> (H_a), (H_F), (H_g) <u>be satisfied. Then</u> (1.5) <u>has a unique (continuous) solution</u> ϕ_0 <u>on</u> $[0,\infty)$ <u>satisfying the following properties</u> :

(2.9) if $y_0 = \phi_0(0)$, then $1 < y_0 < g(0)$;

if $a(t) \not\equiv Ae^{-\lambda t}$, $A > 0$, $\lambda > 0$, then

(2.10) $\phi_0 \in C^1[0,\infty)$, $\phi_0'(t) < 0$ and $1 < \phi_0(t) \le y_0$ $(0 \le t < \infty)$;
if $a(t) \equiv Ae^{-\lambda t}$, $A > 0$, $\lambda > 0$, then $\phi_0(t) \equiv y_0$ $(0 \le t < \infty)$;

(2.11) if g_1, g_2 satisfy (H_g) and if $g_1(t) \ge g_2(t)$ $(-\infty < t \le 0)$, then the corresponding solution $\phi_0^{(1)}$, $\phi_0^{(2)}$ of (1.5) satisfy $\phi_0^{(1)}(t) \ge \phi_0^{(2)}(t)$ $(0 \le t < \infty)$;

$$(2.12)\quad \begin{cases} \text{if } \phi(t,\mu) \text{ is the solution of } (1.1), \text{ for a fixed } \mu > 0 \text{ and if } \phi_0(t) \\ \text{is the solution of } (1.5), \text{ then } \phi_0(t) < \phi(t,\mu) \quad (0 \le t < \infty). \end{cases}$$

From the preceding results one easily has

Corollary 2.5. Let (H_a), (H_F), (H_g) be satisfied. Then

$$(2.13)\qquad \alpha_0 = \lim_{t\to\infty} \phi_0(t) \quad \text{exists and} \quad 1 \le \alpha_0 \le \alpha(\mu);$$

if $a(t) \equiv Ae^{-\lambda t}$, $A > 0$, $\lambda > 0$, then, in fact, $\alpha_0 = y_0 > 1$. If also (2.6) holds, then

$$(2.14)\qquad \lim_{\mu\to 0^+} \alpha(\mu) = \alpha_0$$

and

$$(2.15)\qquad 0 < \phi(t,\mu) - \phi_0(t) \le \alpha(\mu) - \alpha_0 + K\int_t^\infty (s-t)\, a(s)\, ds \quad (0 \le t < \infty;\ 0 < \mu \le \mu_0).$$

In Theorem 2.10 below we establish a more precise result than (2.14), (2.15) under some additional assumptions.

Remark 2.6. An estimate similar to (2.4) holds for the solution ϕ_0 of the reduced problem (1.5), (but, of course, without the term $\frac{K}{\mu}\exp(-K_1 t/\mu)$).

The next task is to establish the physically important fact that the limiting value $\alpha(\mu)$ of the solution $\phi(t,\mu)$ of (1.1) as $t \to \infty$ satisfies $\alpha(\mu) > 1$ $(\mu > 0)$, rather than the weak form $\alpha(\mu) \ge 1$ in (2.5). By properties (2.12) (2.13) it suffices to prove $\alpha_0 > 1$ $(\alpha_0 = \lim_{t\to\infty} \phi_0(t))$.

Theorem 2.7. Let (H_a), (H_F), (H_g) be satisfied. If, in addition, (2.6) is satisfied, then

$$(2.16)\qquad \alpha_0 > 1.$$

We remark that in the special case $a(t) \equiv Ae^{-\lambda t}$, $A > 0$, $\lambda > 0$, there is nothing to prove since $\phi_0(t) \equiv y_0 > 1$. The proof of Theorem 2.7 is sketched in Section 3.

An Abelian argument using Laplace transforms (see [8, Theorem 5]) shows that Theorem 2.7 is best possible in the following sense.

Theorem 2.8. Let (H_a), (H_g) be satisfied and let

$$F(y,z) = y - z . \tag{2.17}$$

If

$$\int_0^\infty s a(s)\, ds = \infty , \tag{2.18}$$

then the solution ϕ_0 of (1.5) satisfies

$$\lim_{t \to \infty} \phi_0(t) = 1 . \tag{2.19}$$

We next discuss the existence of a boundary layer in a neighbourhood of $t = 0$ as $\mu \to 0^+$. For this purpose we consider the following approximation of the problem (1.1) for small $t \geq 0$;

$$-\mu v'(t) = \int_{-\infty}^{0} a(-s) F(v(t), g(s))\, ds \qquad (t > 0 ;\ v(0) = g(0)). \tag{2.20}$$

It will be observed that (2.20) is not a Volterra equation, but acts rather like an ordinary differential equation. Performing the stretching transformation

$$t = \mu \tau \tag{2.21}$$

and setting $w(\tau) = v(t)$ transforms (2.20) to

$$-\frac{dw}{d\tau} = \int_{-\infty}^{0} a(-s) F(w(\tau), g(s))\, ds \qquad (\tau > 0 ;\ w(0) = g(0)). \tag{2.22}$$

The following result is established by appropriate modifications of techniques which are standard in singular perturbation theory for ordinary differential equations.

Theorem 2.9. Let (H_a), (H_F), (H_g) be satisfied. Then the initial value problem (2.22) has a unique solution $w = \xi(\tau)$ existing on $0 \leq \tau < \infty$ and satisfying the following properties :

$$\lim_{\tau \to \infty} \xi(\tau) = y_0 = \phi_0(0) ;\ 0 < \xi(\tau) - y_0 \leq (g(0) - y_0)\, e^{-K\tau} \quad (0 \leq \tau < \infty), \tag{2.23}$$

where ϕ_0 is the solution of (1.5) (see Theorem 2.4) and K is some positive constant.

Moreover, if $\phi(t,\mu)$ is the unique solution of (1.1), in Theorem 2.1 and if $\xi(t/\mu)$ is the unique solution of (2.20) for $\mu > 0$, then for any $t_0 > 0$ there exists a constant $K > 0$ (independent of μ) such that

$$(2.24) \qquad |\phi(t,\mu) - \xi(t/\mu)| \le \overline{K}t \qquad (0 \le t \le t_0;\ \mu > 0).$$

The estimate (2.24) establishes the existence of a boundary layer in a positive neighbourhood of $t = 0$; for details see [3, Theorem 7].

In Corollary 2.5 we showed that $\alpha(\mu) \to \alpha_0$ as $\mu \to 0^+$, so that the solutions $\phi(t,\mu)$ of (1.1) and $\phi_0(t)$ of (1.5) do not differ by much for small $\mu > 0$ and for large t. Our final result makes this precise, under the additional assumptions that $ta(t) \in L^1(0,\infty)$ and $F \in C^2(\mathbb{R}^+ \times \mathbb{R}^+)$.

<u>Theorem 2.10. Let</u> (H_a), (H_F), (H_g) <u>be satisfied. In addition, assume that</u> $F \in C^2(\mathbb{R}^+ \times \mathbb{R}^+)$ <u>and that</u> (2.6) <u>holds. Then there exist constants</u> $K > 0$, $\mu_0 > 0$ <u>and a function</u> $\gamma \in C^1[0,\infty)$, γ <u>positive, bounded and nondecreasing, such that</u>

$$(2.25) \qquad \phi_0(t) < \phi(t,\mu) < \phi_0(t) + (g(0) - \phi_0(0))\exp(-Kt/\mu) + \gamma(t)\mu|\log\mu|$$
$$(0 \le t < \infty;\ 0 < \mu \le \mu_0).$$

<u>In particular, as an immediate consequence of</u> (2.25), <u>there exists a constant</u> $\tilde{K} > 0$ <u>such that</u>

$$(2.26) \qquad 0 < \phi(t,\mu) - \phi_0(t) = O(\mu|\log\mu|), \quad (\mu \to 0^+;\ \tilde{K}\mu|\log\mu| \le t < \infty).$$

The techniquely tedious proof of Theorem 2.10 which is omitted (see [3, Theorem 8] uses the notions of upper and lower solutions of (1.1). The inequality (2.25) is established by observing first that the solution ϕ_0 of the reduced problem (1.5) is a lower solution of (1.1) on $0 \le t < \infty$, and second, by showing that

$$w(t,\mu) = \phi_0(t) + (g(0) - \phi_0(0))\exp(-Kt/\mu) + \gamma(t)\mu|\log\mu|$$

is an upper solution for suitably chosen K and γ (i.e. that

$$-\mu w'(t) < \int_{-\infty}^{t} a(t-s)F(w(t), w(s))ds \qquad \text{for} \quad 0 < t < \infty,$$

where $w(t) = g(t)$ on $-\infty < t < 0$); this is the involved part of the proof.

The question arises whether the order $O(\mu|\log\mu|)$ in (2.26) is best possible. In the linear case it is not; for if $F(y,z) = y - z$, one can establish the inequality

$$\phi_0(t) < \phi(t,\mu) < \phi_0(t) + (g(0) - \phi_0(0))\exp(-Kt/\mu) + \gamma(t)\mu$$

for $0 \le t < \infty$ and $0 < \mu \le \mu_0$, by the method of proof of Theorem 2.10. In addition, one can compute $\alpha(\mu)$ and α_0 in the linear case by the method of Laplace transforms and show that $\alpha(\mu) - \alpha_0$ is precisely of the order μ. In the general case, however, we have been unable to improve the estimate (2.25).

3. Sketches of Proofs. In this section we outline the proofs of those results stated in Section 2 which require further amplification.

a. Proof of Theorem 2.1. The classical Picard successive approximations applied to the integrated form of (1.1) show that for each fixed $\mu > 0$ there is a unique C^1 local solution $\phi(t, \mu)$ existing on some interval $[0, T]$, $T > 0$. To show that this solution can be continued (necessarily uniquely), to the interval $[0, \infty)$, it suffices to establish the inequalities (2.1) on any interval on which the solution $\phi(t, \mu)$ exists. For then the solution satisfies *a priori* upper and lower bounds, independent of T, and hence can be continued to the interval $[0, \infty)$ by a standard result [5].

To establish (2.1) on any interval on which $\phi(t, \mu)$ exists we have from (1.1)

$$(3.1) \qquad -\mu\phi'(0, \mu) = \int_{-\infty}^{0} a(-s)\, F(g(0), g(s))\, ds.$$

The integral clearly exists since $a \in L^1(0, \infty)$ and $F(g(0), g(s))$ is bounded on $(-\infty, 0]$ by (H_F), (H_g). From (H_a), $a(-s) > 0$ $(-\infty < s \le 0)$ and from (H_F), (H_g), $F(g(0), g(s)) \ge 0$ $(-\infty < s \le 0)$, with the strict inequality holding for large negative s; therefore, $\phi'(0, \mu) < 0$. Since $\phi \in C^1$, one has by continuity that $\phi'(t, \mu) < 0$ $(0 \le t < \alpha)$, for some $\alpha > 0$. We claim first that

$$(3.2) \qquad \phi(t, \mu) > 1 \qquad (0 \le t \le \alpha).$$

Indeed, $\phi(0, \mu) = g(0) > 1$, and by continuity (3.2) holds at least on some interval to the right of $t = 0$. Suppose $0 < t_1 \le \alpha$ is the first point at which $\phi(t_1, \mu) = 1$, and $1 < \phi(t, \mu) < g(0)$ $(0 < t < t_1)$. From (1.1) we have

$$(3.3) \qquad -\mu\phi'(t_1, \mu) = \int_{-\infty}^{0} a(t_1 - s)\, F(1, g(s))\, ds + \int_{0}^{t_1} a(t_1 - s)\, F(1, \phi(s, \mu))\, ds.$$

By (H_a), $a(t_1 - s) > 0$ $(-\infty < s \le t_1)$, and by (H_F), (H_g), $F(1, g(s)) \le 0$ on $(-\infty, 0]$, with the strict inequality holding near zero. Since ϕ is strictly decreasing on $[0, t_1]$, we also have $F(1, \phi(s, \mu)) < 0$ $(0 \le s < t_1)$. Therefore each integral in (3.3) is negative and $\phi'(t_1, \mu) > 0$, which, in view of $\phi'(t,\mu) < 0 \quad (0 \le t < \alpha)$, is impossible; this proves (3.2).

We next claim :

$$\phi'(t,\mu) < 0 , \tag{3.4}$$

for as long as the solution exists. Indeed, suppose for contradiction that $\alpha > 0$ is the first point at which

$$\phi'(\alpha,\mu) = 0 \quad \text{and} \quad \phi'(t,\mu) < 0 \quad (0 \le t < \alpha). \tag{3.5}$$

By the argument of the preceding paragraph we have

$$1 < \phi(\alpha,\mu) < g(0) . \tag{3.6}$$

To prove (3.4) we compute $\phi''(\alpha,\mu)$ from (1.1) and we shall obtain an obvious contradiction of (3.5) by showing that

$$\phi''(\alpha,\mu) < 0 ; \tag{3.7}$$

this implies that no such $\alpha > 0$ satisfying (3.5) exists and proves (3.4). Indeed, differentiating (1.1) (justified by (H_a), (H_F), (H_g) - note that by (H_a), $a' \in L^1(0,\infty)$) one has

$$\left\{ \begin{aligned} -\mu\phi''(t,\mu) = & \int_{-\infty}^{t} a'(t-s)F(\phi(t,\mu),\phi(s,\mu))ds \\ & + \phi'(t,\mu)\int_{-\infty}^{t} a(t-s)F_1(\phi(t,\mu),\phi(s,\mu))ds. \end{aligned} \right. \tag{3.8}$$

Putting $t = \alpha$ and using (3.5) gives

$$-\mu\phi''(\alpha,\mu) = \int_{-\infty}^{\alpha} a'(\alpha-s)F(\phi(\alpha,\mu),\phi(s,\mu))\,ds . \tag{3.9}$$

Thus to prove (3.7) we wish to show that

$$I(\alpha) = \int_{-\infty}^{\alpha} a'(\alpha-s)F(\phi(\alpha,\mu),\phi(s,\mu))\,ds > 0 . \tag{3.10}$$

We shall need to consider two cases :

in case (i) $a(t)$ satisfies (H_a) with $a(t) \not\equiv Ae^{-\lambda t}$, $A > 0$, $\lambda > 0$ (i.e. $a'(t)/a(t) \not\equiv -\lambda$);

in case (ii) $a(t)$ satisfies (H_a) with $a(t) \equiv Ae^{-\lambda t}$, $A > 0$, $\lambda > 0$. We shall only deal with (i); in case (ii) (3.4) is established directly from (1.1).

Define a number $-\beta$, $\beta > 0$, by the relation $\phi(\alpha,\mu) = g(-\beta)$; $-\beta$ exists in view of (3.6) and (H_g). Since g may take the constant value $\phi(\alpha,\mu)$ on some interval $J \subset (-\infty, 0]$, we define $-\beta$ uniquely by taking it to be the right-hand end point in such a case. We then have

$$(3.11)\qquad I(\alpha) = \int_{-\infty}^{-\beta} a'(\alpha - s)F(g(-\beta), g(s))\,ds + \int_{-\beta}^{0} a'(\alpha - s)F(g(-\beta), g(s))\,ds$$
$$+ \int_{0}^{\alpha} a'(\alpha - s)F(\phi(\alpha,\mu), \phi(s,\mu))\,ds.$$

Since $\phi'(\alpha,\mu) = 0$ we also have from (1.1)

$$(3.12)\qquad 0 = \int_{-\infty}^{-\beta} a(\alpha - s)F(g(-\beta), g(s))\,ds + \int_{-\beta}^{0} a(\alpha - s)F(g(-\beta), g(s))\,ds$$
$$+ \int_{0}^{\alpha} a(\alpha - s)F(\phi(\alpha,\mu))\,ds.$$

We next define the function σ by the relation

$$\sigma(s) = \frac{a'(\alpha - s)}{a(\alpha - s)} \qquad (-\infty < s \le \alpha),$$

and we observe that the log convexity of a implies that $\sigma(s)$ is negative and nonincreasing; moreover, since $a'(t)/a(t) \not\equiv -\lambda$, $\lambda > 0$, $\sigma(s)$ is strictly decreasing, at least on some interval contained in $(-\infty, \alpha]$.

We rewrite (3.11) in the equivalent form

$$(3.13)\qquad I(\alpha) = \int_{-\infty}^{-\beta} \sigma(s)h(s)\,ds + \int_{-\beta}^{\alpha} \sigma(s)h(s)\,ds$$

where

$$h(s) = \begin{cases} a(\alpha - s)F(g(-\beta), g(s)) & (-\infty < s \le 0) \\ a(\alpha - s)F(\phi(\alpha,\mu), \phi(s,\mu)) & (0 < s \le \alpha)\,; \end{cases}$$

we also write (3.12) in the equivalent form

$$(3.14)\qquad 0 = \int_{-\infty}^{-\beta} h(s)\,ds + \int_{-\beta}^{\alpha} h(s)\,ds.$$

From the definition of $-\beta$ and (3.6) one has $1 < g(-\beta) < g(0)$. Therefore, (H_a), (H_F), (H_g) imply that

$$(3.15)\qquad h(s) \ge 0 \qquad (-\infty < s \le -\beta)$$

with strict inequality for large negative s, and

$$(3.16)\qquad h(s) < 0 \qquad (-\beta < s < \alpha).$$

Combining (3.13), (3.14) yields

$$(3.17)\qquad I(\alpha) = \int_{-\infty}^{-\beta} (\sigma(s) - \sigma(-\beta))h(s)\,ds + \int_{-\beta}^{\alpha} (\sigma(s) - \sigma(-\beta))h(s)\,ds.$$

But

$$(3.18)\qquad \sigma(s) \geq \sigma(-\beta) \qquad (-\infty < s \leq -\beta),$$

$$(3.19)\qquad \sigma(s) \leq \sigma(-\beta) \qquad (-\beta < s \leq \alpha),$$

with strict inequalities holding either for s negative and large or for s near α. Using (3.15), (3.16) and (3.18), (3.19) in (3.17) shows that each integral in (3.17) is nonnegative and at least one of them is positive. This proves (3.10) and hence also (3.7) and (3.4), as well as (2.1).

Properties (2.2), (2.3) are readily proved by contradiction arguments involving the mean value theorem applied to the difference of the appropriate two solutions of (1.1).

Finally, to prove (2.4) let ϕ be the solution of (1.1). We return to equation (3.8) obtained by differentiating (1.1):

$$(3.20)\qquad -\mu\phi''(t,\mu) = G(t,\mu)\,\phi'(t,\mu) + f(t,\mu) \qquad (0 \leq t < \infty),$$

where

$$G(t,\mu) = \int_{-\infty}^{t} a(t-s)\,F_1(\phi(t,\mu), \phi(s,\mu))\,ds,$$

$$f(t,\mu) = \int_{-\infty}^{t} a'(t-s)\,F(\phi(t,\mu), \phi(s,\mu))\,ds.$$

Since $F_1 > 0$, $a \in L^1(0,\infty)$, $a(t) > 0$ $(0 \leq t < \infty)$, and since ϕ satisfies (2.1), we have by (H_g),

$$(3.21)\qquad 0 < \gamma A \leq G(t,\mu) \leq \Gamma A \qquad (0 \leq t < \infty,\ \mu > 0),$$

where

$$\gamma = \inf_S F_1(y,z), \quad \Gamma = \sup_S F_1(y,z), \quad S = [1, g(0)] \times [1, g(0)], \quad A = \int_0^{\infty} a(s)\,ds.$$

One next shows by an argument using the log convexity of a that there exists a constant $K > 0$, independent of μ, such that

$$(3.22) \qquad |f(t,\mu)| \le K \int_t^\infty a(s)\,ds \qquad (0 \le t < \infty, \quad \mu > 0);$$

for details see [3, Theorem 2]. The proof of (2.4) is then accomplished by integrating (3.20) (note that (3.20) is of first order in ϕ') and by using the log convexity of a, (3.21), (3.22), as well as (2.1).

b. <u>Proof of Theorem 2.7</u>. In view of the monotonicity property (2.11) of solutions of (1.5) with respect to the function g, it suffices to prove the result for the function g given by

$$(3.23) \qquad g(t) = \begin{cases} 1+\delta & \text{if} \quad -\eta \le t \le 0 \\ 1 & \text{if} \quad t < -\eta, \quad \delta > 0, \quad \eta > 0. \end{cases}$$

(Strictly speaking, this function g does not satisfy the hypothesis (H_g), being discontinuous, but it is readily verified that the proof below would be essentially unaltered if the given g were replaced by a continuous g approximating sufficiently closely to it.) Since $1 \le g(t) \le 1+\delta$, $-\infty < t \le 0$, property (2.10) implies that the solution ϕ_0 of (1.5) satisfies the inequality $1 \le \phi_0(t) \le 1+\delta$, $0 \le t < \infty$. This means that the arguments of F in (1.5) are close to 1, if $\delta > 0$ is sufficiently small which will be the case in what follows. For this reason and in order to simplify the calculations we assume, consistent with (H_F) and without loss of generality, that

$$F_1(1,1) = 1, \qquad F_2(1,1) = -1;$$

note that $F(x,x) = 0$ $(x > 0)$ implies that $F_1(x,x) = -F_2(x,x)$. Substituting (3.23) into (1.5) yields, after an application of the mean value theorem and some elementary estimation, that the quantity $z(t) = \phi_0(t) - 1$ satisfies the equation

$$(3.24) \qquad z(t)A - a * z(t) = \delta \int_t^{t+\eta} a(s)ds + o\left(\delta \int_t^\infty a(s)ds\right) \qquad (0 \le t < \infty),$$

where $*$ denotes the convolution, $A = \int_0^\infty a(s)ds$, $\eta > 0$ is fixed, and $\delta > 0$ is small.

In [3, Theorem 4] we obtained the conclusion of Theorem 2.7 by applying a Tauberian theorem for Laplace transforms [8, Theorem 4.3, p.192] to the solution z of (6.6). Here we sketch a different elementary argument due to Jordan [1] having the advantage that he has modified it to apply in the nonconvolution case. Define the positive functions

$$\psi(t) = \int_t^{t+\eta} a(s)ds , \qquad \omega(t) = \int_t^{\infty} a(s)\,ds ,$$

and integrate (3.24). A simple calculation shows that

$$(3.25) \qquad \int_0^t z(s)\,\omega(t-s)\,ds = \delta \int_0^t \psi(s)ds + o\left(\delta \int_0^t \omega(s)ds\right) \qquad (0 \le t < \infty).$$

Note that by Theorem 2.4 and Corollary 2.5 $z(t)$ is positive, decreasing, and $z(\infty) = \lim_{t \to \infty} z(t)$ exists, $z(\infty) \ge 0$.

By assumption (2.6) $\psi, \omega \in L^1(0,\infty)$ and thus letting $t \to \infty$ in (3.25) yields

$$(3.26) \qquad z(\infty) \int_0^{\infty} \omega(s)\,ds = \delta \int_0^{\infty} \psi(s)ds + o\left(\delta \int_0^{\infty} \omega(s)\,ds\right),$$

where the integrals in (3.26) are all positive by (H_a); moreover, for small $\delta > 0$ the first term on the right side of (3.26) dominates the second. It is then easy to see from (3.26) that $z(\infty) = \phi_0(\infty) - 1 > 0$, or equivalently $\alpha_0 > 1$, and this completes the proof.

4. Discretization. The objective is to discretize (1.1) using backwards differences such that most of the qualitative properties of the solution are preserved and to give an error bound uniform in μ and t for $0 < \mu \le \mu_0$, $t \ge t_0 > 0$.

Nevanlinna [7] discretizes (1.1) by the implicit scheme

$$(4.1) \quad \begin{cases} \frac{\mu}{h}(y_n - y_{n-1}) + h \sum_{j=-\infty}^{n} a(nh - jh)\, F(y_n, y_j) = 0, & n > 0 \\ y_j = g(jh), & j \le 0 . \end{cases}$$

Let $\phi(t,\mu)$ be the solution of (1.1) on $[0,\infty)$ (see Theorem 2.1 above). The following result is established.

Theorem 4.1. Let (H_a), (H_F), (H_g) hold. Then for all $\mu > 0$, $h > 0$ equation (4.1) has a unique solution $\{y_n\}$ such that $y_n \in [1, g(0)]$ and $y_n \le y_{n-1}$ $(n > 0)$. Moreover, there exist constants $\mu_0 > 0$, $h_0 > 0$, $C_1, C_2, C_3, C_4, C_5 > 0$ such that for any $0 < T < \infty$, $\mu \in (0, \mu_0]$, $h \in (0, h_0)$, and $0 < n \le \frac{T}{h}$ one has

$$(4.2)\quad \begin{cases} |\phi(nh,\mu) - y_n| \le C_1 e^{C_2 T}(1+T)h + C_3\left(\frac{h}{\mu}\right)\left(1+\frac{C_4 h}{\mu}\right)^{-n} \\ \qquad + C_5\left(\frac{h}{\mu}\right)\left\{\left(1+\frac{C_4 h}{\mu}\right)^{1-n} - \left(1+\frac{2C_4 h}{\mu}\right)^{1-n}\right\}. \end{cases}$$

<u>If, in addition</u>, $a(t)$ <u>decays exponentially</u>, <u>then there exists a</u> $\delta \in (0,1)$ <u>and a constant</u> $C_6 > 0$ <u>such that the inequality</u> (4.2) <u>holds for all</u> $n \ge 0$, <u>provided the term</u> $C_1 e^{C_2 T}(1+T)h$ <u>is replaced by</u> $C_6 h^{\delta}$.

The proof of Theorem 4.1 in [7] is lengthy and technical. For this reason we limit ourselves to a few remarks.

(i) It is necessary first to show that solutions of (4.1) have essentially all the properties of those of (1.1) stated in Theorem 2.1 and that all bounds are independent of the stepsize h. The first step is, of course, to show that (4.1) has a unique solution $\{y_n\}$, $1 < y_n < g_0$ for all $n > 0$. This is immediate upon noticing that (4.1) has the form

$$(4.3)\qquad y_n + \frac{h^2}{\mu}\sum_{j=-\infty}^{n} a_{n-j}\, F(y_n, y_j) = y_{n-1} \qquad (n > 0),$$

and that the left side of (4.3) is a continuous increasing function of y_n which maps $[1, g_0]$ onto $[a,b]$, $a < 1$, $b > g(0)$. To show that $\{y_n\}$ is nonincreasing for $n \ge 0$, one puts $\nabla y_n = y_n - y_{n-1}$ and one notes that the analogue of (3.8) is

$$(4.4)\quad \begin{cases} \frac{\mu}{h}\nabla^2 y_m + h\sum_{j=-\infty}^{m-1} a_{m-j}\,[F(y_m, y_j) - F(y_{m-1}, y_j)] \\ \qquad = -h\sum_{j=-\infty}^{m-2} \nabla a_{m-j}\, F(y_{m-1}, y_j). \end{cases}$$

One has $y_1 < y_0$; if $m \ge 2$ is the first index such that $\nabla y_m > 0$, one shows for a contradiction that $\nabla^2 y_m \le 0$. To do this use is made of (4.4) and of the logarithmic convexity of a, now in the form $\left(\frac{\nabla a}{a}\right)_n$ nondecreasing. The analogues of properties (2.2), (2.3) are straightforward. In order to prove the analogue of property (2.4) one has to bound ∇y_n; this is done by "solving" the second order difference equation (4.4). The bound analogous to (2.4) takes the form

$$0 \le -h^{-1} \nabla y_n \le C_1 \mu^{-1} \left(1 + \frac{hG}{\mu}\right)^{1-n} + C_2 h \sum_{j=n}^{\infty} a_j \; ; \quad n > 0 ,$$

where $\mu_0 > 0$, $h_0 > 0$ are given, $\mu \in (0, \mu_0]$, $h \in (0, h_0]$, and C_1, C_2, G are suitable positive constants. From it one obtains as the discrete analogue of property (2.7): if also $\{ j a_j \} \in \ell^1$, then

$$0 < y_n - y_\infty \le C_1 \left(1 + \frac{hG}{\mu}\right)^{-n} + C_2 h^2 \sum_{k=n+1}^{\infty} (k-n) a_k \, .$$

(ii) In order to establish the convergence as $n \to \infty$ of solutions of (4.1) to those of (1.1), one needs to investigate the properties of the difference equation (4.1) under perturbations and then substitute the local truncation error for the perturbation. A complication which space does not permit us to detail here is due to the fact that the local truncation error generally changes sign during time stepping, while it is only fairly simple to control positive perturbations. The local truncation sequence $\{ \tau_n^h \}$ is defined by

$$\tau_n^h = \frac{\mu}{h} \left(\phi(nh, \mu) - \phi((n-1)h, \mu) \right) + h \sum_{j=-\infty}^{n} a_{n-j} F(\phi(nh,\mu), \phi(jh, \mu)) ,$$

where $\phi(\cdot, \mu)$ is the solution of (1.1) for $t > 0$; one shows that there exist constants $C = C(\mu_0, h_0, g(0))$, $K = K(\mu_0, g(0))$ such that

$$|\tau_n^h| \le Ch \left\{ \int_{(n-1)h}^{\infty} a(s)\, ds + \frac{1}{\mu} e^{-K(n-1)h/\mu} + \int_0^{nh} a(nh-s) \left[\int_s^{\infty} a(\xi)\, d\xi + \frac{1}{\mu} e^{-K/\mu} \right] ds \right\} ,$$

for $\mu \in (0, \mu_0]$, $h \in (0, h_0]$, $n > 0$. Note that this estimate implies that $|\tau_n^h|$ is uniformly bounded for $n \ge 2$, $\mu \in (0, \mu_0]$, $h \in (0, h_0]$. In order to prove that the same is true for $|\tau_1^h|$ considerations of the role of the boundary layer (Theorem 2.9, especially (2.24), in the continuous case) come into play; we omit these details. Note that the bound for $|\tau_n^h|$ simplifies considerably if one assumes $a(t) \le \alpha e^{-\beta t}$ $(\alpha, \beta > 0)$.

Supplementary Remarks. (i) The method of discretization used in Theorem 4.1 converges rather slowly. As is understood from the discussion of the continuous model (1.1), the solution $\phi(\cdot, \mu)$ is affected by the boundary layer near $t = 0$ even if one is interested in the solution for large t. While the solution $\phi(\cdot, \mu)$ of (1.1) tends to the limiting value $\alpha(\mu)$, $\mu > 0$, as $t \to \infty$, it need not decay exponentially, so that errors occurring during the numerical time stepping will generally affect the estimated limiting value at ∞. As was seen in Theorem 4.1, the discretized problem (4.1) has qualitative properties very similar to the continuous problem (1.1); this seems to be an essential feature when dealing with Volterra equations numerically. It seems unlikely that similar results can be obtained if one uses a faster numerical technique such as the second order central difference scheme (trapezoidal rule) for the discretization. With such a scheme we could not preserve the pointwise monotonicity of the discretized problem which is essential, and this forces us to use the method of Theorem 4.1; at least, we have not succeeded to do so with any other method. Without this monotonicity the nonlinearities F in (1.1) could cause the numerical solution to run out of control.

(ii) When applying the implicit scheme (4.1) of Theorem 4.1 to the physical problem (1.1) - (1.4), the specific form of F implies that at each step one has to find a real root of a cubic polynomial. The final recursion formula for the specific problem is then essentially an explicit method which generates y_{n+1} from given information $(y_0, y_1, \cdots, y_n)$. Preliminary calculations completed just prior to this conference indicate that the discretized model (4.1) predicts somewhat more recovery of the filament in Meissner's experiments than was obtained experimentally. However, the results become better as μ increases from zero. Surprisingly and unfortunately numerical experiments performed to date indicate that varying μ in the range $(10^{-5}, 1)$ causes only a very small change in the limiting value of the numerical solution of (4.1) (as $t \to \infty$), although these solutions behave very differently for small $t > 0$ because of the boundary layer at zero.

References

[1] G.S. Jordan, A nonlinear singularly perturbed Volterra integrodifferential equation of nonconvolution type. Proc. Royal Soc. of Edinburgh, to appear.

[2] J.J. Levin and J.A. Nohel, Perturbations of a Nonlinear Volterra Equation. Mich. Math. J. 12 (1965), 431-447.

[3] A.S. Lodge, J.B. Mc Leod, and J.A. Nohel, A nonlinear singularly perturbed Volterra integrodifferential equation occurring in polymer rheology. Proc. Royal Soc. of Edinburgh, to appear, (Math. Research Center, Univ. of Wisconsin, T.S.R. # 1694, 73 p).

[4] J. Meissner, Dehnungsverhalten von Polyäthylen-Schmelzen. Rheol. Acta 10 (1971), 230-242.

[5] J.A. Nohel, Some Problems in Nonlinear Volterra Integral Equations. Bull. Amer. Math. Soc. 68 (1962), 323-329.

[6] J.A. Nohel and D.F. Shea, Frequency Domain Methods for Volterra Equations. Advances in Math. 22 (1976), 278-304.

[7] O. Nevanlinna, Numerical solution of a singularly perturbed nonlinear Volterra equation (preprint), Math. Research Center, Univ. of Wisconsin, T.S.R. (to appear).

[8] D.V. Widder, The Laplace Transform. Princeton Univ. Press, 1941.

Periodic Solutions of Nonlinear Autonomous Functional Differential Equations

Roger D. Nussbaum
Rutgers University
New Brunswick, New Jersey, U.S.A.

I would like to discuss in these expository lectures some results concerning periodic solutions in the large for nonlinear, autonomous F.D.E.'s. Questions about such periodic solutions are of interest in mathematical biology and elsewhere, and some applications have been discussed by other lecturers at this conference. Here I shall restrict myself to the purely mathematical side of these questions. I shall describe the blend of general ideas from nonlinear functional analysis and close analysis of specific equations which has proved useful in studying periodic solutions of F.D.E.'s, mention the limitations of known techniques and indicate some of the many intriguing unsolved problems. Indeed, one important reason for studying periodic solutions of F.D.E.'s is that numerical studies suggest they often have some sorts of global stability properties, but very little along these lines has been proved.

It is fair to say that the original impetus for most of the questions we shall discuss came from numerical studies by G.S. Jones [28] of the equation

$$x'(t) = -\alpha x(t-1)[1+x(t)] \tag{1}$$

Here α denotes a positive constant.

Equation (1) came (via a change in variables) from a model in population dynamics, and it was also studied by E.M. Wright [59] because of a connection with the heuristic theory of asymptotic distribution of primes. If φ is a given continuous, real-valued map defined on $[0,1]$, it is not hard to see that there is a unique, continuous, real-valued function $x(t) = x(t;\varphi,\alpha)$ defined on $[0,\infty)$ and such that

$$
\begin{aligned}
x'(t) &= -\alpha x(t-1)[1+x(t)] \quad \text{for} \quad t \geq 1 \\
x|[0,1] &= \varphi
\end{aligned}
\qquad (2)
$$

Indeed, if $\varphi(1) \neq -1$, one can divide both sides of (2) by $1+x(t)$ and solve explicitly for $x(t)$ on $[1,2]$ in terms of φ. One then repeats the procedure on $[2,3]$, etc.

One can show that the solution so obtained never equals -1 for $t \geq 1$; of course if $\varphi(1) = -1$ one has $x(t) = -1$ for all $t \geq 1$.

For a given nonnegative function φ above with $\varphi(1) > 0$, Jones solved (2) numerically on a computer. He found that if $\alpha > (\frac{\pi}{2})$ the solution of (2) looked (for large t) like a periodic function of t, say $x_\alpha(t)$. Furthermore, the appearance of $x(t;\varphi,\alpha)$ seemed to be independent of φ for large t, at least for most φ. The hypothesized periodic solution $x_\alpha(t)$ appeared to have the following properties: $x_\alpha(0) = 0$, $x_\alpha(t) > 0$ on some interval $(0, z_1)$, where $z_1 > 1$, $x_\alpha(t) < 0$ on an interval (z_1, z_2), where $z_2 - z_1 > 1$ and $x_\alpha(t+z_2) = x_\alpha(t)$ for all t. It seemed reasonable to conjecture that for each $\alpha > \frac{\pi}{2}$ there is exactly one periodic solution $x_\alpha(t)$ of the type described above. (We shall see in the next section that there may be other periodic solutions whose zeros may be separated by a distance less than one).

In [27] Jones proved that for $\alpha > \frac{\pi}{2}$ the hypothesized periodic solutions do in fact exist; uniqueness of such periodic solutions remains, fifteen years after Jones's work, an open question. It must be said that Jones's proof is needlessly, complicated and that the details of his argument are somewhat confusing (at least to this writer). Presumably such considerations led R. B. Grafton [20] to give an alternate approach to proving existence of periodic solutions. Nevertheless Jones's basic idea was certainly correct.

I would like to describe here some of the tools which can be used to obtain existence of periodic solutions and then illustrate the use of these ideas for a specific example. The main abstract tools are the fixed point index and certain fixed point theorems, and these we shall have to ask the reader to accept on faith. However, with these tools a typical existence proof involves nothing more advanced than calculus and a smattering of complex variables (so as to locate complex zeros of a transcendental equation).

First, let me describe the fixed point index, Recall that a metric space X is called an ANR (absolute neighborhood retract) if given any closed subset A of a metric space M and a continuous map $f:A \longrightarrow X$, there exists an open neighborhood U of A and a continuous extension $F:U \longrightarrow X$ of f. The precise definition of ANR's is not so important as the properties of this class of spaces: Any closed, convex subset X of a Banach space Y is an ANR. If $X_1, X_2, \ldots, X_n$ are closed, convex subsets of a Banach space Y, then $X = \bigcup_{i=1}^{n} X_i$ is an ANR. If X is a metric space and X_1 and X_2 are closed subsets of X such that $X = X_1 \cup X_2$ and if X_1, X_2 and $X_1 \cap X_2$ are ANR's, then X is an ANR. Any continuous retract of an ANR is an ANR. Any metrizable Banach manifold is an ANR.

Now suppose that X is a complete metric space and an ANR, G is an open subset of X and $f:G \longrightarrow X$ is a continuous map such that if B is any bounded subset of G, then $f(B)$ has compact closure. We shall call any such function f a continuous, "compact" map. Assume in addition that $S = \{x \in G: f(x) = x\}$ is compact or empty. Then there is defined an integer $i_X(f, G)$, which is called the fixed point index of f on G and can be considered an algebraic count of the number of fixed points of f in G. If X is a Banach space, G is a bounded open subset of X and f has a continuous extension to $\bar{G}$ such that $f(x) \neq x$ for $x \in (\bar{G} - G)$, then the Leray-Schauder degree of $I-f$ on G is defined and

$$\deg(I - f, G, 0) = i_X(f, G)$$

The fixed point index has several properties which axiomatically determine it.

1. The additivity property. If G_1 and G_2 are open subsets of a complete metric ANR X, and if $f: G = G_1 \cup G_2 \longrightarrow X$ is a continuous compact map such that $i_X(f, G_j)$ is defined for $j = 1, 2$, then

$$i_X(f, G) = i_X(f, G_1) + i_X(f, G_2)$$

If H is an open subset of X and $f: H \longrightarrow X$ is a continuous map such that $i_X(f, H) \neq 0$, then f has a fixed point in H.

If H_1 is an open subset of H and f has no fixed points in $H - H_1$, then

$$i_X(f, H) = i_X(f, H_1)$$

2. The homotopy property. Suppose that X is a complete metric ANR, G is an open subset of X and $f: G \times [0, 1] \longrightarrow X$ is a continuous map which takes bounded sets to sets which have compact closure and is such that $\Sigma = \{(x, t) \in G \times [0, 1] : f(x, t) = x\}$ is compact or empty. Then if one defines $f_t(x) = f(x, t)$, one has that $i_X(f_t, G)$ is constant for $0 \leq t \leq 1$.

3. The commutativity property. Suppose that X is a complete metric ANR and $Y \subset X$ is a complete metric ANR for which the inclusion map is continuous. Let G be an open subset of X and $f: G \longrightarrow X$ a continuous, compact map such that $i_X(f, G)$ is defined and $f(G) \subset Y$. Then it follows that

$$i_X(f, G) = i_Y(f, G \cap Y)$$

The commutativity property is usually stated more generally as a result relating the fixed point indices of maps $f_1 f_2$ and $f_2 f_1$ and in fact such generality is important for many technical questions in the theory. If g denotes the map f considered as a map from G into Y and i denotes inclusion of Y into X, then $f:G \longrightarrow X$ is simply ig and $f:G \cap Y \longrightarrow Y$ is gi and our version of commutativity relates the fixed point indices of gi and ig.

4. The normalization property. If X is a compact metric ANR and $f:X \longrightarrow X$ a continuous map, then

$$i_X(f, X) = L(f) = \text{the Lefschetz number of } f$$
$$= \sum_{i \geq 0} (-1)^i \operatorname{tr}(f_{*,i})$$

where $f_{*,i}:H_i(X) \longrightarrow H_i(X)$, $H_i(X)$ is the ith singular homology group of X with coefficients in the rationals and $\operatorname{tr}(f_{*,i})$ denotes the trace of the vector space homomorphism $f_{*,i}$.

Typically I shall use the apparatus of the fixed point index in the following situation:

Suppose X is a closed, convex subset of a Banach space Y and $0 \in X$. (Actually, many of the proofs are algebraic topological in nature and convexity is not crucial). Suppose that $f:X - \{0\} \longrightarrow X$ is a continuous, compact map; in the examples I shall consider it will also be true that $f(0) = 0$ and f is continuous at 0, but continuity at 0 may fail in other examples, e.g. Liénard equations with a time lag [40, 46]. I shall want a non-zero fixed point of f. Suppose there exists an open subset G of $X - \{0\}$ (open in the relative topology on X) such that $i_X(f, G)$ is defined and non-zero. Then f will have a fixed point in G. Many fixed point principles for maps of a cone into itself, for example Krasnosel'skii's theorems on expansions and compressions of a cone [34] and Grafton's cone result [20], are most easily proved by a fixed point argument of the general type described above. See [41, Section 1] and [48, Section 5]. The key question, of course, is what verifiable conditions will insure the existence of a set G; for purposes here such a condition is provided by the concept of an "ejective point" introduced by F.E. Browder [6].

Definition. Let X be a topological space, x_0 an element of X, U an open neighborhood of X and $f: U - \{x_0\} \longrightarrow X$ a continuous map. The point x_0 is called an "ejective point of f" if there exists an open neighborhood $U_0 \subset U$ of x_0 such that for any $x \in U_0 - \{x_0\}$ there exists an integer $n = n(x)$ with $f^j(x)$ defined for $1 \leq j \leq n$ and $f^n(x) \notin \overline{U}_0$.

In the original paper [6] f is assumed continuous at x_0 with $f(x_0) = x_0$, but this assumption is unnecessary and undesirable for some applications.

Theorem 1. Suppose that C is a closed, convex subset of a Banach space Y with $x_0 \in C$, $\{x_0\} \neq C$. Assume that C is either infinite dimensional (which here simply means that C is not contained in any finite dimensional linear affine subspace) or that C is finite dimensional and $x_0 \in \tilde{\partial}(C)$, where $\tilde{\partial}(C)$ denotes the boundary of C as a subset of the finite dimensional linear affine subspace which it spans. Assume that $f: C - \{x_0\} \longrightarrow C$ is a continuous, compact map such that (1) x_0 is an ejective point of f and (2) there exists a number $R > 0$ such that $x - tf(x) - (1-t)x_0 \neq 0$ for $x \in C$ with $\|x - x_0\| = R$ and for $0 \leq t \leq 1$. (This second condition is vacuous if C is bounded). Let $r < R$ be any positive number such that $f(x) \neq x$ for $0 < \|x - x_0\| \leq r$. Then if $G = \{x \in C: r < \|x - x_0\| < R\}$, one has

$$i_C(f, G) = 1$$

If f is continuous at x_0 and $B_r = \{x \in X: \|x - x_0\| < r\}$ one has

$$i_C(f, B_r) = 0.$$

In any event f has a fixed point in G different from x_0; and if f is continuous on C and C is infinite dimensional, f has a fixed point x_1 such that $\|x_0-x_1\| < R$ and x_1 is not an ejective point of f. ■

The first version of the above theorem was obtained by F.E. Browder in [6]; the information about the fixed point indices, which is important for certain generalizations I shall describe in the next section, was obtained in [40].

I have restricted my discussion of the abstract theory to compact maps. For certain applications, e.g., for neutral F.D.E.'s, it may be necessary to use the fixed point index as it has been developed for more general classes of maps like condensing operators. See, for example, the last section of [40].

I now wish to study as an example the equation

$$x'(t) = -\alpha\delta^{-1}\int_{-1}^{-1+\delta} f(x(t+\theta))d\theta \tag{3}$$

The function f will always be assumed to satisfy

<u>H 1.</u> $f: \mathbb{R} \longrightarrow \mathbb{R}$ is a continuous function such that $f'(0) = 1$, $xf(x) > 0$ for $x \neq 0$ and $f(x) \geq -B > -\infty$ for all x.

The real numbers α and δ are assumed strictly positive, and we shall have to assume $\delta \leq \frac{1}{2}$. If care is taken one can interpret (3) for $\delta = 0$ to be the equation

$$x'(t) = -\alpha f(x(t-1)) \tag{4}$$

It may be worth emphasizing that the techniques I have described only give weak partial results about periodic solutions for $\frac{1}{2} < \delta < 1$. The reasons for this will become clear later. If it is assumed, in addition to H 1, that f is monotonic increasing and if $\rho:[-1,-1+\delta] \longrightarrow \mathbb{R}^+$ is a given C^1 function with $\rho'(\theta) \geq 0$ for all θ and $\rho(-1) \geq 0$, then for $0 < \delta \leq \frac{1}{2}$, the techniques I have described will give sharp information about periodic solutions of

$$x'(t) = -\alpha I^{-1}\int_{-1}^{-1+\delta} f(x(t+\theta))\,\rho(\theta)\,d\theta \tag{5}$$

where $I = \int_{-1}^{-1+\delta} \rho(\theta)d\theta$. I will not prove this, however.

If ρ is only assumed nonnegative and C^1, the question of existence in the large of periodic solutions of (3) is not well-understood.

Equation (3) was studied by H.-O. Walther [58] for the case $\delta = \frac{1}{2}$. We shall give here a shorter proof of Walther's theorem about existence of periodic solutions of (3); however, it seems that Walther's longer argument may also give other information which we do not obtain.

Theorem 2. (See [58]). Assume that f satisfies H 1 and $0 < \delta \leq \frac{1}{2}$ and define $\omega = \pi(2-\delta)^{-1}$. If $\alpha > \alpha_0 = -\delta\,\omega^2(2\cos\omega)^{-1}$ $(\alpha > (\frac{\pi}{2})$ for $\delta = 0)$ then equation (3) has a periodic solution $x_\alpha(t) = x(t)$ such that $x(0) = 0$, $x(t) > 0$ on an interval $(0, z_1)$, when $z_1 > 1-\delta$, $x(t) < 0$ on an interval (z_1, z_2), where $z_2 - z_1 > 1-\delta$, and $x(t+z_2) = x(t)$ for all t.

Perhaps the most obvious question about Theorem 2 is where the number α_0 comes from. Clearly, $x(t) \equiv 0$ is a solution of (3). One might hope for a nonzero periodic solution if the zero solution is, in some sense, unstable. If one linearizes (3) at 0, one is led to

$$x'(t) = -\alpha\delta^{-1}\int_{-1}^{-1+\delta} x(t+\theta)d\theta \tag{6}$$

and if one seeks solutions of the form $x(t) = e^{zt}$ one is led to the characteristic equation

$$z = -\alpha\delta^{-1}\int_{-1}^{-1+\delta} e^{z\theta}\, d\theta \tag{7}$$

Solutions of (7) with positive real part suggest instability of the zero solution; the first lemma states that such solutions z occur precisely for $\alpha > \alpha_0$.

Lemma 1. Assume that $0 < \delta \leq 1$ and let $\sigma:[-1, -1+\delta] \longrightarrow \mathbb{R}$ be a continuous, real-valued function such that

$$I = \int_{-1}^{-1+\delta} \sigma(\theta)\, d\theta > 0 \tag{8}$$

Let w_j, with $w_j < w_{j+1}$, denote the real positive roots of

$$\int_{-1}^{-1+\delta} \sigma(\theta) \cos w\theta \, d\theta = 0 \tag{9}$$

There will only be countably many roots of (9) with no finite accumulation point. Define numbers β_j by

$$\beta_j = -(w_j I)\left(\int_{-1}^{-1+\delta} \sigma(\theta) \sin w_j \theta \, d\theta\right)^{-1} \tag{10}$$

and define α_0 by

$$\alpha_0 = \inf \{\beta_j : 0 < \beta_j\} \tag{11}$$

Then the equation

$$z = -\alpha I^{-1} \int_{-1}^{-1+\delta} \sigma(\theta)\, e^{z\theta} \, d\theta \tag{12}$$

has no solutions z with $\mathrm{Re}(z) \geq 0$ for $0 < \alpha < \alpha_0$. If $\sigma(\theta) \equiv 1$, $0 < \delta \leq \frac{1}{2}$ and $\theta_0 = \pi(2-\delta)^{-1}$, then

$$\alpha_0 = -\delta \theta_0^2 (2 \cos \theta_0)^{-1} \tag{13}$$

and in this case (12) has, for $\alpha > \alpha_0$, precisely one solution z such

$$\begin{aligned} &\mathrm{Re}(z) > 0 \\ &0 < \mathrm{Im}(z) < 2\pi \end{aligned} \tag{14}$$

Proof. Any solution $z = iw$ of (12) for w real must satisfy

$$\begin{aligned} w &= -\alpha I^{-1} \int_{-1}^{-1+\delta} \sigma(\theta) \sin w\theta \, d\theta \\ 0 &= \int_{-1}^{-1+\delta} \sigma(\theta) \cos w\theta \, d\theta = h(w) \end{aligned} \tag{15}$$

The function $h(w)$ can be interpreted as the Fourier transform of an even function $\sigma_1(\theta)$, where $\sigma_1(\theta) = (\frac{1}{2})\sigma(\theta)$ for $-1 \leq \theta \leq 1-\delta$ and $\sigma_1(\theta) = 0$ for $\theta < -1$ or $-1+\delta < \theta \leq 0$. It follows that $h(w)$ is not a constant function, and since h has an obvious extension to an entire function, the statement about the positive zeros w_j follows. The inequality (8) shows that $z = 0$ is never a solution of (12) for $\alpha \neq 0$.

Now suppose that $0 < \alpha_1 < \alpha_0$. Define $H = \{z \in \mathbb{C}: \operatorname{Re}(z) \geq 0\}$. It is clear that any solution $z \in H$ of (12) for $0 < \alpha \leq \alpha_1$ must satisfy

$$\begin{aligned} &\operatorname{Re} z > 0 \quad \text{and} \\ &|z| \leq M_\alpha = \alpha I^{-1} \int_{-1}^{-1+\delta} |\sigma(\theta)|\, d\theta < M_{\alpha_0} \end{aligned} \tag{16}$$

It follows from Rouché's theorem that (12) has (counting multiplicities) precisely the same number of zeros in H for each α with $0 < \alpha \leq \alpha_1$.

Now define a function $\varphi(z,\alpha)$ for $z \in \mathbb{C}$, $\alpha \in \mathbb{R}$, by

$$\varphi(z,\alpha) = z + \alpha I^{-1} \int_{-1}^{-1+\delta} \sigma(\theta)\, e^{z\theta}\, d\theta \tag{17}$$

The implicit function theorem applies to $\varphi(z,\alpha)$ at $z = 0$, $\alpha = 0$; this combined with the estimate (16), implies that for $\alpha > 0$ and α small enough any solution of (12) in H must be the one insured by the implicit function theorem. If $z(\alpha)$ is the solution of (17) for $|\alpha|$ small, we know that $z(\alpha)$ is a C^1 function of α with $z(0) = 0$. If we can show that

$$\operatorname{Re} \left.\frac{dz}{d\alpha}\right|_{\alpha=0} < 0$$

it will follow that (12) has no solutions z in H for α small and positive and hence for $\alpha < \alpha_0$. However, a calculation shows that

$$\frac{dz}{d\alpha} = -I^{-1} \int_{-1}^{-1+\delta} \sigma(\theta)\, d\theta = -1 \tag{18}$$

If $\sigma(\theta) \equiv 1$, one can solve explicitly for the function $h(w)$ in (15) and find

$$\cos\left(\frac{2w-\delta w}{2}\right) \sin\left(\frac{\delta w}{2}\right) = 0 \tag{19}$$

The only positive solutions for (19) are

$$w = (2n+1)\pi\,(2-\delta)^{-1} \quad \text{or} \quad w = (2m\pi)\,\delta^{-1} \tag{20}$$

where n is a nonnegative integer and m a positive integer, If the second possibility in (20) is chosen one finds that the first equation (15) yields $w = 0$, an impossibility. Thus we can assume $w = (2n+1)\pi\,(2-\delta)^{-1}$ and a calculation yields that the number β in (10) is given by

$$\beta = -(w^2\delta)(2\cos w)^{-1} \tag{21}$$

To show that (12) has no solutions z with $\mathrm{Re}\, z \geq 0$ for $0 < \beta < -(\theta_0^2\delta)(2\cos\theta_0)^{-1}$, it suffices to prove now that

$$\left|\frac{\cos n\,\theta_0}{n\,\theta_0}\right| = \gamma_n$$

is a decreasing function of n for n an odd positive integer. If one sets $\theta_0 = \psi_0 + \frac{\pi}{2}$, one finds (for n odd)

$$\gamma_n = \left|\frac{\sin n\,\psi_0}{n(\psi_0 + \frac{\pi}{2})}\right| \tag{22}$$

and it is an easy inductive exercise (see Lemma 2 in [49]) to show that the right hand side of (22) is a decreasing function of the positive integer n.

It remains to show that if $\alpha > \alpha_0$ (for $\sigma(\theta) \equiv 1$) the equation (12) has exactly one solution z with $\mathrm{Re}\,(z) > 0$ and $0 < \mathrm{Im}(z) < 2\pi$. Define $G = \{z : \mathrm{Re}\, z > 0,\ 0 < \mathrm{Im}\, z < 2\pi\}$. It is clear that the equation

$$f(z,\alpha) = z + \alpha\delta^{-1}\int_{-1}^{-1+\delta} e^{z\theta}\,d\theta = 0 \tag{23}$$

cannot be satisfied for $z = u + 2\pi i$, u real, $\alpha > 0$ and $0 < \delta \leq \frac{1}{2}$, for if it were we would have

$$2\pi = -\alpha\delta^{-1} \int_{-1}^{-1+\delta} e^{u\theta} \sin 2\pi\theta \, d\theta \tag{24}$$

and the right hand side of (24) is negative. Similarly, by using (20) and (21), one can verify that the only solution $z = iw$, $0 \leq w \leq 2\pi$, of (23) for $\alpha > 0$ is for $w = \pi(2-\delta)^{-1}$ and $\alpha = \alpha_0$. Finally, it is clear that (23) has no real, non-negative solutions for $\alpha > 0$. It follows from Rouche's theorem that (23) has the same algebraic number of solutions in G for every $\alpha > \alpha_0$. If one applies the implicit function theorem to $f(z, \alpha)$ at $z = \pi(2-\delta)^{-1} i$ and $\alpha = \alpha_0$, it is not hard to prove that one obtains precisely one solution in G for $\alpha > \alpha_0$. We leave the details to the reader. ∎

Although we have presented a reasonably detailed analysis of the characteristic equation (23), difficulties in proving results like Theorem 2 usually arise elsewhere. The problem is to find a closed, convex set K and a continuous, compact map F whose nonzero fixed points correspond to nontrivial periodic solutions. Of course, to show 0 is an ejective point of F, K cannot be too large.

Our next lemma provides a reasonable choice for K. It is precisely at this point that the condition $\delta \leq \frac{1}{2}$ will enter crucially (although it was also used to a lesser extent in Lemma 1). In the statement of the next lemma we call a function φ monotonic increasing on an interval I if $\varphi(t_1) \leq \varphi(t_2)$ for all $t_1, t_2 \in I$ such that $t_1 < t_2$; φ is strictly monotonic increasing on I if $\varphi(t_1) < \varphi(t_2)$ for t_1, t_2 as before.

<u>Lemma 2.</u> Assume that f satisfies H1, $\alpha > 0$ and $0 < \delta \leq 1$. Let $X = C[-\delta, 1-\delta]$ in the usual sup norm and define $K = \{\varphi \in X : \varphi(t) \leq 0 \;\; -\delta \leq t \leq 0$, $\varphi(0) = 0$ and $\varphi | [0, 1-\delta]$ is monotonic increasing$\}$. For $\varphi \in K - \{0\}$

let $x(t) = x(t;\varphi,\alpha)$ denote the solution of equation (3) for $t \geq 1-\delta$ such that $x|[-\delta, 1-\delta] = \varphi$. Then $x(t)$ is concave down on $[1-\delta, 1]$ and if $x(1-\delta) = 0$, $x'(1-\delta) > 0$. Define $z_1(\varphi) = \inf\{t > 1-\delta : x(t) = 0\}$. If $z_1(\varphi) < \infty$, one has $x'(z_1) < 0$; the set of $\varphi \in K$ for which $z_1(\varphi) < \infty$ is relatively open and $\varphi \longrightarrow z_1(\varphi)$ is continuous on this set. If $z_1(\varphi) < \infty$, the map $x(t)$ is monotonic decreasing on $[z_1, z_1+(1-\delta)]$.

Proof. By changing variables we find that

$$x'(t) = -\alpha\delta^{-1}\int_{t-1}^{t-1+\delta} f(x(\theta))\,d\theta$$

and $x(t)$ is C^2 for $t \geq 1-\delta$ with $x''(t)$ given by

$$x''(t) = -\alpha\delta^{-1}[f(x(t-1+\delta)) - f(x(t-1))] \tag{25}$$

Equation (25) immediately implies that $x''(t) \leq 0$ on $[1-\delta, 1]$. If $x(1-\delta) = 0$, then by the choice of K we have $x(t) = 0$ for $0 \leq t \leq 1-\delta$, and since we assume φ is not identically zero we have $x'(1-\delta) > 0$ and $z_1 > 1-\delta$. If $1-\delta < z_1 \leq 1$, then concavity and the fact that $x(t) > 0$ on $(1-\delta, z_1)$ imply that $x'(z_1) < 0$. If $z_1 > 1$, define $t^* = \sup\{t \in [0, 1-\delta] : \varphi(t) = 0\}$; it is easy to check that $x'(t) \geq 0$ on $[t^*, t^*+(1-\delta)]$ and $x(t) > 0$ for $t^* < t \leq t^*+(1-\delta)$. It follows that we must have $z_1 > t^*+(1-\delta)$, and if we use this fact together with equation (3) and the assumption $z_1 > 1$ we obtain that $x'(z_1) < 0$.

Once one knows that $x'(z_1) < 0$, the facts that $z_1(\varphi) < \infty$ on a relatively open subset of $K - \{0\}$ and that $\varphi \longrightarrow z_1(\varphi)$ is continuous on its domain follow easily. We leave them to the reader.

The fact that x is monotonic decreasing on $[z_1, z_1+(1-\delta)]$ is immediate from equation (3) if $z_1 \geq 1$. If $z_1 < 1$, we know from concavity that x is monotonic decreasing on $[z_1, 1]$ and from (3) that x is monotonic decreasing on $[1, z_1+(1-\delta)]$. ∎

If $\varphi \in K - \{0\}$, $x(t) = x(t;\varphi,\alpha)$ and z_1 are as above, then if $\psi(t) = x(z_1+t)$ for $-\delta \le t \le 1-\delta$, the argument above shows that $\psi(0) = 0$, ψ is monotonic decreasing on $[0, 1-\delta]$ and $\psi(t) \ge 0$ for $-(1-\delta) \le t \le 0$. If one knew that $1-\delta \ge \delta$, i.e. if $\delta \le \frac{1}{2}$, one could repeat the argument of Lemma 2 with ψ as initial function. For this reason assume $0 < \delta \le \frac{1}{2}$ and define $z_2 > z_1$ to be the first time $t > z_1$ such that $x(t) = 0$. If z_2 is finite, define $F_\alpha(\varphi) = \varphi_1$, where $\varphi_1(t) = x(z_2+t)$ for $-\delta \le t \le 1-\delta$. By using the arguments of Lemma 2 one can see that $\varphi_1 \in K - \{0\}$ if $\varphi \in K - \{0\}$, so F_α defines a map of $K-\{0\} \longrightarrow K - \{0\}$, at least for φ with $z_2(\varphi) < \infty$. Extend this map by defining $F_\alpha \varphi = 0$ for φ such that $z_2(\varphi) = \infty$ and for $\varphi = 0$.

At first sight the above definition of F_α seems arbitrary, but it is not hard to prove that F_α so defined is a continuous, compact map. The details of such an argument are given in Section 2 of [41] for a simpler equation. It is also not hard to see that a non-zero fixed point φ of F_α gives the desired periodic solution of (3), i.e. the periodic solution is simply $x(t;\varphi,\alpha)$ for $t \ge -\delta$, and this function can be extended periodically with period $z_2(\varphi)$ to all of $\mathbb{R}$.

For those readers who prefer to be sure that $z_2(\varphi) < \infty$ for $\varphi \in K - \{0\}$, a simple condition can be given. If $\alpha > (1-\delta)^{-1}$, a simple argument (using that $f'(0) = 1$) shows that $z_2(\varphi) < \infty$. In Theorem 2 attention is restricted to $\alpha > \alpha_0$, so it will follow that $z_2(\varphi) < \infty$ for $\varphi \in K - \{0\}$ if $\alpha_0 > (1-\delta)^{-1}$ for $0 < \delta \le \frac{1}{2}$. By the results of Lemma 1, $\alpha_0 > (1-\delta)^{-1}$ is equivalent to

$$\theta_0(1-\delta)(\delta^{-1}\int_{1-\delta}^{1} \sin\theta_0 t\,dt) > 1, \quad \theta_0 = \pi(2-\delta)^{-1} \tag{26}$$

Inequality (26) is certainly valid if $\pi(2-\delta)^{-1}(1-\delta) > 1$, and one can easily check that the latter inequality holds for $0 < \delta \le (\frac{1}{2})$. I should remark that the condition $\alpha > (1-\delta)^{-1}$ is far from best possible.

It will be convenient to replace the unbounded set K by a bounded set K_1.

Recall that $f(x) \geq -B$ for all x. Define constants M_1 and M_2 by

$$M_1 = \alpha B$$

$$M_2 = \sup\{\alpha f(x) \colon 0 \leq x \leq M_1\}$$

and define K_1 by

$$K_1 = \{\varphi \in K \colon -M_2 \leq \varphi(x) \leq M_1 \text{ for all } x\} \tag{27}$$

It is easy to see that $F_\alpha(K_1) \subset K_1$.

If $\varphi \in K$ and $\alpha > (1-\delta)^{-1}$, let z_k denote the successive zeros of $x(t;\varphi,\alpha)$ and let $\tau_k \in (z_k, z_{k+1})$ be the (unique) value of $t \in (z_k, z_{k+1})$ where $|x(t)|$ achieves its maximum on $[z_k, z_{k+1}]$. The next lemma is a technical unpleasantry, but it seems to be necessary. The proof will be deferred until later.

Lemma 3. Suppose $\varphi \in K-\{0\}$, $\alpha > (1-\delta)^{-1}$ and $0 < \delta \leq (\frac{1}{2})$. Define $d > 0$ such that $\frac{1}{2}|x| \leq |f(x)| \leq 2|x|$ for $|x| \leq d$. Then there exist positive constants c_1 and c_2, dependent only on α and δ, such that if $|x(t)| \leq d$ for $z_k \leq t \leq z_{k+1}$, then $|x(z_{k+1} + 1-\delta)| \geq c_1|x(\tau_k)|$ and $|x(\tau_{k+1})| \leq c_2|x(\tau_k)|$.

We are now in a position to prove Theorem 2.

Proof of Theorem 2. By the previous remarks it suffices to prove that the map $F_\alpha \colon K_1 \longrightarrow K_1$ has a nonzero fixed point for $\alpha > \alpha_0$. According to Theorem 1, this will follow if one can prove that 0 is an ejective point of F_α (the condition for $\|x\| = R$ is vacuous, because K_1 is bounded).

For definiteness I shall assume that $\delta = \frac{1}{2}$; the proof is essentially the same for $0 < \delta \leq \frac{1}{2}$. As usual, given $\varphi \in K_1-\{0\}$, let $x(t;\varphi)$ denote the solution of (3) for $t \geq \frac{1}{2}$ such that $x(t) = \varphi(t)$ for $-\frac{1}{2} \leq t \leq \frac{1}{2}$ and let z_k denote the successive zeros of $x(t;\varphi)$. To prove ejectivity it is necessary to find a positive number d_1 such that if $\varphi \in K_1-\{0\}$ and $\|\varphi\| \leq d_1$, then there exists a zero z_{2k} such that

$$\sup\{\,|x(t;\varphi)| : z_{2k}-(\tfrac{1}{2}) \leq t \leq z_{2k}+(\tfrac{1}{2})\} > d_1 \tag{28}$$

Assume first that $d_1 < c_1^2 (c_2)^{-3} d$; d_1 will later be selected even smaller. One can also assume that

$$\sup_{t \geq -\frac{1}{2}} |x(t;\varphi)| < d(c_2)^{-3} \tag{29}$$

For if (29) fails, it is easy to show using Lemma 3 that

$$x(z_{2k} + \tfrac{1}{2}) \geq c_1^2 (c_2)^{-3}\, d$$

for some zero z_{2k}, and (28) has been satisfied.

Thus we can assume that φ is such that (29) holds.

Let z denote the complex root of

$$z = -2\alpha \int_{-1}^{-\frac{1}{2}} e^{z\theta}\, d\theta \tag{30}$$

such that $0 < \operatorname{Re} z$ and $0 < \operatorname{Im}(z) < 2\pi$. Select $\tau \geq \frac{1}{2}$ such that $x'(\tau) = 0$; since $x(t)$ satisfies (29) it makes sense to consider

$$\int_{\tau}^{\infty} x'(t)\, e^{-z(t-\tau)} dt$$

For notational convenience define $y(t) = x(t+\tau)$ and note that $y(t)$ also satisfies (3). If we integrate by parts we find that

$$\int_0^{\infty} y'(t)\, e^{-zt}\, dt = z\int_0^{\infty} y(t)\, e^{-zt} dt - y(0) \tag{31}$$

If we define $g(t)$ by

$$g(t) = \int_{-1}^{-\frac{1}{2}} (f(y(t+\theta)) - y(t+\theta))d\theta \tag{32}$$

and R by

$$R = \int_0^{\infty} g(t)\, e^{-zt}\, dt \tag{33}$$

then if we substitute for $y'(t)$ from equation (3) we find

$$\int_0^\infty y'(t)e^{-zt}dt = -2\alpha \int_0^\infty \int_{-1}^{-\frac{1}{2}} y(t+\theta)d\theta\, e^{-zt}dt + R \tag{34}$$

Simplifying (34) yields

$$\int_0^\infty y'(t)\,e^{-zt}dt = R - 2\alpha \int_{-1}^{-\frac{1}{2}} e^{z\theta}\left(\int_0^\infty y(s)e^{-zs}ds\right)d\theta - 2\alpha \int_{-1}^{-\frac{1}{2}} \int_\theta^0 y(s)e^{-z(s-\theta)}ds\,d\theta \tag{35}$$

If one equates (35) and (31) and uses (30) one obtains

$$-y(0) + 2\alpha \int_{-1}^{-\frac{1}{2}} \left(\int_\theta^0 y(s)\,e^{-z(s-\theta)}\,ds\right)d\theta = R \tag{36}$$

If one writes $\varphi(s) = y(s)$ for $-1 \leq s \leq 0$, so $\varphi \in C([-1,0])$, notice that the left hand side of (36) is just a continuous linear functional of φ, independent of the values of y on $[0,\infty)$. As we shall indicate more precisely later, this is a general phenomenon for such calculations.

If $Y(t)$ and $W(t)$ are defined by

$$Y(t) = \int_{-1}^{t} y(s)\,ds$$
$$W(t) = \int_{-1}^{t} f(y(s))\,ds \tag{37}$$

then the selection of τ implies

$$x'(\tau) = y'(0) = -2\alpha W\left(-\frac{1}{2}\right) = 0 \tag{38}$$

A further integration by parts in (36) together with (38) yields

$$-\int_{-\frac{1}{2}}^{0} y'(s)e^{-zs}ds - y\left(-\frac{1}{2}\right)e^{\frac{1}{2}z} - 2\alpha \int_{-1}^{-\frac{1}{2}} W(s)\,e^{-z(1+s)}\,ds =$$
$$R - z\,e^{\frac{1}{2}z}\left[W\left(-\frac{1}{2}\right) - Y\left(-\frac{1}{2}\right)\right] - 2\alpha \int_{-1}^{-\frac{1}{2}} [W(s) - Y(s)]e^{-z(1+s)}\,ds \tag{39}$$

Define ζ to be the zero of $y(t)$ which immediately precedes 0, so that $\zeta \leq -\frac{1}{2}$.

Assuming for definiteness that $y(0) = x(\tau) > 0$ (the proof is the same if $x(\tau)<0$), one can see that $y'(s) \geq 0$ for $-\frac{1}{2} \leq s \leq 0$ and $y(-\frac{1}{2}) \geq 0$. It is also clear that $W(t) \leq 0$ for $-1 \leq t \leq \zeta$; and since $W'(t) \geq 0$ for $\zeta \leq t \leq 0$ and $W(-\frac{1}{2}) = 0$, it follows that $W(t) \leq 0$ for $-1 \leq t \leq -\frac{1}{2}$. The point of these observations is that if one takes the imaginary part of the left hand side of (39) one finds that all three terms on the left hand side are negative.

Now define $m = \sup\{|x(t)| : t \geq 0\}$ and denote by LHS and RHS the left hand side and right hand side respectively of (39). Select $\tau > 0$ to be a local extremum of $x(t)$ such that $|x(\tau)| \geq (\frac{1}{2})m$. It is easy to see that there exists a function $\epsilon(m)$ (independent of φ), with $\lim_{m \to 0} \epsilon(m) = 0$, such that the imaginary part of RHS is estimated by

$$|\operatorname{Im}(\text{RHS})| \leq \epsilon(m)m \tag{40}$$

By Lemma 3 one has that (assuming $x(\tau) > 0$)

$$y'(s) \leq 2\,\alpha(c_1)^{-1}\,y(0) \tag{41}$$

for $-\frac{1}{2} \leq s \leq 0$. Using this fact and the observations above, it is not hard to see that there exists a positive constant c such that

$$|\operatorname{Im}(\text{LHS})| \geq |\operatorname{Im}\left(\int_{-\frac{1}{2}}^{0} y'(s)e^{-zs}\,ds\right) + \operatorname{Im}\left(y(-\tfrac{1}{2})\,e^{\frac{1}{2}z}\right)| \geq c\,m \tag{42}$$

It follows from (39), (40) and (42) that

$$c\,m \leq \epsilon(m)\,m \tag{43}$$

If $m_0 > 0$ is selected so that $\epsilon(m) < c$ for $0 < m \leq m_0$, it follows that

$$m = \sup_{t \geq 0} |x(t)| \geq m_0 \tag{44}$$

If one uses Lemma 3 again, there is a positive constant c_3 and a zero z_{2k}

of $x(t)$ such that

$$x(z_{2k} + \frac{1}{2}) \geq c_3 m_0 \tag{45}$$

If d_1 is selected less than $c_3 m_0$, the proof of ejectivity is finished. ∎

The basic idea of the above proof of ejectivity can be traced to E.M. Wright [59]. Wright proved that if $x(t)$ satisfies eq. (1) for $t \geq 0$ and $x|[-1,0] = \varphi$ is a nonnegative continuous function (not identically zero) then if $\alpha > \pi/2$ there exists a positive constant $c = c(\alpha)$ (independent of φ) such that

$$\limsup_{t \to \infty} |x(t;\varphi)| \geq c$$

We should point out that the method of proof used in Theorem 2 actually applies in all instances where cone methods have been used to prove existence of periodic solutions. By this we mean the following: Consider the vector F.D.E.

$$\begin{aligned} x'(t) &= L(x_t) + N(x_t) \quad \text{for } t \geq 0 \\ x|[-r,0] &= \varphi \end{aligned} \tag{46}$$

Here $x(t)$ takes values in $\mathbb{R}^n$, $L: X = C([-r,0]; \mathbb{R}^n) \longrightarrow \mathbb{R}^n$ is a continuous linear map and $N: X \longrightarrow \mathbb{R}^n$ is continuous and takes bounded sets to bounded sets. As in [24], $x_t = \psi \in X$ is defined by $\psi(s) = x(t+s)$ for $-r \leq s \leq 0$. The maps L and N extend in the obvious way to complex-valued functions.

For each complex number z, one obtains a linear map $\Lambda(z): \mathbb{C}^n \longrightarrow \mathbb{C}^n$ where $\Lambda(z)$ is defined by the formula

$$\begin{aligned} (\Lambda(z))(w) &= L(\varphi) \\ \varphi(t) &= e^{zt} w \quad \text{for } -r \leq t \leq 0 . \end{aligned} \tag{47}$$

$\Lambda(z)$ can be considered an $n \times n$ matrix, so it makes sense to consider $\Lambda(z)^t$, the transpose matrix. Let $z \in \mathbb{C}$ be a solution of the characteristic equation

$$\det (zI - \Lambda(z)) = 0 \tag{48}$$

and let $w \in \mathbb{C}^n$ satisfy

$$z\,w = \Lambda(z)^t\, w \tag{49}$$

If $<u, v>$ denotes the standard Hermitian inner product on $\mathbb{C}^n$ and if one assumes $x(t)$ is bounded and $\mathrm{Re}\, z > 0$ one finds

$$\int_0^\infty <e^{-zt} w,\ x'(t)>dt = \int_0^\infty <z\,w,\ x(t)>e^{-zt}dt - <w, \varphi(0)> \tag{50}$$

On the other hand one can prove that

$$\begin{aligned} &\int_0^\infty <e^{-zt} w,\ L(x_t)>dt + \int_0^\infty <e^{-zt} w,\ N(x_t)>dt = \\ &l(\varphi) + \int_0^\infty <z\ w,\ x(t)>e^{-zt}dt + \int_0^\infty <e^{-zt}\ w,\ N(x_t)>dt \end{aligned} \tag{51}$$

where $l : X \longrightarrow \mathbb{C}$ is a continuous linear functional dependent on z, w and L. Equating (50) and (51) gives

$$l(\varphi) + <w, \varphi(0)> = -\int_0^\infty <e^{-zt}\ w,\ N(x_t)> dt \tag{52}$$

which is the sort of identity obtained in (36).

A gap still remains in the proof of Theorem 2 since Lemma 3 has not been proved.

Proof of Lemma 3. The ideas underlying the proof are geometrical and the reader may find it helpful to draw some pictures. For simplicity we assume $\delta = \frac{1}{2}$; the proof is the same for $0 < \delta \leq \frac{1}{2}$.

The existence of the constant c_2 is easy; details are left to the reader. To prove existence of c_1, suppose $\varphi \in K - \{0\}$ and $x(t; \varphi, \alpha) = x(t)$ is the corresponding solution of (3) and $\zeta_1 > \frac{1}{2}$ and ζ_2 are, respectively, the first and second zeros of $x(t)$. Define $\tau_0 \in [\frac{1}{2}, \zeta_1)$ by $x'(\tau_0) = 0$ and $\tau_1 \in (\zeta_1, \zeta_2)$ by $x'(\tau_1) = 0$, so x has a local maximum at τ_0 and a local minimum at τ_1. Assume $x(\tau_0) \leq d$. To prove Lemma 3 it suffices to find a constant c_1, independent of φ, such that $|x(\tau_1)| \leq c_1 x(\tau_0)$. The proof consists of a laborious case-by-case analysis depending on the size of ζ_1.

Case 1. Assume $\zeta_1 \leq 1$. Recall that $x(t)$ is known to be concave down for $(\frac{1}{2}) \leq t \leq 1$, so it follows that

$$
\begin{aligned}
x'(\zeta_1) &\leq (x(\zeta_1) - x(\tau_0))(\zeta_1 - \tau_0)^{-1} = -x(\tau_0)(\zeta_1 - \tau_0)^{-1} \\
x'(t) &\leq -x(\tau_0)(\zeta_1 - \tau_0)^{-1} \quad \text{for } \zeta_1 \leq t \leq 1
\end{aligned}
\tag{53}
$$

Equation (53) implies that

$$
x(1) \leq -2x(\tau_0)(1 - \zeta_1) \tag{54}
$$

If $\zeta_1 \leq (\frac{7}{8})$, equation (54) gives the lemma, so one can assume $(\frac{7}{8}) \leq \zeta_1 \leq 1$. For $1 \leq t \leq (\frac{11}{8})$ one has the estimate

$$
\begin{aligned}
x'(t) &= -2\alpha \int_{t-1}^{t-\frac{1}{2}} f(x(s))\,ds \\
&\leq -\alpha \int_{\frac{1}{2}}^{t-\frac{1}{2}} x(s)\,ds
\end{aligned}
\tag{55}
$$

Let $u_1(s)$ denote the piecewise linear function whose graph is a straight line segment connecting the point $(\frac{1}{2}, 0)$ to the point $(\tau_0, x(\tau_0))$ and a straight line segment connecting the point $(\tau_0, x(\tau_0))$ to the point $(\zeta_1, 0)$. The concavity of $x(s)$ and (55) yield (for $1 \leq t \leq (\frac{11}{8})$)

$$
x'(t) \leq -\alpha \int_{\frac{1}{2}}^{t-\frac{1}{2}} u_1(s)\,ds \tag{56}
$$

It is a simple geometry exercise to verify that if $v_1(s)$ is the function whose graph is a straight line segment connecting the points $(0, \frac{1}{2})$ and $(\zeta_1, x(\tau_0))$ then (for $1 \le t \le \frac{11}{8}$)

$$\int_{\frac{1}{2}}^{t-\frac{1}{2}} u_1(s)\,ds \ge \int_{\frac{1}{2}}^{t-\frac{1}{2}} v_1(s)\,ds \tag{57}$$

One can verify using (56) and (57) that for $1 \le t \le \frac{11}{8}$ one has

$$x'(t) \le -\alpha x(\tau_0)(t-1)^2 \tag{58}$$

Equation (58) implies that (for $\zeta_1 \ge (\frac{7}{8})$)

$$x(\tau_1) \le x(\tfrac{11}{8}) - x(1) \le -(\tfrac{\alpha}{3})(\tfrac{3}{8})^3 x(\tau_0) \tag{59}$$

and (59) gives the desired estimate in case 1.

Case 2. Assume that $1 < \zeta_1 \le (\frac{3}{2})$. As before $x(t)$ is concave down on $[\frac{1}{2}, 1]$. It is necessary to estimate $x'(t)$ for $\zeta_1 \le t \le \zeta_1 + (\frac{1}{2})$. First suppose that $(\frac{3}{2}) \le t \le \zeta_1 + (\frac{1}{2})$ and let $u_2(s)$ denote the function whose graph is a straight line segment connecting the points $(\frac{1}{2}, x(\tau_0))$ and $(1, 0)$. Then one finds that (for t as above)

$$\begin{aligned} x'(t) &= -2\alpha \int_{t-1}^{t-\frac{1}{2}} f(x(s))\,ds \\ &\le -\alpha \int_{t-1}^{1} u_2(s)\,ds \\ &= -\alpha(2-t)^2\, x(\tau_0) \end{aligned} \tag{60}$$

We have omitted an intermediate step here in which (just as in case 1) one observes that if $v_2(s)$ is the piecewise linear function whose graph is a straight line segment from $(\frac{1}{2}, 0)$ to $(\tau_0, x(\tau_0))$ and a straight line segment from $(\tau_0, x(\tau_0))$ to $(1, 0)$, then

$$\int_{t-1}^{1} v_2(s)\,ds \geq \int_{t-1}^{1} u_2(s)\,ds\,.$$

If $\zeta_1 \leq t \leq (\frac{3}{2})$ and if $u_3(s)$ denotes the function whose graph is a straight line segment from $(\frac{1}{2}, 0)$ to $(1, x(\tau_0))$, one finds that

$$\begin{aligned} x'(t) &= -2\alpha \int_{t-1}^{t-\frac{1}{2}} f(x(s))\,ds \\ &\leq -\alpha \int_{\frac{1}{2}}^{t-\frac{1}{2}} u_3(s)\,ds \qquad (61) \\ &= -\alpha x(\tau_0)\,(t-1)^2 \end{aligned}$$

If one uses the estimates (60) and (61) one obtains

$$\begin{aligned} x(\zeta_1 + \tfrac{1}{2}) &= -\alpha \int_{\zeta_1}^{(3/2)} x'(t)\,dt - \alpha \int_{(3/2)}^{\zeta_1 + \frac{1}{2}} x'(t)\,dt \qquad (62) \\ &\leq -(\tfrac{\alpha}{3}) x(\tau_0) \left[\tfrac{1}{4} - (\zeta - 1)^3 - (\tfrac{3}{2} - \zeta_1)^3\right] \end{aligned}$$

If ζ_1 in (62) is considered a variable with $1 \leq \zeta_1 \leq (\frac{3}{2})$, the maximum of the right hand side of (62) is achieved at $\zeta_1 = 1$ or $\zeta_1 = (\frac{3}{2})$, and one obtains

$$x(\zeta_1 + \tfrac{1}{2}) \leq -(\tfrac{\alpha}{24})\, x(\tau_0) \qquad (63)$$

The estimate (63) completes the proof for case 2.

<u>Case 3.</u> Assume that $\zeta_1 \geq (\frac{3}{2})$. If $1 \leq t \leq (\frac{3}{2})$, the same argument used to obtain (61) still applies and one obtains

$$x'(t) \leq -\alpha x(\tau_0)\,(t-1)^2 \qquad (64)$$

Since it is being assumed that $\zeta_1 \geq \frac{3}{2}$, one must have for $1 \leq t \leq (\frac{3}{2})$

$$x(t) - \alpha\, x(\tau_0) \int_{t}^{(\frac{3}{2})} (s-1)^2 ds \geq 0 \quad \text{or} \qquad (65)$$

$$x(t) \geq (\frac{\alpha}{3}) x(\tau_0) [(\frac{1}{8}) - (t-1)^3] = \psi(t)$$

Since $\psi(t)$ is monotonic decreasing and concave down on $[1, \frac{3}{2}]$ one finds that for $1 \leq t \leq (\frac{3}{2})$

$$x(t) \geq (\frac{\alpha}{12}) x(\tau_0) (\frac{3}{2} - t) \tag{66}$$

As in the proof of equation (60) one still has (if $t \leq 2$)

$$\int_{t-1}^{1} x(s)\, ds \geq x(\tau_0)(2-t)^2 \tag{67}$$

If $(\frac{3}{2}) \leq t \leq 2$ one obtains from (66) and (67) that

$$\begin{aligned} x'(t) &\leq -\alpha \int_{t-1}^{1} x(s)\, ds - \alpha \int_{1}^{t-\frac{1}{2}} x(s)\, ds \\ &\leq -\alpha x(\tau_0)[(2-t)^2 + (\frac{\alpha}{24})((\frac{1}{4}) - (2-t)^2)] \end{aligned} \tag{68}$$

Equation (66) implies that $\alpha \leq 24$ (otherwise $x(1) > x(\tau_0)$), and one then obtains from (68) that for $\frac{3}{2} \leq t \leq 2$

$$x'(t) \leq -(\frac{\alpha^2}{96}) x(\tau_0) \tag{69}$$

If $(\frac{3}{2}) < \zeta_1 \leq 2$ one obtains from (66) and (69) and the fact that $x(t)$ is concave down on $[\frac{1}{2}, 1]$ that $x(t) \geq \theta(t)$ for $\frac{1}{2} \leq t \leq \zeta_1$, where

$$\begin{aligned} \theta(t) &= (\frac{\alpha}{12}) x(\tau_0)(t - \frac{1}{2}) & \frac{1}{2} \leq t \leq 1 \\ \theta(t) &= (\frac{\alpha}{12}) x(\tau_0)(\frac{3}{2} - t) & 1 \leq t \leq \frac{3}{2} \\ \theta(t) &= (\frac{\alpha^2}{96}) x(\tau_0)(\zeta_1 - t) & \frac{3}{2} \leq t \leq \zeta_1 \end{aligned} \tag{70}$$

Equation (70) implies that for $\zeta_1 \leq t \leq \zeta_1 + \frac{1}{2}$

$$x'(t) \leq -\alpha \int_{t-1}^{t-\frac{1}{2}} \theta(s)\,ds$$
$$\leq -\alpha k_1 x(\tau_0) \tag{71}$$

where k_1 is a positive constant independent of ζ_1, $(\frac{3}{2}) \leq \zeta_1 \leq 2$. Equation (71) implies that

$$x(\zeta_1 + \tfrac{1}{2}) \leq -(\tfrac{1}{2})\alpha k_1 x(\tau_0) \tag{72}$$

which proves the lemma. Thus one can assume $\zeta_1 \geq 2$.

Actually the above argument proves more. Assume $x(2) \geq 0$ and let $\theta(t)$ be defined as in (70) for $\zeta_1 = 2$. Then equation (71) is still valid and one obtains

$$x(\tfrac{5}{2}) \leq x(2) - (\tfrac{1}{2})\alpha k_1 x(\tau_0) \tag{73}$$

Thus one can also assume that $x(2) \geq (\frac{1}{4})\alpha k_1 x(\tau_0)$ or the lemma is proved.

Now select a positive constant $\gamma < 1$ such that $\frac{1}{2}\alpha\gamma > 1$ (in general, so that $(1-\delta)\alpha\gamma > 1$) and define $\epsilon > 0$ to be such that for $|x| \leq 2\epsilon d$

$$|f(x)| \geq \gamma |x|$$

and $\epsilon < (\frac{1}{4})\alpha k_1$. Define T to be the first $t \geq 2$ such that $x(t) = \epsilon x(\tau_0)$; $T > 2$ by the above remarks. Then using the fact that x is decreasing on $[T-1, T]$ one obtains for $T \leq t \leq T + (\frac{1}{2})$

$$x'(t) = -2\alpha \int_{t-1}^{t-(\frac{1}{2})} f(x(s))\,ds$$
$$\leq -(2\alpha)(\tfrac{1}{2})\gamma x(t - \tfrac{1}{2}) \tag{74}$$
$$\leq -\alpha\gamma x(T - \tfrac{1}{2})$$

Equation (74) implies that $\zeta_1 - T \leq (\alpha\gamma)^{-1} < \frac{1}{2}$ and

$$x(T + \tfrac{1}{2}) \leq \epsilon x(\tau_0)\,[1 - (\tfrac{1}{2})(\alpha\gamma)] \tag{75}$$

Equation (75) completes the proof of Lemma 3. ∎

2. The equation $y'(t) = -\alpha f(y(t-1))$.

In this section I would like to specialize to the equation

$$y'(t) = -\alpha f(y(t-1)) \tag{76}$$

and describe some known results and some open questions. Because of its simple form it is tempting to try to understand as well as possible the structure of the set of periodic solutions of (76). Equation (76) serves as a model: if one cannot understand it well, the prospect for more complicated looking equations is bleak.

Equation (76) already contains a variety of interesting examples. For instance, the substitution $x(t) = \exp(y(t)) - 1$ transformes (1) to the form of (76). In general suppose $N: [-b, a] \longrightarrow \mathbb{R}$ $(b > 0,\ a > 0)$ is a continuous map such that $N(x) > 0$ for $-b < x < a$ and $N(-b) = 0$ and $N(a) = 0$. Consider the equation

$$x'(t) = -\alpha x(t-1) N(x(t)) \tag{77}$$

If $f(y)$ is a solution of the ordinary differential equation

$$f'(y) = N(f(y))$$

$$f(0) = 0$$

then the substitution $x(t) = f(y(t))$ transforms (77) to (76).

It is necessary first to recall what the techniques of the previous section yield for the equation (76). Define $X = C[0, 1]$, the Banach space of continuous, real-valued functions in the usual sup norm. Define a cone K in X by

$$K = \{\varphi \in X: \varphi(0) = 0 \text{ and } \varphi \text{ is monotonic increasing}\}$$

Assume that f satisfies H1 and given $\varphi \in K$ and $\alpha > 0$ let $x(t; \varphi, \alpha) = x(t)$

denote the solution of

$$
\begin{aligned}
x'(t) &= -\alpha f(x(t-1)) \quad \text{for } t \geq 1 \\
x(t) &= \varphi(t) \quad \text{for } 0 \leq t \leq 1
\end{aligned}
\tag{78}
$$

Denote by $z_1(\varphi, \alpha) = z_1$ the first time $t > 0$ such that $x(t) = 0$ (if such a time exists) and by $z_2(\varphi, \alpha) = z_2$ the first time $t > z_1(\varphi, \alpha)$ such that $x(t) = 0$. Define a map $F_\alpha: K \longrightarrow K$ by $F_\alpha(\varphi) = 0$ if $z_2(\varphi, \alpha)$ is not defined or if $\varphi = 0$ and $F_\alpha(\varphi) = \psi$, where $\psi(t) = x(z_2 + t)$ for $0 \leq t \leq 1$ if $z_2(\varphi, \alpha) < \infty$ Then one has the following theorem.

Theorem 3. Assume that f satisfies H1 and that $\alpha > (\frac{\pi}{2})$. Then the map $F_\alpha: K \longrightarrow K$ has a nonzero fixed point, so that (76) has a periodic solution $x(t)$ such that $x(0) = 0$, $x(t) > 0$ for $0 < t < z_1$, where $z_1 > 1$, $x(t) < 0$ for $z_1 < t < z_2$, where $z_2 - z_1 > 1$, and $x(t+z_2) = x(t)$ for all t.

It will be convenient to call a periodic solution of the type described in Theorem 3 a "slowly oscillating periodic solution".

Actually one can obtain sharper results than are given in Theorem 2 or Theorem 3. To describe these a definition is needed.

Definition. Suppose Z is a topological space and $g: Z \longrightarrow Z$ a continuous map with fixed point z_0. The point z_0 is called an "attractive point for g" if there exists an open neighborhood U_0 of z_0 such that for any open neighborhood V of z_0 there is an integer $N(V)$ such that $g^n(U_0) \subset V$ for $n \geq N(V)$.

The following theorem is proved (in greater generality) in [43]; in some important applications the function F defined below is not continuous at all points $(0, \alpha)$, but for simplicity I have ignored this point.

Theorem 4. Suppose that C is a closed, convex subset of a Banach space Y and that $0 \in C$. Assume either that C is infinite dimensional or that C is finite dimensional and $0 \in \tilde{\partial}(C)$ = the boundary of C in the affine subspace spanned by C. Let $I = (a, \infty)$ be an interval of reals $(a \geq -\infty)$ and suppose that $F: C \times I \longrightarrow C$ is a continuous, compact map such that $F(0, \alpha) = 0$ for all $\alpha \in I$. Define $F_\alpha(x) = F(x, \alpha)$ and assume that there exists $\alpha_0 \in I$ such that 0 is an attractive point for F_α for $\alpha < \alpha_0$ and an ejective point for F_α for $\alpha > \alpha_0$ or vice-versa. Finally, if $a > -\infty$ and if $F(x_k, \alpha_k) = x_k$ for any sequence $x_k \in C - \{0\}$ and $\alpha_k \longrightarrow a$ suppose that $\|x_k\| \longrightarrow \infty$. Define a set S by

$$S = \text{closure}\{(x, \alpha) \in C \times I : x \neq 0 \text{ and } F(x, \alpha) = x\}$$

and let S_0 denote the maximal connected component of S which contains $(0, \alpha_0)$. Then S_0 is nonempty and unbounded and contains no points $(0, \beta)$ for $\beta \neq \alpha_0$.

Theorem 4 is directly applicable to equation (3) or equation (76) if one takes $C = K$ = the set defined in Section 1, $I = (0, \infty)$, F_α the map already defined, and $\alpha_0 = -\delta\omega^2(2\cos\omega)^{-1}$ for $\omega = \pi(2-\delta)$ $(\alpha_0 = (\frac{\pi}{2})$ for eqn. (76)). When applied to (76) Theorem 4 yields

<u>Theorem 5.</u> (See [43]). Assume that f satisfies H1, let F_α, and K be as defined for equation (76), and let $I = (0, \infty)$. Define $S = \text{closure}\{(\varphi, \alpha) \in K \times I : \varphi \neq 0$ and $F_\alpha(\varphi) = \varphi\}$. Then there exists a constant $\epsilon > 0$ such that S contains no points (φ, α) with $0 < \alpha \leq \epsilon$ and there exists a continuous function $M(\alpha)$ for $\alpha \geq 0$ such that $\|\varphi\| \leq M(\alpha)$ if $(\varphi, \alpha) \in S$. If S_0 denotes the maximal connected component of S which contains the point $(0, \frac{\pi}{2})$, S_0 is nonempty and unbounded and contains no points $(0, \beta)$ for $\beta \neq (\frac{\pi}{2})$.

Theorem 5 provides more information than Theorem 3, but its real interest is to provide a method for obtaining information (e.g. , about the periods of solutions of (76)) which until now has proved inaccessible by other means.

Notice that the map $(\varphi, \alpha) \rightarrow z_2(\varphi, \alpha)$ = the minimal period of $x(t; \varphi, \alpha)$ is clearly a continuous function on $S - \{(0, \frac{\pi}{2})\}$, but that it is not immediately obvious that z_2 can be defined continuously at $(0, \frac{\pi}{2})$. The next lemma shows how to remedy this problem.

Lemma 4. (See [43]). Assume f satisfies H 1 and let $z_2(\varphi, \alpha)$ be defined as before for $(\varphi, \alpha) \in S - \{(0, \frac{\pi}{2})\}$. If one defines $z_2(0, \frac{\pi}{2}) = 4$, then z_2 is a continuous function on S.

The following lemma describes the behaviour of $z_2(\varphi, \alpha)$ for $(\varphi, \alpha) \in S$ and α large. The proof is considerably more difficult than that of Lemma 4 and is given in [44].

Lemma 5. Assume that f satisfies H 1. In addition assume that $\lim_{x \to -\infty} f(x) = -b$, where $0 < b < \infty$, and $\lim_{x \to \infty} f(x) = a$, where $0 < a$ but the possibility $a = \infty$ is allowed. If $a = \infty$ assume that f is monotonic increasing (much less is necessary). Then one has

$$\lim_{\beta \to \infty} (\inf\{z_2(\varphi, \alpha) : (\varphi, \alpha) \in S \text{ and } \alpha \geq \beta\}) =$$

$$\lim_{\beta \to \infty} (\sup\{z_2(\varphi, \alpha) : (\varphi, \alpha) \in S \text{ and } \alpha \geq \beta\}) = \tag{79}$$

$$= p_\infty = 2 + ab^{-1} + ba^{-1}$$

Actually, more than Lemma 5 is proved in [44]: it is shown that if $x(t)$ is a periodic solution of (76) corresponding to $(\varphi, \alpha) \in S$ and α large, then $f(x(t))$ looks essentially like a step function alternating between the values +a and -b.

As an immediate consequence of Theorem 5 and Lemma 4 and 5 one obtains

Corollary 1. Assume that f satisfies H 1 and the additional assumptions of Lemma 5. Then for every number p with $4 < p < 2 + ab^{-1} + ba^{-1}$ there exists a $(\varphi, \alpha) \in S_0 - \{(0, \frac{\pi}{2})\}$ with $z_2(\varphi, \alpha) = p$. In particular, for every such p there is an $\alpha = \alpha_p > 0$ and a periodic solution $x(t)$ of (76) with $x(0) = 0$, $x(t) > 0$

on $(0, z_1)$ with $z_1 > 1$, $x(t) < 0$ on (z_1, p) and $x(t+p) = x(t)$ for all t.

Proof. By Lemmas 4 and 5 the function $z_2(\varphi, \alpha)$ is continuous on the connected set S_0 and

$$\inf\{z_2(\varphi,\alpha) : (\varphi,\alpha) \in S_0\} \leq 4$$

$$\sup\{z_2(\varphi,\alpha) : (\varphi,\alpha) \in S_0\} \geq 2 + ab^{-1} + ba^{-1} = p_\infty \tag{80}$$

By connectedness $z_2(\varphi,\alpha) = p$ for some $(\varphi,\alpha) \in S_0 - \{(0,\frac{\pi}{2})\}$. ■

Notice that Corollary 1 provides no information if f is an odd function, for then $p_\infty = 4$. In fact in this case it is known [30] that for every $\alpha > (\frac{\pi}{2})$ there is a slowly oscillating periodic solution of (76) of period 4.

I have restricted myself until now to slowly oscillating periodic solutions. In fact, as was observed formally by K. Cooke [29], there are other periodic solutions of (76). For suppose $x(t)$ is a slowly oscillating periodic solution of

$$x'(t) = -\alpha f(x(t-1))$$

of minimal period p. If n is an integer, define $y(t) = x((np+1)t)$ and $\beta = (np+1)\alpha$. Then one can verify that $y(t)$ satisfies

$$y'(t) = -\beta f(y(t-1))$$

and $y(t)$ is of minimal period $p(np+1)^{-1}$. However, one needs results like those above above to obtain any real information from Cooke's observation, and in fact this was the original motivation for proving Lemmas 4 and 5 and Corollary 1. With these results one can proceed as follows: assume f satisfies H 1 and let Y denote the Banach space of bounded continuous functions $x: \mathbb{R} \to \mathbb{R}$. One is interested in pairs $(x,\alpha) \in Y \times I$ such that x is a periodic solution of $x'(t) = -\alpha f(x(t-1))$; call the set of such pairs T. For $(\varphi,\alpha) \in S_0$ the function

$x(t;\varphi,\alpha)$ can be extended by periodicity to all of $\mathbb{R}$, and one obtains a connected set $T_0 \subseteq T$ by

$$T_0 = \{(x(t;\varphi,\alpha),\alpha) : (\varphi,\alpha) \in S_0\}$$

The set T_0 of periodic solutions bifurcates from the zero solution at $\alpha = \frac{\pi}{2}$. For each integer n one obtains another connected subset T_n of T by

$$T_n = \{(y(t),\beta) : y(t) = x((np+1)t;\varphi,\alpha) \quad \text{for some } (\varphi,\alpha) \in S_0$$

$$p = z_2(\varphi,\alpha) \text{ and } \beta = (np+1)\alpha\}$$

The set T_n bifurcates from the zero solution at $\beta = (4n+1)(\frac{\pi}{2})$, n an integer. It is not hard to prove that any bifurcation of periodic solutions from the zero solution must occur at $\alpha = (4n+1)(\frac{\pi}{2})$, n an integer. If further assumptions are made on f, one can use Lemma 5 and obtain more detailed information about T_n.

It should be noted that Chow and Mallet-Paret [10] have also proved the exsistence of "rapidly-oscillating" periodic solutions. Their results give less information than the methods above for equation (76), but their techniques also apply to some other equations.

The results I have described so far are not limited to equation (76). Similar theorems about the qualitative properties of periodic solutions can be proved for other F. D. E.'s; see, for instance, [46]. When one turns to the question of global stability, however, the known results are almost entirely limited to (76) (and under stronger assumptions on f than H1) or to

$$x'(t) = -\alpha f(x(t-1), x(t))$$

An important and difficult problem is to obtain global stability results for periodic solutions of other F. D. E.'s. For equation (76), J. Kaplan and

J. Yorke have proved

Theorem 6. [31]. Assume that $\alpha > (\frac{\pi}{2})$ and that $f: \mathbb{R} \longrightarrow \mathbb{R}$ is C^1, bounded below, $f'(x) > 0$ for all x, $f(0) = 0$ and $f'(0) = 1$.

Then there exist two (possibly identical) slowly oscillating periodic solutions $x_0(t)$ and $x_1(t)$ of (76) such that if $X_0(t) = (x_0(t), x_0'(t))$ and $X_1(t) = (x_1(t), x_1'(t))$ then the simple closed curve $X_0(t)$ lies inside the simple closed curve $X_1(t)$ (both being considered as curves in the Cartesian plane). Furthermore, if $\varphi \in K - \{0\}$, $x(t) = x(t; \varphi, \alpha)$ is the corresponding solution of (78) and $X(t) = (x(t), x'(t))$, then

$$d(X(t), Q) \longrightarrow 0$$

as $t \longrightarrow \infty$, where Q is the region between $X_0(t)$ and $X_1(t)$ and $d(P, Q)$ denotes the distance of a point P to the region Q.

Theorem 6 is proved by a type of phase plane analysis applied to $X(t) = X(t; \varphi, \alpha)$ in the theorem.

The difficulty is that trajectories for different φ in K can cross. Related ideas were used by Myschkis [36] in studying linear, scalar F. D. E.'s .

The conclusions one can draw from Theorem 6 become very strong if one knows in addition that for a given f equation (76) has precisely one slowly oscillating periodic solution. The following proposition, whose proof is omitted, is an example of the consequences of uniqueness.

Proposition 1. Assume that $\alpha > (\frac{\pi}{2})$, that f satisfies the hypotheses of Theorem 6, and that equation (76) (for this α and f) has precisely one slowly oscillating periodic solution $x_0(t)$. Assume that $x_0(t) > 0$ for $0 < t < \zeta_1$, $\zeta_1 > 1$, $x_0(t) < 0$ for $\zeta_1 < t < \zeta_2$, $\zeta_2 - \zeta_1 > 1$, and $x_0(t + \zeta_2) = x_0(t)$ for all t. Then if $F_\alpha^j(\varphi) = \varphi$ for some $\varphi \in K$, it follows that $F_\alpha(\varphi) = \varphi$.

If $\varphi \in K - \{0\}$, $x(t)$ is the corresponding solution of (78) and $z_i = z_i(\varphi, \alpha)$ is the ith zero of $x(t)$, then

$$\lim_{i \to \infty} z_{2i} - z_{2i-1} = \zeta_2 - \zeta_1$$

$$\lim_{i \to \infty} z_{2i+1} - z_{2i} = \zeta_1$$

In view of the above remarks it would be very desirable to find conditions on f which insure uniqueness of slowly oscillating periodic solutions. The following recent theorem provides a close-to-definitive answer to this question within the class of odd functions f (f is odd if $f(-x) = -f(x)$ for all x).

Theorem 7. See [51]. Suppose that $a > 0$ (possibly $a = +\infty$) and $f: [-a, a] \to \mathbb{R}$ is an odd C^1 function which is bounded and such that $f'(x) > 0$ for $-a \leq x \leq a$ and $f'(0) = 1$. Assume that $f'(x)$ is monotonic decreasing (not necessarily strictly) on $[0, a]$ and that $\Phi(x) = f(x)x^{-1}$ is strictly monotonic decreasing for $0 < x \leq a$. Then if $0 < \alpha \leq (\frac{\pi}{2})$ equation (76) has no slowly oscillating periodic solution $x(t)$ such that $-a \leq x(t) \leq a$ for all t. If $\alpha > (\frac{\pi}{2})$, equation (76) has at most one slowly oscillating periodic solution such that $-a \leq x(t) \leq a$ for all t.

As an immediate consequence of Theorem 7 one obtains a uniqueness result for an equation which has been studied for many years (see [29]).

Corollary 2. The equation $x'(t) = -\alpha x(t-1)[1-(x(t))^2]$ has no slowly oscillating periodic solutions for $0 < \alpha \leq (\frac{\pi}{2})$ and exactly one slowly oscillating periodic solution for $\alpha > (\frac{\pi}{2})$.

The assumption that f is odd is a very strong one. If f is odd and satisfies H 1, it is known [30, 49] that for $\alpha > \frac{\pi}{2}$ equation (76) has a periodic solution $x(t)$ such that $x(t+2) = -x(t)$ for all t, $x(-t) = -x(t)$ for all t and $x(t) > 0$ for $0 < t < 2$. Nevertheless, numerical studies for some special odd functions f suggest that the dynamics of equation (78) may be complicated, e.g.

F_α^j may have fixed points in K which are not fixed points of F_α (contrast Proposition 1). J. Yorke has, for example, modified an equation of Mackey and Glass [19], and numerically found complicated behaviour of (78) for $f(x) = x(1+x^8)^{-1}$.

Until now very little rigorous has been done for examples like those of Mackey and Glass or Yorke. The next theorem sheds some light on the dynamics of (78) for an f similar to Yorke's and for large α. It also shows how close Lemma 5 and Theorem 7 are to being best-possible.

Theorem 8. [51]. For $r > 0$ define $f_r(x)$ to be an odd function such that

$$\begin{aligned} f_r(x) &= x \quad \text{for } 0 \le x \le 1 \\ f_r(x) &= x^{-r} \quad \text{for } x \ge 1 \end{aligned} \tag{81}$$

and consider the equation

$$x'(t) = -\alpha f_r(x(t-1)) \tag{82}$$

For $\alpha > \frac{\pi}{2}$ equation (82) has precisely one solution $x_\alpha(t)$ such that (1) $x_\alpha(t)>0$ for $0 < t < 2$, (2) $x_\alpha(t+2) = -x_\alpha(t)$ for all t and (3) $x_\alpha(-t) = -x_\alpha(t)$ for all t. Nevertheless, if $r > 2$ there exists a number $\alpha_r > 0$ such that for $\alpha \ge \alpha_r$ equation (82) has a second slowly oscillating periodic solution $y_\alpha(t)$ and a number $q_\alpha > 2$ such that $y_\alpha(t) > 0$ for $0 < t < q_\alpha$, $y_\alpha(t+q_\alpha) = -y_\alpha(t)$ for all t and $\lim_{\alpha \to \infty} q_\alpha = \infty$.

Theorem 8 gives an example in which uniqueness of slowly oscillating periodic solutions fails. If one does not have the condition that $f(x)x^{-1}$ is monotonic decreasing it is easy to find examples of functions f for which one has an arbitrarily large number of periodic solutions of period 4 for an $\alpha > \frac{\pi}{2}$.

The example in Theorem 8 is of a subtler type.

One might conjecture in Lemma 5 that if f is an odd function and $a = b = 0$, then the various limits defined in equation (79) should all equal 4. Theorem 8 shows that this conjecture is false.

Theorem 8 was proved in [51] for the example $f_r(x)$ only for simplicity. It seems likely that a modification of the proof will prove the existence of a slowly oscillating periodic solution like $y_\alpha(t)$ for an odd function $f(x)$ such that $xf(x) > 0$ for $x \neq 0$, $\lim_{x \to \infty} x^2 f(x) = 0$, and $f(x)$ is monotonic increasing on $[0, x_0]$, $x_0 > 0$, and monotonic decreasing on $[x_0, \infty)$. The functions $f(x) = x \exp(-bx^2)$ provide examples. The cut-off $r = 2$ is precise in the sense that the period of slowly oscillating period solutions of (82) must remain bounded as $\alpha \to \infty$ for $0 < r < 2$.

3. Differential - delay equations with two time lags.

There are many examples for which the techniques of Section 1 seem inapplicable. Perhaps the simplest of these is the equation

$$x'(t) = -\alpha f(x(t-1)) - \beta f(x(t-\gamma)) \tag{83}$$

We shall always assume that $\alpha \geq 0$, $\beta \geq 0$, $\gamma \geq 1$ and f satisfies H 1, and we are interested in studying the set of periodic solutions of (83). At present there are only incomplete results about (83), but what is known suggests that even for $\gamma = 3$ the structure of the set of periodic solutions may be complicated. Some confusion has arisen because of incorrect claims in the literature. For example, Jones suggests in [29] that his techniques prove existence of periodic solutions for equations like (83); he mentions in particular

$$x'(t) = [-\alpha x(t-1) - \beta x(t-2)][1-(x(t))^2] \tag{84}$$

which can be transformed to the form of (83) by the substitution $x(t) = f(y(t))$, $f(y) = (\exp(2y) - 1)(\exp(2y) + 1)^{-1}$. In fact, even (84) has still not been completely analyzed; and, as I shall indicate, some of Jones's statements about it are simply wrong.

I want to describe now some results from [50]. Most of the theorems below were motivated by computer studies and provide some explanation for numerically observed results.

Theorem 9. [50]. Assume that $f: \mathbb{R} \longrightarrow \mathbb{R}$ is continuous, $f(0) = 0$, f is bounded below and f is monotonic increasing and set $-b = \lim_{x \to -\infty} f(x) > -\infty$ and $a = \lim_{x \to \infty} f(x) \leq +\infty$. Assume that $\alpha \geq 0$, $\beta \geq 0$ and $1 < \gamma \leq 2$. Suppose that $\epsilon > 0$, $0 < a_1 < a$ and $0 < b_1 < b$. Then there exists a bounded open neighborhood $U = U(\epsilon, a_1, b_1, f, \gamma)$ of 0 in $\mathbb{R}^2$ such that if $(\alpha, \beta) \notin U$ equation (83) has a periodic solution $x(t)$ such that (1) $x(t) > 0$ for $0 < t < z_1$ where $z_1 \geq 1$

(z_1 depends on α, β, γ and f) and $f(x(t)) \geq a_1$ for $\epsilon \leq t \leq \max(1, z_1 - \epsilon)$, (2) $x(t) < 0$ for $z_1 < t < z_1 + z_2$ where $z_2 \geq 1$ (z_2 depends on α, β, γ and f) and $f(x(t)) \leq -b_1$ for $z_1 + \epsilon \leq t \leq \max(z_1 + 1, z_1 + z_2 - \epsilon)$ and (3) $x(t + z_1 + z_2) = x(t)$ for all t. If $\alpha_0 \geq 0$ and $\beta_0 \geq 0$ are selected such that $\alpha_0^2 + \beta_0^2 = 1$ and if $\alpha = r\alpha_0$ and $\beta = r\beta_0$, the periodic solution x(t) can be chosen so that $\lim_{r \to \infty} z_1 = \zeta_1$ and $\lim_{r \to \infty} z_2 = \zeta_2$, where one can give explicit formulas (see [50]) for ζ_1 and ζ_2 in terms of α_0, β_0, a, b and γ.

Theorem 9 is a crude result; periodic solutions exist for $\alpha \geq 0$, $\beta \geq 0$, $(\alpha, \beta) \notin U$. For some functions f the techniques of Section 1 can be applied for a limited range of α and β; and these results, when combined with Theorem 9, may be sharp. For equation (84) one can prove

Proposition 2. Assume that $\alpha_0 = \cos\theta$ and $\beta_0 = \sin\theta$ for θ such that $0 \leq \theta \leq (\frac{\pi}{2})$ and for $r > 0$ consider the equation

$$x'(t) = -r[\alpha_0 x(t-1) + \beta_0 x(t-2)][1 - (x(t))^2] \tag{85}$$

Define ν_0 to be the unique solution of the equation

$$\alpha_0 \cos\nu + \beta_0 \cos(2\nu) = 0 \tag{86}$$

such that $0 \leq \nu \leq (\frac{\pi}{2})$ and define $r_0 = \nu_0(\alpha_0 \sin\nu_0 + \beta_0 \cos\nu_0)^{-1}$. Then if $\beta_0 \geq (\frac{5}{2})\alpha_0$ and if $r > r_0$, equation (86) has a periodic solution x(t) such that $x(t) > 0$ for $0 < t < z_1$, where $z_1 \geq 2$ and $x(t + z_1) = -x(t)$ for all t.

If one only demands that $z_1 > 1$, it is likely that Proposition 2 is true without the restriction on $\beta_0 \alpha_0^{-1}$. The periodic solution one obtains above is bifurcating from the zero solution at $r = r_0$; locally, Hopf bifurcation is applicable [11].

Jones claims in [29] that the periodic solution of (85) can be taken to have period independent of r. If $\alpha_0 = 1$ and $\beta_0 = 0$ or $\alpha_0 = \beta_0$ or $\alpha_0 = 0$ and

$\beta_0 = 1$, this is true: see [29, 30, 49]. However, if α_0 and β_0 do not fall in one of these three special cases, it is proved in [50] that there exist values of r, say r_1 and r_2, for which there are corresponding periodic solutions x_1 and x_2 of (85) of minimal periods p_1 and p_2, with $p_1 > 4$, $p_2 > 4$ and $p_1 \neq p_2$.

Once one allows $\gamma > 2$ in (83) a variety of intriguing new phenomena appear. For instance, if $\alpha = r\alpha_0$ and $\beta = r\beta_0$, number ζ_1 and ζ_2 can be defined as in Theorem 9, but ζ_1 and ζ_2 are no longer continuous functions of α_0 and β_0 (see Theorem 2.1 in [50]). If f is as in Theorem 9 and $a = b$ and $\gamma = 3$, it is proved in [50] that for $(\frac{1}{3})\alpha_0 < \beta_0 < \alpha_0$ and r large enough, equation (83) has at least two distinct periodic solutions whose zeros are separated by a distance greater than $(\frac{3}{2})$. Although this is not proved in [50], it seems likely that there is a third periodic solution (for the same range of α_0 and β_0 and for r large) whose zeros are separated by a distance between 1 and $(\frac{3}{2})$. A better understanding of equation (83) seems both desirable and difficult. Even the discussion of the characteristic equation for (83) for $\gamma = 3$

$$
\begin{gathered}
g(z, r) = z + r\alpha_0 e^{-z} + r\beta_0 e^{-3z} = 0 \\
r > 0,\ \alpha_0 \geq 0,\ \beta_0 \geq 0,\ \alpha_0^2 + \beta_0^2 = 1
\end{gathered}
\tag{87}
$$

is nontrivial. One can obtain a reasonably complete analysis of those roots z of (87) for which $|\mathrm{Im}(z)| < \pi$ — how they vary with r and when $\mathrm{Re}(z) > 0$ — but for reasons of length I omit this.

A partial bibliography for fixed point theory, the fixed point index, and periodic solutions of F.D.E.'s .

[1] J. Alexander and J. Yorke, "Global bifurcation of periodic orbits", Amer. J. Math. 100 (1978), 263-292

[2] W. Alt, "Some periodicity criteria for functional differential equations", Manuscripta Math. 23 (1978), 295 - 318.

[3] K. Borsuk, Theory of Retracts, Polish Sci. Publ., Warsaw, 1967.

[4] F.E. Browder, "Local and global properties of nonlinear mappings in Banach spaces", Institute Naz. di Alta Math. Symposia Math. 2(1968), 13 - 35.

[5] F.E. Browder, "On a generalization of the Schauder fixed point theorem", Duke Math. J. 26 (1959), 291 - 303.

[6] F.E. Browder, "A further generalization of the Schauder fixed point theorem", Duke Math. J. 32 (1965), 575 - 578.

[7] F.E. Browder, "Asymptotic fixed point theorems", Math. Ann. 185 (1970), 38 - 60.

[8] Shui-Nee Chow, "Existence of periodic solutions of autonomous functional differential equations", J. Differential Equations 15 (1974), 350 - 378.

[9] Shui-Nee Chow and J. Hale, "Periodic solutions of autonomous equations", J. Math. Anal. Appl. 66 (1978),495-506

[10] Shui-Nee Chow and J. Mallet-Paret, "Fuller's index and global Hopf's bifurcation", to appear.

[11] Shui-Nee Chow and J. Mallet-Paret, "Integral averaging and bifurcation", J. Differential Equations 26 (1977), 112 - 159.

[12] D. Darbo, "Punti uniti in trasformazioni a condiminio non compatto", Rend. Sem. Math. Univ. Padova 24 (1955), 84 - 92.

[13] R.F. Brown, The Lefschetz Fixed Point Theorem, Scott Foresman Co., Henview, Illinois, (1971).

[14] A. Dold, "Fixed point index and fixed point theorems for Euclidean neighborhood retracts", Topology 4 (1965), 1 - 8.

[15] J. Eells and G. Fournier, "La théorie des points fixes des applications a itérée condensante", Bull. Soc. Math. France, Memoire 45 (1976), 91 - 120.

[16] C. Fenske, "Analytische Theorie des Abbildungsgrades für Abbildungen in Banachräumen", Math. Nachr. 48 (1971), 279 - 290.

[17] C. Fenske and H.-O. Peitgen, "On fixed points of zero index in asymptotic fixed point theory", Pacific J. Math.66 (1976) 391-410

[18] G. Fournier and H.-O. Peitgen, "On some fixed point principles for cones in linear normed spaces", Math. Ann. 225 (1977), 205 - 218

[19] L. Glass and M. Mackey, "Oscillation and chaos in physiological control systems", Science 197 (1977), 287 - 289.

[20] R.B. Grafton, "A periodicity theorem for autonomous functional differential equations", J. Differential Equations 6 (1969), 87 - 109.

[21] R.B. Grafton, "Periodic solutions of certain Liénard equations with delay", J. Differential Equations 11 (1972) 519 - 527.

[22] K.P. Hadeler and J. Tomiuk, "Periodic solutions of difference-differential equations", Arch. Rat. Mech. Anal. 65 (1977), 87 - 95.

[23] U. an der Heiden, "Periodic solutions of a nonlinear second order differential equation with delay", to appear.

[24] J. Hale, Theory of Functional Differential Equations, Springer Verlag, New York, Heidelberg, Berlin, (1977).

[25] J. Hale, "Nonlinear oscillations in equations with delays", to appear.

[26] J. Ize, "Bifurcation theory for Fredholm operators", Memoirs Amer. Math. Soc. 174 (1976).

[27] G.S. Jones, "The existence of periodic solutions of $f'(x) = -\alpha f(x-1)[1 + f(x)]$", J. Math. Anal. Appl. 5 (1962), 435 - 450.

[28] G.S. Jones, "On the nonlinear differential difference equation $f'(x) = -\alpha f(x-1)[1 + f(x)]$", J. Math. Anal. Appl. 4 (1962), 440 - 469.

[29] G.S. Jones, "Periodic motions in Banach space and applications to functional differential equations", Contributions to Differential Equations 3 (1964), 75 - 106.

[30] J. Kaplan and J. Yorke, "Ordinary differential equations which yield periodic solutions of differential-delay equations", J. Math. Anal. Appl. 48 (1974), 317 - 324.

[31] J. Kaplan and J. Yorke, "On the stability of a periodic solution of a differential delay equation", SIAM J. Math. Anal. 6 (1975), 268 - 282.

[32] J. Kaplan and J. Yorke, "On the nonlinear differential delay equation $x'(t) = -f(x(t), x(t-1))$", J. Differential Equations 23 (1977), 293 - 314.

[33] M. Krasnosel'skii and P.P. Zabreiko, "Iterations of operators and the fixed point index", Soviet Math. Doklady 12 (1971), 294 - 298.

[34] M. Krasnosel'skii, Positive Solutions of Operator Equations, P. Noordhoff Ltd., Groningen, The Netherlands, (1964).

[35] J. Leray, "Théorie des points fixes: indice total et nombre de Lefschetz", Bull. Soc. Math. France 87 (1959), 221 - 233.

[36] A.D. Myschkis, "Lineare Differentialgleichungen mit nacheilendem Argument", German translation of Russian original, VEB Deutscher Verlag der Wissenschaften, volume 17, Berlin, (1955).

[37] R. D. Nussbaum, "The fixed point index for local condensing maps", Ann. Mat. Pura Appl. 89 (1971), 217 -258.

[38] R. D. Nussbaum, "Some asymptotic fixed point theorems", Transactions AMS 171 (1972), 349 - 375.

[39] R. D. Nussbaum, "Generalizing the fixed point index", Math. Ann. 228 (1977), 259 -278.

[40] R. D. Nussbaum, "Periodic solutions of some nonlinear autonomous functional differential equations", Annali di Math. pura e appl. 101 (1974), 263 - 306.

[41] R. D. Nussbaum, "Periodic solutions of some nonlinear autonomous functional differential equations,II ", J. Differential Equations 14 (1973), 360 - 394.

[42] R. D. Nussbaum, " A correction of 'Periodic solutions of some nonlinear autonomous functional differential equations, II`,", J. Differential Equations 16 (1974), 548 - 549.

[43] R. D. Nussbaum, "A global bifurcation theorem with applications to functional differential equations", J. Functional Analysis 19 (1975), 319 - 338.

[44] R. D. Nussbaum, "The range of periods of periodic solutions of $x'(t) = -\alpha f(x(t-1))$", J. Math. Anal. Appl. 58 (1977), 280 - 292.

[45] R. D. Nussbaum, "Periodic solutions of analytic functional differential equations are analytic", Michigan Math. J. 20 (1973), 249 - 255.

[46] R. D. Nussbaum, "Global bifurcation of periodic solutions of some autonomous functional differential equations", J. Math. Anal. Appl. 55 (1976), 699 - 725.

[47] R. D. Nussbaum, "A Hopf global bifurcation theorem for retarded functional differential equations", Transactions of Amer. Math. Soc. 238 (1978), 139 - 163.

[48] R. D. Nussbaum, "Integral equations from the theory of epidemics", in Nonlinear Systems and Applications, edited by

V. Lakshmikantham, Academic Press, New York, (1977), pp. 235 - 255.

[49] R.D. Nussbaum, "Periodic solutions of special differential-delay equations: an example in nonlinear functional analysis", to appear in Proc. Royal Soc. Edinburgh.

[50] R.D. Nussbaum, "Differential-delay equations with two time lags", to appear in Memoirs of the Amer. Math Soc..

[51] R.D. Nussbaum, "Uniqueness and nonuniqueness for periodic solutions of $x'(t) = -g(x(t-1))$", to appear.

[52] H.-O. Peitgen, "On continua of periodic solutions for functional differential equations", Rocky Mountain J. 7 (1977), 609 - 617.

[53] B.N. Sadovskii, "Limit compact and condensing operators", Russian Math. Surveys 27 (1972), 85 - 155.

[54] A. Somolinos, "Periodic solutions of the sunflower equation", Quart. Applied Math. 35 (January, 1978), 465 - 478.

[55] H. Steinlein, "Ein Satz über den Leray-Schauderschen Abbildungsgrad", Math. Z. 120 (1972), 176 - 208.

[56] H. Steinlein, "Über die verallgemeinerten Fixpunktindizes von Iterierten verdichtender Abbildungen", Manuscripta Math. 8 (1972), 251 - 266.

[57] H.-O. Walther, "A theorem on the amplitudes of periodic solutions of differential delay equations with applications to bifurcation", J. Diff. Equat. 29 (1978), 396 - 404

[58] H.-O. Walther, "Über Ejektivität und periodische Lösungen bei autonomen Funktionaldifferentialgleichungen mit verteilter Verzögerung", Habilitationsschrift, Univ. München, (1977).

[59] E.M. Wright, "A nonlinear difference-differential equation", J. Reine Angew. Math. 194 (1955), 66 - 87.

THE LERAY-SCHAUDER CONTINUATION METHOD IS A CONSTRUCTIVE ELEMENT IN THE NUMERICAL STUDY OF NONLINEAR EIGENVALUE AND BIFURCATION PROBLEMS*

BY

HEINZ-OTTO PEITGEN AND MICHAEL PRÜFER

Fachbereich Mathematik
Universität Bremen
2800 Bremen 33

CONTENTS

---

* This work was partially supported by Sonderforschungsbereich 72 "Approximation und Optimierung in einer anwendungsbezogenen Mathematik", Universität Bonn and Forschungsschwerpunkt "Dynamische Systeme" Universität Bremen.

I. INTRODUCTION

It has been typical for the study of nonlinear phenomena in anal· ysis that they must by analyzed with one technique to obtain quanti· tative information (numerical values of solutions) and by another tc obtain qualitative information (existence, uniqueness-multiplicity, stability, bifurcation). In a sense this paper is devoted to a unified approach, i.e. for a number of topological methods which have been used in the past only to obtain qualitative insight we will show how to exploit them at the same time also to provide numerical knowledge up to implementable and tested procedures.

Topological degree (Brouwer and Leray-Schauder) is at the basis of many of these methods and in view of a constructive approach it seems to be noteworthy that one of the early forerunners of topologi cal degree which certainly is the Kronecker characteristic [34] was directly motivated by a classical numerical tool, i.e. Sturm's Theorem (cf. H.W. Siegberg [54]).

Since the fundamental work of J. Leray and J.P. Schauder [37] degree theory and what is called the Leray-Schauder Continuation Method has become one of the most prominent topological tools in the study of existence of solutions of both linear and nonlinear prc blems (integral equations, ordinary and partial differential equations, problems of optimal control, variational inequalities, e.t.c.

It is an observation of M.A. Krasnosel'skii [32] (local form) and of P.H. Rabinowitz [50] (global form) that certain bifurcation phenomena can be detected by degree and this actually means thε bifurcation can be considered as a negation of the Leray-Schauder Continuation Method.

In the following we restrict all considerations to finite dimensions, i.e. we study problems

$$F(x,\lambda) = 0$$

$$F : X \times R \to X \quad \text{continuous}$$

($X = R^n$ or some triangulable subset of R^n, e.g. a cone in R^n).

Typical problems that arise with the numerical computation of bifurcation problems are the following (figure 1):

(1): How to follow a piece of a bifurcation branch, i.e. of $F^{-1}(0)$? (the continua $F^{-1}(0)$ might be non smooth, or higher dimensional, or solutions might be unstable) (see example (10.2.4))

(2): How to follow a piece of $F^{-1}(0)$, which contains a "turning point"? (see example (10.2.4))

(3): How to detect and how to pass bifurcation points? (see examples (10.2.4), (10.3.3).

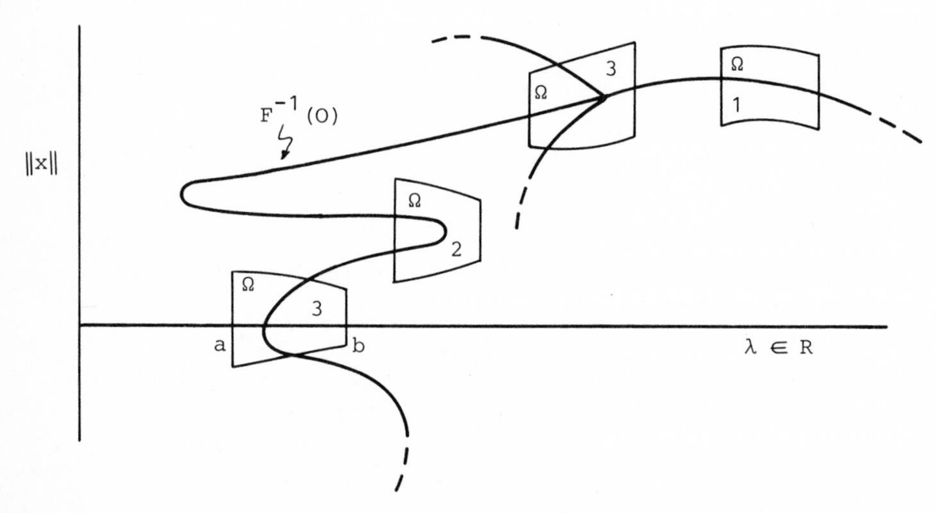

Figure 1.

For our discussion of problems (1)-(3) it will be crucial that they can be characterized by a topological invariant: degree (see figure 2).

Let $\Omega \subset X \times [a,b]$ ($[a,b]$ a suitable interval in R) be open, bounded and connected and such that the projection onto the λ-axis fills $[a,b]$ and such that it isolates situations (1)-(3). ($F_\lambda(x) := F(x,\lambda)$)

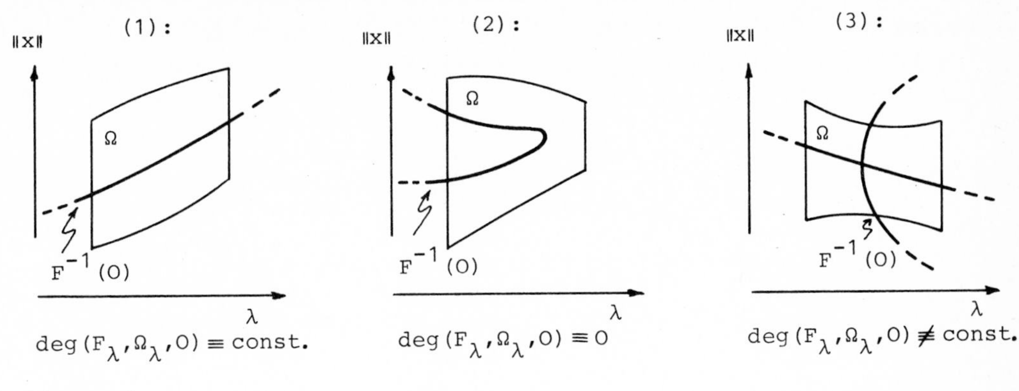

Figure 2.

The central idea of the following will be to exploit the topological information contained in (1)-(3) for numerical procedures. At the heart of our approach are three ideas which are essentially motivated by Sperner's Lemma [55] and the Pontryagin Construction [40]:

1° labeling, triangulation, the notion of a completely labeled simplex and orientation;

2° finite representation of Brouwer degree (or fixed point index) by completely labeled oriented simplices;

3° Door-In/Door-Out Principle (codimension one completely labeled simplices) and homotopy.

The Pontryagin Construction together with the Brown-Sard theorem and the classification of smooth 1-manifolds is a modern way to define Brouwer degree and to prove its basic properties [7, 22]. One can consider the recent work of J.C. Alexander, J.A. Yorke, T.-Y. Li, S.N. Chow, J. Mallet-Paret and B. Kellog as an exploitation of this approach to degree.

Labeling (integer-, integer-L-, and vector-labeling) and triangulation will provide a simplicial or piecewise linear approximation F_T (T for triangulation) to F such that the relevant topological information in F will be inherited to F_T and such that $F_T^{-1}(0)$

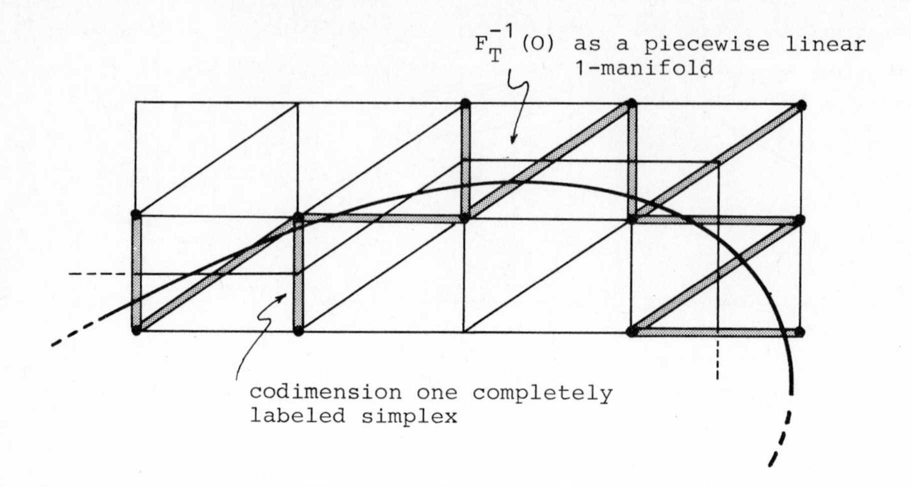

Figure 3.

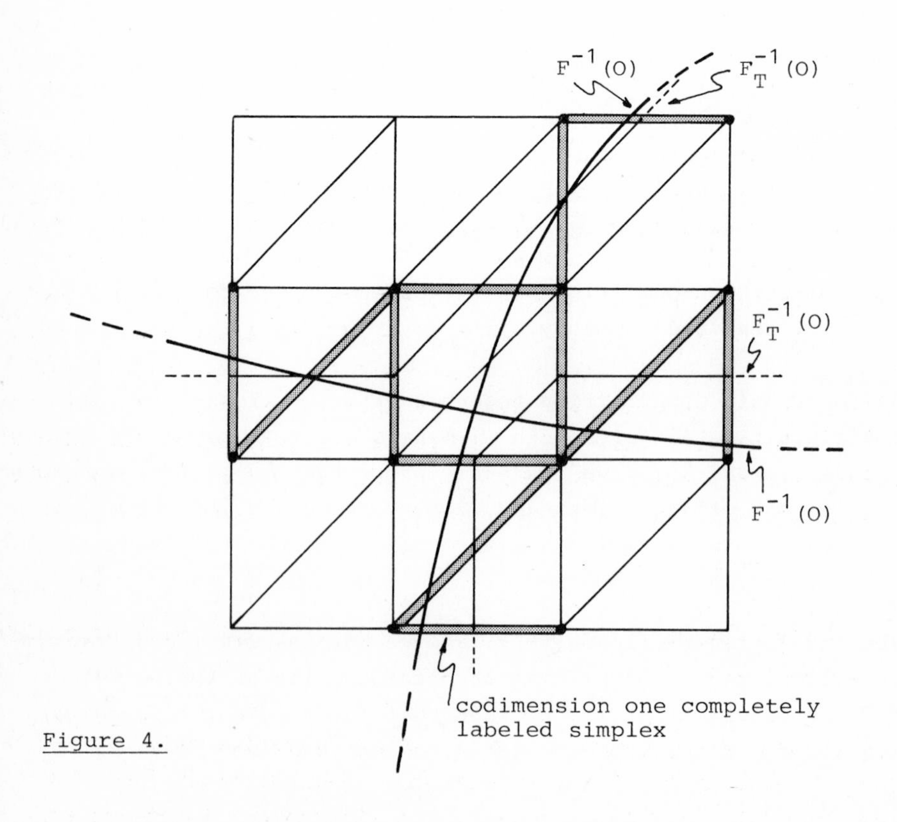

Figure 4.

is a piecewise linear or simplicial 1-manifold which approximates $F^{-1}(0)$ (no branching, bifurcation disappears) (see figure 4). Observe that in contrast to $F_T^{-1}(0)$ being a manifold even in concrete problems almost nothing is known about the structure of $F^{-1}(0)$: bifurcation destroys the manifold structure. The Door-In/Door-Out Principle (this is a term due to Ky Fan) will mean that given an (n+1)-simplex in a triangulation of $X \times R$, than either <u>two</u> or <u>no</u> n-face have a certain property (are codimension one completely labeled).

This will be the constructive and repeatable element in the design of a numerical procedure to compute, i.e. to follow $F_T^{-1}(0)$ (see figure 3). Finally, whether and how this constructive step can be initiated and carried through is completely determined by the topological invariant carried by $F_T^{-1}(0)$: the Brouwer degree, resp. its representation by completely labeled simplices.

Two types of labeling, i.e. associating F_T with F are known

- integer-labeling and
- vector-labeling .

Given a linear Isomorphism L of X ($X = R^n$ or R^n_+) we will introduce a new integer labeling with respect to L, called integer-L-labeling. Numerical experience (see XI) indicates that procedures based on vector labeling should be given preference in competition with those based on classical integer labeling. However, our new integer-L-labeling will share the good properties of both integer and vector labeling. Moreover, by introducing a new variable, i.e. the isomorphism L, into the concept we will create more flexibility (predictor-corrector) and another possibility to pilot the algorithm (artificial bifurcation and connections, see X). Especially, this will provide new possibilities to

- handle secondary bifurcation;
- find new pieces of $F^{-1}(0)$ from known parts;
- to give estimates of how good actually $F_T^{-1}(0)$ will approximate $F^{-1}(0)$.

In many cases (see example (12.3) the qualitative analysis of $F(x,\lambda) = 0$ can only be made successfully (see H. Amann [8] and M.A. Krasnosel'skii [33]) if one uses more structure: F leaves a cone invariant, i.e. solutions are positive. In this case degree is replaced by fixed point index and it seems to be interesting that our constructive approach can be made to adapt the cone situation completely.

The monograph of J. Mawhin and B. Gaines [25] provides a number of nonlinear problems, which can be studied more adequately by looking at them as coincidence problems and there a coincidence degree is applied successfully. Again, our approach can be extended to a coincidence set up.

For a recent survey on the field we recommend E. Allgower and K. Georg [5].

We acknowledge with pleasure the invaluable help of H. Jürgens and D. Saupe who where patient enough to implement the ideas of this paper on a computer as part of their Diploma Thesis [28]. Their enthusiastic cooperation provided us with most of the examples given at the end of the paper. These examples have had, in fact, a great impact on sections IX and X.

II. TRIANGULATION

A k-simplex σ in R^n $(k \leq n)$ is the closed convex hull of $(k+1)$ affinely independent elements $a^o,\dots,a^k \in R^n$ which are called <u>vertices</u> of σ and we sometimes write $\sigma = [a^o,\dots,a^k]$. To indicate that σ is of dimension k we write k as a superscript: σ^k. Given a simplex $\sigma = [a^o,\dots,a^k]$, a simplex σ' is called a <u>face</u> of σ provided all its vertices are taken from $\{a^o,\dots,a^k\}$.

(2.1.1) DEFINITION

Let $T_n = \{\sigma_1,\sigma_2,\dots\}$ be a (possibly infinite) set of n-simplices in R^n. The set

$$M = \bigcup_{\sigma \in T} \sigma$$

is called a <u>triangulable set of homogeneous dimension n</u> provided

(T_1) For all $\sigma, \sigma' \in T_n$ the intersection $\sigma \cap \sigma'$ is a common face of both σ and σ';

(T_2) for every $x \in M$ there is a neighborhood U_x which intersects only a finite number of simplices $\sigma \in T_n$.

Any simplicial subdivision T_n of M of type (T_1) and (T_2) together with the collection of all faces of simplices in T_n is called a <u>triangulation</u> T of M.

The set of k-simplices $(0 \leq k \leq n)$ in T is called the <u>k-skeleton</u> of T and is denoted by T_k. The 0-skeleton of T is the set of all vertices T_0 of T.

III. LABELING

In this section M will always denote a triangulable set of homogeneous dimension n and T will be a fixed triangulation of M.

The introduction indicates that labeling is a means to associate with $f: M \to R^n$ a piecewise linear (PL) or simplicial approximation $f_T: M \to R^n$. As a problem of topology this is classical and well understood. Here, we focus, however, on the computability of $f^{-1}(0)$ via $f_T^{-1}(0)$ and it is there that labeling becomes important:

(3.1) VECTOR-LABELING

Let $f: M \to R^n$ be continuous. Being interested in $f^{-1}(0)$ in a numerical sense the obvious first candidate for f_T is this:

Let f_T be the map which is affine on each $\sigma \in T_n$ and which coincides with f on T_0.

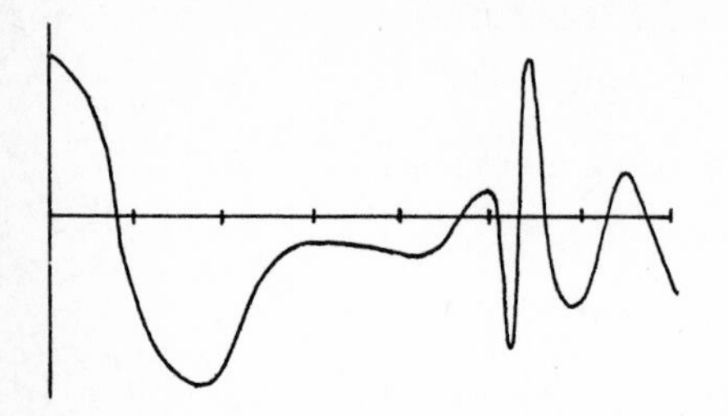

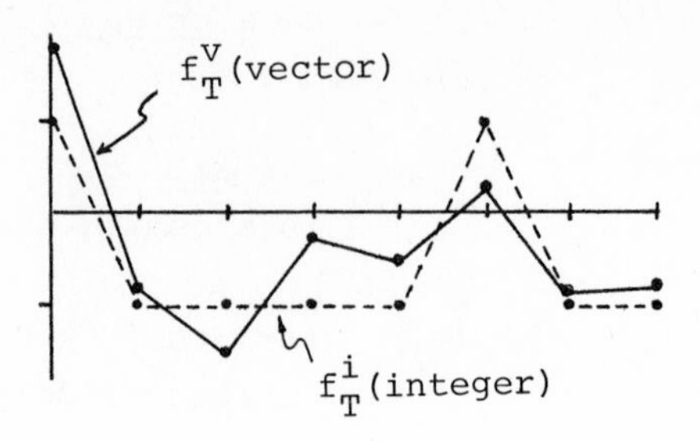

Figure 5.

Vector-labeling is formally any map $\ell: M \to R^n$. A special vector-labeling to investigate the zeros of $f: M \to R^n$ is $\ell^f(x) := f(x)$. To discuss $f_T^{-1}(0)$ as an approximation of $f^{-1}(0)$ one introduces completely labeled simplices. A first definition could be: a simplex $\sigma \in T_n$ is called completely labeled provided $0 \in f_T(\sigma)$. However, this definition does not provide a Door-In/ Door-Out Principal (see figure 9). In view of this the following definition is appropriate (cf. [21]):

(3.1.1) DEFINITION

Let $f: M \to R^n$ be continuous. A simplex $\sigma = [a^0,\dots,a^n] \in T_n$ is called <u>completely labeled</u> provided Λ^{-1} , the inverse of the $(n+1) \times (n+1)$ matrix

$$\Lambda = \begin{pmatrix} 1 & \dots & 1 \\ f(a^0) & \dots & f(a^n) \end{pmatrix}$$

exists and is lexicographically positive, i.e. the first non-zero element of each row of Λ^{-1} is positive.

One shows that this definition has the following geometric interpretation, i.e. it is equivalent with: there exists $\varepsilon_0 > 0$ such that for all $0 \le \varepsilon \le \varepsilon_0$ the n-vector $(\varepsilon,\varepsilon^2,\dots,\varepsilon^n)$ is in $f_T(\sigma)$.

To link the set up with topology one introduces an <u>orientation</u> $or(\sigma) \in \{-1,+1\}$ for completely labeled simplices $\sigma \in T_n$:

(3.1.2) $$or(\sigma) = \text{sign det} \begin{pmatrix} 1 & \cdots & 1 \\ a^o & \cdots & a^n \end{pmatrix} \circ \begin{pmatrix} 1 & \cdots & 1 \\ f(a^o) & \cdots & f(a^n) \end{pmatrix}$$

(3.2) INTEGER-LABELING

Let L be a linear isomorphism of R^n. We will associate with $f: M \longrightarrow R^n$ (continuous) and L labeling sets $A_o,\ldots,A_n$ depending on f and L in such a way that

$$N = \bigcap_{i=0}^{n} A_i$$

consists of zeroes of f. These sets are strongly motivated by coverings in the sense of Knaster-Kuratowski-Mazurkiewicz (see [31] and [19] for a related application in the context of variational inequalities). We set

$$B^o = \{x \in R^n: x_1 \leq 0\}$$
$$\vdots$$
$$B^i = \{x \in R^n: x_1 > 0,\ldots,x_i > 0, x_{i+1} \leq 0\}$$
$$\vdots$$
$$B^n = \{x \in R^n: x_1 > 0,\ldots,x_n > 0\}$$

(3.2.1) DEFINITION

Let $f: M \to R^n$ be continuous and let L be a linear isomorphism of R^n. The function $\ell_L^f: M \to \{0,\ldots,n\}$ defined by

$$\ell_L^f(x) = i \quad \text{provided} \quad f(x) \in L(B^i)$$

is called an <u>integer-L-labeling</u> of M.

Setting now $A_i(f,L) := \text{cl } f^{-1}(L(B^i))$ one has the immediate consequence:

(3.2.2) $$x \in \bigcap_{i=0}^{n} A_i \quad \text{then} \quad f(x) = 0.$$

Observe that $L = Id$ provides the classical integer-labeling as studied e.g. in [6 ,27,47].

For a numerical computation of $\ell_L^f(x)$ observe that Cramer's rule implies that

$$(3.2.3) \quad \ell_L^f(x) = \begin{cases} \min\{i-1 : z_i \leq 0\} , & \text{if this set is nonempty} \\ n , & \text{otherwise} \end{cases}$$

where z is the solution of the linear system $Lz = f(x)$.

(for each x the solution z could be easily computed by forward- and back-substitution from a fixed triangular factorization of L).

In sections VI and XI we will discuss the effect of introducing L into the labeling concept in detail along concrete numerical examples which have been studied using an implementation of the forthcoming procedures. In the very special case where $f = L$, however, the effect is clear immediately and is visualized in figure 6.

(a) integer-Id-labeling for f = L :

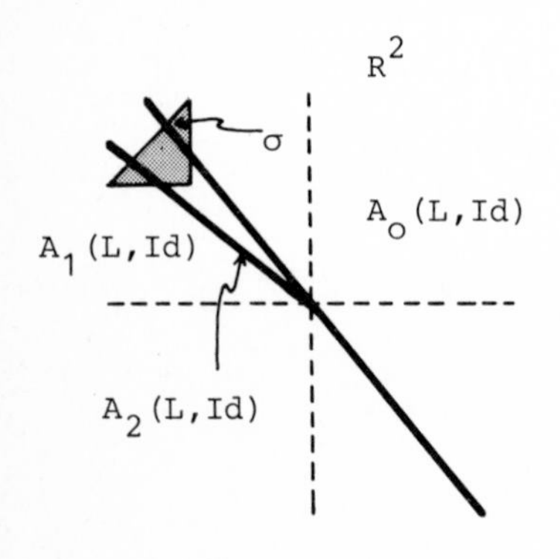

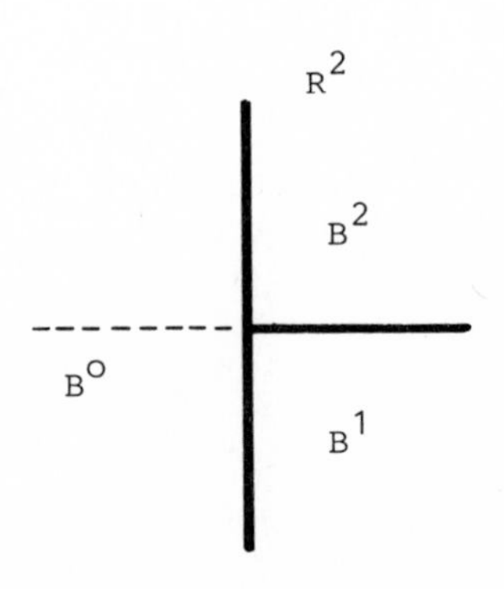

(b) integer-L-labeling for f = L :

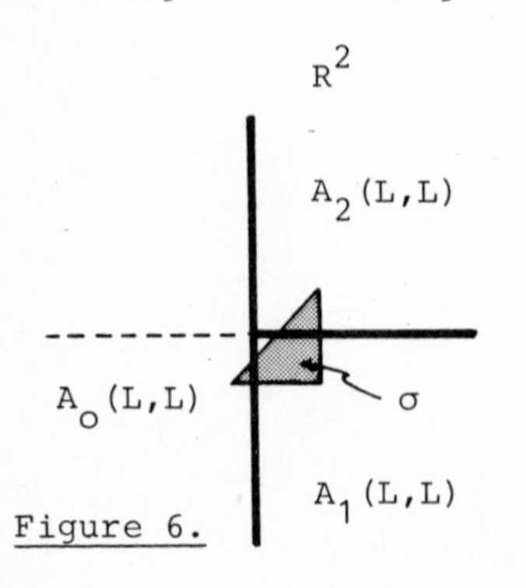

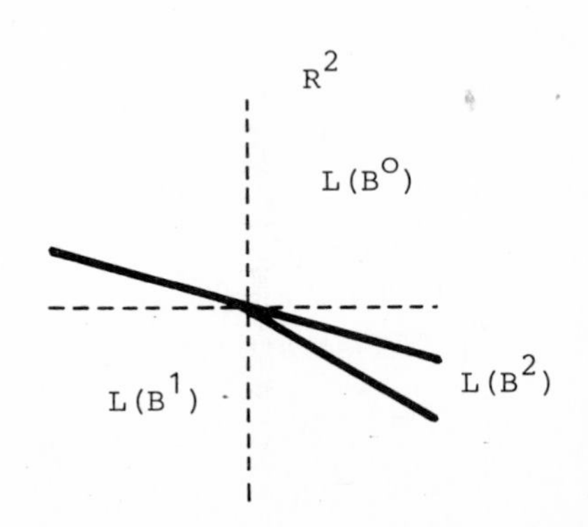

Figure 6.

As with vector-labeling one introduces completely labeled simplices and an orientation:

(3.2.4) DEFINITION

A simplex $\sigma = [a^o,\dots,a^n] \in T_n$ is called L-completely labeled or L-Sperner simplex provided

$$\{\ell_L^f(a^o),\dots,\ell_L^f(a^n)\} = \{0,\dots,n\} .$$

Let $\sigma = [a^o,\dots,a^n]$ be an L-Sperner simplex. Assume that $\ell_L^f(a^i) = i$. Let $C: R^n \to R^n$ be the unique affine isomorphism such that

$$C(a^i) = e^i$$

($e^o = 0$, e^i = i-th unit vector in R^n). Define an L-orientation $or_L(\sigma) \in \{-1,+1\}$ of σ by

$$or_L(\sigma) = \text{sign det } (C-C(0)) .$$

Set $\mathcal{S}(f;L) = \{\sigma \in T_n: \sigma$ is an L-completely labeled n-simplex of a triangulation T for $M\}$.

One advantage for introducing L into the concept of labeling (see also sections VI and XI) is immediately clear from the pictures in figure 6: integer-Id-labeling may admit completely labeled simplices σ of small diameter which are far from a zero of f; whereas a small diameter L-completely labeled simplex σ for a suitable L should be always close to a zero of f.

The PL-approximation f_T which is suitable for integer-labeling is defined as follows:

(3.2.5) (The PL-map f_T)

Let $b^o,\dots,b^n$ be the n-vectors in R^n:

$$b^o = (-1,0,\dots,0)$$
$$\vdots$$
$$b^i = (0,\dots,1,-1,0,\dots,0)\ , \quad \text{for} \quad 1 \le i \le n-1,$$
$$\vdots$$
$$b^n = (0,\dots,0,1).$$

For $x \in T_o$ (the vertices of T) set

$$f_T(x) := L(b^i) \quad \text{provided} \quad \ell_L^f(x) = i.$$

Then extend f_T to the map which is affine on each $\sigma \in T_n$ (see figure 5).

(3.2.6) $(f_T^{-1}(0))$

Obviously, since L is an isomorphism, $f_T^{-1}(0) =$ $\{x_\sigma : \sigma \in \mathcal{S}(f;L)$ and x_σ is the barycenter of $\sigma\}$ (for $\sigma = [a^o,\dots,a^n] \in T_n$ one has $x_\sigma = \sum_{i=0}^{n} \lambda_i a^i$ with $\lambda_i = 1/n+1)$.

So far we have worked on the computation of $f^{-1}(0)$ (a possibly very inaccessible and wild set) in two steps:

(3.2.7) (1) $f^{-1}(0) \supsetneqq \bigcap_{i=0}^{n} A_i(f,L) = N(f,L)$;

N(f,L) selects the "important" zeroes of f, i.e. those of degree different from zero (see section IV).

(2) $f_T^{-1}(0)$ approximates $N(f,L)$ and this will be made precise in section V.

$$f_T^{-1}(0) \approx \bigcap_{i=0}^{n} A_i(f,L) \approx f^{-1}(0)$$

Comparing vector-labeling with integer-labeling it seems to be interesting to observe that vector-labeling can be considered as a secant method and integer-labeling as a bisection method in dimensions greater than one (see figure 7; there the superscript v respectively i in f_T specifies the PL-map assigned to vector-respectively integer-labeling):

σ compl. lab. implies $0 \in f_T^v(\sigma)$, but does not imply $0 \in f_T^i(\sigma)$

σ compl. lab. implies $0 \in f_T^i(\sigma)$, but does not imply $0 \in f_T^v(\sigma)$

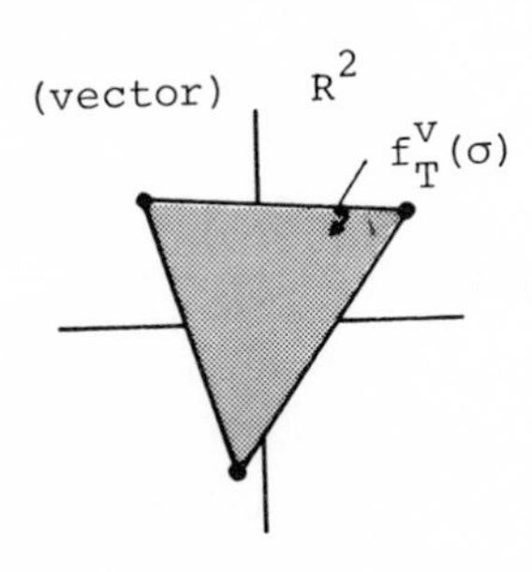

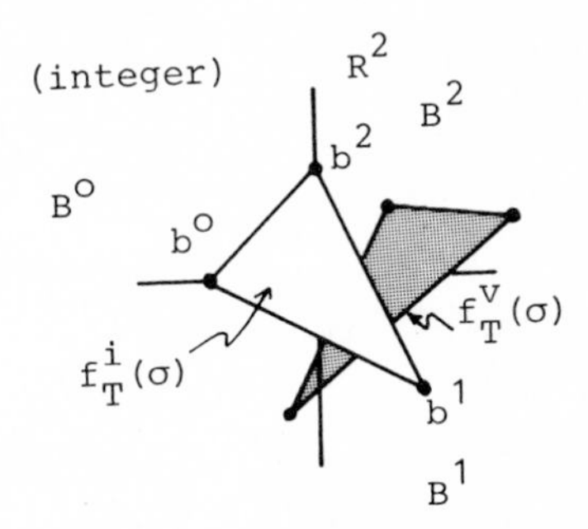

Figure 7.

In fact, in dimension one vector-labeling is the secant method and integer-labeling is the bisection method (L = Id). Vector - labeling has been considered superior to classical integer-labeling since it uses an approximation to a linearization of f near a zero. Observe that this drawback is removed from integer-labeling when using integer-L-labeling (e.g. L could be chosen to be the derivative of f in a zero).

IV. Brouwer Degree and Fixed Point Index

In this section we will give the crucial link between a new numerical study of $f^{-1}(0)$ and the classical study of $f^{-1}(0)$ as a problem of existence. Brouwer degree and fixed point index [8,15,16,22,25,32,37,41,51] so far have been considered as topological tools in problems of existence mainly. In fact, the following will show that with a suitable representation both are an important numerical tool to compute $f^{-1}(0)$ and other problems.

The following result is due to M. Prüfer ([46] see also [48]).

(4.1.1) THEOREM (Degree and Integer-L-Labeling)

Let $M \subset R^n$ be bounded and triangulable of homogeneous dimension n, and let $f: M \longrightarrow R^n$ be continuous such that $f(x) \neq 0$ for all $x \in \partial M$. Let L be a linear isomorphism of R^n.

Then one has

(1) $\{\partial M \setminus A_i(f,L)\}_{i=0}^{n} = \mathfrak{U}$ is an open covering of ∂M;

(2) if $\lambda > 0$ is the Lebesgue number of $\mathfrak{U}$ and T is any triangulation of M with $\text{mesh}(T_{n-1} \cap \partial M) \leq \lambda$ then

$$\deg(F,\text{int } M,0) = (\text{sign det } L) \cdot \sum_{\sigma \in \mathcal{S}(f,L)} or_L(\sigma) .$$

($\text{mesh}(T)$ is the maximal diameter of simplices in T).

We sketch a proof from [48]: Observe, that $f_T^i(x) = 0$ if and only if $x = x_\sigma$ is the barycenter of an L-completely labeled simplex $\sigma \in \mathcal{S}(f;L)$. However, the definition of deg for C^∞-maps and regular values implies

$$\deg(f_T^i,\text{int } \sigma,0) = (\text{sign det } L) \cdot or_L(\sigma),\ \sigma \in \mathcal{S}(f;L).$$

Thus,

$$\deg(f_T^i,\text{int } M,0) = \Big(\sum_{\sigma \in \mathcal{S}(f,L)} or_L(\sigma) \Big) \text{ sign det } L$$

is a consequence of the additivity property of degree. Finally, the assumption on mesh $(T_{n-1} \cap \partial M)$ implies that one can construct a homotopy h for f and f_T^i with no solutions on ∂M and the conclusion follows from the homotopy invariance of deg.

A detailed proof of (4.1.1) for the case $L = Id$ is given in [48], where the connection with other recently obtained formulas for the Brouwer degree due to F. Stenger [56], B. Kearfott [29] and M. Stynes [57] is discussed.

It is clear that a corresponding formula will be true for the concept of vector-labeling (see [5], where Prüfer's result and proof is transfered to that case and see [28] for a more elementary proof).

(4.2.1) THEOREM (Degree and Vector-Labeling)

Let $M \subset R^n$ be bounded and triangulable of homeogenous dimension n, and let $f: M \to R^n$ be continuous and such that $f(x) \neq 0$ for all $x \in \partial M$. Then there exists $\delta_o > 0$ such that for any triangulation T for M with $\mathrm{mesh}(T) \leq \delta_o$ one has

$$\deg(f, \mathrm{int}M, 0) = \sum_{\substack{\sigma \in T_n \\ \sigma \text{ completely labeled}}} \mathrm{or}(\sigma)$$

Finally, let X be a triangulable set in R^n of homogeneous dimension n and let $f: X \to X$ be continuous. Let $M \subset X$ be open in X and a triangulable set such that any triangulation T of X induces a triangulation for M. Since X is an ANR (see [13]) there exists an open neighborhood $\mathcal{O}$ of X in R^n and a retraction $r: \mathcal{O} \to X$ (i.e. $r_{|X} = Id_X$). Note that the fixed point index can be defined by

$$\text{(4.1.3)} \quad \mathrm{ind}(X,f,M) := \deg(Id - fr, r^{-1}(M), 0) \quad \text{and}$$

this definition does not depend on the choices $(\mathcal{O}, r)$ involved. In this generality formulas in the sense of (4.1.1) and (4.1.2) exist for ind but will depend on $r^{-1}(M)$ and r. However, in the special case where X is a special cone in R^n, i.e. $X = R^n_+$, one can avoid the dependence from r introducing a suitable L-labeling. This is considered in section XII.

V. Homotopy and Orientation, the Door-In/Door-Out Principle

In this section M denotes a triangulable subset of $R^{n+1} = R^n \times R$ of homogeneous dimension $(n+1)$. Let $F: M \to R^n$ $((x,\lambda) \mapsto F(x,\lambda))$ be continuous. We want to study a problem

$$F(x,\lambda) = 0 \ .$$

To implement such a problem into the setting we have described before we define

$$(5.1.1) \qquad \ell^F: M \to \begin{cases} R^n & \text{(vector-labeling)} \\ \{0,1,2,\dots,n\} & \text{(integer-L-labeling)} \end{cases}$$

by

$$\ell^F(x,\lambda) := \ell(x)^{F(x,\lambda)} \ .$$

Note that according to this definition for any triangulation T of M completely labeled simplices will be in the n-skeleton T_n of T_{n+1} i.e. they are codimension one simplices in T_{n+1}.

The following lemma has the quality of a constructive element in the forthcoming procedures:

(5.1.2) LEMMA (None or Two)

Let $F: M \to R^n$ be continuous and assume that M is labeled according to (5.1.1). Let T be any triangulation of M and let $\sigma \in T_{n+1}$. Then σ has

either two completely labeled faces
or no completely labeled face .

The proof is trivial in the case of integer-L-labeling and follows from an exercise in linear algebra in the case of vector-labeling. Note, however, that it is here and in (4.1.2) that definition (3.1.1) has to be used in its full sharpeness. Lemma (5.1.2) provides us both in the case of vector- and integer-L-labeling with the crucial

(5.1.3) DOOR-IN/DOOR-OUT PRINCIPLE

Let $F: M \to R^n$ be continuous, and let ℓ^F be a labeling according to (5.1.1). Let T be any triangulation of M and assume that $\sigma_o \in T_n$ is a (codimension one) completely labeled simplex.

Then σ_o determines a unique "n-chain"

$$ch_F(\sigma_o) = (\ldots,\sigma_{-2},\sigma_{-1},\sigma_o,\sigma_1,\sigma_2,\ldots)$$

of completely labeled, i.e. codimension one, n-simplices in the n-skeleton of T and a connected simplicial 1-manifold

$$m_F(\sigma_o) = \bigcup_i s_i^{i+1}$$

where s_i^{i+1} is determined by $\sigma_i, \sigma_{i+1} \in ch_F(\sigma_o)$ as follows

$$s_i^{i+1} = \begin{cases} t\,x_{\sigma_i} + (1-t)x_{\sigma_{i+1}}, & 0 \le t \le 1,\ x_{\sigma_i} \text{ and } \\ \text{(integer-L-labeling)} & x_{\sigma_{i+1}} \text{ the barycenters} \\ & \text{of } \sigma_i \text{ and } \sigma_{i+1} \\ & \\ tx_i + (1-t)x_{i+1}, & 0 \le t \le 1,\ \bar{\varepsilon}=(\varepsilon,\varepsilon^2,\ldots,\varepsilon^n) \\ \text{(vector-labeling)} & \varepsilon > 0 \text{ and sufficiently} \\ & \text{small, } F_T \text{ according to} \\ & (3.1) \text{ and} \\ & x_j = F_T^{-1}(\bar{\varepsilon}) \cap \sigma_j, j=i,i+1. \end{cases}$$

Moreover, $ch_F(\sigma_o)$ and $m_F(\sigma_o)$ have the following properties (see figure 8):

(1) Each $\sigma_i \in ch_F(\sigma_o)$ which is not in ∂M is a common face of exactly two (n+1)-simplices of T, hence, $ch_F(\sigma_o)$ is "connected".

(2) There are two possibilities for $ch_F(\sigma_o)$:

(a) $ch_F(\sigma_o)$ is a cycle (i.e. for some i one has $\sigma_i = \sigma_o$ and $ch_F(\sigma_o)$ is completely determined by $\{\sigma_o,\ldots,\sigma_{i-1}\}$) and $m_F(\sigma_o)$ is homeomorphic with S^1 and

$$m_F(\sigma_o) \cap \partial M = \phi \quad .$$

(b) $ch_F(\sigma_o)$ contains no cycle (i.e. each σ_i in $ch_F(\sigma_o)$ occurs only once) and $m_F(\sigma_o)$ is homeomorphic with $[0,1]$, $(0,1)$, or $[0,1)$ and

$$m_F(\sigma_o) \cap \partial M = \partial m_F(\sigma_o)$$

($\partial m_F(\sigma_o)$ denotes the boundary of $m_F(\sigma_o)$ in the manifold sense).

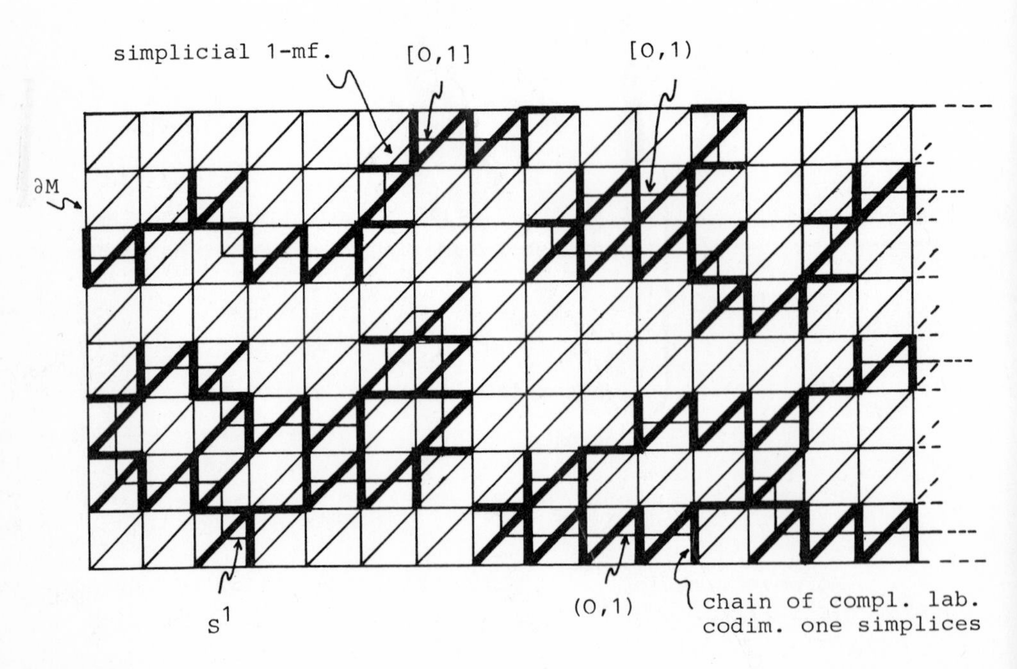

Figure 8.

Observe that if M is compact then $m_F(\sigma_o)$ is homeomorphic with $[0,1]$ or S^1 and thus

$$\partial m_F(\sigma_o) = \begin{cases} \{x_{\sigma_{-i}}, x_{\sigma_j}\} \; ; & \sigma_{-i}, \sigma_j \subset \partial M \\ \emptyset \; . & \end{cases}$$

In this formulation (5.1.2) is a direct analogue to the classification of the smooth manifold $F^{-1}(0)$ when M is a smooth manifold, F is C^∞ and 0 is a regular value. In the case of vector-labeling it is not true in general that $F_T^{-1}(0)$ is a simplicial 1-manifold (see figure 9):

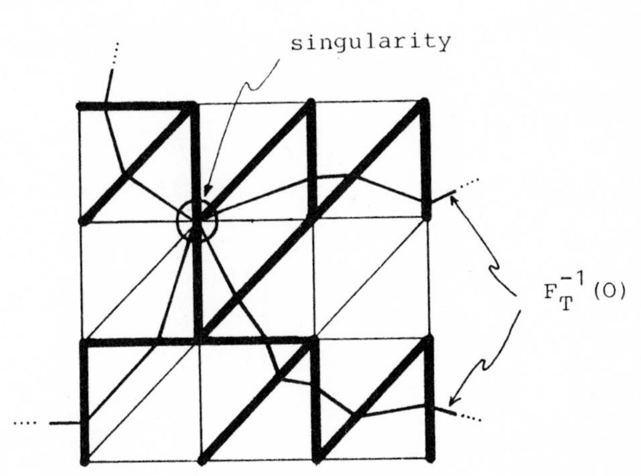

Figure 9.

However, for $\bar{\varepsilon} = (\varepsilon, \varepsilon^2, \ldots, \varepsilon^n)$, $\varepsilon > 0$ and sufficiently small, one can show easily from definition (3.1.1) and the remarks following that definition that $F_T^{-1}(\bar{\varepsilon})$ (F_T PL-map according to vector-labeling) is as well as $F_T^{-1}(0)$ (F_T PL-map according to integer-L-labeling) a collection of simplicial 1-manifolds which we denote in the following by

$$(5.1.4) \qquad \mathcal{M}_F = \begin{cases} F_T^{-1}(0) & \text{(integer-L-labeling)} \\ F_T^{-1}(\bar{\varepsilon}) & \text{(vector-labeling)} \; . \end{cases}$$

Our next observation together with (5.1.3) is the crucial step for a simplicial analogue to the Pontryagin Construction (see for example [40]) in differentiable topology:

(5.1.4) LEMMA (Orientation and Homotopy)

Let $[\lambda_o,\lambda_1]$ be an interval in R and let $M = N \times [\lambda_1,\lambda_2]$, where $N \subset R^n$ is a triangulable set of homogeneous dimension n. Let $F: M \to R^n$ be continuous and let ℓ^F be a labeling (vector- or integer-L-labeling) determined by F according to (5.1.1). Let T be any triangulation of M and let $\sigma_o \in T_n$ be a completely labeled simplex in $N \times \{\lambda_o\}$ Assume that

$$\text{ch}_F(\sigma_o) \text{ has an } n\text{-simplex } \sigma_s \in \text{ch}_F(\sigma_o)$$
$$\text{with } \sigma_s \subset N \times \{\lambda_o\} \cup N \times \{\lambda_1\}$$

Then one has

$$\sigma_s \subset N \times \{\lambda_1\} \quad \text{implies that } or(\sigma_o) = or(\sigma_s)$$
and
$$\sigma_s \subset N \times \{\lambda_o\} \quad \text{implies that } or(\sigma_o) = -or(\sigma_s).$$

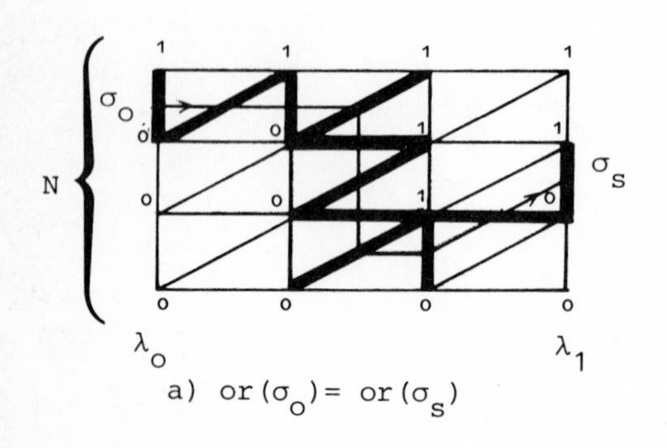

a) $or(\sigma_o) = or(\sigma_s)$

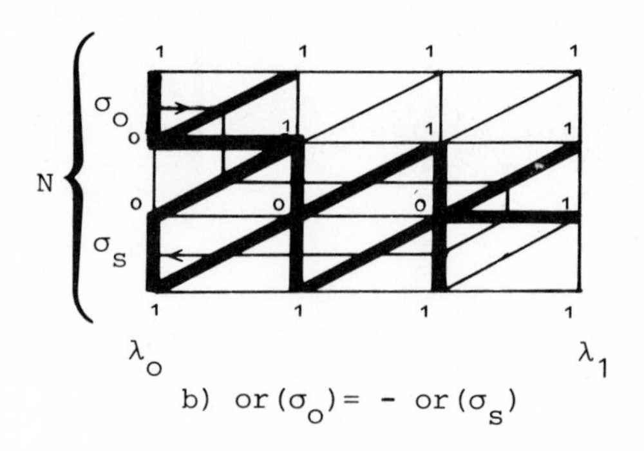

b) $or(\sigma_o) = -\,or(\sigma_s)$

Figure 10.

Using lemma (5.1.4) it is now possible to assign an orientation to the components (which are simplicial 1-manifolds) of $\mathcal{M}_F$ in an obvious manner (see figure 10).

Recall that the essential ingredients from differentiable topology or geometry which are used

- to define the Brouwer degree and
- to show its most important property: the homotopy invariance

are the following (see [7 ,22 ,40]:

(1) Mollifier (to reduce to the C^∞ case)

(2) Brown-Sard Theorem (to reduce to regular values)

(3) Implicit Function Theorem (to prove that $F^{-1}(0)$ is a smooth 1-manifold and a characterization of $\partial F^{-1}(0)$)

(4) Classification of Smooth 1-Manifolds

(5) The Pontryagin Construction

Observe that we have established so far "simplicial substitutes" for each of these ingredients in a constructive sense, except for (5). An analogue for the Pontryagin Construction using (5.1.2) and (5.1.4) will be given in VIII.

VI. Integer-Labeling Versus Integer-L-Labeling

With respect to theory theorem (4.1.1) gives a complete answer to the question: which zeroes of $f:M \to R^n$ can be approximated by a computation of $f_T^{-1}(0)$? Numerically, i.e. in view of a concrete computation, the choice of L will have a tremendous influence on the quality of the approximation process (see figure 6). We demonstrate this by a first example (see also section XI).

$$(6.1.1) \quad \begin{cases} f: R^2 \to R^2 \\ f(x,y) = (x+y+2x^2,\ 10x+11y+21y^2) \end{cases}$$

In figure 11 we give a picture of the labeling sets with $L = Id$:

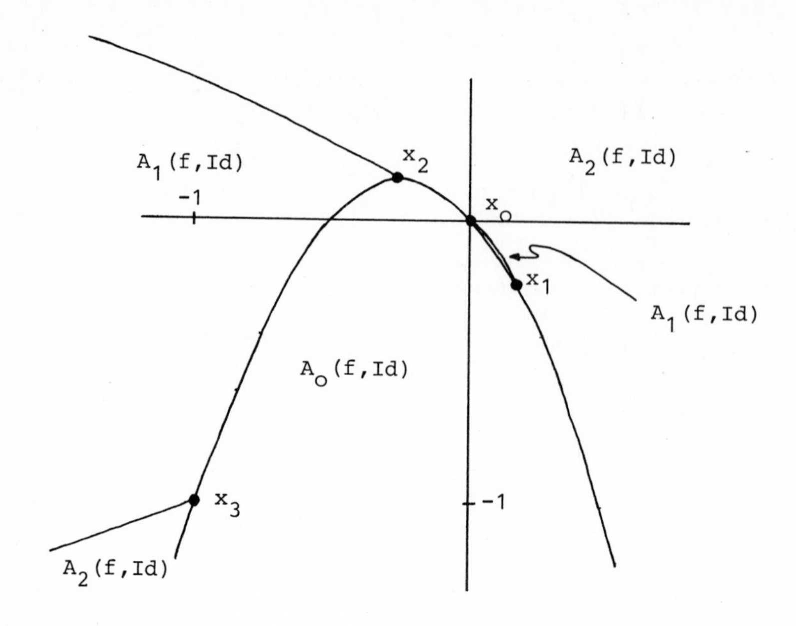

Figure 11.

Note that $\bigcap_{i=0}^{2} A_i(f,Id) = \{x_o,x_1,x_2,x_3\}$, with $x_o = 0$, $x_3 = (-1,-1)$.
Figure 12 microscopes a neighborhood of 0 in figure 11:

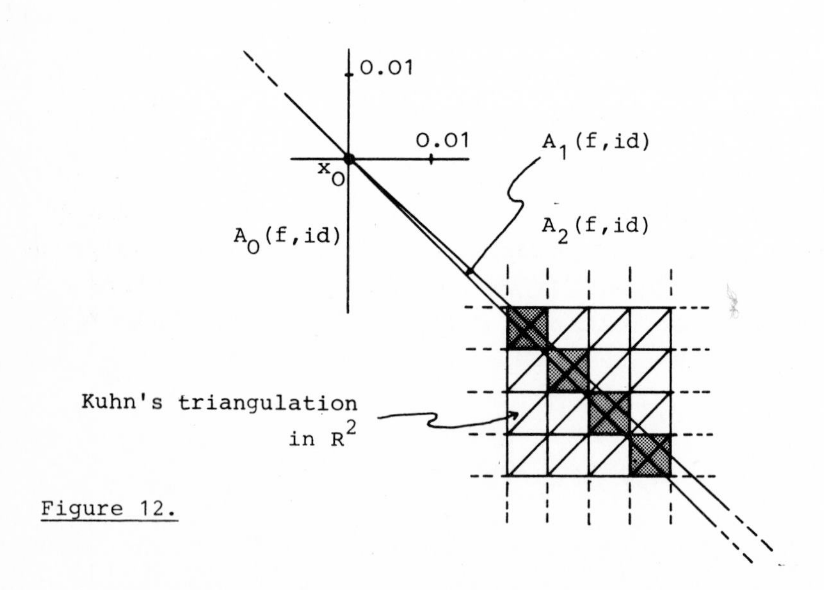

Figure 12.

Observe that an approximation of x_o with ℓ^f_{Id}-Sperner-simplices will be disasterous: Endowing R^2 with Kuhn's triangulation T^K [59] (see also figure 12) with mesh $T^K = 0.005$ we will find a huge number of Sperner-simplices within a relatively large distance from x_o. The reason for this pathology is, of course, the skinny shape of $A_1(f,Id)$ near x_o.

In figure 13 we have drawn the labeling sets with $L = df(x_o)$:

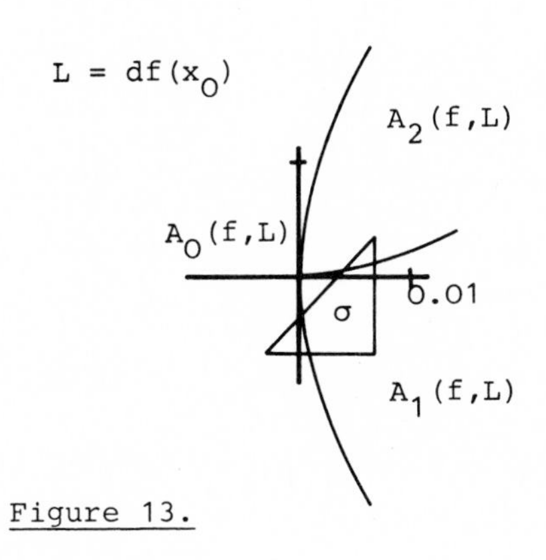

Figure 13.

Obviously, the approximation problem for x_o has improved tremendously. For example, any Kuhn type triangulation T^K with $\text{mesh} T^K = 0.01$ contains exactly one ℓ^f_L-Sperner-simplex σ which approximates x_o and for the barycenter x_σ we can estimate $||x_\sigma||_\infty \leq 0.01$ ($||\cdot||_\infty$ = sup-norm).

Note, however, that replacing $L = Id$ by $L = df(x_o)$ in the labeling is a double edged sword: the zero x_3 having fairly reasonable labeling sets in figure 11 inherits the pathology under L-labeling ($L = df(x_o)$) which before was with x_o (see figure 14).

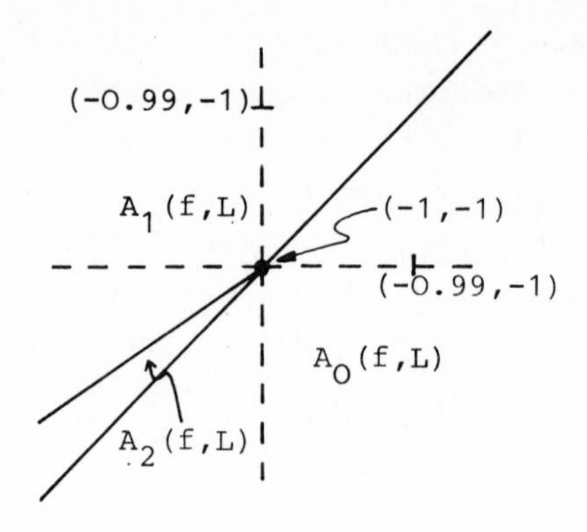

Figure 14.

Figures 11-13 suggest the following lemma describing the effect of integer-L-labeling: (a cone in R^n is a closed convex set which contains with x also the ray tx, $t \geq 0$).

(6.1.2) LEMMA

Let $U \subset R^n$ be open, $x_o \in U$, and let $f: U \to R^n$ be sufficiently smooth. Assume that $f(x_o) = 0$ and $df(x_o)$ is an isomorphism.

(1) The tangential cones for the labeling sets $A(f;Id)$ (i.e. $L = Id$) in x_o are given by the labeling sets $A(df(x_o), Id)$ which are cones in R^n.

(2) The tangential cones for the labeling sets $A(f,df(x_o))$ (i.e. $L = df(x_o)$) are given by the cones $clB^o, clB^1, \dots, clB^n$ (see section III for the definition of B^i).

VII. $f_T^{-1}(0)$ Versus $f^{-1}(0)$

It has become clear in the preceding sections that completely labeled simplices provide approximations for zeroes of a continuous mapping (i.e. $f_T^{-1}(0) \approx f^{-1}(0)$). In this section we make precise by some standard estimates how good $f_T^{-1}(0)$ approximates $f^{-1}(0)$.

As before M denotes a triangulable subset of R^n of homogeneous dimension n.

(7.1.1) LEMMA (vector- and integer-L-labeling)

Let $f: M \to R^n$ be continuous and let $\{T^k\}_{k\in N}$ be a sequence of triangulations for M such that $\lim_{k\to\infty} \text{mesh}(T^k)=0$. Let $\{\sigma_k \in T_n^k\}$ be a sequence of completely labeled n-simplices. Then any clusterpoint of $\{\sigma_k\}_{k\in N}$ is a zero of f.

In the following two lemmata we restrict to integer-L-labeling:

(7.1.2) LEMMA

Let $f: M \to R^n$ be continuous and assume that M is bounded. Let $\varepsilon, \delta > 0$ be such that

$$||x-y||_\infty \leq \delta \text{ implies } ||f(x)-f(y)||_\infty \leq \varepsilon$$

for all $x,y \in M$. Let T be any triangulation for M with mesh $(T) \leq \delta$ and let $\sigma \in T_n$ be ℓ_L^f-completely labeled.

Then

$$||f(x)||_\infty \leq \varepsilon \cdot \text{cond}_\infty(L)$$

for all $x \in \sigma$ $(\text{cond}_\infty(L) = ||L||_\infty \cdot ||L^{-1}||_\infty)$.

Our last lemma is suggested by figure 13 and lemma (6.1.2) and provides an a priori estimate for a zero:

(7.1.3) LEMMA

Let $f: M \to R^n$ be continuous and let $x_o \in \text{int} M$ be a zero of f. Assume that $df(x_o)$ exists and that $df(x_o)$ is an isomorphism. Let T be a triangulation of M which is Kuhn's triangulation in a neighborhood of x_o.
Then there exists a unique ℓ_L^f-Sperner-simplex σ with $L = df(x_o)$ which approximates x_o provided mesh(T) is sufficiently small and then one has the estimate

$$||x_\sigma - x_o||_\infty \le \text{mesh}(T),$$

where x_σ is the barycenter of σ.

Thus, if ε and δ are as in (7.1.2) and mesh(T) $\le \delta$ then

$$||f(x_\sigma)||_\infty \le \varepsilon .$$

Lemma (7.1.3) suggests x_σ as a candidate for an approximate zero x^* of f. Note, however, that (3.2.3) provides a better one:
Let $\sigma = [a^o,\dots,a^n]$ be the ℓ_L^f-Sperner-simplex given by lemma (7.1.3) $(L = df(x_o))$. Assume that $\ell_L^f(a^i) = i$ for $0 \le i \le n$, and let z^i be the solution of $Lz^i = f(a^i)$, $0 \le i \le n$, cf. (3.2.3). Then $\ell_L^f(a^i) = i$ means:

$$\begin{cases} z_1^o \le 0 & \\ z_j^i \begin{cases} > 0 \\ \le 0 \end{cases} \begin{cases} 1 \le j \le i \\ j = i+1 \end{cases} & \text{for } 1 \le i \le n-1 \\ z_j^n > 0 \quad 1 \le j \le n & \end{cases}$$

Define the components of x^* (which might not be in σ) by

$$(7.1.4) \qquad x_i^* = a_i^{i-1} - z_i^{i-1} \cdot \frac{a_i^i - a_i^{i-1}}{z_i^i - z_i^{i-1}}, \quad 1 \leq i \leq n .$$

Observe that $x^* = x^o$ in the case where f is an affine homeomorphism $f = L + b$, which actually was our motivation to consider x^* as the "natural" guess for a zero of f.

VIII. Leray-Schauder Continuation Method

Continuation, embedding or homotopy methods have a long history in modern mathematics. Certainly, H. Poincaré [45] and F. Klein [30] and S. Bernstein [12], have to be mentioned. However, the great success of continuation methods in analysis is certainly due to J. Leray and J.P. Schauder [37] who presented the method free from all unnecessary restrictions and assumptions as a basic and global result in topology: (in the language of degree theory: the homotopy invariance of degree).

(8.1.1) Let $J = [\lambda_o, \lambda_1]$ be a compact interval in R and let $\Omega \subset R^n \times J$ be an open and bounded connected subset whose projection onto the second component fills J. Let $F: \mathrm{cl}\Omega \to R^n$ be continuous and assume that

(A PRIORI ESTIMATES) $F(x,\lambda) \neq 0$, for all $(x,\lambda) \in \partial\Omega$, and

(TRIVIAL PROBLEM) $\deg(F_{\lambda_o}, \Omega_{\lambda_o}, 0) \neq 0$

Then there exists a continuum $\mathscr{C} \subset \Omega$ (closed, connected set) such that

(1) $F(x,\lambda) = 0$ for all $(x,\lambda) \in \mathscr{C}$

(2) $\mathscr{C} \cap \Omega_{\lambda_o} \neq \phi \neq \mathscr{C} \cap \Omega_{\lambda_1}$.

($F_\lambda := F(\cdot,\lambda)$ and $\Omega_\lambda = \{x \in R^n: (x,\lambda) \in \Omega\}$).

Actually, in view of applications to existence problems in analysis it is important to have the continuation method in an ∞-dimensional Banach space for completely continuous mappings [15,32,37,49,51] Observe that the Leray-Schauder Continuation Method is of global nature, i.e. it guarantees a global branch in $F^{-1}(0)$:

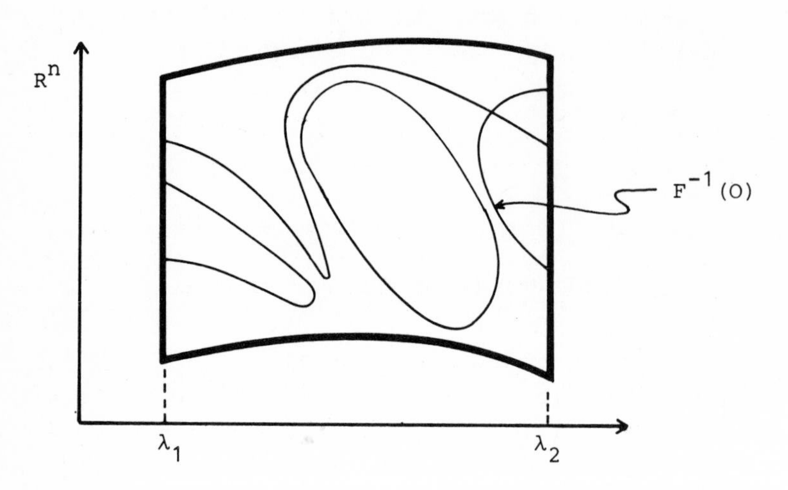

Figure 15.

Observe that $F^{-1}(0)$ can have branching points and turning points and it seems to be therefore that the principle resisted any numerical accessibility for such a long time though it became one of the most prominent tools in non-linear analysis. Local continuation methods, however, are well known in numerical analysis for quite a long time [5 ,44,61] and it is due to J.C. Alexander, S.N. Chow, M. Hirsch, J. Mallet-Paret and J.A. Yorke (none of them is a numerical analyst) [1 , 2 , 3 , 4 , 5 ,14,61] who observed that one can obtain global continuation methods in a constructive sense using the Pontryagin Construction [40] and some generalisations from transversality theory. They obtain "with probability one" smooth 1-manifolds as global solution sets which ought to be followed by solving the so called Davidenko differential equation (which is an ODE whose trajectory is the 1-manifold in question).

Instead, our approach will guarantee by passing from F to F_T a simplicial 1-manifold connecting Ω_{λ_0} and Ω_{λ_1} as a component of $\mathcal{M}_F$, and a numerical algorithm to follow this is provided by an implementation of the Door-In/Door-Out Principle (5.1.3).

(8.1.2) THEOREM (Leray-Schauder Continuation Method)

Let $J \subset R$ and $\Omega \subset R^n \times J$ be as above and let $F: \mathrm{cl}\Omega \to R^n$ be continuous. Assume that

(A PRIORI ESTIMATES) $F^{-1}(0) \cap \partial\Omega = \phi$, and,

(TRIVIAL PROBLEM) $\deg(F_{\lambda_0}, \Omega_{\lambda_0}, 0) \neq 0$.

(i) Then there exists $\delta_1 > 0$ such that for any triangulation T of a neighborhood of $\mathrm{cl}\Omega$ with $\mathrm{mesh}(T) \leq \delta_1$

$$\mathcal{M}_F = \begin{cases} F_T^{-1}(0) & \text{(integer-L-labeling), respectively} \\ F_T^{-1}(\bar{\varepsilon}) & \text{(vector-labeling and } \bar{\varepsilon} = (\varepsilon, \varepsilon^2, \ldots, \varepsilon^n), \\ & \varepsilon \text{ small)} \end{cases}$$

contains a connected simplicial 1-manifold m_F with

$$\Omega_{\lambda_0} \cap m_F \neq \phi \neq \Omega_{\lambda_1} \cap m_F .$$

Furthermore, the manifold m_F can be traced numerically by a chain of completely labeled simplices in the n-skeleton T_n of T by an implementation of the Door-In/Door-Out Pinciple (5.1.3).

(ii) $\deg(F_\lambda, \Omega_\lambda, 0) \equiv$ constant for all $\lambda \in J$.

<u>Proof:</u> a. Observe that $\inf\{\|F(x,\lambda)\|: (x,\lambda) \in \partial\Omega\} = \delta_0 > 0$. Hence, Lemma (7.1.1) will imply that for any sufficiently fine triangulation of a neighborhood of $\mathrm{cl}\Omega$ no completely labeled n-simplex intersects $\partial\Omega$.

b. Theorems (4.1.1), respectively (4.1.2), imply for any sufficiently fine triangulation T of a neighborhood of $\mathrm{cl}\Omega$

$$0 \neq \deg(F_{\lambda_0}, \Omega_{\lambda_0}, 0) = \sum_{\substack{\sigma \in T_n \cap \Omega_{\lambda_0} \\ \sigma \text{ completely labeled}}} \mathrm{or}(\sigma)$$

c. Now let σ_o be any completely labeled n-simplex in Ω_{λ_o} (b. guarantees at least one). According to lemma (5.1.3) σ_o initiates a chain $ch_F(\sigma_o)$ which is not a cycle. Hence, if $ch_F(\sigma_o) \cap \Omega_{\lambda_1} = \phi$ then a. and lemma (5.1.4) will imply that

$$ch_F(\sigma_o) \cap \Omega_{\lambda_o} = \{\sigma_o, \sigma_i\}, \quad \text{and}$$

$$or(\sigma_o) = - or(\sigma_i)$$

Thus, the conclusion follows form $\deg(F_{\lambda_o}, \Omega_{\lambda_o}, O) \neq O$, and a. - c.

Observe that this proof is strictly analogous to the Pontryagin Construction used by H. Amann [7] and C. Fenske/G. Eisenack [22] and that an obvious limit argument provides a proof of the classical Leray-Schauder Continuation Method in finite dimensions.

In view of problem (1) (see introduction) we remark that lemma (7.1.1) will imply:

(8.1.3) REMARK: (Approximating a continuum in $F^{-1}(O)$)

Let $S = \{(x,\lambda) \in \Omega: F(x,\lambda) = O\}$ and let Ω^S be any open neighborhood of S. Then any sufficiently fine triangulation of a neighborhood of $cl\Omega$ will have all completely labeled simplices in Ω^S, i.e. the chains which are guaranteed by (8.1.2) can be forced to follow the components of $F^{-1}(O)$.

Figure 16 illustrates (8.1.3) for integer-L-labeling. In figure 16a the mesh size is small enough to force all chains to follow the different components of $F^{-1}(O)$. In figure 16b the triangulation is too coarse and this is to the effect that a chain may "switch" between several continua, which, however, is not in contradiction with (8.1.2). In fact, this is a valuable observation in view of the original background of the Leray-Schauder Continuation Method:

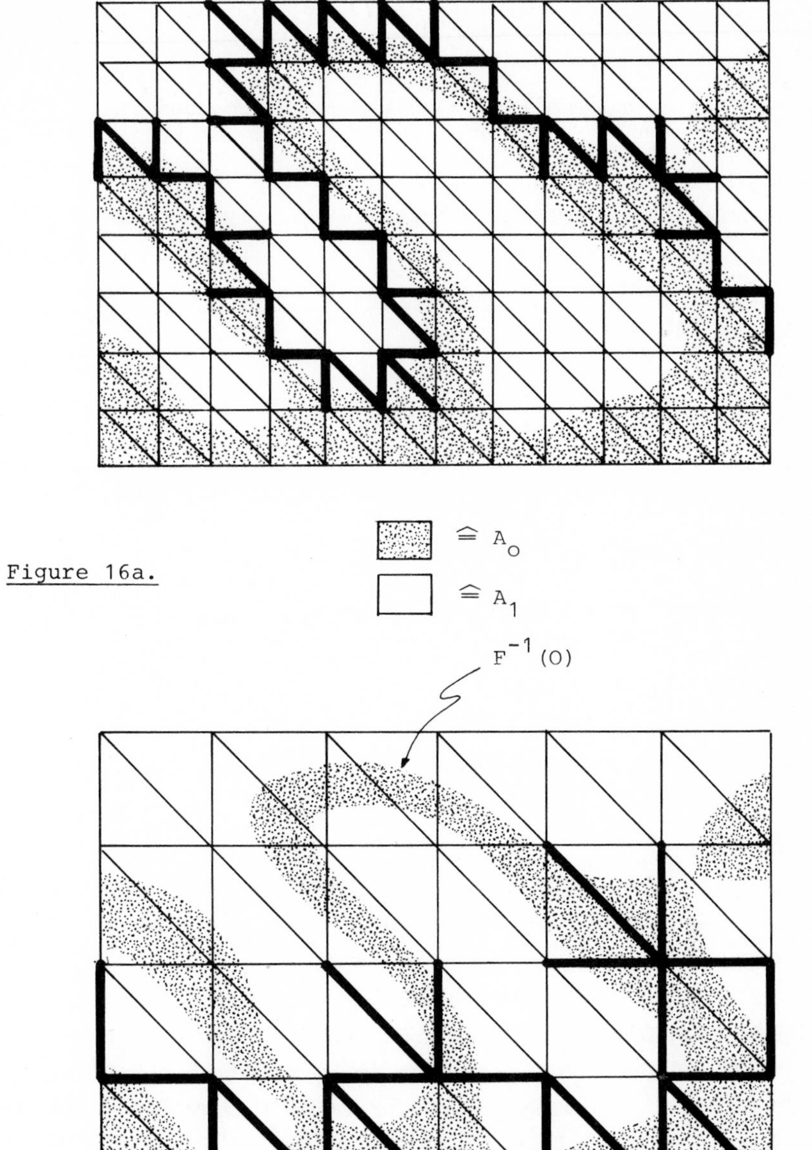

Figure 16a.

Figure 16b.

(8.1.4) REMARK:

(i) (Appropriate triangulations)

Assume that a problem $f(x) = 0$ has been embedded into $F(x,\lambda) = 0$ with $F(x,\lambda_1) = f(x)$ and such that (8.1.2) can be applied. In view of this $F(x,\lambda) = 0$ is an artificial problem for all $\lambda_o \leq \lambda < \lambda_1$, i.e. the numerical problem to approximate a given continuum in $F^{-1}(0)$ is of secondary interest. In such a situation a favourable triangulation will be one which is coarse near λ_o and becomes finer as λ approaches λ_1, (cf. [20]) see figure 17.

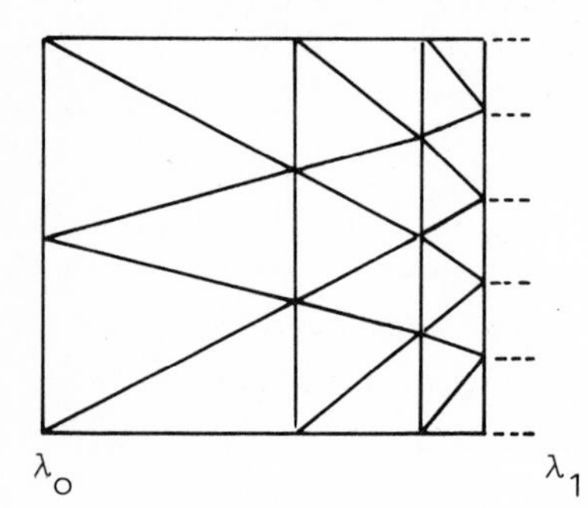

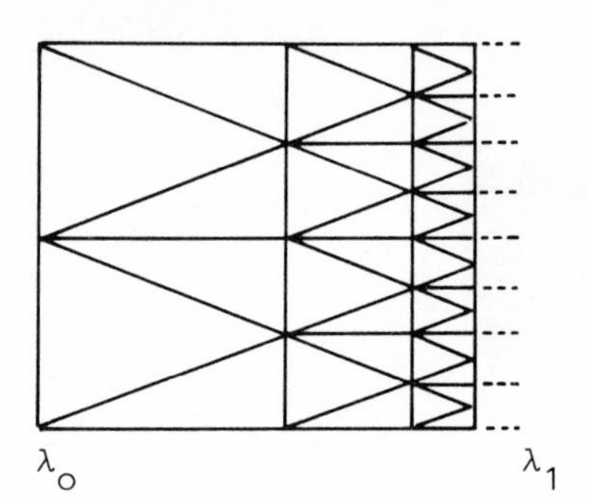

Figure 17.

So far it has become clear that the topological information in a problem of analysis which has made an existence proof work along the Leray-Schauder Continuation Method can now also be exploited for a "stable" numerical procedure to approximate solutions or at least to provide starting values for rapidly converging procedures.

(ii) (To start a chain in the trivial problem)

Observe that in many cases the Leray-Schauder Continuation Method is applied with

$F(x,\lambda_o) \equiv \mathrm{Id}$ or

$F(x,\lambda_o) \equiv \mathrm{Id} - c.$

In both cases (integer-L-labeling) and vector-labeling) Ω_{λ_o} will contain a unique completely labeled simplex to start with.

(iii) (Non-linear eigenvalue problems)

If F is for example derived from the discretization of a non-linear boundary value problem

$$\text{(BVP)} \quad \begin{cases} u'' + \lambda f(u) = 0 \\ u(0) = u(1) = 0 , \end{cases}$$

then $F(x,\lambda) = \mathscr{L}x + \mathscr{N}(x,\lambda)$, where $\mathscr{L}$ is linear and $\mathscr{N}$ is non-linear. In this case one might be interested to follow a continuum of solutions of (BVP) as good as possible. For integer-L-labeling the choice of L will be crucial. In XI we will discuss (BVP) for $f(u) = \sin u$ and $f(u) = \exp(u)$. In both cases numerical experiments have shown that the choice $L = Id$ provides very bad numerical results, whereas the choice $L = \mathscr{L}$ provides results which are comparable to those obtained using vector-labeling. In general one will try to update L as λ changes according to (6.1.2). Then, however, $\operatorname{sign}(\det L_\lambda)$ has to be kept constant.

IX. BIFURCATION I (Local and Global Bifurcation in the sense of M.A. Krasnosel'skii and P.H. Rabinowitz)

Bifurcation is a term used in several parts of mathematics. It generally refers to a qualitative change in the objects being studied due to a change in the parameters on which they depend. In view of the analysis of differential equations we are interested in functional equations

$$F(x,\lambda) = 0$$

where F is a nonlinear operator, x is the solution vector and λ is a parameter. Here bifurcation theory is the study of the branching of solutions of $F(x,\lambda) = 0$. It is of particular interest to study how the solutions $x(\lambda)$ and their multiplicities change as λ varies. Let E be a Banach space, $\Omega \subset E \times R$, and $F: \Omega \to E$. Suppose there is a one-to-one curve or branch

$$\mathscr{S} = \{(x(\lambda),\lambda) \in \Omega: \quad \lambda \in (0,1) \quad \text{and} \quad F(x(\lambda),\lambda) = 0\}$$

of "known solutions". A <u>bifurcation point for F</u> with <u>respect to</u> $\mathscr{S}$ is a point $(x(\lambda_0),\lambda_0) \in \mathscr{S}$ such that every neighborhood of $(x(\lambda_0),\lambda_0)$ contains zeros of F not on $\mathscr{S}$. In applications one usually has $\mathscr{S} = \{(0,\lambda): \lambda \in (a,b)\}$, called "trivial solutions" of $F(x,\lambda) = 0$.

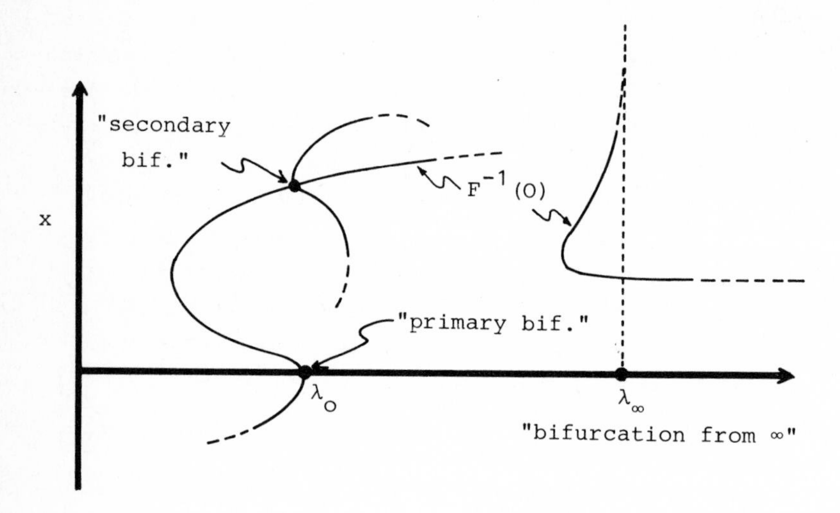

<u>Figure 18.</u>

More precisely we will use the terms "primary bifurcation, secondary bifurcation, and bifurcation from ∞", see figure 18 and [8 ,9 ,32,39,51,52,58].

A quarter century ago Krasnosel'skii [32] introduced topological methods, especially degree theory, into the subject and obtained a local existence result (eigenvalues of odd multiplicities imply bifurcation). Later Rabinowitz [50] observed that the hypotheses in Krasnosel'skii's theorem imply something global rather than local about the behaviour of nontrivial zeros of F. In both cases it is essential that the proof is by contradiction and this seems to be noteworthy in view of our constructive approaches.

Observe that the hypotheses (eigenvalue of odd multiplicity) in [32] and [50] is only to guarantee from the theorem of Leray and Schauder [37] that the degree or index changes passing through a crucial parameter λ_o. Therefore the following is a finite-dimensional version of their theorems.

(9.1.1) THEOREM (Bifurcation and Degree)

Let $F: R^n \times R \to R^n$ be continuous and such that $F(0,\lambda) = 0$ for all $\lambda \in R$. Let $[\lambda_1,\lambda_2] \subset R$ be an interval and let U_i be open neighborhoods of zero in $R^n \times \{\lambda_i\}$, $i = 1,2$, such that

$$F^{-1}(0) \cap \mathrm{cl}U_i = \{0\} \quad , \quad i=1,2.$$

Assume that

$$\deg(F_{\lambda_1};U_1,0) \neq \deg(F_{\lambda_2};U_2,0).$$

Then one has:

(9.1.1) LOCAL BIFURCATION (Krasnosel'skii)

There exists a bifurcation point $(0,\lambda^*)$ for $F(x,\lambda) = 0$ with $\lambda^* \in [\lambda_1,\lambda_2]$.

(9.1.2) GLOBAL BIFURCATION (Rabinowitz)

There exists a continuum $\mathscr{C} \subset F^{-1}(0)$ of nontrivial solutions of $F(x,\lambda) = 0$ with

$$\mathscr{C} \cap (\{0\} \times [\lambda_1,\lambda_2]) \neq \phi$$

(i.e. $\mathscr{C}$ emanates from 0 in $[\lambda_1,\lambda_2]$) and $\mathscr{C}$ has at least one of the following properties:

- $\mathscr{C}$ is unbounded in $R^n \times R$
- There exists $\mu \in R \setminus [\lambda_1,\lambda_2]$ such that $(0,\mu) \in \mathscr{C}$.

Observe that topological bifurcation is simply the "negation" of the Leray-Schauder Continuation Method for a suitably chosen neighborhood Ω:

$$\left\{\begin{array}{c}\text{A PRIORI ESTIMATES}\\ (F^{-1}(O) \cap \partial\Omega = \phi)\end{array}\right\} \quad \Rightarrow \quad \deg(\lambda) \equiv \text{constant}$$

$$\left\{\begin{array}{c}\text{No A PRIORI ESTIMATES}\\ (F^{-1}(O) \cap \partial\Omega \neq \phi)\end{array}\right\} \quad \Leftarrow \quad \deg(\lambda) \not\equiv \text{constant}$$

In the proof of (8.1.2) we have shown the existence of a chain of completely labeled simplices, resp. the existence of a simplicial 1-manifold, which due to the a priori estimates could not leave the bounded set $\Omega \subset R^n \times R$. Here we shall have that at least one chain of completely labeled simplices, resp. simplicial 1-manifold, which intersects $U_1 \cup U_2$ must leave any bounded set Ω through $\partial\Omega \setminus (U_1 \cup U_2)$ which is a "proper connection" between U_1 and U_2:

(9.1.2) We say that $\Omega \subset R^n \times R$ is a <u>proper connection</u> for U_1 and U_2 (see figure 19) provided: $(\Omega_\lambda := \{x \in R^n : (x,\lambda) \in \Omega\})$

(i) $\Omega \cap \{O\} \times R = \{O\} \times [\lambda_1,\lambda_2]$;

(ii) Ω is connected;

(iii) $\Omega = \Omega^t \cup \Omega^n$, (t = trivial, n = non-trivial) where Ω^t is open and bounded in $R^n \times [\lambda_1,\lambda_2]$ and where Ω^n is open and bounded in $R^n \times R$;

(iv) $\Omega^t_{\lambda_1} = U_1$ and $\Omega^t_{\lambda_2} = U_2$ and $\mathrm{cl}\Omega^t_{\lambda_1} \cap \mathrm{cl}\Omega^n_{\lambda_1} = \phi = \mathrm{cl}\Omega^t_{\lambda_2} \cap \mathrm{cl}\Omega^n_{\lambda_2}$.

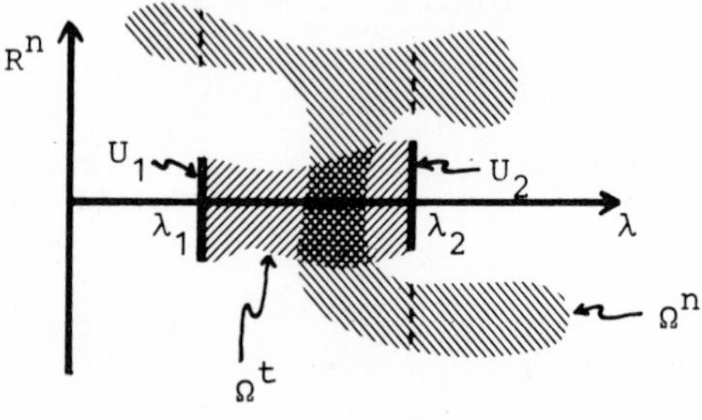

<u>Figure 19.</u>

The following theorem subsumes local as well as global bifurcation and its proof is obtained simply by "reversing" the arguments in the proof of (8.1.2):

(9.1.3) THEOREM (Bifurcation Constructive)

Let $F: R^{n+1} \to R^n$ be continuous and such that $F(0,\lambda) = 0$ for all $\lambda \in R$. Let $[\lambda_1,\lambda_2] \subset R$ be an interval and let U_i be open neighborhoods of zero in $R^n \times \{\lambda_i\}$, $i = 1,2$, such that

$$F^{-1}(0) \cap \mathrm{cl}\, U_i = \{0\} \quad , \quad i = 1,2.$$

Let $\Omega \subset R^n \times R$ be any proper connection for U_1 and U_2 and assume that

$$\deg(F_{\lambda_1},U_1,0) \neq \deg(F_{\lambda_2},U_2,0) .$$

Then there exists $\delta_o > 0$ such that any triangulation T of a neighborhood of Ω which is compatible with $R^n \times [\lambda_1,\lambda_2]$ and for which $\mathrm{mesh}(T) \le \delta_o$ has the following property:

There is at least one completely labeled n-simplex $\sigma^n \subset T_n \cap (U_1 \cup U_2)$ such that $ch_F(\sigma^n)$ leaves Ω through $\partial\Omega \setminus (U_1 \cup U_2)$. This means that

$$\mathcal{M}_F = \begin{cases} F_T^{-1}(0) & \text{(integer-L-labeling), respectively} \\ F_T^{-1}(\bar{\varepsilon}) & \text{(vector-labeling, } \bar{\varepsilon} = (\varepsilon,\varepsilon^2,\ldots,\varepsilon^n) \text{ and } \varepsilon \text{ small),} \end{cases}$$

contains a simplicial 1-manifold which connects

$$U_o \cup U_1 \quad \text{with} \quad \partial\Omega \setminus (U_o \cup U_1) .$$

(A triangulation T of a neighborhood of $\Omega \subset R^n \times R$ is called compatible with $R^n \times [\lambda_1,\lambda_2]$ provided $T_n \cap R^n \times \{\lambda_1\}$ and $T_n \cap R^n \times \{\lambda_2\}$ is a triangulation of a neighborhood of Ω_{λ_1} and Ω_{λ_2}).

Proof: The proof is immediate from the arguments in the proof of (8.1.2) and the pictures in figure 20.

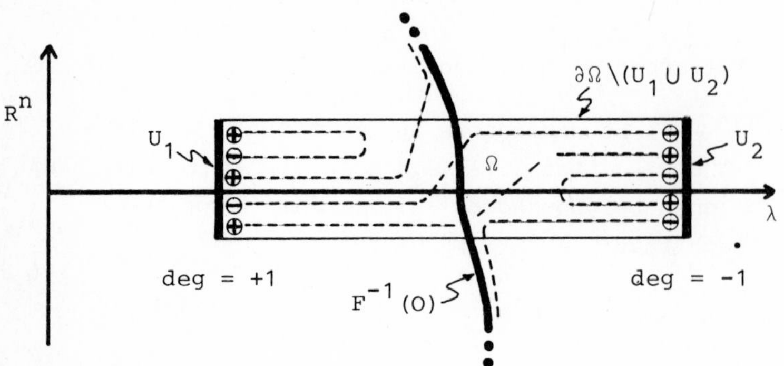

Figure 20a. (local bifurcation : Ω "small")

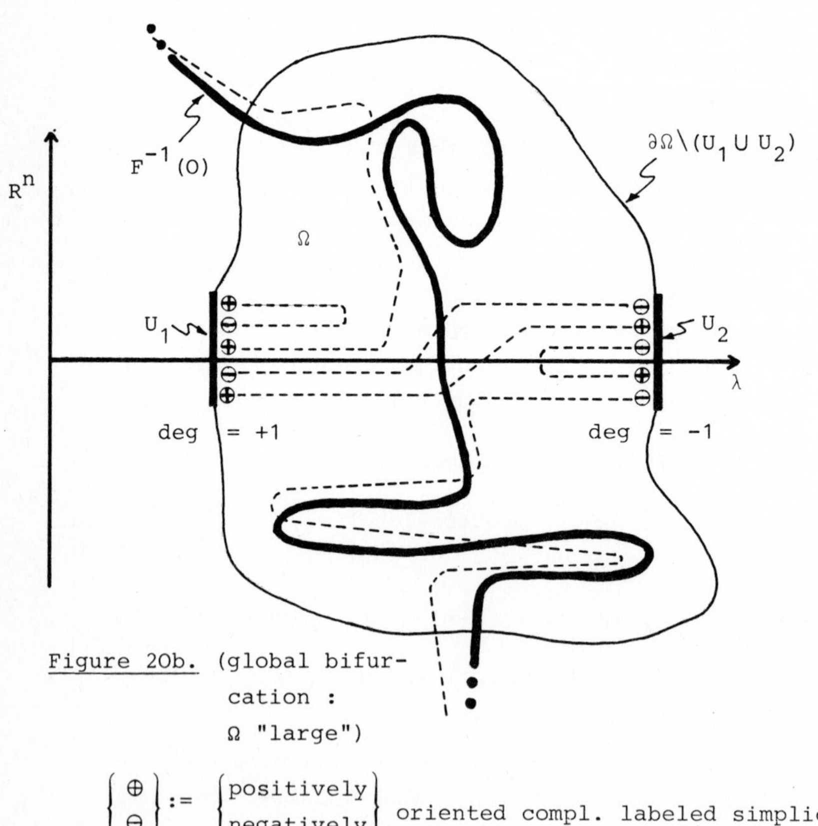

Figure 20b. (global bifurcation : Ω "large")

$\begin{Bmatrix} \oplus \\ \ominus \end{Bmatrix}$:= $\begin{Bmatrix} \text{positively} \\ \text{negatively} \end{Bmatrix}$ oriented compl. labeled simplices

------- := chains of compl. labeled simplices

As a consequence of lemma (7.1.1) a chain of completely labeled simplices as given in theorem (9.1.3) approximates a continuum of nontrivial solutions if mesh(T) is sufficiently small. However, as is indicated in figure 20, to select the emanating chain from those chains which start in $U_1 \cup U_2$ might be another problem:

(9.1.4) REMARK (Start for an emanating chain)

(i) In many applications (see examples in X-XII) λ_1 can be chosen such that F_{λ_1} is linear. In that case

- vector-labeling generates a unique completely labeled simplex σ_1 approximating $(0,\lambda_1)$ (cf. [5]), and
- integer-L-labeling generates a unique completely labeled simplex σ_1 approximating $(0,\lambda_1)$ if one chooses $L := F_{\lambda_1}$.

In these cases the emanating chain is often determined by σ_1.

(ii) In general it might be necessary in order to find the emanating chain to check with any completely labeled simplex in $U_1 \cup U_2$. Then a systematic search in $U_1 \cup U_2$ seems to be appropriate and this can be achieved by an implementation of techniques by M. Todd [60].

X. BIFURCATION II (Artificial Bifurcation and Connections, Secondary Bifurcation, Bifurcation from ∞, Even Multiplicities)

Our approach here will be to describe "artificial bifurcation" as a new technique which in combination with previous considerations will provide a tool to study

- secondary bifurcation (see example (10.3.3)
- bifurcation from ∞ (see example (10.2.4)
- bifurcation from eigenvalues of even multiplicity (see example (10.3.3)
- problems with nontrivial solutions which do not bifurcate - variational methods (see example (10.2.4)).

In this section $F: R^n \times R \to R^n$ will always denote a continuous map. As in the previous sections all results here are true both for vector-labeling and integer-L-labeling. Therefore, F_T denotes either the PL-map according to vector- or integer-L-labeling. Note that its solution set $\mathcal{M}_F$ (cf. (5.1.4)) is a collection of simplicial 1-manifolds which approximate $F^{-1}(0)$. In the following we prefer often to graph these manifolds instead of $F^{-1}(0)$ or the chains of completely labeled simplices which generate $\mathcal{M}_F$. We use the following notation:

$$GL(R^n) = GL_+ \cup GL_- , \quad \text{where}$$

$$GL_\pm = \{P \text{ linear isomorphism of } R^n\text{: sign det}(P) = \pm 1\}$$

Let Γ be a subset of $R^n \times R$. Then Γ and any triangulation T of $R^n \times R$ determine a decomposition $\Gamma_+ \cup \Gamma_- \cup \Gamma_a$ of $R^n \times R$ (see figure 21):

$$\Gamma_+ = \bigcup_{\sigma \in T_+} \sigma , \quad \text{where} \quad T_+ = \{\sigma \in T_{n+1}\text{: } \sigma \cap \Gamma = \phi\}$$

$$\Gamma_- = \bigcup_{\sigma \in T_-} \sigma , \quad \text{where} \quad T_- = \{\sigma \in T_{n+1}\text{: } \sigma \subset \Gamma\}$$

$$\Gamma_a = (R^n \times R) \setminus (\Gamma_+ \cup \Gamma_-)$$

The subscript a indicates the region in $R^n \times R$ where artificial bifurcation takes place.

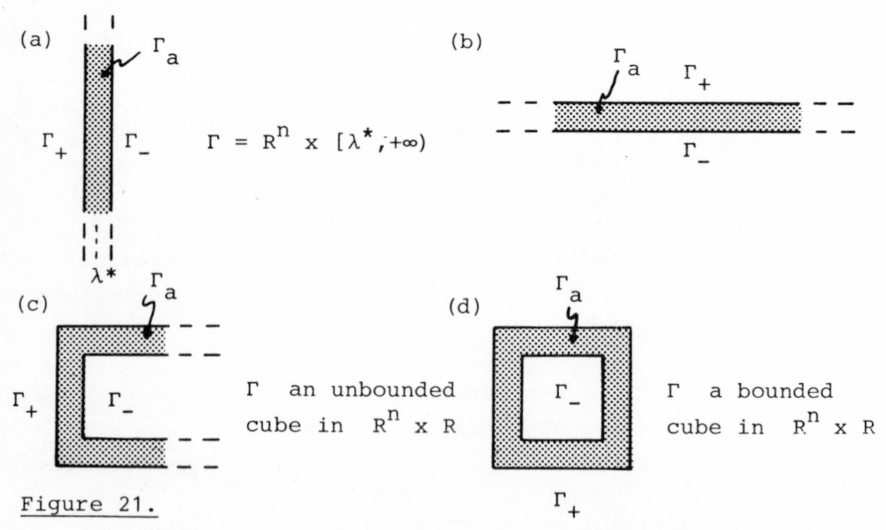

Figure 21.

(10.1) ARTIFICIAL BIFURCATION

Let $P \in GL_-$ (P plays the role of an orientation reversing "switch" in Γ_-). We now associate with F_T a new PL-map $G_T: R^n \times R \longrightarrow R^n$ defined by

$$(10.1.1)\quad \begin{cases} G_T(x,\lambda) = \begin{cases} F_T(x,\lambda)\,, & (x,\lambda) \in \Gamma_+ \\ (PF)_T(x,\lambda)\,, & (x,\lambda) \in \Gamma_- \end{cases} \\ \text{On } \Gamma_a \text{ let } G_T \text{ be the unique PL-extension of} \\ G_T: \Gamma_+ \cup \Gamma_- \to R^n\,. \end{cases}$$

This manipulation roughly means that in the interior of Γ the map F has been "switched" by $P \in GL_-$. To obtain a similar manipulation in the exterior of Γ our notations and definition are such that this can simply be achieved by replacing Γ by its complement $c(\Gamma) := R^n \times R \setminus \Gamma$.

Passing from F_T to G_T along the above definition can be considered to be a formal operation

$$F_T \xmapsto{S(F;P,\Gamma)} G_T$$

which we call switch (in forthcoming figures a switch is indicated by Ⓢ).

WARNING: In the case of vector-labeling one has that $(PF)_T = PF_T$. Moreover, observe that if $F_T^{-1}(0) \cap T_{n-1} = \phi$ (i.e. the zeros of F_T are nondegenerate with respect to T) then any completely labeled n-simplex for F is also completely labeled for PF and vice versa. However, in the case of integer-L-labeling one has in general $(PF)_T \neq PF_T$ and, furthermore,

$$\mathcal{S}(F(\cdot,\lambda);L) \neq \mathcal{S}(PF(\cdot,\lambda);L) = \mathcal{S}(F(\cdot,\lambda);P^{-1}L)\,.$$

Therefore, in both cases

$$\mathcal{M}_F \cap \Gamma_- \neq \mathcal{M}_G \cap \Gamma_-\,. \qquad (\mathcal{M}_G = G_T^{-1}(0),\ \text{resp.} = G_T^{-1}(\bar{\varepsilon}))$$

However, one has the following trivial result:

(10.1.2) LEMMA

Let F,P and Γ be as above. Then $\mathcal{M}_F \cap \Gamma_-$ and $\mathcal{M}_G \cap \Gamma_-$ are approximations of $F^{-1}(0) \cap \Gamma_-$ (i.e. the manipulation (10.1.1) does not affect the numerical approximation procedure outside of Γ_a (see figures 22,23).

To understand the local effect of (10.1.1) it suffices (up to an appropriate change of parameters) to restrict ourselves to the situation indicated in figure 21a. We assume that T is a triangulation of $R^n \times R$ which is "parallel to R^n", i.e. given T there exists a triangulation T^R of R such that for any vertex $\lambda \in T_o^R$ one has that $T_n \cap R^n \times \{\lambda\}$ is a triangulation of $R^n \times \{\lambda\}$. In this case λ_1, λ_2 will always denote two successive vertices in T_o^R, i.e. two successive λ-levels in T.

Now, from our discussion in IX and IV we have

(10.1.3) LEMMA

Let F and P be as above and let $\Gamma = R^n \times [\lambda^*, \infty)$, $\lambda_1 < \lambda^* < \lambda_2$. Assume that there exists $\Omega \subset R^n \times [\lambda_1, \lambda_2]$ open and bounded and such that

$$\partial\Omega \cap F^{-1}(0) = \phi .$$

Then for T sufficiently fine one has that

(i) $\deg(G_T(\cdot,\lambda_1),\Omega_{\lambda_1},0) = (-1)\deg(G_T(\cdot,\lambda_2),\Omega_{\lambda_2},0)$ and,

(ii) hence, if $\deg(F(\cdot,\lambda_1),\Omega_{\lambda_1},0) \neq 0$ then $S(F;P,\Gamma)$ creates artificial bifurcation in $[\lambda_1,\lambda_2]$, i.e. $\mathcal{M}_G \cap \partial\Omega \neq \phi$ for any T, which numerically is pursued by the chains of completely labeled simplices which correspond to G_T (see figure 22 and 23).

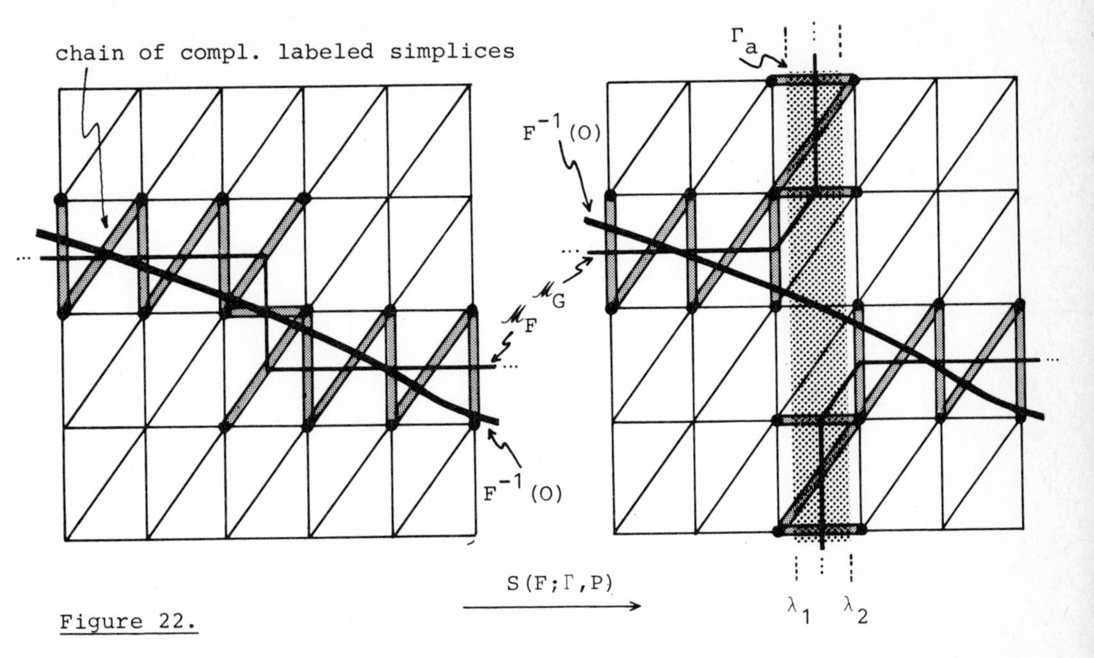

Figure 22.

WARNING: Parallel to our considerations in IX it is clear that again $\mathcal{M}_G$ is a collection of simplicial 1-manifolds, i.e. $\mathcal{M}_G$ is without bifurcation. The term artificial bifurcation in (10.1.3) refers merely to a nonlinear model for G_T which is given in (10.1.5).

We discuss a simplified example for the case of vector-labeling. A similar discussion is possible for integer-L-labeling.

(10.1.4) EXAMPLE

Choose $P = I(i) \in GL_-$, where

$$I(i) = \begin{pmatrix} 1 & & & & & \\ & \ddots & & & 0 & \\ & & 1 & & & \\ & & & -1 & & \\ & & & & 1 & \\ & 0 & & & & \ddots \\ & & & & & & 1 \end{pmatrix} \text{.........i-th row}$$

Let $[\underline{\lambda},\overline{\lambda}]$ be an interval containing $[\lambda_1,\lambda_2]$ and let $\Omega \subset R^n \times [\underline{\lambda}\ \overline{\lambda}]$ be open and bounded and such that $\mathcal{M}_F \cap \partial\Omega = \phi$ and $\mathcal{M}_F \cap \Omega$ is a one-to-one continuous curve $(x_T(\lambda),\lambda)$. Then we know that $\deg(F_T(\cdot,\lambda),\Omega_\lambda,\bar{\varepsilon}) \equiv \text{constant} = \pm 1$ for all

$\lambda \in [\underline{\lambda},\overline{\lambda}]$ ($\overline{\varepsilon} = (\varepsilon,\varepsilon^2,\ldots,\varepsilon^n)$, ε small). However, since $(PF)_T = PF_T$, we have that $\deg(G_T(\cdot,\lambda_1),\Omega_{\lambda_1},\overline{\varepsilon}) = (-1)\deg(G_T(\cdot,\lambda_2),\Omega_{\lambda_2}),\overline{\varepsilon})$ and, thus, it follows from (9.1.1) or (9.1.3) that $\mathcal{M}_G$ bifurcates off in $[\lambda_1,\lambda_2]$ vertically (see figure 23).

$\Gamma = R^n x\ [\lambda^*,\infty)$

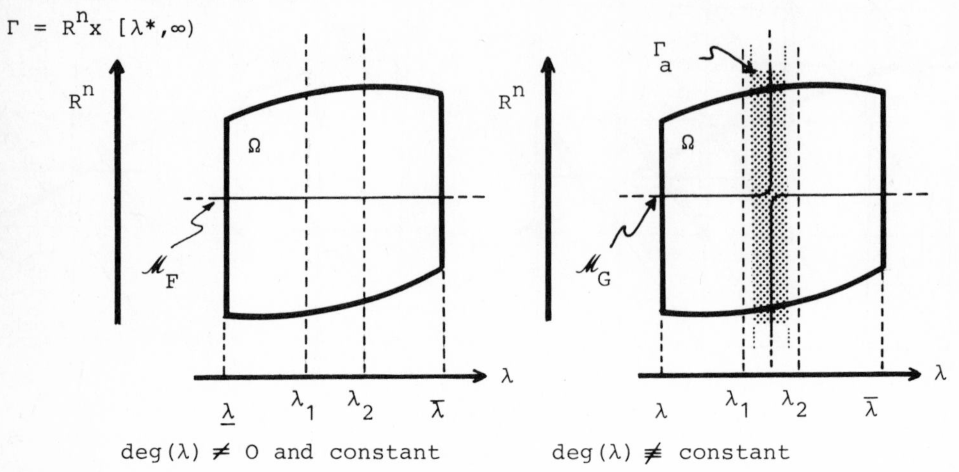

Figure 23.

Our forthcoming considerations will be to describe how to use the artificial "bifurcation branch" $\mathcal{M}_G \cap \Gamma_a$ of G_T to study $F^{-1}(0)$ in a numerical sense.

To understand what we have done so far we consider a nonlinear model G for G_T:

(10.1.5) MODEL FOR G_T

Let $P \in GL_-$ (e.g. $P = I(i)$). Choose a continuous curve $[\lambda_1,\lambda_2] \ni \lambda \mapsto P(\lambda)$, such that

$$\begin{cases} P(\lambda_1) = Id \\ P(\lambda_2) = P \\ P(\lambda) \in \begin{cases} GL_+ , & \lambda \in [\lambda_1,\lambda^*) \\ GL_- , & \lambda \in (\lambda^*,\lambda_2] \end{cases} \\ \dim \ker P(\lambda^*) = 1 \end{cases}$$

Assume that $F^{-1}(0) \cap R^n \times [\lambda_1,\lambda_2]$ is a set of disjoint curves

$$[\lambda_1,\lambda_2] \ni \lambda \mapsto (x_i(\lambda),\lambda)$$

and assume that F is Fréchet-differentiable and that

$$T_i(\lambda) := \frac{\partial F}{\partial x}(x_i(\lambda),\lambda) \in GL(R^n), \quad \text{i.e.}$$

$$\deg(F(\cdot,\lambda),\ x_i(\lambda),0) = \text{sign}\ \det T_i(\lambda).$$

We describe a nonlinear model G for G_T:

$$G(x,\lambda) = \begin{cases} F(x,\lambda) & , \quad \lambda \le \lambda_1 \\ P(\lambda)(F(x,\lambda)) & , \quad \lambda_1 \le \lambda \le \lambda_2 \\ P(F(x,\lambda)) & , \quad \lambda \ge \lambda_2 \end{cases}$$

(e.g. $P(\lambda) = \frac{\lambda-\lambda_1}{\lambda_2-\lambda_1} P + \frac{\lambda_2-\lambda}{\lambda_2-\lambda_1} \text{Id}$)

Our assumptions imply now: (see figure 24)

- F has no bifurcation in $[\lambda_1,\lambda_2]$;
- G has "vertical" bifurcations in $(x_i(\lambda^*),\lambda^*)$ according to (9.1.1) or (9.1.3)
- the artificial bifurcation branches are

  $$\mathscr{C}_a := G^{-1}(0) \cap R^n \times \{\lambda^*\} \neq \phi$$

  given by $\mathscr{C}_a = F^{-1}_{\lambda^*}(\ker P(\lambda^*))$ which are locally curves emanating from $(x_i(\lambda^*),\lambda^*)$ and which in a small neighborhood of these bifurcation points "look like"

  $$T_i(\lambda^*)^{-1}(\ker P(\lambda^*)) =: t_i \cdot R, \quad t_i \in R^n.$$

(10.1.6) PROBLEM: Is it possible to arrange artificial bifurcation such that disconnected continua become connected?

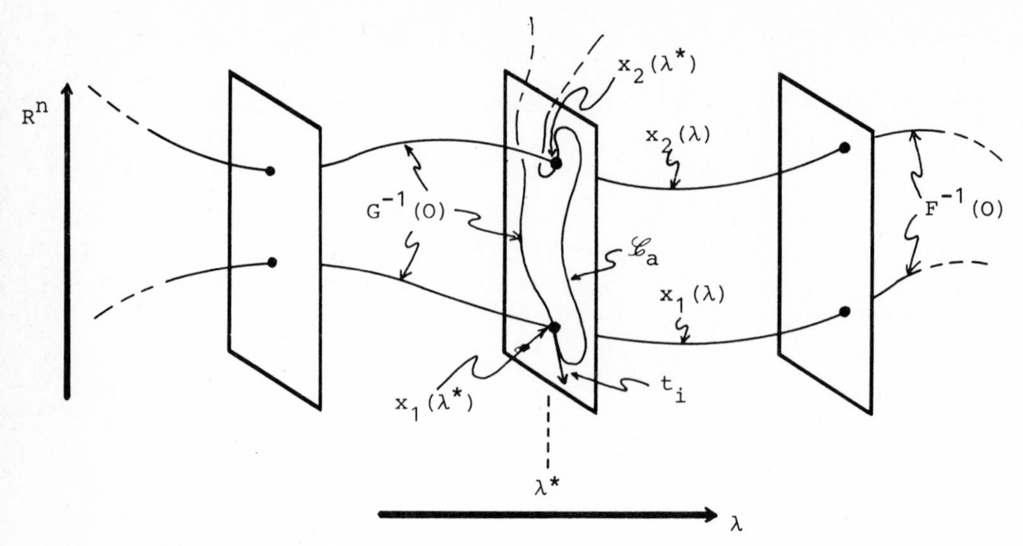

Figure 24.

(10.1.7) REMARK

In the case of a general Γ similar considerations as in (10.1.3) and (10.1.5) yield locally artificial bifurcation provided $F^{-1}(0)$ intersects Γ_a "transversal" and the solutions $F^{-1}(0) \cap \Gamma_a$ carry nonzero degree (see figure 25).

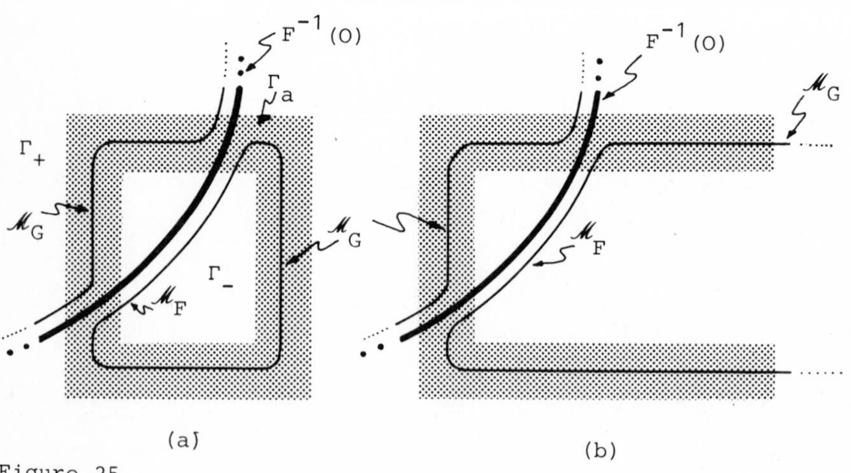

Figure 25.

(10.1.8) REMARK (Global properties of artificial bifurcation)

Globally artificial bifurcation will highly depend on the choice of Γ :

(i) If Γ is unbounded (see figures 21 a,b and c) then $\mathcal{M}_G$ may have unbounded components in Γ_a. This effect will be exploited to construct artificial "connections" between disjoint continua of $F^{-1}(0)$.

(ii) If Γ is bounded (see figure 21 d) then $\mathcal{M}_G$ may contain homeomorphs of S^1 (see figure 25 a) which are a consequence of artificial bifurcation. However, in that case, any component of $\mathcal{M}_G \cap (\Gamma_+ \cup \Gamma_-)$ (which approximates solutions of $F^{-1}(0)$) which enters Γ_a must leave again Γ_a (i.e. "real" solutions in $\mathcal{M}_G$ cannot vanish into "artificial" solutions in $\mathcal{M}_G \cap \Gamma_a$). This effect will be exploited for the complete study of bifurcation.

(10.2) ARTIFICIAL CONNECTIONS, BIFURCATION FROM ∞

Assume that we have situations as indicated in figure 26.

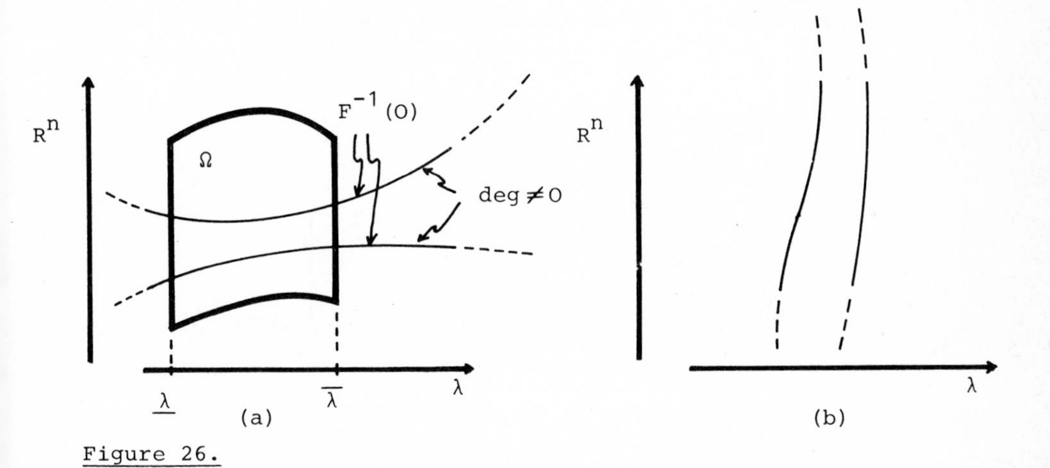

Figure 26.

(10.2.1) We discuss the situation in figure 26 a in details:

Let $\Omega \subset R^n \times [\underline{\lambda},\bar{\lambda}]$ be a connected open and bounded subset such that $F^{-1}(0) \cap R^n \times [\underline{\lambda},\bar{\lambda}]$ consists locally of two continuous curves

$$[\underline{\lambda},\bar{\lambda}] \ni \lambda \begin{array}{c} \nearrow \\ \searrow \end{array} \left.\begin{array}{c} (x_1(\lambda),\lambda) \\ (x_2(\lambda),\lambda) \end{array}\right\} \in \Omega \qquad \text{both carrying nontrivial degree}$$

Then $\deg(F_\lambda,\Omega_\lambda,0) \equiv$ constant for all $\lambda \in [\underline{\lambda},\bar{\lambda}]$. We want to use artificial bifurcation between two successive λ-levels $\lambda_1,\lambda_2 \in (\underline{\lambda},\bar{\lambda})$ in a given triangulation to connect $x_1(\lambda)$ with $x_2(\lambda)$ (see also figure 24), i.e. we choose

$$\Gamma = R^n \times [\lambda^*,\infty)$$

with $\lambda_1 < \lambda^* < \lambda_2$ and T parallel to R^n. Using a switch which is indicated in the forthcoming figures by Ⓢ we can have the following cases (see figure 26) passing from F_T to G_T:

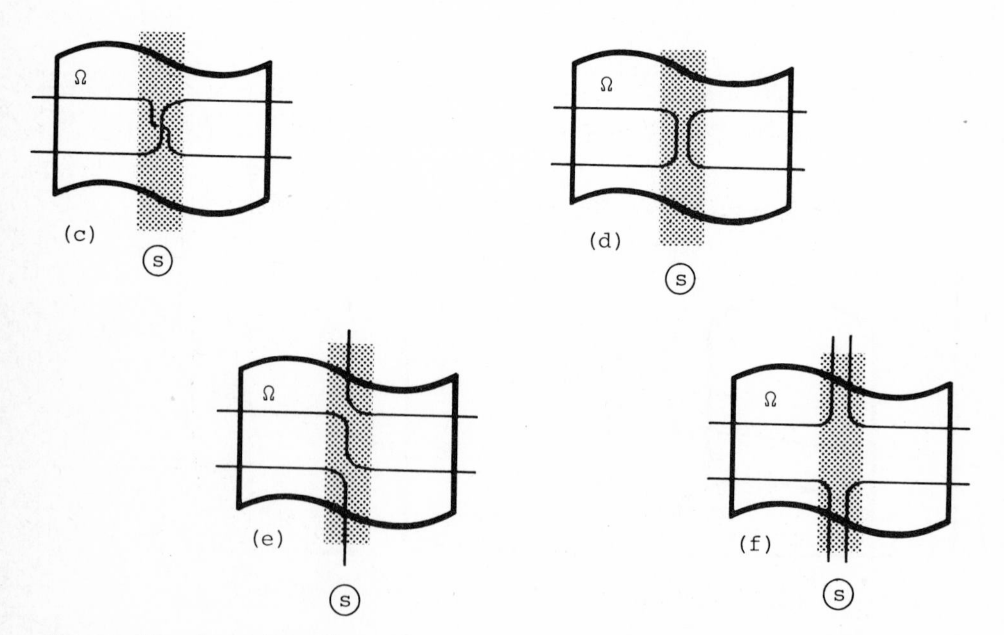

Figure 26. (continued)

Cases (c),(d) provide a perfect connection between the continua $x_1(\lambda)$ and $x_2(\lambda)$. This is possible, however, only if $\deg(F_\lambda,\Omega_\lambda,0) \equiv 0$ and if the local degrees of $x_1(\lambda)$ and $x_2(\lambda)$ are different from zero, thus,add up to be zero. Then one can prove cases (c),(d) provided $G_T^{-1}(0) \cap R^n \times [\lambda_1,\lambda_2]$ is bounded (a priori bounds for $G_T(x,\lambda) = 0$) analoguous to theorem (8.1.2).

In case $\deg(F_\lambda,\Omega_\lambda,0) \neq 0$ any switch $S(F;\Gamma,P)$ will create bifurcation through $\partial\Omega$ for G_T because then $\deg(G_T(\cdot,\lambda),\Omega_\lambda,0) \not\equiv$ constant. This is typical for cases (e) and (f).

(10.2.2) In a situation which is indicated in figure 26 b the following choice for Γ is appropriate (cf. figure 21 b):

Let H be some half-space in R^n and set

$$\Gamma = H \times R$$

Again one can obtain a perfect connection similar to 26 a. But also an effect parallel to 26 e and 26 f is possible.

(10.2.3) However, it is possible to avoid the unpleasant effects of 26 e and 26 f a priori by choosing Γ a bounded cube in R^{n+1} as indicated in figure 21 d. Then according to (10.1.8) the artificial branches are always bounded (Γ_a is bounded). To obtain connections between disjoint continua it is necessary, however, to guess an appropriate cube (see figure 27).

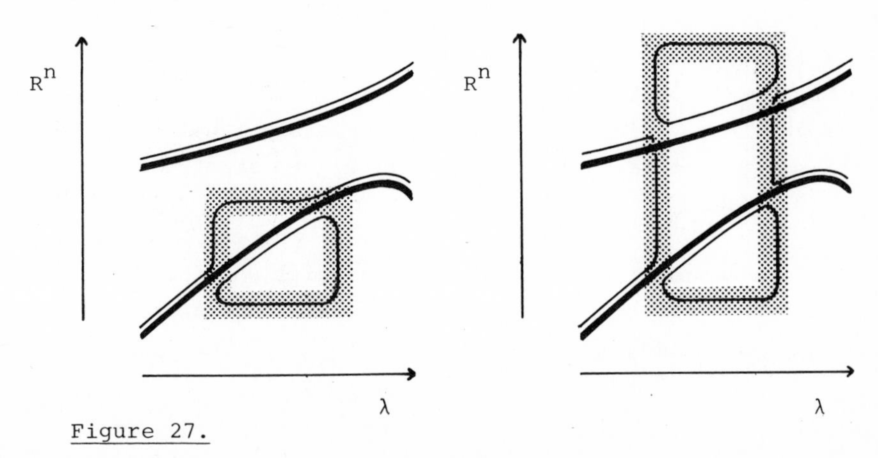

Figure 27.

(10.2.4) NUMERICAL EXAMPLES

A numerical realization of artificial bifurcation is simply obtained by

vector-labeling: replace Λ by $\hat{P}\Lambda$, $\hat{P} = \begin{bmatrix} 1 & 0 \\ 0 & P \end{bmatrix}$

integer-L-labeling: replace L by $P^{-1}L$.

Most of our numerical experiments simply used switches of type $P = I(i)$. In view of the discussion of this section the following two-point boundary value problems have been solved using vector-labeling and a triangulation by reflection [5].

$$\text{BVP} \quad \begin{cases} u'' + \lambda f(u) = 0 \\ u(0) = u(a) = 0\,, \quad a > 0\,. \end{cases}$$

BIFURCATION FROM ∞

A. Ambrosetti and P. Hess [9] by use of degree methods recently have proved bifurcation from "0" and "∞" of positive solutions for BVP provided f is a nonlinearity as indicated in figure 28 a (λ_1 denotes the first eigenvalue of the corresponding linear BVP) and $a = \pi$.

Figure 28 c shows how artificial bifurcation has created a connection between the branches emanating from "0" and "∞" and, thus, has made both of them computable.More precisely, the algorithm proceeds as follows:

- BVP for $\lambda = 0$ provides a unique completely labeled simplex approximating the trivial solution (0,0) to start a chain (cf. (9.1.4));

- the chain started like so follows the trivial solutions until it bifurcates off from "0" at λ_0 and then approximates nontrivial solutions; ($\lambda_0 = \lambda_1/f'(0)$)

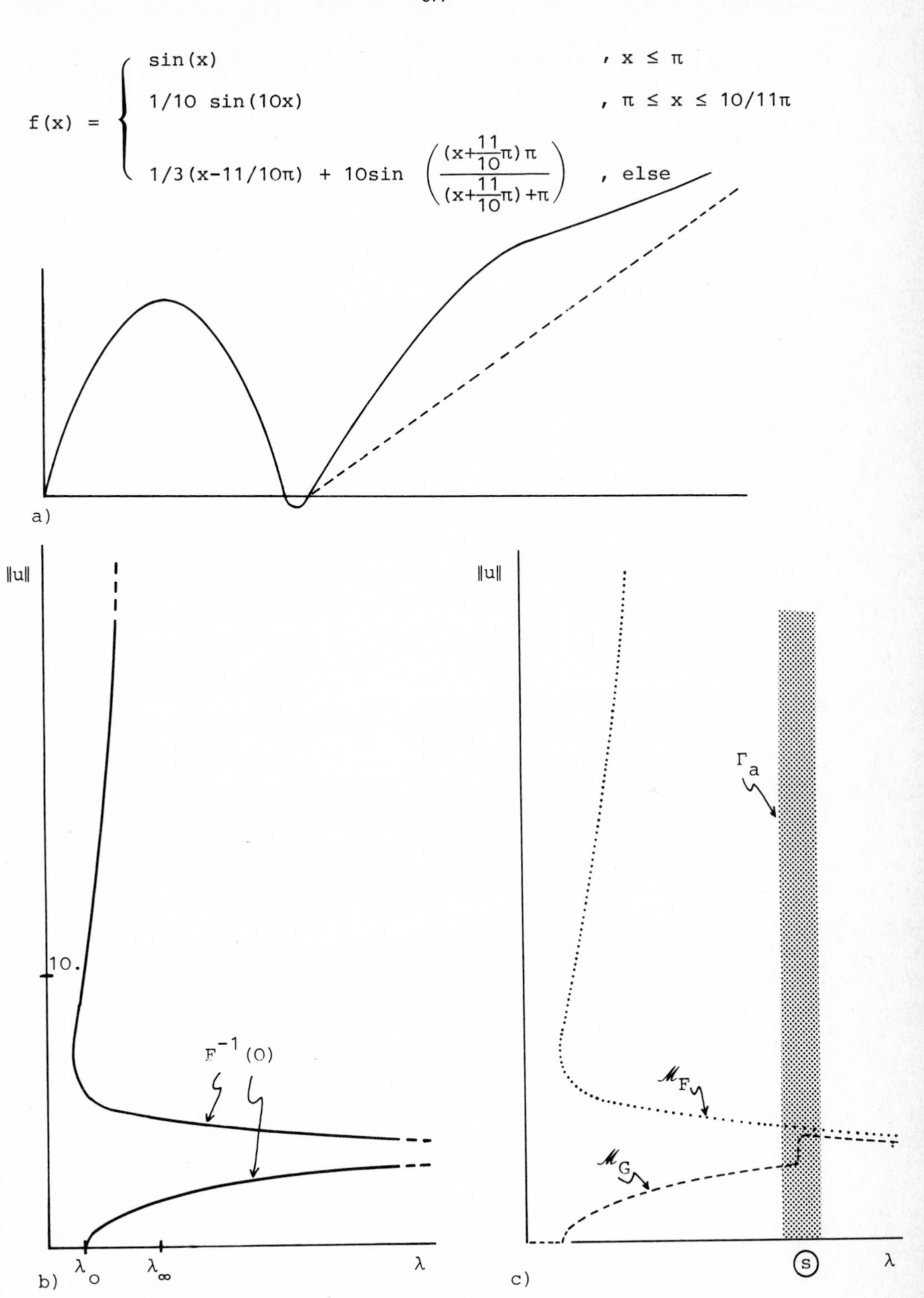

f(x) = sin(x) , x ≤ π
1/10 sin(10x) , π ≤ x ≤ 10/11π
1/3(x-11/10π) + 10sin ((x+11/10π)π / (x+11/10π)+π) , else
a)
‖u‖
10.
F⁻¹(O)
λo
λ∞
λ
b)
‖u‖
Γa
MF
MG
s
λ
c)

Figure 28.

- switching artificial bifurcation on at Ⓢ with $P \in GL_-$ and Γ as in figure 21 a one creates an artificial connecti[on] towards the branch emanating from "∞" and then one follows the "∞"-branch according to the different possibilities in figure 26 c,d;

- to obtain also the solutions on the "∞"-branch beyond the switching point one distinguishes two cases:

  (a) the chain is in the unswitched region Γ_+: then one "reverses" the chain and switches off artificial bifurcation at Ⓢ;

  (b) the chain is in the switched region Γ_-: then one "reverses" the chain and one switches on artificial bifurcation at Ⓢ with the same P but with $c(\Gamma)$, i.e. one simply passes from F to PF also in Γ_+.

NON BIFURCATING CONTINUA

A. Ambrosetti and P.H. Rabinowitz [10] using variational methods have proved the existence of infinitely many solutions for each $\lambda > 0$ provided the nonlinearity is of type $f(u) = u^3$, $a = \pi$. Figure 29 a shows what one can expect for $F^{-1}(0)$ from their paper. Figure 29 b indicates the trace of

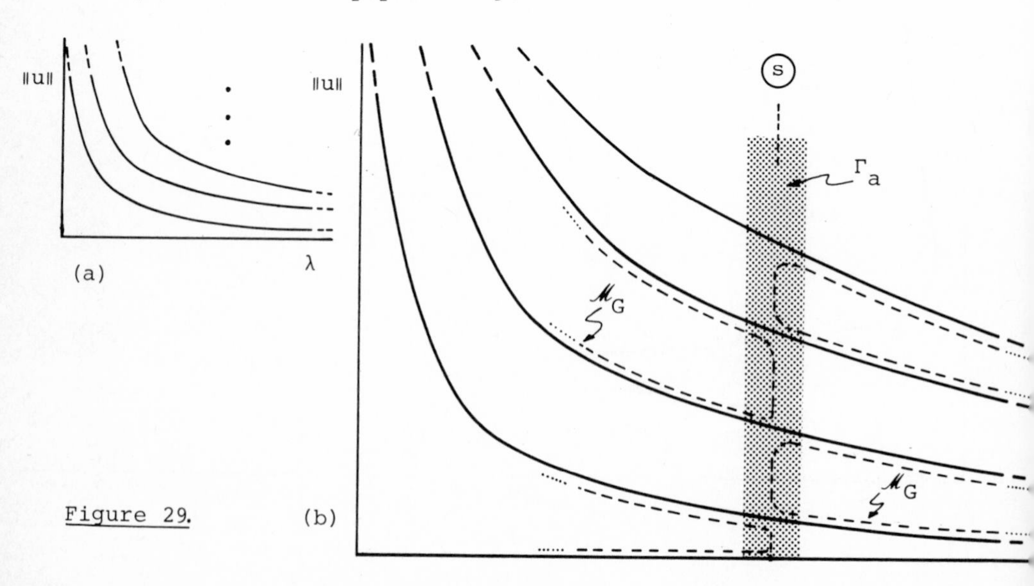

Figure 29.

the numerical procedure and the switching point.

Our numerical results together with the topological background of our procedures suggest that $F^{-1}(0)$ actually is a family of continua each of which carries nonzero degree.
The differential equations have been discretized on 12 points, i.e. the problems have been computed in R^{12}. The mesh size for T has been varied between 0.05-0.001. All possible switches S of type I(i) have been tested and, surprisingly, it turned out that in all numerical examples studied here and in later paragraphs almost every artificial bifurcation created an artificial connection to some "new" continuum.
Figures 28-29 indicate that artificial bifurcation is indeed an appropriate numerical tool to connect disconnected continua in $F^{-1}(0)$ in order to make them accessible for our algorithms which are designed to follow continua. However, it might be necessary to test with several choices $P \in GL_-$ to find an artificial connection. Numerical experience has suggested that when working with artificial bifurcation one can have a simple indicator from $||F(x,\lambda)||$ being small or large ((x,λ) in the current completely labeled simplex) to decide whether one is approximating "real" or "artificial" solutions in Γ_a.

(10.3) SECONDARY BIFURCATION

Here we restrict attention to bifurcations of the following type (supercritical case)

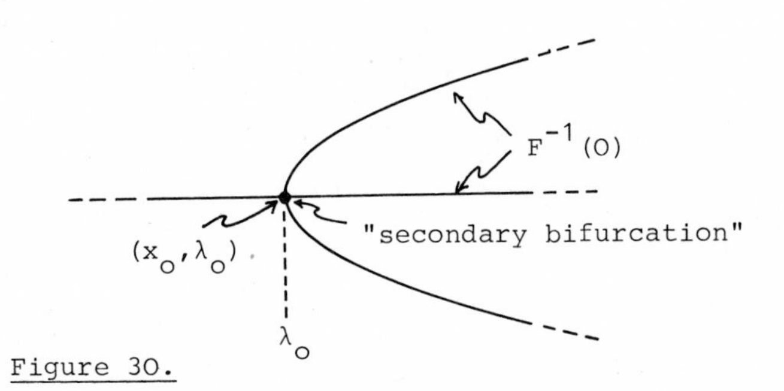

Figure 30.

We assume that every solution (except at bifurcation points) carries nonzero degree. Bifurcations of the types given in figure 31 can have a similar analysis.

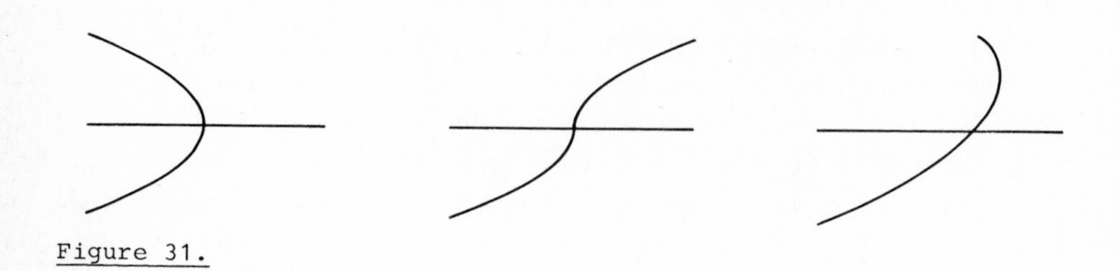

Figure 31.

In figure 32 we graph once again the simplicial 1-manifolds $\mathcal{M}_F$ ($\oplus,\ominus$ represent examples of possible local orientation, i.e. degree):

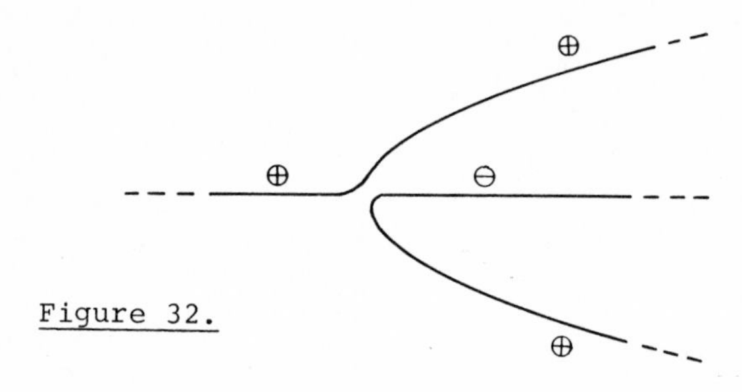

Figure 32.

(10.3.1) PASSING SECONDARY BIFURCATION

Here we discuss how to use artificial bifurcation such that the algorithm "ignores" secondary bifurcation, i.e. continues to follow the "known" solutions: The idea is to switch on artificial bifurcation with Γ as in figure 21 a in an interval $[\lambda_1,\lambda_2]$ where $\lambda_o < \lambda_1$.

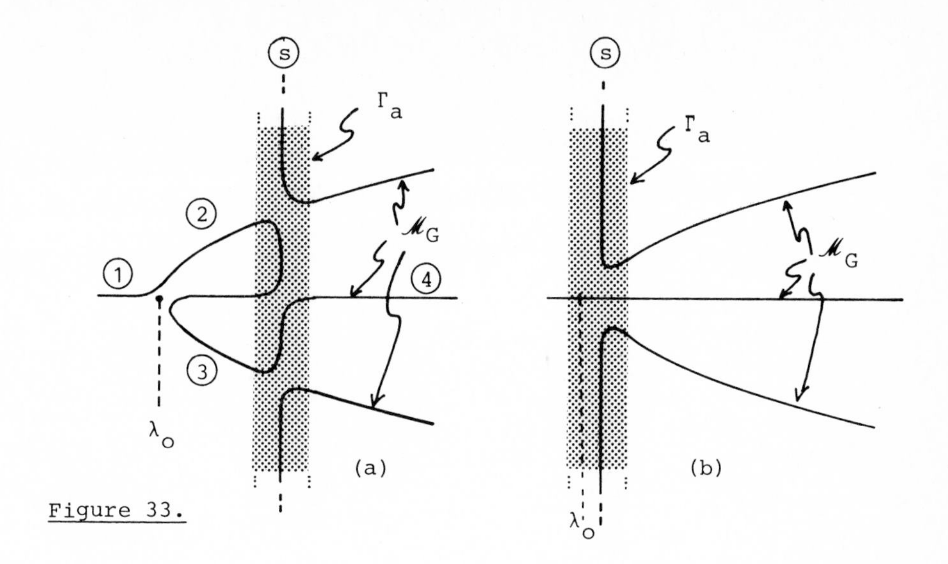

Figure 33.

Different choices of $P \in GL_-$ will provide various artificial bifurcation types. However, each of them must necessarily have bifurcation "towards ∞" provided the degrees are as in figure 32 or vice versa, i.e. the total sum is different from zero. Comparing figures 32 and 33 a it is shown how artificial bifurcation can link "known" solutions. In a sense figure 33 b is the limit case of figure 33 a, i.e. λ_o , the bifurcation point, is in the switching interval $[\lambda_1, \lambda_2]$ and to this effect we refer as to "passing secondary bifurcation". Actually, numerical experience with many examples has verified that the effect described above is manageable. However, to perform this technique one should know (x_o, λ_o) the point of secondary bifurcation.

Numerically, secondary bifurcation is indicated by the following changes (see figure 32), none of which is necessary, of course:

- drastic change of the norm of solutions $(x(\lambda), \lambda)$;
- $\frac{\partial F}{\partial x}(x(\lambda), \lambda)$ becomes singular (provided F is differentiable)
- drastic change of the structure of $x(\lambda)$ (symmetry change, node change, e.t.c.).

(10.3.2) HOW TO FIND ALL BRANCHES

Having found a switch which creates artificial bifurcation along figure 33 a one can easily approximate three different branches:

Assume the procedure is in position ①, ② or ③ (see figure 33 a):

①: then "switch on" provides

②: then "switch off" provides

③: then "switch off" provides

However, it might be advantageous, especially in presence of more branches of different degrees to insert more switches. The most interesting and important cases which might occur are given in figure 34. All of them are a combination of figure 33 b and a second switch to the right of the secondary bifurcation point.

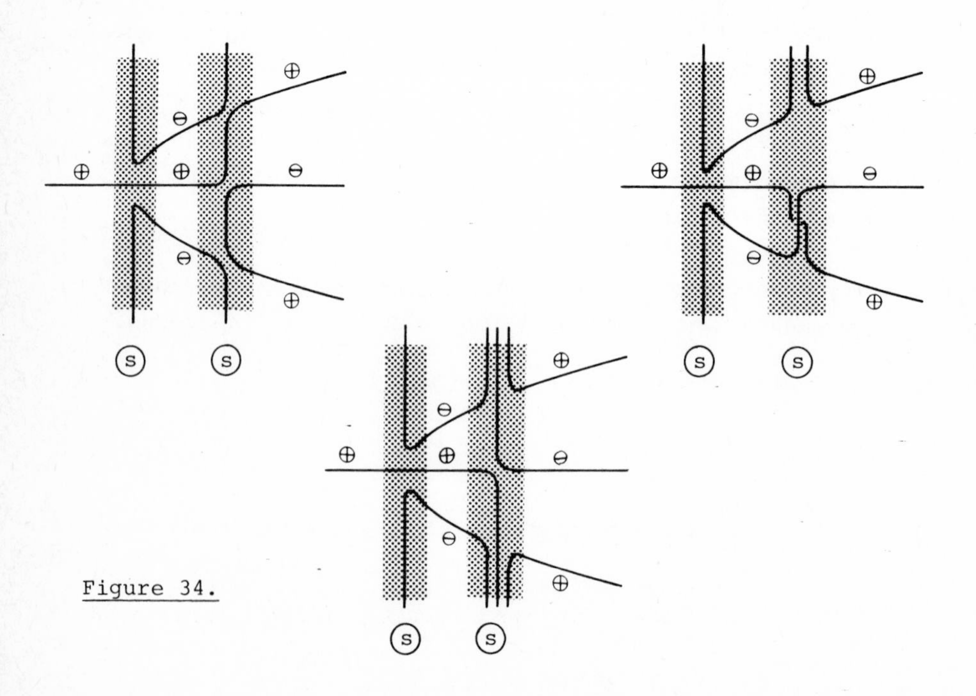

Figure 34.

So far we have used unbounded $\Gamma \subset R^n \times R$ to create artificial bifurcation in the study of secondary bifurcation. Finally, we briefly describe the effect of using the technique when Γ is <u>bounded</u>. This might be advantageous for several reasons:

- artificial solutions are bounded;
- precise knowledge about the secondary bifurcation point is not necessary;
- the technique is independent of the possible "directions" of the bifurcating branches.

Figure 35 suggests that it is possible performing a suitable "switch on" "switch off" technique to obtain all branches ① - ④ from one single chain. In a sense, using a bounded Γ which contains the secondary bifurcation point (x_o,λ_o) creates an effect which can be understood as a "blow up" of (x_o,λ_o).

(10.3.3) NUMERICAL EXAMPLES

Two types of bifurcation problems have been considered:

- bifurcations which are characterized by a change of degree (bifurcation in the sense of Krasnosel'skii and Rabinowitz)
- bifurcations in presence of eigenvalues of even multiplicity (no change of degree).

For the first type of problem we have studied two-point boundary value problems

$$\text{BVP} \quad \begin{cases} u'' + \lambda f(u) = 0 \\ u(0) = u(\pi) = 0 \end{cases}$$

$$\text{with} \quad \begin{aligned} f_1(u) &= \sin(u) \\ f_2(u) &= u + u^3 \end{aligned}$$

and as a model problem the equation $F(x,\lambda) = 0$ (see [58], p. 81)

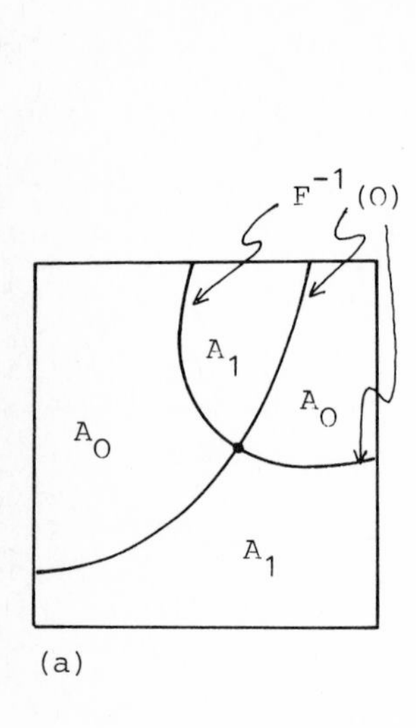

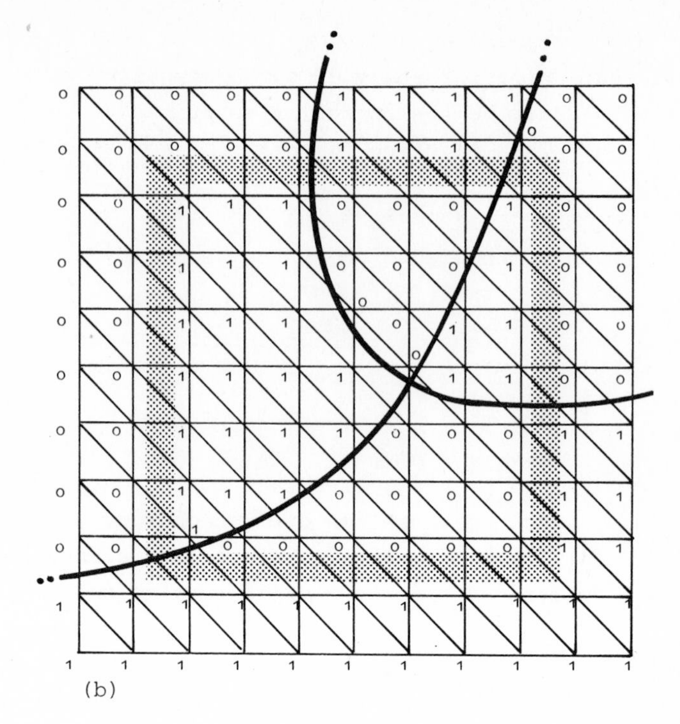

Figure 35.

a) Labeling sets A_0, A_1 in the neighborhood of a bifurcation point.

b) The labeling after switching - Γ_a = [shaded].

c) The simplicial one-manifold $\mathcal{M}_G$ approximates all bifurcating branches.

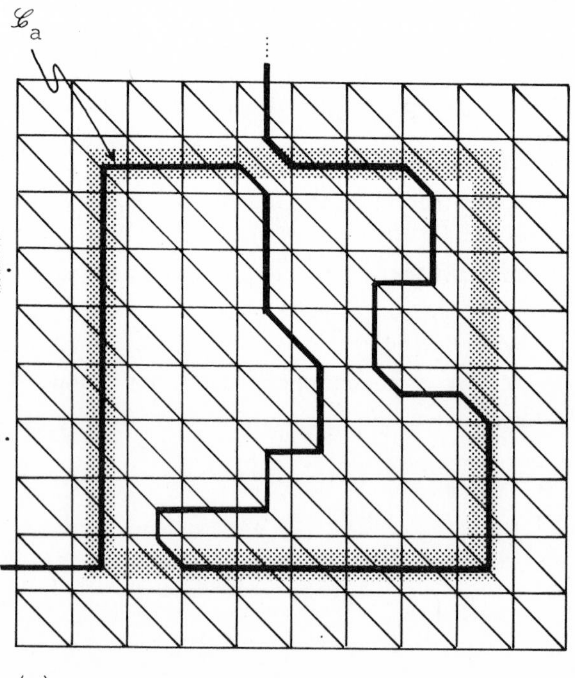

$$F: R^2 \times R \to R^2$$

$$F((x_1,x_2),\lambda) := (16x_1 - \lambda x_1 + 12x_1^3 + 24\, x_1 x_2^2,\ 12x_2 - \lambda x_2 + 9x_2^3 + 18x_2 x_1^2)$$

The problem $F(x,\lambda) = 0$, as one easily verifies, has the following solutions: $F^{-1}(0)$ has bifurcation from zero in

$$\lambda = 12 \quad \text{and} \quad \lambda = 16$$

and the bifurcating branches are given by

$$\text{I}: \quad x_1 = 0, \quad x_2^2 = 1/9\ (\lambda - 12)$$

and

$$\text{II}: \quad x_1^2 = 1/12\ (\lambda - 16), \quad x_2 = 0$$

respectively. But for $\lambda > 24$ we also have the non-trivial solution

$$x_1^2 = 4/9\ (\frac{5\lambda}{48} - 1), \quad x_2^2 = \frac{4}{9}\ (\frac{\lambda}{24} - 1)\ .$$

The latter solution bifurcates from the solution curve II, thus, we have secondary bifurcation in $((\sqrt{\frac{2}{3}},0),24)$.

Figures 36-38 show the numerical results and switching points for artificial bifurcation. The BVP has been discretized on 12 points. The mesh size for T was varied between $1/20$ - $1/100$.

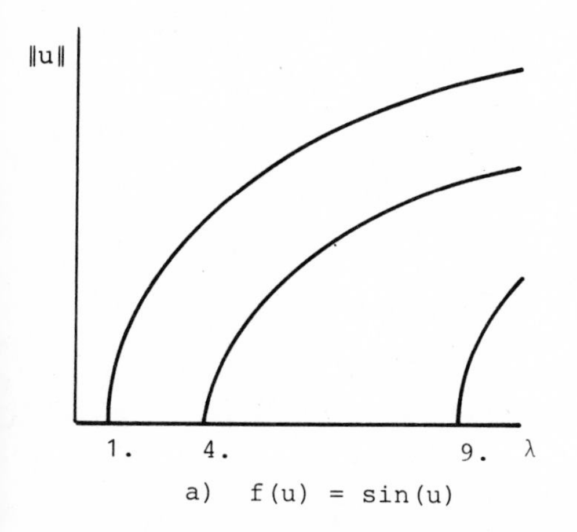

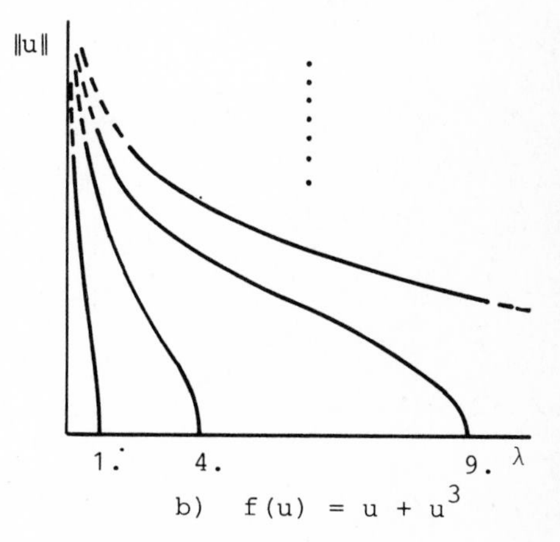

Figure 36. Continua of solutions of (BVP)

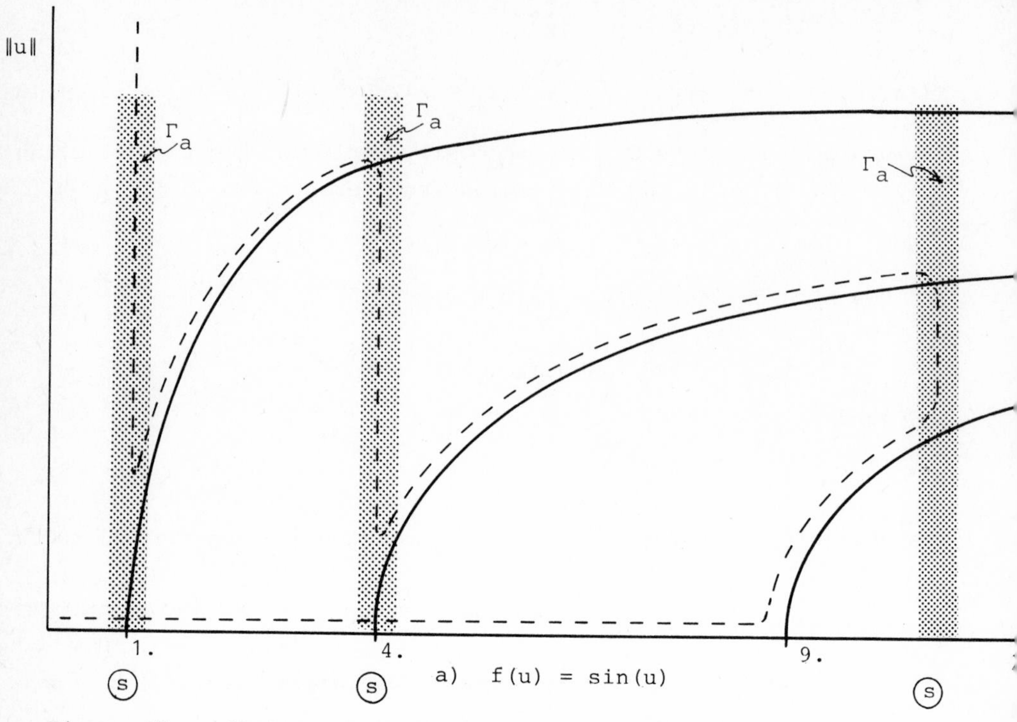

Figure 37. Chains approximating solutions of (BVP)

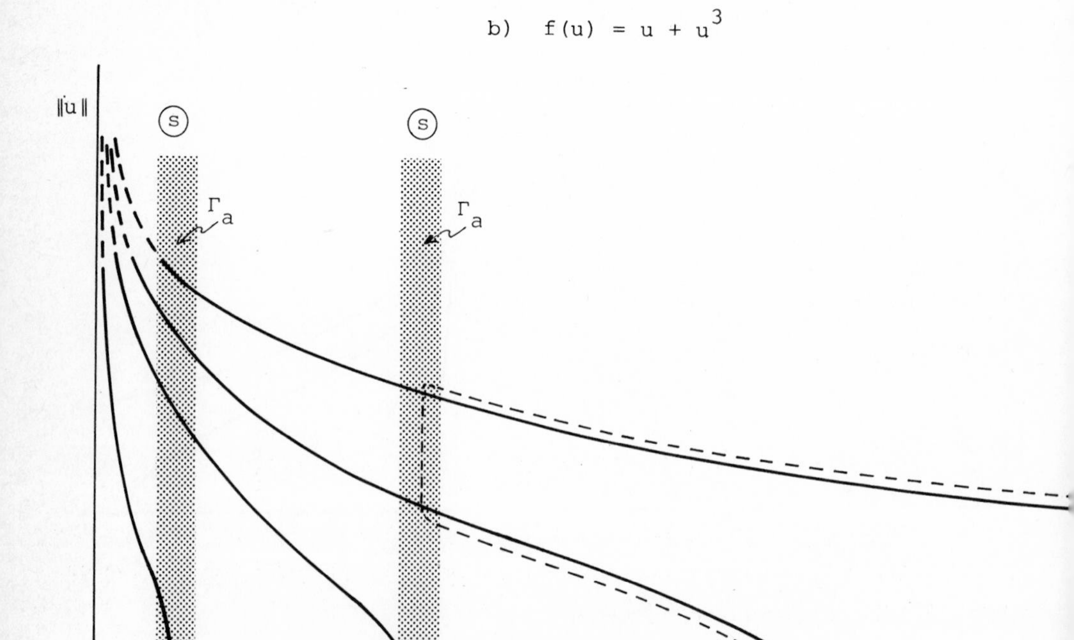

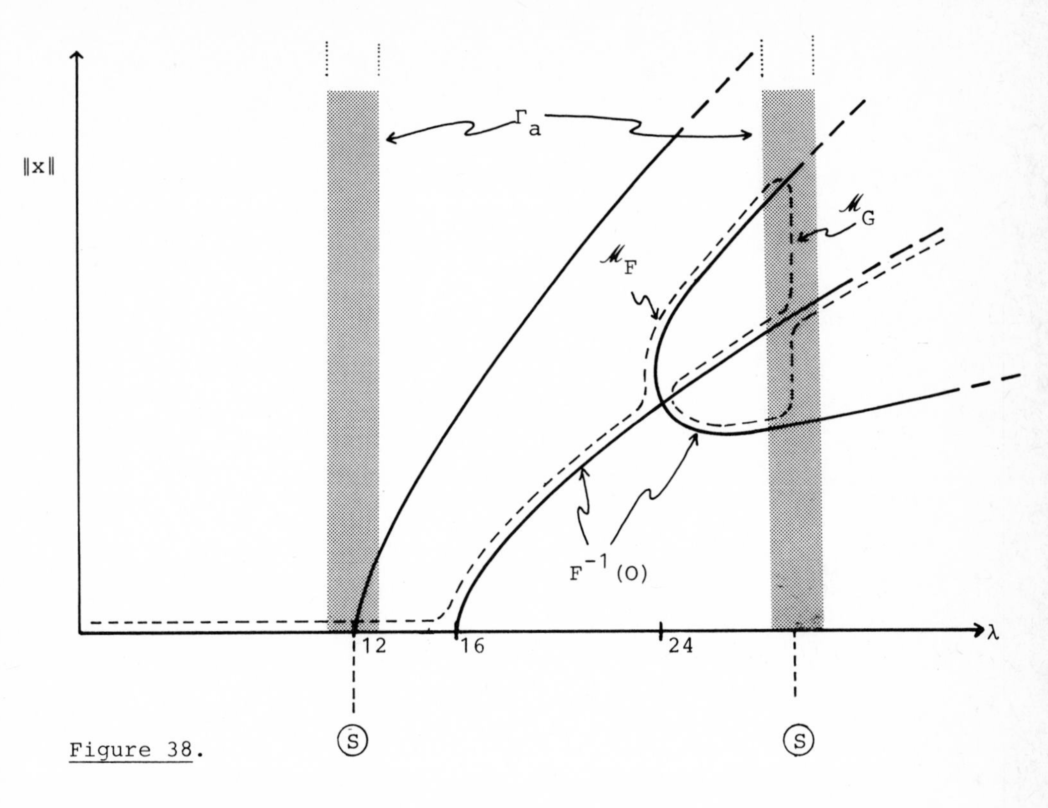

Figure 38.

For the second type of problem we considered two model problems (see [58], p. 83):

$$F: R^2 \times R \to R^2$$

$$F(x,\lambda) = 0$$

with , $a > 1$,

$$F_1((x_1,x_2),\lambda) = (x_1-\lambda x_1+ax_1(x_1^2+x_2^2),\ x_2-\lambda x_2+x_2(x_1^2+x_2^2))$$

and

$$F_2((x_1,x_2),\lambda) = (x_1-\lambda x_1+2x_1x_2,\ x_2-\lambda x_2+x_1^2+2x_2^2)$$

One observes that in both examples $\lambda = 1$ is the only eigenvalue of the linearized equation, thus , $\lambda = 1$ is an eigenvalue of <u>even</u> multiplicity.

However, in both cases $F^{-1}_{1,2}(0)$ has bifurcation from zero in $\lambda = 1$. The solution curves are given by:

$$F_1^{-1}(0): \quad \begin{cases} x_1^2 = \frac{1}{a}(\lambda-1), \; x_2 = 0 \\ x_1 = 0, \quad x_2^2 = \lambda-1. \end{cases}$$

$$F_2^{-1}(0): \quad x_1 = 0, \; x_2 = \frac{1}{2}(\lambda-1) \; .$$

The indices of the different solution curves are given in figures 39 and 40 . Observe that typically in both cases there is <u>no change of the index</u> in the bifurcation point.

Figures 39 - 40 show numerical results and switching points for artificial bifurcation. Most surprisingly these two examples sugges

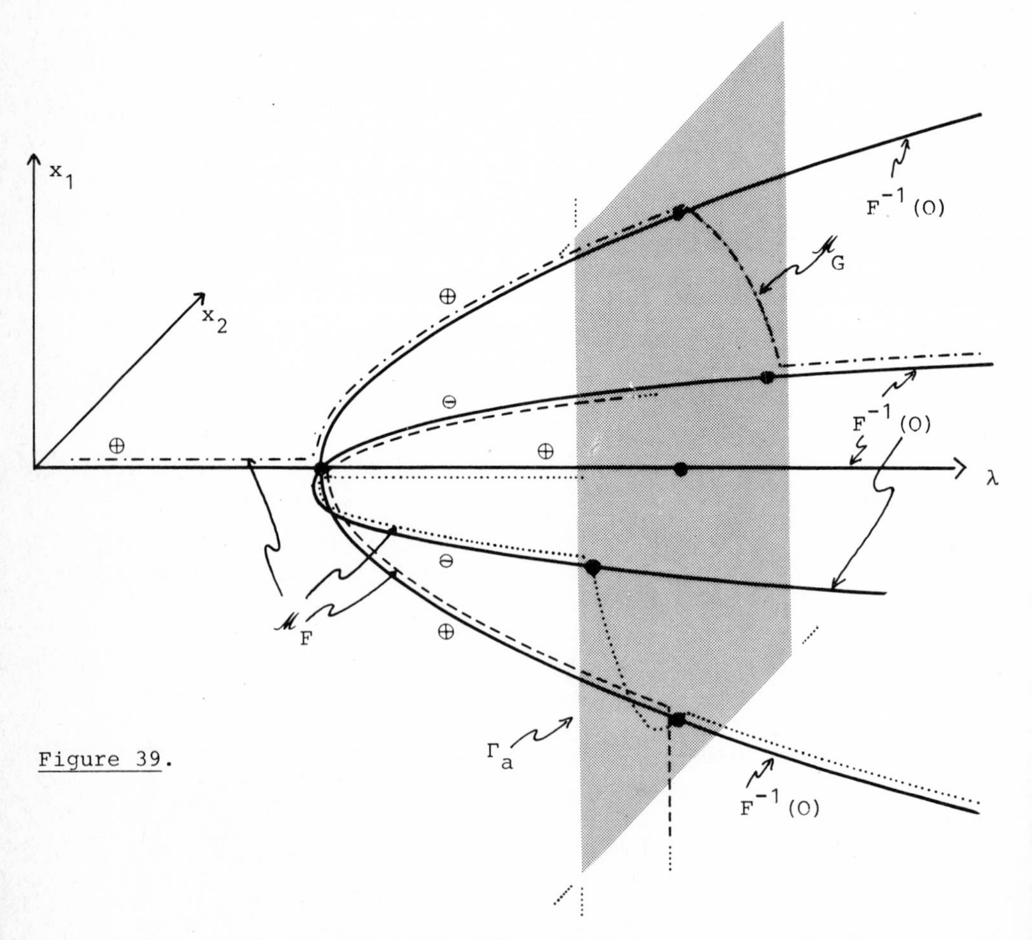

Figure 39.

that our numerical procedures to study bifurcation problems are not restricted to problems which are characterized by a change of degree.

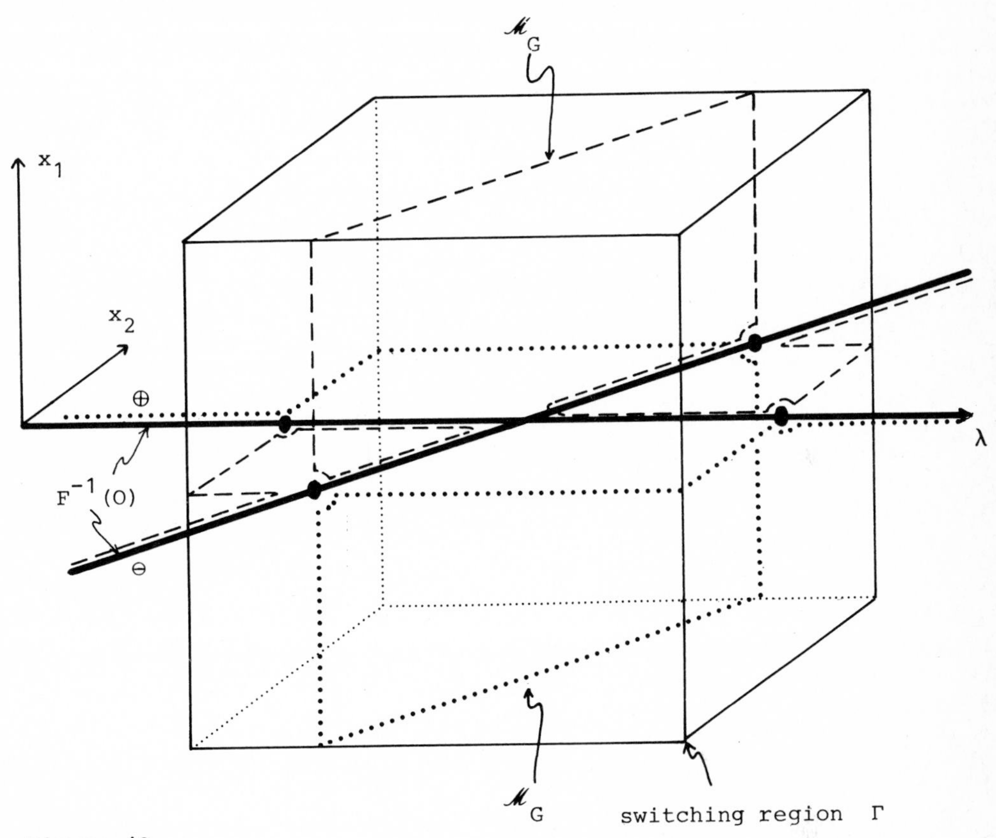

Figure 40.

XI. Integer-Labeling versus Integer-L-Labeling and Coincidence Problems

In the study of nonlinear problems one often has the alternative to consider the problem as a coincidence problem

$$\mathcal{L}(x) = \mathcal{N}(x,\lambda)$$

or a fixed point problem

$$x = \mathcal{L}^{-1}\mathcal{N}(x,\lambda)$$

(assume that $\mathscr{L}$ is a linear isomorphism and $\mathscr{N}$ is nonlinear)

In both cases in order to apply our numerical procedures, one has to put them into the form of a zero problem $F(x,\lambda) = 0$. From the definition of vector-labeling and the definition of a completely labeled simplex in that context it follows trivially that both problems are equivalent. However, for integer-labeling this is not the case.

(11.1) TWO EXAMPLES

In this paragraph we discuss two numerical examples to demonstrate the effect of introducing integer-L-labeling.

We consider the two-point boundary value problem

$$\text{(BVP)} \quad \begin{cases} u'' + f(u) = 0 \\ u(0) = u(a) = 0 \end{cases}$$

where $a > 0$ is fixed. Using central-difference approximations on a grid of $(n+2)$ equidistant points in $[0,a]$ one can easily put (BVP) in

$$(\text{BVP})_{\text{appr}} \quad F(x,\lambda) = 0$$

where $F(x,\lambda) = \mathscr{L}(x) + \mathscr{N}_f(x,\lambda)$ and
$\mathscr{L}: R^n \to R^n$ is a linear isomorphism and
$\mathscr{N}_f: R^n \times R \to R^n$ is continuous and nonlinear

We restrict attention to two specific examples:

$$f_1(u) = \exp(u), \quad a_1 = 1$$

$$f_2(u) = \sin(u), \quad a_2 = \pi$$

It is known that $(\text{BVP})_2$ has bifurcation for all $\lambda \in \{n^2 : n \in N\}$. Moreover, all solutions of $(\text{BVP})_1$ are symmetric with respect to $t = 1/2$. The same is true for $(\text{BVP})_2$ for $\lambda \leq 4$ ant $t = \pi/2$. This symmetry is used to provide 2-dimensional pictures for the labeling sets A_i of F_λ.

Figure 41 shows the "classical" integer-Id-labeling sets $A_i(F_\lambda, Id)$ for typical values of λ :

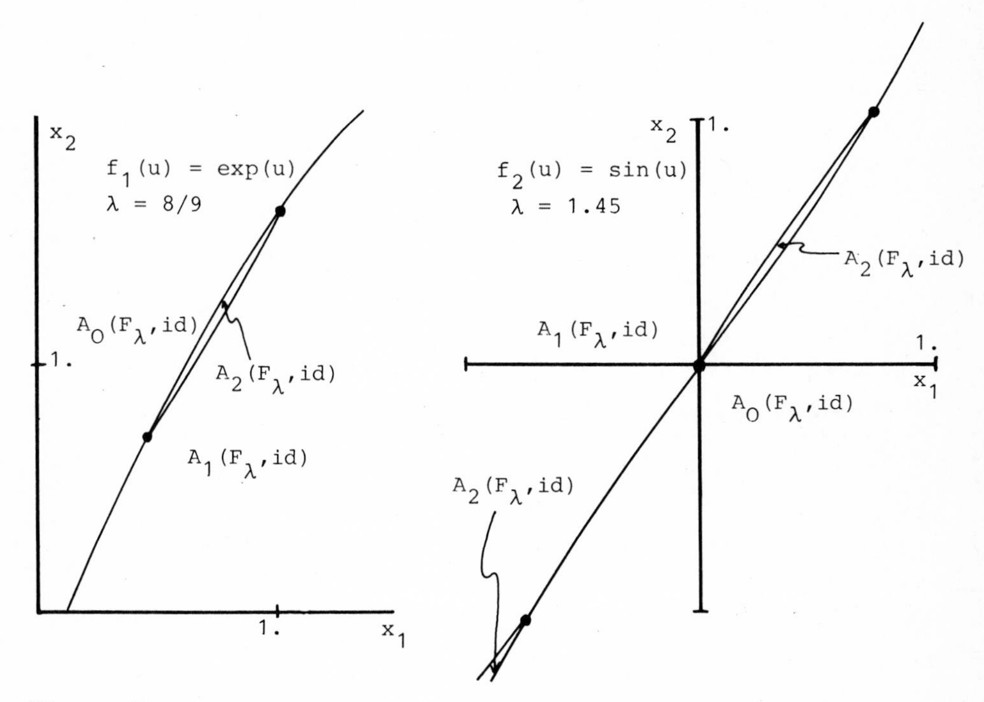

Figure 41.

In both cases we observe that A_2 has the skinny shape discussed in example (6.1.1) and this means that one can find completely labeled simplices in large distance from zeros of F even if the mesh size of a triangulation is small. Furthermore, if σ is a completely labeled simplex in some λ-level of a triangulation of $R^n \times R$ it will turn out that copies of σ in the λ-levels nearby are also completely labeled. This will have an effect for the chains of completely labeled simplices as shown in figure 42 and this has been observed by computer experiments.

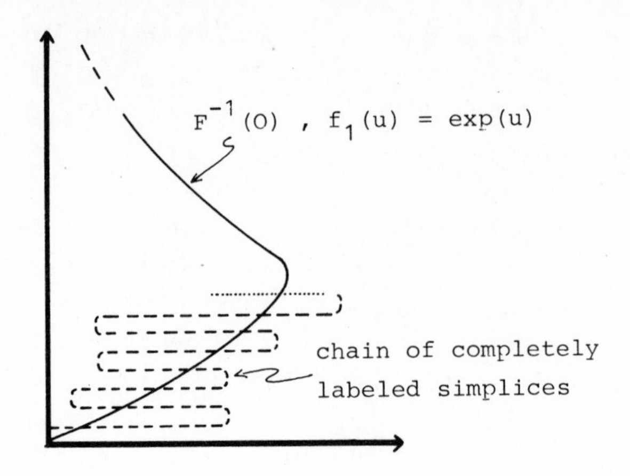

Figure 42.

In figure 43 we show the labeling sets A_i for integer-L-labeling with

$$L = \mathscr{L}$$

i.e. the linear differential operator is incorporated into the labeling process.

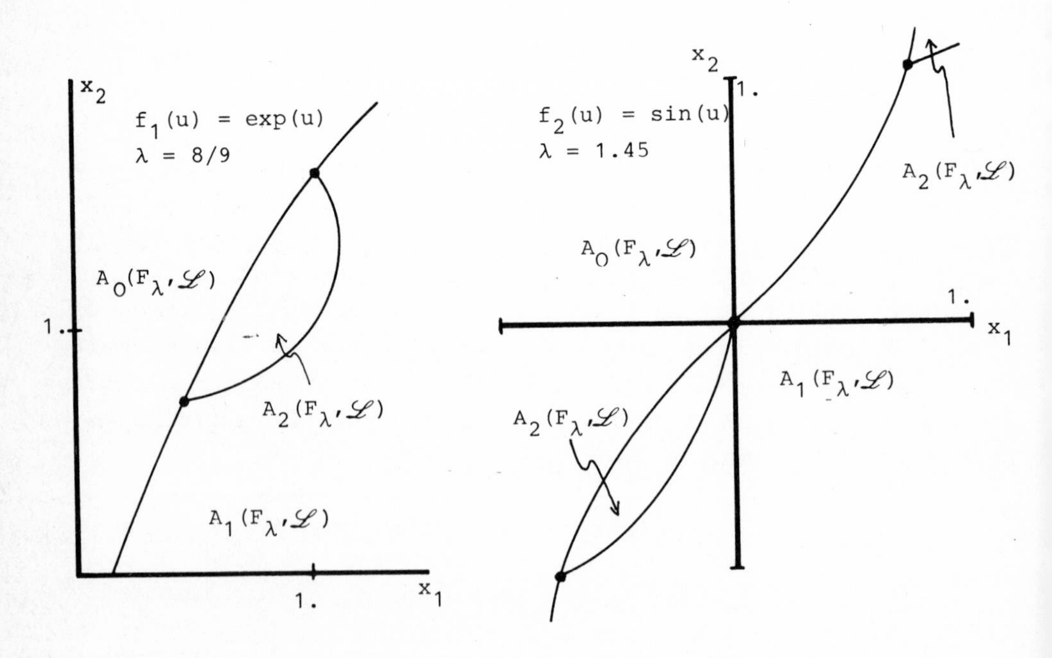

Figure 43.

Obviously, the geometry of the labeling sets has improved considerably in view of our simplicial algorithms. In fact, computer experiments have verified that integer-L-labeling with $L = \mathcal{L}$ generates efficient algorithms to approximate solutions of boundary value problems. Both for vector- and integer-L-labeling the fact has been used that $F(\cdot,0)$ is a linear problem; thus, in view of (9.1.4) the start for an emanating chain was easy to find. As for a comparison between integer-L-labeling and vector-labeling computer experiments verified that Door-In/Door-Out steps are easier and faster (by a factor of 0.2) to perform with integer-L-labeling than with vector-labeling. However, to approximate a given piece of a continuum vector-labeling needs much less (by a factor of 0.1) Door-In/Door-Out steps than integer-L-labeling.

(11.2) COINCIDENCES

A number of nonlinear problems (e.g. nonlinear boundary value problems with non-invertible left hand side) can be studied more adequately as coincidence problems, see [25]. There, instead of Leray-Schauder degree, a coincidence degree for infinite dimensional spaces is developed and successfully applied. After a suitable discretization typically these problems have the form

$$\mathcal{L}x = \mathcal{N}(x,\lambda), \quad \text{where}$$

$\mathcal{L}: R^n \to R^n$ (possibly non-invertible) and $\mathcal{N}: R^n \times R \to R^n$ (nonlinear). Then all the difficulties which arise in the infinite dimensional case with the definition of a suitable degree vanish and one can simply define the coincidence degree by

$$(11.2.1) \quad \mathrm{DEG}(\mathcal{L},\mathcal{N}(\cdot,\lambda),\Omega) := \deg(\mathcal{L}-\mathcal{N}(\cdot,\lambda),\Omega,0).$$

In fact, the boundary value problems in (11.1) have been studied as coincidence problems, i.e. as a zero problem for

$$F(x,\lambda) = \mathcal{L}x - \mathcal{N}(x,\lambda) .$$

(11.2.2) REMARK

Assume that $F(x,\lambda) = \mathcal{L}x - \mathcal{N}(x,\lambda)$ and that $\mathcal{L}$ is a linear isomorphism. Then the following are equivalent for inter-labeling:

(i) the problem $\mathcal{L}x - \mathcal{N}(x,\lambda) = 0$ and integer-L-labeling with $L = \mathcal{L}$;

(ii) the problem $x - \mathcal{L}^{-1}\mathcal{N}(x,\lambda) = 0$ and integer-id-labeling.

Hence, whenever one has that in a problem of the form $\mathcal{L}x - \mathcal{N}(x,\lambda) = 0$ the linear part is invertible then integer-L-labeling in (i) can provide better results than integer-id-labeling in (ii) only if one chooses L more carefully (see (8.1.4)).

In general, coincidence problems of type $\mathcal{L}x = \mathcal{N}(x,\lambda)$ are included in our previous discussion by use of (11.2.1).

XII. Positive Operators-Cones

It is typical for many nonlinear problems, e.g.

- second order elliptic boundary value problems
- functional differential equations,

that a successful use of topological tools like those discussed in previous sections requires additional structure: abstractly speeking this means that the operators which are in a natural way induced by the differential equations are compatible with the natural ordering of the underlying function spaces, i.e. these operators leave some cone invariant (see [8 , 33]. However, in such a case the Leray-Schauder degree is no longer an appropriate measure of solutions and has to be replaced by a kind of "relative" degree and that is provided by the topological fixed point index (see [17, 18, 22] for a definition). Our goal here is to point out how a cone situation can be adapted numerically in the line of our previous discussions. For reasons of length we restrict ourselves to the standard cone in R^n which is $K := R^n_+$. It will be clear, however, that our discussion here can be easily generalized to any linear isomorphic image of R^n_+. In the following all point set topological notions are to be understood in K .

Similar to the previous sections we consider now mappings

$$F: K \times R \to K$$

and the problem

$$F(x,\lambda) = x .$$

We associate an appropriate integer-L-labeling by setting

$$G(x,\lambda) := x - F(x,\lambda)$$

and

$$\ell_L^F: K \times R \to \{0,1,\dots,n\}$$

$$\ell_L^F := \ell_L^G , \quad \text{where}$$

ℓ_L^G is defined according to (3.2.1).

Now, it is most important that we can have a Door-In/Door-Out principle (cf. (5.1.3)) which is K-invariant, i.e. any chain of completely labeled simplices will not leave $K \times R$.

For the remaining part of this section we will assume that

$$L \in GL \quad \text{and} \quad L(K) \subset K.$$

We have the following

(12.1.1) LEMMA

Let F, K and L be as above and let ∂K denote the boundary of K in R^n.
Then

$$\ell_L^F(\partial K \times R) \subset \{0,1,2,\dots,n-1\},$$

i.e. no n-simplex in $\partial K \times R$ is completely labeled.

<u>Proof</u>: Observe that $\mathrm{int}K = B^n$ (definition of B^n: (3.2)). Let $y \in \partial K$, i.e. $y_i = 0$ for some $1 \le i \le n$. Then $F(y,\lambda) \in K$ implies $y_i - F_i(y,\lambda) \le 0$, thus, $y - F(y,\lambda) \notin B^n$. However, L being an open mapping leaves also B^n invariant and this proves the lemma.

One consequence of (12.1.1) is that integer-L-labeling provides a K-invariant Door-In/Door-Out principle provided L is K-invariant. Another consequence is the following paragraph:

(12.2) FIXED POINT INDEX

Let $U \subset K$ be open and bounded and let $f\colon \mathrm{cl}U \to K$ be continuous such that $\mathrm{Fix}(f) \cap \partial U = \phi$ where $\mathrm{Fix}(f) = \{x \in \mathrm{cl}U\colon f(x) = x\}$. Then the fixed point index

$$\mathrm{ind}(K,f,U) \in Z = \{\text{integers}\}$$

can be defined, for example along (4.1.3). A numerical exploitation of (4.1.3), however, would require a concrete knowledge of a retraction

$$r\colon \mathcal{O} \to K.$$

It is therefore that for the case of integer-L-labeling we present a useful alternative which avoids this unpleasant situation.

(12.2.1) THEOREM (Fixed Point Index and Integer-L-Labeling)

Let $U \subset K$ be open and bounded and such that clU is triangulable. Let $f: clU \to K$ be continuous and assume that $Fix(f) \cap \partial U = \phi$. Furthermore, let L be a linear isomorphism of R^n such that

- L is K-invariant and
- L has only positive eigenvalues.

Then we have for any triangulation T of clU sufficiently fine

$$ind(K,f,U) = \sum_{\sigma \in \mathcal{S}(id-f;L)} or(\sigma).$$

$(\mathcal{S}(id-f;L) = \{\sigma \in T_n: \sigma \text{ is a } \ell_L^{id-f}\text{-Sperner-simplex}\})$

Proof: We sketch only those steps which are needed to make possible a reference to [48]. First, we consider the case where $L = id$:

1) One defines a PL-map f_T which is K-invariant:

If $x \in T_o$ (a vertex of T) and $\ell_{id}^{id-f}(x) = i$

then

$$f_T(x) := x - \varepsilon_x b^i,$$

where $b^i = (0,\ldots,0,\underset{i}{1},\underset{i+1}{-1},0,\ldots,0)$ and

$\varepsilon_x > 0$ is chosen such that $f_T(x) \in K$.

Now set $\varepsilon = \min \{\varepsilon_x : x \in T_o\}$ and define f_T to be the PL-extension on each $\sigma \in T_n$.

2) If T is sufficiently fine then f and f_T are homotopic maps such that the homotopy is fixed point free on ∂U. Thus, the homotopy invariance for ind implies

$$\text{ind}(K,f,U) = \text{ind}(K,f_T,U).$$

3) Now, $\text{Fix}(f_T)$ is already in $\text{int}_{R^n}(U)$ and therefore

$$\text{ind}(K,f_T,U) = \deg(\text{id}-f_T,\text{int}_{R^n}U,0).$$

Hence the case $L = \text{id}$ is a consequence of theorem (4.1.1).

Finally, we consider a general L:

4) Define $L_t := t\,\text{id} + (1-t)L$, $0 \leq t \leq 1$, and observe that $L_t(K) \subset K$ and that $L_t \in GL_+$ for all t. Indeed, $tx + (1-t)Lx = 0$ would imply that $Lx = -\frac{t}{1-t}x$ and this contradicts our assumption that L has only positive eigenvalues.

5) Consider the homotopy

$$H: \text{cl}U \times [0,1] \to R^n$$

$$H(x,t) = L_t(\text{id}-f)(x)$$

and observe that $H(x,t) \neq 0$ for all $x \in \partial U$ and $t \in [0,1]$. Apply the Door-In/Door-Out principle (integer-L-labeling) to H in $\text{cl}U \times [0,1]$ and conclude that for T sufficiently fine lemma (12.1.1) implies

$$\sum_{\sigma \in \mathcal{S}(\text{id}-f;L)} \text{or}(\sigma) = \sum_{\sigma \in \mathcal{S}(L(\text{id}-f);L)} \text{or}(\sigma).$$

6) However, $\mathcal{S}(id-f;id) = \mathcal{S}(L(id-f);L)$ and this proves the theorem.

In view of artificial bifurcation in the context of a problem in K, of course, "switches" $P \in GL_-$ have to be chosen such that they are K-invariant, e.g. P a permutation of rows or columns. For such $P \in GL_-$ one can easily show that

$$ind(K,f,U) = (-1) \sum_{\sigma \in \mathcal{S}(id-f,PL)} or(\sigma) \quad .$$

Equipped with (12.1.1) and (12.2.1) one immediately can have results parallel to VIII, IX and X for cones simply by replacing degree arguments by index arguments.

(12.3) NUMERICAL EXAMPLE

To discuss an example where the order structure has been used significantly in the theoretical study (see [42,43]), and also in the numerical approximation we present the problem of finding non-trivial periodic solutions of

$$\text{(FDE)} \quad \begin{cases} \dot{u}(t) = -\lambda f(u(t-1)), & t \geq 0 \\ u(t) = \phi(t) & -1 \leq t \leq 0 . \end{cases}$$
$$\lambda \in R \ , \ \phi \in C[-1,0]$$

Equations like the above have been used to model growth phenomena in biology (see [26]).

Assume that $f: R \to R$ is continuous, $f(t) \cdot t > 0$ for all $t \neq 0$ and that f is differentiable in a neighborhood of zero with $f'(0) = 1$.

Periodic solutions of (FDE) can be obtained from fixed points of

$$\mathcal{F}(\phi,\lambda) = \phi$$

where $\mathcal{F}$ is the "shift operator" defined on the cone $C_m = \{\phi \in C[-1,0]: \phi(-1) = 0$ and ϕ is monotonic increasing$\}$ (see [42,43]). Since $f(0) = 0$ one has that

$$\mathcal{F}(0,\lambda) = 0$$

and the problem of finding periodic solutions for equation (FDE) fits into the framework of bifurcation in cones (see [42]). Indeed, it can be shown that

$$\mathrm{ind}(C_m, \mathcal{F}_\lambda, 0)$$

changes its value from +1 to 0 when λ passes through $\pi/2$, i.e. $(0,\pi/2)$ is a point of bifurcation for $\mathcal{F}$ (see [42]). Moreover, in view of our simplicial approach it seems to be noteworthy that it is not known that $\mathcal{F}$ has any differentiability properties.

After a suitable discretization of $\mathcal{F}$ we obtain an operator

$$F: R^n_+ \times R \longrightarrow R^n_+$$

and the index change of $\mathcal{F}$ described above suggests that our bifurcation algorithm in a cone version should be applicable.

Indeed, a computer program using integer-Id-labeling implementing the bifurcation algorithm has been successfully used to approximate periodic solutions of (FDE) for many nonlinearities f. Two examples of special interest in the context of multiplicity results (see [11]) are pictured in figure 44. In fact, these numerical experiments have motivated the results given in [11].

The differential equation was discretized on 11 points. The mesh size for T was 10^{-2}. cpu-time per example was less than 2 min.

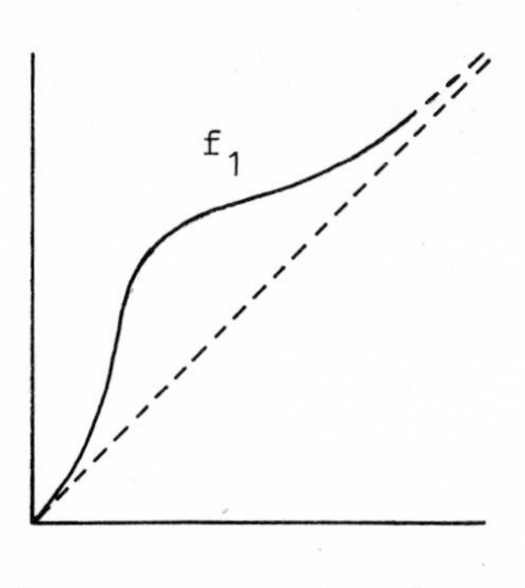

(i) asymptotically linear

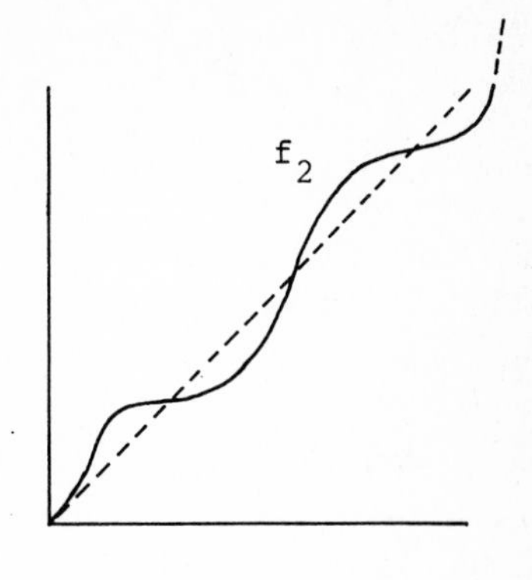

(ii) oscillating

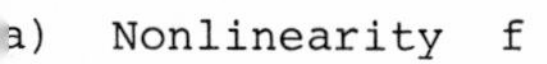

a) Nonlinearity f

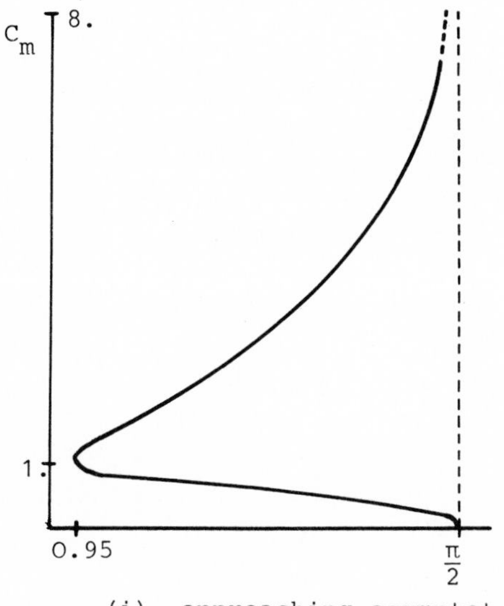

(i) approaching asymptotically the $\frac{\pi}{2}$ - level

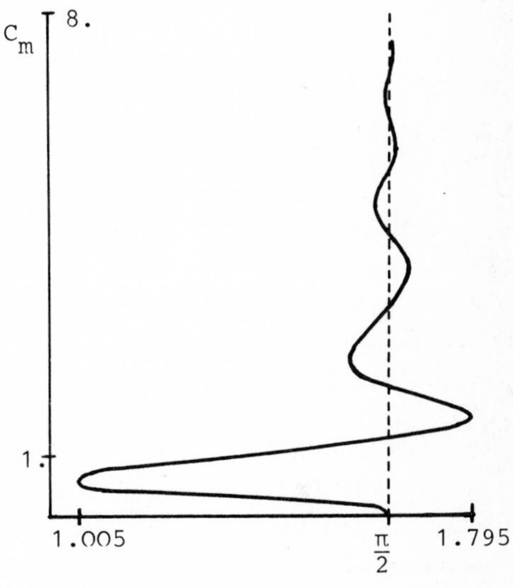

(ii) oscillating around the $\frac{\pi}{2}$ - level

b) Continua of periodic solutions of (FDE)

Figure 44.

XIII. Concluding Remarks

Several questions and problems arise at the end of our discussion:

(13.1) INFINITE DIMENSIONS versus FINITE DIMENSIONS

Though the following problem seems to be less important for concrete computations one should recall, however, that from the point of view of theory it was most important in our discussions that a finite dimensional approximation

$$F(x,\lambda) = 0$$

to an infinite dimensional problem

$$\mathcal{F}(x,\lambda) = 0$$

($\mathcal{F}$ completely continuous) could be chosen such that the topological degree is preserved. One knows from the original paper of Leray-Schauder [37] as well as from the Ritz-Galerkin method [32] how and that this can be achieved abstractly.

(13.1.1) PROBLEM

Given, for example, a certain type of differential equation. Find for a favourable discretization concrete sufficient conditions which guarantee that topological degree is inherited to the corresponding finite dimensional approximation.

(13.2) DEFINITION OF DEGREE

Brouwer degree has been accepted to be a powerful tool in many branches of modern mathematics. Accordingly, it can and actually is defined in many different settings, as for example differential topology, differential geometry, algebraic topology, functional analysis, e.t.c. In view of section V and the remarks there (Pontryagin Construction) it seems to be adequate to give also a definition for the needs of numerical analysis. Such a definition is of course implicit in V: Define the degree for PL-maps according to (4.1.2). Observe that such a degree satisfies all the properties of a degree. Especially, (5.1.3) the Door-In/Door-out principle provides the homotopy property for PL-homotopies. Then define the degree for continuous maps by a suitable PL-approximation (cf. (4.1.1) and (4.1.2)) and verify that such a definition does not depend on the various choices involved.

(13.3) MULTIVALUED PROBLEMS

Control problems [24], problems of optimal control [35] and problems in terms of variational inequalities [19] are studied in a typical manner by means of multivalued fixed point theory [36], i.e. degree and fixed point index for multivalued mappings [38,23], to obtain existence results. It should be noted that our methods can be easily extended to multivalued mappings G provided $G(x)$ is a compact and convex set for all x: A suitable vector-labeling is simply obtained by replacing G by any selection $g \in G$ (cf. [5,59]. Thus, problems modeled with multivalued mappings are naturally included in our discussion, whereas any numerical method which relies on differentiable structure is not applicable in presence of multivalued problems.

<u>NOTE ADDED IN PROOF</u>:

For an extensive study of PL-homotopies and orientation (index) of PL-curves we refer to [EAVES, B.C., SCARF, H. : The Solution of Systems of Piecewise Linear Equations, Cowles Foundation Paper No. 434, New Haven, Connecticut, 1976]. Certainly, our approach in sections II, III and V is very much in their spirit. Especially, their results concerning the curve index could be used to define the Brouwer Degree for PL-maps and prove its homotopy invariance. For a proof of the homotopy property of the Brouwer degree using PL-methods see [GARCIA, C.B. : Computation of Solutions to Nonlinear Equations under Homotopy Invariance, Mathematics of Operations Research, 2 1 (1977)].

References

[1] ALEXANDER, J.C., YORKE, J.A.: The homotopy continuation method, Numerically implementable topolgogical procedures, Trans. AMS, 242 (1978),271-284

[2] ALEXANDER, J.C.: The topolgocial theory of an embedding method, in: "Continuation Methods", H.J. Wacker, ed., New York: Academic Press, 1978.

[3] ALEXANDER, J.C.: Numerical continuation methods and bifurcation, in "Functional differential equations and approximation of fixed points", H.O. Peitgen and H.O. Walther, eds., Berlin, Heidelberg, New York: Springer Lecture Notes, 1979.

[4] ALEXANDER, J.C., YORKE, J.A.: A numerical continuation method that works generically, University of Maryland, Dept. of Math., MD 77-9-JA, TR 77-9.

[5] ALLGOWER, E.L., GEORG,K.: Simplicial and Continuation Methods for Approximating Fixed Points and Solutions to Systems of Equations, SIAM Review (to appear).

[6] ALLGOWER, E.L., KELLER, C.L., REEVES, T.E.: A program for the numerical approximation of a fixed point of an arbitrary continuous mapping of the n-cube or n-simplex into itself, Aerospace Research Laborations, Report ARL 71-0257, 1971.

[7] AMANN, H.: Lectures on some fixed point theorems, Monografias de Matemâtica, Instituto de matemâtica pura e aplicada, Rio de Janeiro.

[8] AMANN, H.: Fixed point equations and nonlinear eigenvalue problems in ordered Banach spaces, SIAM Review, 18.4 (1976).

[9] AMBROSETTI, A., HESS, P.: Positive solutions of asymptotically linear elliptic eigenvalue problems, to appear.

[10] AMBROSETTI,S., RABINOWITZ, P.H.: Dual variational methods in critical point theory and applications, J. Functional Analysis, 14.4 (1973), 349-381.

[11] ANGELSTORF, N.: Global branching and multiplicity results for functional differential equations, in "Functional differential equations and approximation of fixed points", H.O. Peitgen and H.O. Walther, eds., Berlin, Heidelberg, New York: Springer Lecture Notes, 1979.

[12] BERNSTEIN, S.: Sur la généralisation du problème de Dirichlet, Math. Ann. 69 (1910), 82-136.

[13] BORSUK, K.: Theory of Retracts, Warszawa: PWN, Polish Scientific Publishers, 1967.

[14] CHOW, S.N., MALLET-PARET, J., YORKE, J.A.: Finding zeros of maps: Homotopy methods that are constructive with probability one, Math. Comp. 32 (1978), 887-899.

[15] CRONIN, J.: Fixed Points and Topological Degree in Nonlinear Analysis, Providence: Amer. Math. Soc., 1964.

[16] DEIMLING, K.: Nichtlineare Gleichungen und Abbildungsgrade, Berlin, Heidelberg, New York: Springer, 1974.

[17] DOLD, A.: Fixed point index and fixed point theorems for Euclidean neighborhood retracts,Topology 4 (1965), 1-8.

[18] DOLD, A.: Lectures on Algebraic Topology, Berlin, Heidelberg, New York: Springer Verlag, 1972.

[19] DUGUNDJI, J., GRANAS, A.: KKM maps and Variational Inequalities, Annali della Scuola Norm. Sup. di Pisa, 5.4 (1978), 679-682.

[20] EAVES, B.C.: Homotopies for computation of fixed points, Mathematical Programming 3 (1972), 1-22.

[21] EAVES, B.C.: Properly labeled simplices, Studies in Optimization, 10, MAA, Studies in Mathematics, G.B. Dantzig and B.C. Eaves, (eds.), (1974) 71-93.

[22] EISENACK, G., FENSKE, C.C.: Fixpunkttheorie, Mannheim, Wien, Zürich: Bibliographisches Institut, 1978.

[23] FENSKE, C.C., PEITGEN, H.O.: Repulsive fixed points of multivalued transformations and the fixed point index, Math. Ann. 218 (1975), 9-18.

[24] FILIPPOV, A.F.: Differential equations with many valued discontinuous right hand side, Dok. Akad. Nauk SSSR 151 (1963), 65-68.

[25] GAINES, R.E., MAWHIN, J.L.: Coincidence Degree and Nonlinear Differential Equations, Berlin, Heidelberg, New York: Springer Verlag, 1977.

[26] HADELER, K.P.: Delay equations in biology, in "Functional differential equations and approximation of fixed points", H.O. Peitgen and H.O. Walther, eds., Berlin, Heidelberg, New York: Springer Lecture Notes, 1979.

[27] JEPPSON, M.M.: A search for the fixed points of a continuous mapping, Mathematical Topics in Economics Theory and Computation, R.H. Day and S.M. Robinson, eds., Philadelphia: SIAM, (1972) 122-129.

[28] JÜRGENS, H., SAUPE, D.: Methoden der simplizialen Topologie zur numerischen Behandlung von nichtlinearen Eigenwert- und Verzweigungsproblemen, Diplomarbeit, Bremen, 1979.

[29] KEARFOTT, B.: An efficient degree-computation method for a generalized method of bisection, Numer. Math., to appear.

[30] KLEIN, F.: Neue Beiträge zur Riemannschen Funktionentheorie, Math. Annalen 21, 1882/3

[31] KNASTER, B., KURATOWSKI, C., MAZURKIEWICZ, S.: Ein Beweis des Fixpunktsatzes für n-dimensionale Simplexe, Fund. Math. 14, (1929),132-137.

[32] KRASNOSEL'SKII, M.A.: Topological Methods in the Theory of Nonlinear Integral Equations, Oxford: Pergamon, 1963.

[33] KRASNOSEL'SKII, M.A.: Positive solutions of operator equations, Groningen: Noordhoff, 1964.

[34] KRONECKER, L.: Über Systeme von Funktionen mehrerer Variablen I (1869), Ges. Werke Bd. I, Teubner, Leibzig (1895), 177-234.

[35] LASOTA A., OPIAL, Z.: Fixed point theorems for multi-valued mappings and optimal control problems, Bull. Acad. Pol. Sci. 16 (1968), 645-649.

[36] LASRY, J.M., ROBERT, R.: Analyse non lineaire multivoque, Cahiers de Matématiques de la Decision, Université Paris IX Dauphine, N° 7611.

[37] LERAY, J., SCHAUDER, J.P.: Topologie et équations fonctionelles, Ann. Ecole Norm. Sup. (3) 51 (1934), 45-78.

[38] MA, T.W.: Topological degree for set-valued compact vector fields in locally convex spaces, Dissertations Math. 92 (1972), 1-43.

[39] MARSDEN, J.E.: Qualitative Methods in Bifurcation Theory, Bull. Amer. Math. Soc. 84.6 (1978), 1125-1148.

[40] MILNOR, J.: Topology from the differentiable viewpoint, 2nd printing, Charlottesville: The University of Virginia Press, 1969.

[41] NIRENBERG, L.: Topics in Nonlinear Functional Analysis, New York University Lecture Notes, 1973-74.

[42] NUSSBAUM, R.D.: A global bifurcation theorem with applications to functional differential equations, J. Functional Analysis 19, (1975), 319-338.

[43] NUSSBAUM, R.D.: Periodic Solutions of Nonlinear Autonomous Functional Differential Equations, in "Functional Differential Equations and Approximation of Fixed Points", H.O. Peitgen and H.O. Walther, eds., Berlin, Heidelberg, New York: Springer Lecture Notes, 1979.

[44] ORTEGA, J.M., RHEINBOLDT, W.C.: Iterative Solution of Nonlinear Equations in Several Variables, New York: Academic Press, 1970.

[45] POINCARÉ, H.: Sur les Groupes des Equationes Lineaires. Acta Mathem. 4, (1884).

[46] PRÜFER, M.: Sperner simplices and the topological fixed point index, Universität Bonn, SFB 72, preprint no. 134, 1977.

[47] PRÜFER, M.: Simpliziale Topologie und globale Verzweigung, Dissertation, Bonn 1978.

[48] PRÜFER, M., SIEGBERG, H.W.: On computational aspects of degree in R^n, in "Functional Differential Equations and Approximations of Fixed Points", H.O. Peitgen and H.O. Walther, eds., Heidelberg, Berlin, New York: Springer Lecture Notes, 1979.

[49] PEITGEN, H.O.: Methoden der topologischen Fixpunkttheorie in der nichtlinearen Funktionalanalysis, Habilitationsschrift, Universität Bonn, 1976.

[50] RABINOWITZ, P.H.: Some global results for nonlinear eigenvalue problems, J. Functional Analysis 7 (1971), 487-513.

[51] RABINOWITZ, P.H.: Some aspects of nonlinear eigenvalue problems, Rocky Mountain J. Math. 3 (1973), 162-202.

[52] RABINOWITZ, P.H.: On bifurcation from infinity, J. Differential Equations 14 (1973), 462-475.

[53] RABINOWITZ, P.H.: A survey on bifurcation theory, in "Dynamical Systems: An International Symposium", vol. I, New York: Academic Press, 1976.

[54] SIEGBERG, H.W.: Abbildungsgrade in Analysis und Topologie, Diplomarbeit, Bonn 1977.

[55] SPERNER, E.: Neuer Beweis für die Invarianz der Dimensionszahl und des Gebietes, Abh. Math. Sem. Hamburg 6 (1928), 265-272.

[56] STENGER, F.: Computing the topological degree of a mapping in R^n, Numer. Math. 25 (1975), 23-38.

[57] STYNES, M.J.: An algorithm for the numerical calculation of the degree of a mapping, Ph. D. Thesis, Oregon, State University, Corvallis, 1977.

[58] TEMME, M.M. (ed.): Nonlinear Analysis, Vol. I, II, Amsterdam: Mathematisch Zentrum, 1976.

[59] TODD, M.J.: The Computation of Fixed Points and Applications, Berlin, Heidelberg, New York: Springer Lecture Notes in Economics and Mathematical Systems, 1976.

[60] TODD, M.J.: Hamiltonian triangulations of R^n, in "Functional Differential Equations and Approximations of Fixed Points", H.O. Peitgen and H.O. Walther, eds., Berlin,Heidelberg, New York: Springer Lecture Notes, 1979.

[61] WACKER, H.J.: A Summary of the Developments in Imbedding Methods, in "Continuation Methods", H.J. Wacker, ed., New York: Academic Press, 1978.

[62] ZEIDLER, E.: Existenz, Eindeutigkeit, Eigenschaften und Anwendungen des Abbildungsgrades in R^n, in "Theory of Nonlinear Operators; Proceedings of a Summer School", Berlin: Akademie-Verlag, 1974 , 259-311.

On Computational Aspects of Topological Degree in $\mathbb{R}^n$

Michael Prüfer - Hans Willi Siegberg

I. INTRODUCTION : Topological degrees are an important tool in topology and its applications to nonlinear functional analysis, e.g. fixed point theory and bifurcation theory. In most of these applications degree arguments are used to prove existence.
In this note we discuss two computational formulas for the topological degree, deg(f,int P,o), of a continuous map $f: (P,\partial P) \to (\mathbb{R}^n, \mathbb{R}^n - \{o\})$ where P is a homogeneous n-dimensional polyhedron in $\mathbb{R}^n$.

The first formula was given by F.Stenger [15] in 1975 :

$$(1.1) \qquad \deg(f,\text{int } P,o) = (2^n n!)^{-1} \sum_{j=1}^{m} \eta_j . \det(\text{ sign } f(t_j^o), \ldots, \text{sign } f(t_j^{n-1})),$$

where $\sum_{j=1}^{m} \eta_j . < t_j^o, \ldots, t_j^{n-1} >$, $\eta_j \in \{-1,1\}$, is the oriented boundary of P.

The other formula given in [11] calculates the degree in terms of oriented Sperner n-simplices associated to a labeling function $\ell^f: P \to \{o, \ldots, n\}$ induced by f :

$$(1.2) \qquad \deg(f,\text{int } P,o) = \sum_{\sigma \ \ell^f\text{-Sperner n-simplex}} \text{or}(\sigma) ,$$

where σ is a simplex of the triangulation of P.

For a definition and basic properties of the degree see, e.g. [3], [4], [5], [7], [12].

A proof of formula (1.2) is given in section II . In section III we show how (1.2) can be replaced by a formula counting certain oriented Sperner (n-1)-simplices on ∂P. In sections IV - VII we give a new proof of Stenger's formula (1.1) revealing its geometry and its topological background. The course of this proof will detect a relation between (1.1) and (1.2) ; moreover, it

---

This work was supported by the "Deutsche Forschungsgemeinschaft, Sonderforschungsbereich 72, Universität Bonn" and by the "Forschungsschwerpunkt 'Dynamische Systeme', Universität Bremen" .

turns out that Stenger's crucial assumption ("sufficient refinement relative to sign F") can be replaced by a simpler one, see also Stynes [16]. Finally, in section VIII we discuss numerical aspects.

II. DEGREE AND SPERNER SIMPLICES : Let $Q \subseteq \mathbb{R}^n$ be a polyhedron. A continuous mapping $f: Q \to \mathbb{R}^n$ induces a labeling function $\ell^f: Q \to \{o, \dots n\}$ in the following way:

First, define the following sets in $\mathbb{R}^n$:

$$B^o := \{x \in \mathbb{R}^n : x_1 \leq o\}$$

$$B^i := \{x \in \mathbb{R}^n : x_1 > o, \dots, x_i > o, x_{i+1} \leq o\} , \quad 1 \leq i \leq n-1$$

$$B^n := \{x \in \mathbb{R}^n : x_1 > o, \dots, x_n > o\} .$$

Define

$$(2.1) \qquad \ell^f(x) = i \iff f(x) \in B^i \quad \text{and}$$

$$(2.2) \qquad A_i^f := cl\{x \in Q : \ell^f(x) = i\} \quad \text{for } o \leq i \leq n .$$

As an immediate consequence we have

$$(2.3) \qquad x \in \bigcap_{i=o}^{n} A_i^f \implies f(x) = o .$$

Definition 2.4 : i) Let $\{\sigma_1, \dots, \sigma_r\}$ be a finite set of n-simplices in $\mathbb{R}^n$. The set $P = \{x \in \mathbb{R}^n : x \in \sigma_j$ for some $j, 1 \leq j \leq r\}$ is called a homogeneous n-dimensional (hom. n-dim.) polyhedron iff the simplices $\sigma_1, \dots, \sigma_r$ together with their faces form a simplicial complex T . T is called a triangulation of P.

ii) Let $f: P \to \mathbb{R}^n$ be continuous. An n-simplex $\sigma = co\{a^o, \dots, a^n\} \in T$ is called an ℓ^f-Sperner simplex iff $\{\ell^f(a^o), \dots, \ell^f(a^n)\} = \{o, \dots, n\}$.

Let $\sigma := co\{a^o, \dots, a^n\} \subseteq \mathbb{R}^n$ be an n-simplex. Recall that an orientation of σ is defined by choosing some ordering (modulo even permutations) of its vertices. The oriented simplex determined by the ordering $a^o < a^1 < \dots < a^n$ is denoted by $< a^o, \dots, a^n >$. (Thus we have $< a^o, \dots, a^n > = < a^{\pi(o)}, \dots, a^{\pi(n)} >$ for an even permutation π.) An oriented n-simplex $< a^o, \dots, a^n >$ in $\mathbb{R}^n$ will be called positive oriented provided

$$\text{or}< a^o, \dots, a^n > := \text{sign det } M = 1 ,$$

where M is the matrix belonging to the affine homeomorphism which maps a^i into e^i ($e^o = o, e^1, \dots, e^n$ the standard basis in $\mathbb{R}^n$) for $o \leq i \leq n$. Observe that

$$\text{or}< a^o, \dots, a^n > = \text{sign det} \begin{pmatrix} 1 & \dots & 1 \\ a^o & \dots & a^n \end{pmatrix} .$$

Let $\sigma := \text{co}\{a^o, \dots, a^n\}$ be an ℓ^f-Sperner simplex, and assume without loss of generality that $\ell^f(a^i) = i$ for $o \leq i \leq n$. Define the orientation of σ by $\text{or}(\sigma) := \text{or}< a^o, \dots, a^n >$.

We can now formulate the main result of this section :

Theorem 2.5 : Let $P \subseteq \mathbb{R}^n$ be a hom. n-dim. polyhedron, and let $f: (P, \partial P) \to (\mathbb{R}^n, \mathbb{R}^n - \{o\})$ be a continuous map. Let T be a triangulation of P which satisfies the following "smallness condition" on ∂P:

(SC I) $\begin{cases} \text{For every (n-1)-simplex } \tau \in T, \tau \subseteq \partial P, \text{ there exists } i \in \{o, \dots, n\} \\ \text{such that } \ell^f(\tau) \subseteq \{o, \dots, \hat{i}, \dots, n\} . \end{cases}$

Under these assumptions the following formula holds :

$$\deg(f, \text{int } P, o) = \sum_{\substack{\sigma \in T \\ \sigma \ \ell^f\text{-Sperner simplex}}} \text{or}(\sigma) .$$

Remark : It is readily verified that condition (SC I) is satisfied by every triangulation T which is sufficiently small on ∂P : together with (2.3) the assumption $o \notin f(\partial P)$ implies that the sets $\partial P - A_i^f$, $o \leq i \leq n$, form an open covering $\mathcal{U}$ for ∂P. If $\lambda > o$ is any Lebesgue number (cf. [9]) of $\mathcal{U}$, and T is any triangulation of P such that $\text{diam } \tau \leq \lambda$ for all $\tau \in T, \tau \subseteq \partial P$, then T satisfies condition (SC I).

Proof of theorem 2.5 : We give a sketch of the proof :

In the first part we will construct a piecewise linear mapping $f_T : (P, \partial P) \to (\mathbb{R}^n, \mathbb{R}^n - \{o\})$ which satisfies

$$\deg(f_T, \text{int } P, o) = \sum_{\substack{\sigma \in T \\ \sigma \ \ell^f\text{-Sperner simplex}}} \text{or}(\sigma) .$$

In the second part we will construct a homotopy $\check{H}: (P, \partial P) \times [o,1] \to (\mathbb{R}^n, \mathbb{R}^n - \{o\})$ between f and f_T . The homotopy invariance of degree will then yield the assertion.

PART 1 : The Mapping f_T

Define the following vectors in $\mathbb{R}^n$:

$$b^o := (-1, o, \ldots, o)$$

$$b^i := (1, \ldots, 1, -1, o, \ldots, o) \qquad \text{for} \quad o < i < n$$

$$\uparrow$$
$$i\text{-th component}$$

$$b^n := (1, \ldots, 1)$$

We define f_T on the vertices of T : let y be such a vertex and assume $\ell^f(y) = i$. Define $f_T(y) := b^i$, and extend f_T to an affine mapping on each n-simplex $\sigma \in T$.

We conclude from the definition of f_T that $f_T(x) = o$ if and only if

$$x = \sum_{i=o}^{n-1} 2^{-(i+1)} . a^i + 2^{-n} . a^n ,$$

where $\sigma = \text{co}\{a^o, \ldots, a^n\}$ is an ℓ^f-Sperner simplex ($\ell^f(a^i) = i$ for $o \leq i \leq n$).

This means that $\deg(f_T, \text{int } \sigma, o)$ is defined for every n-simplex $\sigma \in T$, and equals zero, whenever σ is not an ℓ^f-Sperner simplex. Furthermore, one calculates $\deg(f_T, \text{int } \sigma, o) = \text{or}(\sigma)$, if σ is a Sperner simplex. By the additivity of degree we have

$$\deg(f_T, \text{int } P, o) = \sum_{\substack{\sigma \in T \\ \sigma \ \ell^f\text{-Sperner simplex}}} \text{or}(\sigma) .$$

PART 2 : The Homotopy $\check{H}$

It will be sufficient to construct $\check{H}$ on $\partial P \times [o,1]$. By Tietzes theorem [4] such a homotopy can be extended to a homotopy joining f and f_T on the whole of P.

PART 2a : Definition of $\check{H}$ on the vertices of T

If $y \in T$ is a vertex, $y \in \partial P$, define

$$\check{H}(y,t) := t\, f(y) + (1-t)\, f_T(y) \quad , \quad t \in [o,1]$$

We have immediately that $\check{H}(y,t) \neq o$ for all $t \in [o,1]$. Furthermore, one has that $\{\check{H}(y,t) : o \leq t \leq 1\} \subseteq B^i$ for some $o \leq i \leq n$.

PART 2b :

In the following any k-simplex ($1 \leq k \leq n-1$) of the triangulation T will be denoted by σ^k.

$\check{H}$ will now be defined on the simplices $\sigma^k \in T$, $\sigma^k \subseteq \partial P$, $1 \leq k \leq n-1$. The construction of $\check{H}$ will be complete if we have the following lemma:

Lemma 2.6 (Homotopy extension): Let $\sigma^k := co\{a^o, \ldots, a^k\} \subseteq \mathbb{R}^n$ be a k-simplex ($1 \leq k \leq n-1$), and let $f_1, f_2 : \sigma^k \to \mathbb{R}^n$ be continuous mappings which have no zeros on σ^k and which satisfy the following condition:

(C) There exists $\{i_o, \ldots, i_r\} \subsetneqq \{o, \ldots, n\}$ such that
$f_i(\sigma^k) \subseteq cl \bigcup_{o \leq j \leq r} B^{i_j}$ for $i = 1,2$.

Assume further that there exists a continuous homotopy $H: \partial_c \sigma^k \times [o,1] \to \mathbb{R}^n$ (∂_c = combinatorial boundary) with

(i) $H(x,o) = f_1(x)$, $H(x,1) = f_2(x)$ for $x \in \partial_c \sigma^k$

and

(ii) $H(.,t)(\partial_c \sigma^k) \subseteq (cl \bigcup_{o \leq j \leq r} B^{i_j}) - \{o\}$ for all $t \in [o,1]$.

Then there exists a continuous homotopy $\bar{H}: \sigma^k \times [o,1] \to \mathbb{R}^n$ with

$$\text{(iii)} \quad \bar{H}(x,t) = H(x,t) \quad \text{for all } (x,t) \in \partial_c \sigma^k \times [o,1]$$

$$\text{(iv)} \quad \left.\begin{array}{l} \bar{H}(x,o) = f_1(x) \\ \bar{H}(x,1) = f_2(x) \end{array}\right\} \quad \text{for all } x \in \sigma^k$$

and

$$\text{(v)} \quad \bar{H}(.,t)(\sigma^k) \subseteq \Big(\text{cl} \bigcup_{o \leq j \leq r} B^{i_j} \Big) - \{o\} \quad \text{for all } t \in [o,1].$$

We will apply this lemma in our situation with $f_1 = f$ and $f_2 = f_T$. Since T satisfies the smallness condition (SC I), condition (C) is fulfilled on every k-simplex $\sigma^k \in T$, $\sigma^k \subseteq \partial P$. Thus, applying lemma (2.6) (n-1) times to the homotopy $\check{H}$ as defined in PART 2a we obtain the desired homotopy $\check{H}: \partial P \times [o,1] \to \mathbb{R}^n - \{o\}$.

<u>Proof of lemma 2.6</u> : Define a homotopy $H': \sigma^k \times [o,1] \to \mathbb{R}^n$ by

a) $H'(x,t) := H(x,t)$ for $(x,t) \in \partial_c \sigma^k \times [o,1]$

b) $H'(x,o) := f_1(x)$, $H'(x,1) := f_2(x)$ for $x \in \sigma^k$.

(Observe that by a) and b) H' is defined on $\partial_c(\sigma^k \times [o,1])$.)

c) $H'(b(\sigma^k),1/2) := o$, where $b(\sigma^k)$ denotes the barycenter of σ^k.

d) If $(x,t) \in (\sigma^k \times [o,1]) - (b(\sigma^k),1/2)$, let $(\bar{x},\bar{t})$ denote the projection of (x,t) onto $\partial_c(\sigma^k \times [o,1])$ along the ray given by $(b(\sigma^k),1/2)$ and (x,t). We have a unique representation

$$(x,t) = \theta(\bar{x},\bar{t}) + (1-\theta)(b(\sigma^k),1/2) \quad \text{with some } \theta \in (o,1] ,$$

and we define

$$H'(x,t) = \theta H'(\bar{x},\bar{t}) + (1-\theta)H'(b(\sigma^k),1/2) = \theta H'(\bar{x},\bar{t}) .$$

By construction H' has properties (iii) and (iv) of lemma 2.6 . Since the sets B^i are convex, we have furthermore

$$H'(x,t) \in \Big(\text{cl} \bigcup_{o \leq j \leq r} B^{i_j} \Big) - \{o\} \quad \text{for } (x,t) \neq (b(\sigma^k),1/2) .$$

Thus, H' satisfies (v) on $(\sigma^k \times [o,1]) - (b(\sigma^k),1/2)$.

However, $H'(b(\sigma^k),1/2) = o \in B^o$.

This zero will be removed in the next step by a slight perturbation of H' :

Let w be an n-vector satisfying

1) $w_j > o$ if $j = i_k + 1$ for some $i_k \varepsilon \{i_o, \dots, i_r\}$

2) $w_j < o$ otherwise

Define $\bar{H}(x,t) := H'(x,t) - \delta(x)w$,

where $\delta(x)$ denotes the distance from x to $\partial_c(\sigma^k \times [o,1])$. It is easily verified that $\bar{H}$ satisfies conditions (iii) - (v) of lemma 2.6 .
This completes the proof of the lemma . ■

Remark: A closer analysis shows that the piecewise linear map $f_T: P \to S$, $S := co\{b^o, \dots, b^n\}$, is actually (provided the triangulation T is a bit smaller than (SC I)) a simplicial approximation (in the sense of simplicial topology, see e.g. [1]) of the mapping $R \circ f: P \to S$, where $R: \mathbb{R}^n \to S$ is the radial retraction onto S. Thus in formula (1.2) a renaissance of Brouwer's ideas is reflected: indeed, Brouwer defined the degree at first for simplicial mappings, and then he extended the concept to continuous mappings using simplicial approximation , see [14] .

III. DEGREE AND CODIMENSION ONE SPERNER SIMPLICES IN ∂P :

For numerical purposes it is important to remark that the degree may be calculated by considering codimension one completely labeled simplices in ∂P.

Let $f:(P,\partial P) \to (\mathbb{R}^n, \mathbb{R}^n - \{o\})$ be a continuous mapping, P a hom. n-dim. polyhedron, and let T be a triangulation of P. Let $j \varepsilon \{o, \dots, n\}$ be fixed.

Definition 3.1 : An (n-1)-simplex $\tau := co\{t^o, \dots, t^{n-1}\} \varepsilon T, \tau \subseteq \partial P$, is called a $\hat{j}$ - Sperner simplex, iff $\{\ell^f(t^o), \dots, \ell^f(t^{n-1})\} = \{o, \dots, \hat{j}, \dots, n\}$.

Let $\tau = co\{t^o, \dots, t^{n-1}\}$ be such a Sperner simplex and assume without loss of generality that $\ell^f(t^o) = o, \dots, \ell^f(t^{j-1}) = j-1, \ell^f(t^j) = j+1, \dots, \ell^f(t^{n-1}) = n$.
There exists a unique point $b \varepsilon$ int P such that the n-simplex $co\{t^o, \dots, t^{j-1}, b, t^j, \dots, t^{n-1}\}$ belongs to T .

Define $or(\tau) := \text{sign} \det M$, where M is the matrix belonging to the affine homeomorphism which maps t^i into e^i for $i < j$, b into e^j, and t^i into e^{i+1} for $i > j$.

Using the familiar "door in - door out" pivoting scheme [2],[6],[1o], we observe that Sperner simplices and $\hat{j}$ - Sperner simplices are connected by chains of simplices (cf. fig. 3.2):

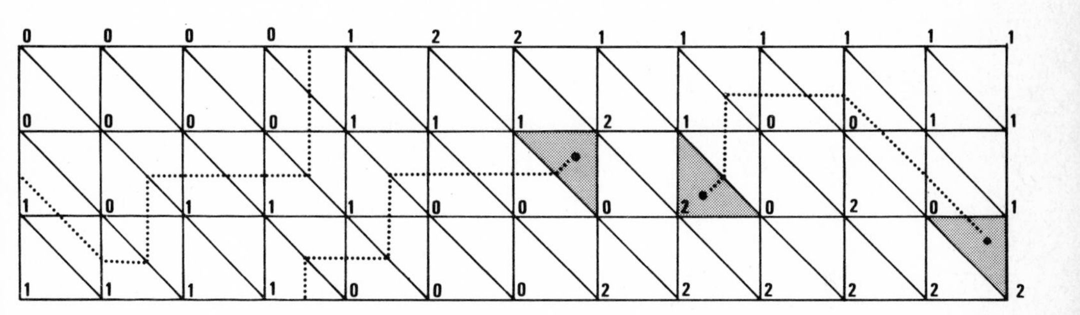

3.2 figure : example for $n = 2$, $j = 2$.

There are exactly three types of simplicial chains:

(3.3) Chains starting with a $\hat{j}$ - Sperner simplex $\tau \subseteq \partial P$ and ending with a Sperner simplex σ ; in this case we have $or(\tau) = or(\sigma)$.

(3.4) Chains joining two $\hat{j}$ - Sperner simplices τ_o, $\tau_1 \subseteq \partial P$; in this case we have $or(\tau_o) = -or(\tau_1)$.

(3.5) Chains joining two Sperner simplices σ_o, $\sigma_1 \subseteq P$; in this case we have $or(\sigma_o) = -or(\sigma_1)$.

From (3.3) - (3.5) we conclude

Theorem 3.6 : Let $P \subseteq \mathbb{R}^n$ be a hom. n-dim. polyhedron, and let $f:(P,\partial P) \to (\mathbb{R}^n, \mathbb{R}^n - \{o\})$ be a continuous mapping. Let T be any triangulation of P and let $j \in \{o, \dots ,n\}$ be arbitrary but fixed.

Then $$\sum_{\substack{\sigma \in T \\ \sigma\ \ell^f\text{-Sperner simplex}}} \mathrm{or}(\sigma) = \sum_{\substack{\tau \in T \\ \tau\ \ell^f\text{-}\hat{j}\text{ - Sperner simplex}}} \mathrm{or}(\tau) .$$

Note that theorem 3.6 together with theorem 2.5 proves the well known fact that $\deg(f,\mathrm{int}\ P,o)$ is already determined by the values of f on ∂P.

IV. A HOMOLOGICAL APPROACH TO DEGREE THEORY :

In order to deduce Stenger's formula (1.1) we give an appropriate definition of $\deg(f,\mathrm{int}\ P,o)$ along [13]. For that purpose we introduce homology classes in $H_n(P,\partial P)$ and in $H_n(\mathbb{R}^n,\mathbb{R}^n-\{o\})$: By an affine singular q-simplex

$$(w^o, \dots ,w^q): S_q \to V \quad , \quad V = \text{vector space}$$

we mean the affine mapping that maps e^i into w^i, $i = o, \dots ,q$ ($S_q := \mathrm{co}\{e^o, \dots ,e^q\}$ the standard simplex in $\mathbb{R}^q$); $w^o, \dots ,w^q$ are arbitrary points in V.

(4.1) For a (geometric) n-simplex $\sigma = \mathrm{co}\{a^o, \dots ,a^n\}$ in $\mathbb{R}^n$ we will denote each corresponding affine singular n-simplex of positive orientation (i.e. $(a^{\pi(o)}, \dots ,a^{\pi(n)}): S_n \to \mathbb{R}^n$ such that $\mathrm{or}< a^{\pi(o)}, \dots ,a^{\pi(n)}> = 1$) by the corresponding Latin letter s .

<u>Lemma and Definition 4.2</u> : Let P be a hom. n-dim. polyhedron in $\mathbb{R}^n$ with triangulation T , and let $\sigma_1, \dots ,\sigma_r$ be the n-simplices in T . Then $\sum_{i=1}^{r} s_i$ is a cycle mod ∂P , and its homology class $\left[\sum_{i=1}^{r} s_i\right] \in H_n(P,\partial P)$ (H = sing. homology with integer coefficients) is independent of the choice of the s_i .

$\mathcal{O}(P;T) := \left[\sum_{i=1}^{r} s_i\right]$ is called the fundamental class of $(P,\partial P)$ induced by T.

<u>Proof</u>: The lemma follows from the equivalence of singular and simplicial homology and an analogous lemma in simplicial homology ,cf. [1],[9]. ■

We need the following observations:

(4.3) Let $\sigma = co\{a^o, \dots, a^n\}$ be an n-simplex in $\mathbb{R}^n$ such that $o \in int\ \sigma$. Then the affine singular n-simplex $(a^o, \dots, a^n)$ is a cycle mod $\mathbb{R}^n - \{o\}$, and its homology class $[(a^o, \dots, a^n)] \in H_n(\mathbb{R}^n, \mathbb{R}^n - \{o\})$ generates $H_n(\mathbb{R}^n, \mathbb{R}^n - \{o\})$.

(4.4) Let $(a^o, \dots, a^n)$ and $(b^o, \dots, b^n)$ be two affine singular n-simplices in $\mathbb{R}^n$ such that $or\langle a^o, \dots, a^n\rangle = 1 = or\langle b^o, \dots, b^n\rangle$, and such that $o \in \mathbb{R}^n$ is contained in $int\ co\{a^o, \dots, a^n\}$ and also in $int\ co\{b^o, \dots, b^n\}$. Then

$$[(a^o, \dots, a^n)] = [(b^o, \dots, b^n)] \in H_n(\mathbb{R}^n, \mathbb{R}^n - \{o\}) .$$

The generator of $H_n(\mathbb{R}^n, \mathbb{R}^n - \{o\})$ induced by these n-simplices of positive orientation will be denoted by $\mathcal{O}$ in the following.

Using these facts we can have the topological degree as follows:

Theorem 4.5 : Let P be a hom. n-dim. polyhedron in $\mathbb{R}^n$ with triangulation T, and let $f: (P, \partial P) \to (\mathbb{R}^n, \mathbb{R}^n - \{o\})$ be a continuous map. Then for the induced map $f_* : H_n(P, \partial P) \to H_n(\mathbb{R}^n, \mathbb{R}^n - \{o\})$ the following identity holds

$$f_* (\mathcal{O}(P;T)) = deg(f, int\ P, o) \cdot \mathcal{O} . \tag{4.6}$$

In particular, $f_*(\mathcal{O}(P;T))$ does not depend on the triangulation T of P, and identity (4.6) may be used as a definition of the degree for f .

Let $\|\cdot\|$ be the ℓ_∞-norm in $\mathbb{R}^n$, $\|x\| := max\{|x_i| : i = 1, \dots, n\}$ and Δ^n, Σ^{n-1} the corresponding unit ball rsp. unit sphere.

(4.7) The map $r: \mathbb{R}^n - \{o\} \to \Sigma^{n-1}$, $r(x) := x/\|x\|$, is a deformation retraction; therefore, the induced map $r_*: \tilde{H}_{n-1}(\mathbb{R}^n - \{o\}) \to \tilde{H}_{n-1}(\Sigma^{n-1})$ is an isomorphism ($\tilde{H}$ = reduced homology).

(4.8) Since $\mathbb{R}^n$ is contractible the connecting homomorphism $d_*: H_n(\mathbb{R}^n, \mathbb{R}^n - \{o\}) \to \tilde{H}_{n-1}(\mathbb{R}^n - \{o\})$ is an isomorphism.

Therefore we have the following relation

$$\deg(f,\text{int } P,o) = N \iff f_*(\mathcal{O}(P;T)) - N.\mathcal{O} = o \in H_n(\mathbb{R}^n,\mathbb{R}^n-\{o\})$$
$$\iff r_*\circ d_*(\, f_*(\mathcal{O}(P;T)) - N.\mathcal{O}\,) = o \in \tilde{H}_{n-1}(\Sigma^{n-1})$$
$$(4.9) \qquad \iff (r\circ f_{|\partial P})_*(d_*\mathcal{O}(P;T)) - N\cdot r_* d_* \mathcal{O} = o \in \tilde{H}_{n-1}(\Sigma^{n-1})$$

Using (4.9) formula (1.1) will be proved in the following way :

If the triangulation of P is small enough we can replace $(r\circ f_{|\partial P})_*(d_*\mathcal{O}(P;T))$ by the homology class of a cycle (on Σ^{n-1}) which is composed only of affine singular (n-1)-simplices; the same holds for $r_*d_*\mathcal{O}$ (see section V and VI). Then in section VII we will show how an affine cycle on Σ^{n-1} whose homology class vanishes can be splitted into $2n$ affine cycles (on C_j) mod ∂C_j ($C_1, \dots ,C_{2n}$ are the $2n$ faces of Σ^{n-1}) such that their homology class in $H_{n-1}(C_j,\partial C_j)$ vanish. Using induction this will imply (1.1).

V. A TRIANGULATION OF THE n-CUBE :

An appropriate representation of $r_*d_*\mathcal{O}$ - in terms of affine singular (n-1)-simplices - can be obtained by a special triangulation of Δ^n introduced in [17]. For our purposes it is useful to describe this triangulation in a way slightly different from the usual one.

(5.1) For $n > 1$ and $o \le i \le n-1$ let $A_i^n, B_i^n : \mathbb{R}^{n-1} \to \mathbb{R}^n$ be the maps

$$A_i^n : \mathbb{R}^{n-1} \to \mathbb{R}^n \;,\quad x = (x_1, \dots ,x_{n-1}) \mapsto \begin{cases} (1,x_1, \dots ,x_{n-1}) & , i = o \\ (x_1, \dots ,x_i,1,x_{i+1}, \dots ,x_{n-1}) & , i > o \end{cases}$$

$$B_i^n : \mathbb{R}^{n-1} \to \mathbb{R}^n \;,\quad x = (x_1, \dots ,x_{n-1}) \mapsto \begin{cases} (-1,x_1, \dots ,x_{n-1}) & , i = o \\ (x_1, \dots ,x_i,-1,x_{i+1}, \dots ,x_{n-1}) & , i > o \end{cases}$$

(If possible we will omit the index n in A_i^n , B_i^n in the following.)

Let $\mathcal{D}^n$ be the following set of n-simplices

$$\mathcal{D}^1 := \{\beta_1,\beta_2\} \;;\quad \beta_1 = \text{co}\{-1,o\} \;,\quad \beta_2 = \text{co}\{o,1\}$$

$$\mathcal{D}^n := \{C_i[\beta] \;:\; C_i \in \{A_i,B_i\};\ \beta \in \mathcal{D}^{n-1} \;;\ o \le i \le n-1\} \qquad n > 1$$

where $C_i[\beta]$ is the n-simplex

$$C_i[\beta] := \mathrm{co}\{o, C_i(v^o), \ldots, C_i(v^{n-1})\} \qquad \text{for} \quad \beta := \mathrm{co}\{v^o, \ldots, v^{n-1}\} .$$

The n-simplices in $\mathcal{D}^n$ together with their faces form a simplicial complex which will be denoted by $\mathcal{D}^n$, too; Δ^n is a hom. n-dim. polyhedron with triangulation $\mathcal{D}^n$.

The following remarks are easily verified:

(5.2) The triangulation $\mathcal{D}^n$ of Δ^n is composed of $k(n) := 2^n n!$ n-simplices.

(5.3) Let $s = (w^o, \ldots, w^{n-1}) \colon S_{n-1} \to \mathbb{R}^{n-1}$ be an affine singular (n-1)-simplex. If we denote the affine singular n-simplex $(o, C_i(w^o), \ldots, C_i(w^{n-1}))$ by $C_i[s]$, for $C_i \in \{A_i, B_i\}$ and $o \leq i \leq n-1$, then the fundamental class $\mathcal{O}(\Delta^{n-1};\mathcal{D}^{n-1}) = \left[\sum_{\beta \in \mathcal{D}^{n-1}} b\right]$ $(n > 1)$ can be used to describe the fundamental class $\mathcal{O}(\Delta^n;\mathcal{D}^n)$ as follows :

$$\mathcal{O}(\Delta^n;\mathcal{D}^n) = \left[\sum_{\substack{o \leq i \leq n-1 \\ \beta \in \mathcal{D}^{n-1}}} (-1)^i A_i[b] - \sum_{\substack{o \leq i \leq n-1 \\ \beta \in \mathcal{D}^{n-1}}} (-1)^i B_i[b]\right]$$

In order to prove the main result of this section we need the following relation:

(5.4) $$r_* d_* \mathcal{O} = d_*(\mathcal{O}(\Delta^n;\mathcal{D}^n))$$

Proof: Since the mapping $i_* \colon H_n(\Delta^n, \Sigma^{n-1}) \to H_n(\mathbb{R}^n, \mathbb{R}^n - \{o\})$, i = inclusion , maps $\mathcal{O}(\Delta^n;\mathcal{D}^n)$ into $\mathcal{O}$, and since r is a deformation retraction, identity (5.4) follows from the commutativity of the diagram

$$\begin{array}{ccc} H_n(\Delta^n, \Sigma^{n-1}) & \xrightarrow{\;i_*\;} & H_n(\mathbb{R}^n, \mathbb{R}^n - \{o\}) \\ \downarrow d_* & & \downarrow d_* \\ \tilde{H}_{n-1}(\Sigma^{n-1}) & \xrightarrow[\;i_* = (r)_*^{-1}\;]{} & \tilde{H}_{n-1}(\mathbb{R}^n - \{o\}) \end{array}$$ ∎

Using (5.4) we can prove the desired representation of $r_* d_* \mathcal{O}$:

Lemma 5.5 : The class $r_* d_* \mathcal{O} \in \tilde{H}_{n-1}(\Sigma^{n-1})$, $n \geq 1$, can be represented as follows:

$$r_* d_* \mathcal{O} = \left[\sum_{j=1}^{k(n)} \varepsilon_j a_j \right], \quad \varepsilon_j \in \{-1,1\}$$

such that

i) each a_j is an affine singular (n-1)-simplex

$$a_j = (a_j^o, \ldots, a_j^{n-1}) : S_{n-1} \to \Sigma^{n-1} ,$$

and each vertex of a_j is contained in $\{-1,o,1\}^n$

ii) for each $a_j = (a_j^o, \ldots, a_j^{n-1})$: $\det(a_j^o, \ldots, a_j^{n-1}) = \varepsilon_j$.

Sketch of proof:

$n = 1$: $\mathcal{O}(\Delta^1;\mathcal{D}^1) = [(-1,o) + (o,1)]$. Therefore, by (5.4), $r_* d_* \mathcal{O} = [-(-1) + (1)]$. Set $\varepsilon_1 := -1$, $\varepsilon_2 := 1$, $a_1 := (-1)$, $a_2 := (1)$.

$n > 1$: Using (5.3) and (5.4) we conclude

$$r_* d_* \mathcal{O} = \left[\sum_{\substack{o \leq i \leq n-1 \\ \beta \in \mathcal{D}^{n-1}}} (-1)^i \delta A_i[b] - \sum_{\substack{o \leq i \leq n-1 \\ \beta \in \mathcal{D}^{n-1}}} (-1)^i \delta B_i[b] \right] .$$

(δ = usual boundary operator in homology)

Since $\mathcal{O}(\Delta^n;\mathcal{D}^n)$ is a cycle mod Σ^{n-1}, by some cancellations this expression reduces to

$$\left[\sum_{\substack{o \leq i \leq n-1 \\ \beta \in \mathcal{D}^{n-1}}} (-1)^i A_i(b) - \sum_{\substack{o \leq i \leq n-1 \\ \beta \in \mathcal{D}^{n-1}}} (-1)^i B_i(b) \right] ,$$

where $C_i(b)$, $C_i \in \{A_i, B_i\}$, denotes the affine singular (n-1)-simplex

$$C_i(b) := (C_i(v^o), \ldots, C_i(v^{n-1})) \quad \text{for } b = (v^o, \ldots, v^{n-1}) .$$

Finally, using induction one shows

$\det A_i(b) = (-1)^i$ and $\det B_i(b) = -(-1)^i$ for $\beta \in \mathcal{D}^{n-1}$, $o \leq i \leq n-1$, and the lemma follows. ∎

VI. THE STENGER FORMULA : Let P be a hom. n-dim. polyhedron in $\mathbb{R}^n$ with triangulation T, and let $\sigma_1, \dots, \sigma_r$ be the n-simplices in T.
One can write the "boundary" $d_* \mathcal{O}(P;T)$ of the fundamental class $\mathcal{O}(P;T) = \left[\sum_{i=1}^{r} s_i\right]$ in the following way :

$$d_* \mathcal{O}(P;T) = \left[\sum_{i=1}^{r} \delta s_i\right] = \left[\sum_{j=1}^{m} \eta_j t_j\right] , \quad \eta_j \in \{-1,1\} .$$

Here t_j is an affine singular (n-1)-simplex

$$t_j = (t_j^o, \dots, t_j^{n-1}) : S_{n-1} \to \partial P$$

Observe that the (topological) boundary ∂P of P is composed of the geometric (n-1)-simplices $\tau_j := \mathrm{co}\{t_j^o, \dots, t_j^{n-1}\}$ corresponding to the affine singular simplices t_j.

Before we proceed to the main result of this section we state a basic lemma which will be proved in the following section .

<u>Lemma 6.1</u> : Let $Z := \sum_{i=1}^{q} \gamma_i z_i \in Z_{n-1}(\Sigma^{n-1})$, $\gamma_i \in \{-1,1\}$, $n \geq 1$,

be a cycle on Σ^{n-1} of following type:

Each z_i is an affine singular (n-1)-simplex

$z_i = (z_i^o, \dots, z_i^{n-1}) : S_{n-1} \to \Sigma^{n-1}$,

and each vertex of z_i is contained in $\{-1,o,1\}^n$.

If the (reduced) homology class of Z equals zero, i.e. $[Z] = o \in \tilde{H}_{n-1}(\Sigma^{n-1})$, then

$$\sum_{i=1}^{q} \gamma_i \det(z_i^o, \dots, z_i^{n-1}) = o .$$

<u>Theorem 6.2</u> : Let $P \subset \mathbb{R}^n$ be a hom. n-dim. polyhedron, and let

$f:(P,\partial P) \to (\mathbb{R}^n, \mathbb{R}^n - \{o\})$ be a continuous map.

Let T be a triangulation of P which satisfies the following "smallness condition" on ∂P:

(SC II) $\begin{cases} \text{For every } (n-1)\text{-simplex } \tau \in T, \tau \subseteq \partial P, \text{ there exists } i \in \{1, \ldots, n\} \\ \text{such that } \operatorname{sign}(f_{i|\tau}) = \text{const } (\neq o). \quad (f = (f_1, \ldots, f_n)) \end{cases}$

Further let $d_* \mathcal{O}(P;T)$ have the representation $d_* \mathcal{O}(P;T) = \left[\sum_{j=1}^{m} \eta_j t_j \right]$
$\eta_j \in \{-1,1\}$.

Under these assumptions the following formula holds :

$$\deg(f, \text{int } P, o) = (2^n n!)^{-1} \sum_{j=1}^{m} \eta_j \det(\operatorname{sign} f(t_j^o), \ldots, \operatorname{sign} f(t_j^{n-1}))$$

$$(\operatorname{sign} f = (\operatorname{sign} f_1, \ldots, \operatorname{sign} f_n))$$

Remark : If the boundary of P is "sufficiently refined rel. to sign f" in the sense of Stenger [15] , then, in particular, T satisfies condition (SC II). Thus, conditions (c) and (d) in Stenger's definition 4.4 [15] can be dropped; see also Stynes [16] .

Proof: By associating to each vertex y of the (n-1)-simplices τ_j lying on the boundary ∂P the point sign f(y) we obtain a piecewise linear mapping E by linear extension : $E: \partial P \to \mathbb{R}^n$.

Observe that the image $E(\partial P)$ is contained in Σ^{n-1} (by virtue of (SC II)). Again by virtue of condition (SC II) the image $h(\partial P \times [o,1])$ of the homotopy $h(x,t) := t\, f(x) + (1-t)\, E(x)$ is contained in $\mathbb{R}^n - \{o\}$.

Therefore, we have $(f_{|\partial P})_* = E_* : \tilde{H}_{n-1}(\partial P) \to \tilde{H}_{n-1}(\Sigma^{n-1})$, and thus,

$$\deg(f, \text{int } p, o) = N \iff (r \circ E)_* d_* \mathcal{O}(P;T) - N.\, r_* d_* \mathcal{O} = o \in \tilde{H}_{n-1}(\Sigma^{n-1}).$$

This means, however, that

$$\left[\sum_{j=1}^{m} \eta_j (E(t_j^o), \ldots, E(t_j^{n-1})) - \deg(f, \text{int } P, o).(\sum_{i=1}^{k(n)} \varepsilon_i a_i) \right] = o \in \tilde{H}_{n-1}(\Sigma^{n-1}).$$

Now, the cycle

$$\sum_{j=1}^{m} \eta_j (E(t_j^o), \ldots, E(t_j^{n-1})) - \deg(f, \text{int } P, o).(\sum_{i=1}^{k(n)} \varepsilon_i a_i) =: Z$$

satisfies the hypothesis of lemma 6.1. Hence, it follows

$$\sum_{j=1}^{m} \eta_j \det(E(t_j^o), \ldots, E(t_j^{n-1})) - \deg(f, \text{int } P, o).\, 2^n n! = o \ . \ \blacksquare$$

Remark : It is worthwile, to trace back the historical roots of Stenger's formula. An early manifestation of degree theory was provided by means of the Kronecker integral which was discovered in 1869 [14] :

$f: (M, \partial M) \to (\mathbb{R}^n, \mathbb{R}^n - \{o\})$ smooth, $M \subseteq \mathbb{R}^n$ compact n-manifold with boundary .

$$\deg(f, \text{int } M, o) = (\text{vol } S^{n-1})^{-1} \int_{\partial M} ||f||^{-n} \det(f, \frac{\partial f}{\partial x_1}, \ldots, \frac{\partial f}{\partial x_{n-1}}) dx_1 \ldots dx_{n-1}$$

$$=: K(f)$$

($x = (x_1, \ldots, x_{n-1})$ local coordinates of ∂M ; $||\cdot||$ euclidean norm)

To prove that $K(f)$ is always an integer, one can show that $K(f)$ coincides with the intersection number of $f(\partial M)$ with an half ray emanating from the origin. This was done by Hadamard in 1910. A closer look at Stenger's inductive degree relation [15 ; ch. 4.2], which is crucial for his proof, shows that it describes the same property:

$2n \deg(f, \text{int } M, o)$ = intersection number of $f(\partial M)$ with the coordinate axes of $\mathbb{R}^n$,

and this intersection number can be represented as a sum of certain (n-1)-dim. degrees. Thus, roughly speaking, one can say that Stenger's formula roots in the early work of Hadamard; this was observed at first in [16] .

VII. A RELATION BETWEEN FORMULA (1.1) AND FORMULA (1.2) : First we will prove lemma 6.1 stated in the previous section.

Proof of lemma 6.1 : The proof will proceed by induction on n .

$n = 1$: Each z_i in the cycle Z has the form $z_i = (-1)$ or $z_i = (1)$; thus, we can write

$$o = [Z] = \left[\sum_{j_1=1}^{R_1} \gamma_{j_1} \cdot (1) + \sum_{j_2=1}^{R_2} \gamma_{j_2} \cdot (-1) \right] = \sum_{j_1=1}^{R_1} \gamma_{j_1} [(1)] - \sum_{j_2=1}^{R_2} -\gamma_{j_2} [(-1)]$$

Recall that $H_o(\Sigma^o) \cong \mathbb{Z}$ is generated by $[(1)] - [(-1)]$. Therefore,

$$\sum_{j_1=1}^{R_1} \gamma_{j_1} = o = \sum_{j_2=1}^{R_2} -\gamma_{j_2}$$

$n \geq 2$: Fix a face $C_j(\Delta^{n-1})$ of Σ^{n-1}, $C_j \in \{A_j, B_j\}$, and let $z_1, \dots, z_R$ be the set of those simplices in Z whose image is contained $C_j(\Delta^{n-1})$ (if necessary, renumber the simplices).

We claim that $\sum_{i=1}^{R} \gamma_i z_i$ is a cycle mod $\partial C_j(\Delta^{n-1})$. Then, in a following step we prove that $\sum_{i=1}^{R} \gamma_i z_i$ is homologous to zero. Using the connecting homomorphism d_* we are ready to proceed by induction.

<u>STEP I</u> : $\sum_{i=1}^{R} \gamma_i z_i \in Z_{n-1}(\, C_j(\Delta^{n-1}), \partial C_j(\Delta^{n-1})\,)$

<u>Proof of STEP I</u> : Assume that $\sum_{i=1}^{R} \gamma_i z_i$ is not an (n-1)-cycle mod $\partial C_j(\Delta^{n-1})$. Then there exists a singular (n-2)-simplex c in the chain $\delta(\sum_{i=1}^{R} \gamma_i z_i)$ such that $c(S_{n-2}) \not\subseteq \partial C_j(\Delta^{n-1})$. This means that $c(S_{n-2}) \cap C_j(\text{int } \Delta^{n-1}) \neq \emptyset$. However, Z is a cycle, i.e. $\delta Z = \sum_{i=1}^{R} \gamma_i \delta z_i + \sum_{i=R+1}^{q} \gamma_i \delta z_i = o$; hence,

there is a singular (n-2)-simplex c' in the chain $\sum_{i=R+1}^{q} \gamma_i \delta z_i$ such that $\lambda_1 c + \lambda_2 c' = o$ for certain $\lambda_1, \lambda_2 \in \{-1,1\}$, but this is impossible .

<u>STEP II</u> : $\left[\sum_{i=1}^{R} \gamma_i z_i\right] = o \in H_{n-1}(\, C_j(\Delta^{n-1}), \partial C_j(\Delta^{n-1})\,)$

<u>Proof of STEP II</u> : Define $d_j := \begin{cases} -e^{j+1} & , C_j = A_j \\ e^{j+1} & , C_j = B_j \end{cases}$, and let

$d_j * C_j(\Delta^{n-1}) := \{t \cdot d_j + (1-t) \cdot x : x \in C_j(\Delta^{n-1}),\ o \leq t \leq 1\}$ be the cone over $C_j(\Delta^{n-1})$ with vertex d_j. Then $d_j * C_j(\Delta^{n-1})$ is a hom.n-dim. polyhedron [9] and we have a simplicial map L_j

$$L_j : (\Delta^n, \Sigma^{n-1}) \rightarrow (d_j * C_j(\Delta^{n-1}), \partial(d_j * C_j(\Delta^{n-1}))\,)$$

$$\text{vertex } y \mapsto \begin{cases} y & , \quad y \in C_j(\Delta^{n-1}) \\ d_j & , \quad \text{otherwise} \end{cases}$$

L_j induces an isomorphism in homology

$$(L_j)_*: \tilde{H}_{n-1}(\Sigma^{n-1}) \to \tilde{H}_{n-1}(\ \partial(d_j * C_j(\Delta^{n-1}))\ ),$$

and it can be shown that

$$(L_j)_*\left(\left[\sum_{i=1}^{q} \gamma_i z_i\right]\right) = \left[\sum_{i=1}^{q} \gamma_i (L_j(z_i^o), \ \dots \ , L_j(z_i^{n-1}))\right]$$

(This is a variant of the homotopy invariance in homology; the techniques used for a proof can be found in [9; ch.4.3] .)

Now let D be the map

$$D: H_{n-1}(C_j(\Delta^{n-1}), \partial(C_j(\Delta^{n-1}))) \to H_n(d_j * C_j(\Delta^{n-1}), \partial(d_j * C_j(\Delta^{n-1})))$$
$$[\Sigma\, \alpha_i c_i] \qquad \mapsto \qquad [\Sigma\, \alpha_i \tilde{c}_i]$$

where $\tilde{c}_i$ is the n-simplex

$$\tilde{c}_i: S_n \to d_j * C_j(\Delta^{n-1}) \quad , \quad x = \sum_{i=o}^{n} \lambda_i e^i \mapsto \begin{cases} d_j & , \lambda_o = 1 \\ (1-\lambda_o)c_i(y) + \lambda_o d_j, & \text{otherwise} \end{cases}$$

$$(\ y := \sum_{i=1}^{n} (\lambda_i/(1-\lambda_o))e^{i-1} \ \varepsilon \ S_{n-1} \ )$$

We conclude that for an affine singular (n-1)-simplex $c_i = (c_i^o, \ \dots \ , c_i^{n-1})$ one has $\tilde{c}_i = (d_j, c_i^o, \ \dots \ , c_i^{n-1})$, and that D is an isomorphism.

Now consider the following diagram

$$\begin{array}{ccccc} \tilde{H}_{n-1}(\Sigma^{n-1}) & \xrightarrow[(L_j)_*]{} & \tilde{H}_{n-1}(\partial(d_j * C_j(\Delta^{n-1}))\ ) & \underset{d_*}{\overset{\simeq}{\leftarrow}} & H_n(d_j * C_j(\Delta^{n-1}), \partial(d_j * C_j(\Delta^{n-1}))\ ) \\ & & & & \simeq \uparrow D \\ & & & & H_{n-1}(C_j(\Delta^{n-1}), \partial(C_j(\Delta^{n-1}))\ ) \end{array}$$

Chasing the diagram one has

$$[z] = \left[\sum_{i=1}^{q} \gamma_i z_i\right] \mapsto \left[\sum_{i=1}^{q} \gamma_i (L_j(z_i^o), \dots, L_j(z_i^{n-1}))\right] \mapsto \left[\sum_{i=1}^{q} \gamma_i (d_j, L_j(z_i^o), \dots, L_j(z_i^{n-1}))\right] =$$

$$= \left[\sum_{i=1}^{R} \gamma_i (d_j, L_j(z_i^o), \dots, L_j(z_i^{n-1}))\right] \mapsto \left[\sum_{i=1}^{R} \gamma_i (L_j(z_i^o), \dots, L_j(z_i^{n-1}))\right] = \left[\sum_{i=1}^{R} \gamma_i z_i\right]$$

Since $[Z] = o \in \tilde{H}_{n-1}(\Sigma^{n-1})$, the proof of STEP II is finished.

Before we proceed we observe that the map $C_j : \Delta^{n-1} \to C_j(\Delta^{n-1})$ is bijective, hence, each vertex z_i^k of $z_i = (z_i^o, \ldots, z_i^{n-1})$, $1 \leq i \leq R$ and $o \leq k \leq n-1$, has a unique preimage, say $\bar{z}_i^k \in \Delta^{n-1}$, under the mapping C_j. This yields an (n-1)-simplex $\bar{z}_i$

$$\bar{z}_i = (\bar{z}_i^o, \ldots, \bar{z}_i^{n-1}) : S_{n-1} \to \Delta^{n-1} \quad .$$

In view of STEP I and STEP II it is clear that $\sum_{i=1}^{R} \gamma_i \bar{z}_i$ is a cycle mod Σ^{n-2}, and that $\left[\sum_{i=1}^{R} \gamma_i \bar{z}_i\right] = o \in H_{n-1}(\Delta^{n-1}, \Sigma^{n-2})$.

Because of the linearity of the connecting homomorphism d_* it follows immediately

$$d_*\left[\sum_{i=1}^{R} \gamma_i \bar{z}_i\right] = \left[\sum_{i=1}^{R} \gamma_i \delta\bar{z}_i\right] = o \in \tilde{H}_{n-2}(\Sigma^{n-2}).$$

Moreover, after possible cancellations the cycle

$$\sum_{i=1}^{R} \gamma_i \delta\bar{z}_i = \sum_{i=1}^{R} \gamma_i \Big(\sum_{l=o}^{n-1} (-1)^l \, (\bar{z}_i^o, \ldots, \hat{\bar{z}}_i^l, \ldots, \bar{z}_i^{n-1}) \Big)$$

satisfies the hypothesis of the lemma. Hence, by induction hypothesis

$$\sum_{i=1}^{R} \gamma_i \Big(\sum_{l=o}^{n-1} (-1)^l \det(\bar{z}_i^o, \ldots, \hat{\bar{z}}_i^l, \ldots, \bar{z}_i^{n-1}) \Big) = o \,.$$

In particular, we have

$$\sum_{i=1}^{R} \gamma_i \det(z_i^o, \ldots, z_i^{n-1}) = \sum_{i=1}^{R} \gamma_i \det(C_j(\bar{z}_i^o), \ldots, C_j(\bar{z}_i^{n-1}))$$

$$= \pm (-1)^j \sum_{i=1}^{R} \gamma_i \det\begin{pmatrix} 1 & \ldots & 1 \\ \bar{z}_i^o & \ldots & \bar{z}_i^{n-1} \end{pmatrix}, \quad \text{according to the choice of } C_j \in \{A_j, B_j\}$$

$$= o \,.$$

Now, observe that each affine singular (n-1)-simplex z_i in Z, such that $z_i(S_{n-1})$ is contained in the intersection of two faces of Σ^{n-1}, fulfilles

$$\det(z_i^o, \ldots, z_i^{n-1}) = o.$$

Hence, summing up over the $2n$ faces of Σ^{n-1} we obtain the desired result:

$$\sum_{i=1}^{q} \gamma_i \det(z_i^o, \ldots, z_i^{n-1})$$

$$= \sum_{j=o}^{n-1} \Big(\sum_{\substack{z_i \text{ in } Z \\ z_i(S_{n-1}) \subseteq A_j(\Delta^{n-1})}} \gamma_i \det(z_i^o, \ldots, z_i^{n-1}) + \sum_{\substack{z_i \text{ in } Z \\ z_i(S_{n-1}) \subseteq B_j(\Delta^{n-1})}} \gamma_i \det(z_i^o, \ldots, z_i^{n-1}) \Big)$$

$$= \quad o \; . \; \blacksquare$$

Using local homology, $H_{n-1}(\Sigma^{n-1}, \Sigma^{n-1} - \{p\})$ for an appropriate $p \in \Sigma^{n-1}$, the proof of the (crucial) STEP II can be replaced by a simpler one [13]. However, the proof we gave above may be used to see a relation between (1.1) and (1.2) .

Instead of a rigorous proof we confine ourselves to some pictures :

As we have seen by collapsing the face $B_o^n(\Delta^{n-1})$ to the point $b^o = (-1, o, \ldots, o)$, i.e. via the map

$$L_o^n = L_o : (\Delta^n, \Sigma^{n-1}) \to (b^o * A_o^n(\Delta^{n-1}), \partial(b^o * A_o^n(\Delta^{n-1})))$$

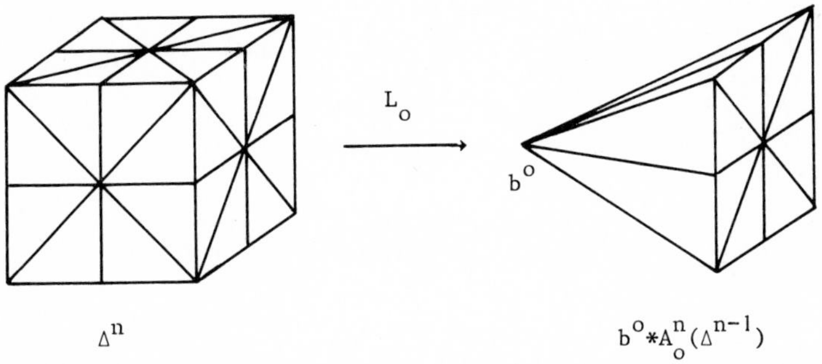

L_o b^o

Δ^n $b^o * A_o^n(\Delta^{n-1})$

the cycle $\sum_{i=1}^{q} \gamma_i z_i \in Z_{n-1}(\Sigma^{n-1})$ changes into $\sum_{i=1}^{q} \gamma_i (L_o^n(z_i^o), \ldots, L_o^n(z_i^{n-1})) \in$

$\in Z_{n-1}(\partial(b^o * A_o^n(\Delta^{n-1})))$

Now the map $L_o^{n-1} : (\Delta^{n-1},\Sigma^{n-2}) \to (b^o * A_o^{n-1}(\Delta^{n-2}), \partial(b^o * A_o^{n-1}(\Delta^{n-2})))$ which collapses the face $B_o^{n-1}(\Delta^{n-2})$ of Σ^{n-2} to the point $b^o = (-1,o, \dots ,o)$ in $\mathbb{R}^{n-1}$ may be seen also as a map

$$(A_o^n(\Delta^{n-1}),A_o^n(\Sigma^{n-2})) \to (A_o^n(b^o) * A_o^n A_o^{n-1}(\Delta^{n-2}), \partial(A_o^n(b^o) * A_o^n A_o^{n-1}(\Delta^{n-2})))$$

Joining with $b^o \varepsilon \mathbb{R}^n$ we get a simplicial map

$$(\Delta^n,\Sigma^{n-1}) \xrightarrow[L_o^n]{} (b^o * A_o^n(\Delta^{n-1}), \partial(b^o * A_o^n(\Delta^{n-1}))) \xrightarrow{b^o * L_o^{n-1}} (b^o * b^1 * A_o^n A_o^{n-1}(\Delta^{n-2}), \partial(b^o * b^1 * A_o^n A_o^{n-1}(\Delta^{n-2})))$$

$$\text{vertex } y \mapsto \begin{cases} b^o & , \ y \ \varepsilon \ B^o \\ b^1 & , \ y \ \varepsilon \ B^1 \\ y & , \ \text{otherwise} \end{cases}$$

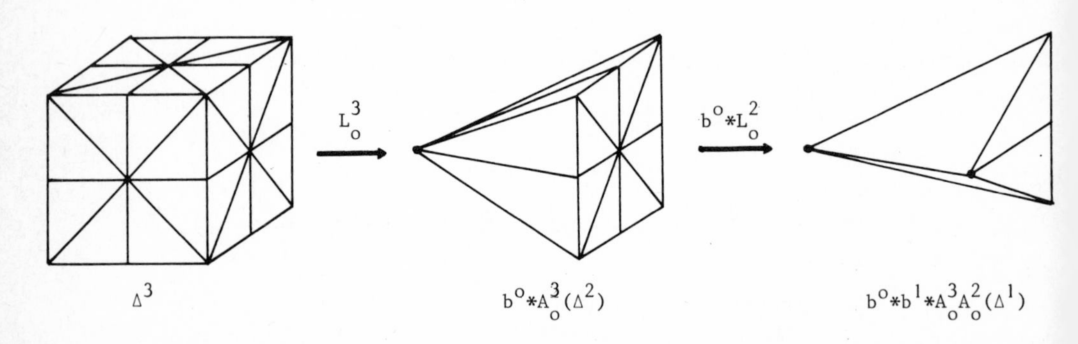

Repeating this procedure in each dimension, finally, one gets a simplicial map

$$L : (\Delta^n,\Sigma^{n-1}) \to (S,\partial S) \quad , \quad S = co\{b^o, \dots ,b^n\}$$
$$\text{vertex } y \mapsto b^{\ell(y)}$$

L induces an isomorphism in homology, and again we have the identity

$$L_*\left(\left[\sum_{i=1}^{q} \gamma_i z_i \right] \right) = \left[\sum_{i=1}^{q} \gamma_i (L(z_i^o), \dots ,L(z_i^{n-1})) \right] \quad .$$

Observe that L_* maps

$$\left[\sum_{i=1}^{k(n)} \varepsilon_i a_i\right] \varepsilon \tilde{H}_{n-1}(\Sigma^{n-1}) \quad \text{into} \quad \left[\sum_{i=o}^{n} (-1)^i (b^o, \ldots, \hat{b}^i, \ldots, b^n)\right] \varepsilon \tilde{H}_{n-1}(\partial S) .$$

Therefore we conclude

$$\left[\sum_{j=1}^{m} \eta_j (L\circ E(t_j^o), \ldots, L\circ E(t_j^{n-1}))\right] = \deg(f, \text{int } P, o) . \left[\sum_{i=o}^{n} (-1)^i (b^o, \ldots, \hat{b}^i, \ldots, b^n)\right]$$

Using notation of section III this means (comp. 3.6):

(7.2) If the triangulation T of the hom. n-dim. polyhedron P satisfies condition (SC II) , for each $j \varepsilon \{o, \ldots, n\}$ the degree $\deg(f, \text{int } P, o)$ is determined by the identity

$$\deg(f, \text{int } P, o) = \sum_{\substack{\tau \varepsilon T \\ \tau\ \ell^f\text{-}\hat{j}\text{ - Sperner simplex}}} \text{or}(\tau) .$$

<u>Remark</u> : Observe that for $j=o$ (7.2) coincides with B. Kearfotts "Parity theorem" which he deduces from Stenger's formula (1.1) by combinatorial arguments, provided the boundary of P is "sufficiently refined rel to sign f" [8] .

VIII. NUMERICAL ASPECTS :

In the previous section we have pointed out, how (1.1) and (1.2) are related, provided the triangulation of ∂P is fine enough. However, the smallness condition (SC I) is much less restrictive than condition (SC II). This is demonstrated by the following two examples:

(8.1) Consider $P := \Delta^2$ and a continuous mapping $f = (f_1, f_2) : (P, \partial P) \to (\mathbb{R}^2, \mathbb{R}^2 - \{o\})$ with the following behavior :

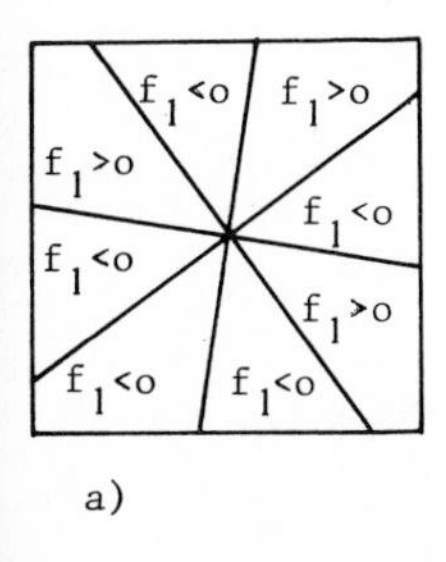

a)

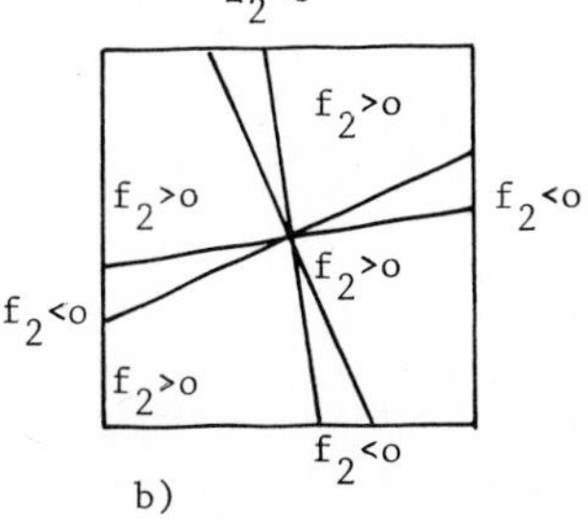

b)

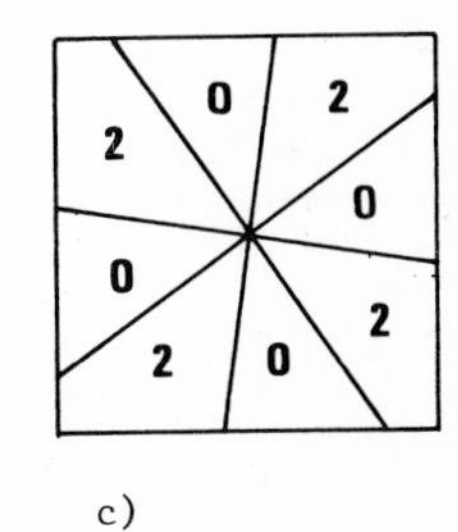

c)

The labeling induced by f is described in fig. c) . The following triangulations of ∂P are the coarsest possible according to (SC I) and (SC II) :

Following the pattern indicated in the above figures one can construct mappings $f: (P,\partial P) \to (\mathbb{R}^2, \mathbb{R}^2-\{o\})$ such that the mesh size of a triangulation satisfying (SC II) is any multiple of the mesh size of a triangulation satisfying (SC I) .

(8.2) Another example is provided by complex multiplication on the unit ball in $\mathbb{C}$: $f: P = \{z \in \mathbb{R}^2 : z_1^2 + z_2^2 \leq 1\} \to \mathbb{R}^2$, $f(z) := z^n$ $(n \neq o)$

It is easy to verify that any triangulation of ∂P satisfying (SC II) consists of at least $4|n|$ 1-simplices ; however, a triangulation satisfying (SC I) with $3|n|$ 1-simplices is easy to construct.

A numerical procedure based on theorem 3.6 was implemented on a computer and tested for $P = [o,1]^n$, $n \leq 5$ and mesh $T \geq o.2$. In all examples the cpu-time was below 1 min.

IX. ACKNOWLEDGEMENT : The problems considered in this paper were brought to our attention by Prof. Heinz-Otto Peitgen. We are indebted to him for advice and many valuable discussions.

REFERENCES

1 : P. ALEXANDROFF and H. HOPF : Topologie I , Springer, Berlin (1974) Berichtigter Reprint

2 : D.I.A. COHEN : On the Sperner lemma , J. Comb. Theory 2 (1967), 585 - 587

3 : J. CRONIN : Fixed points and topological degree in nonlinear analysis, AMS Math. Surveys No. 11 , Providence Rhode Island (1964)

4 : K. DEIMLING : Nichtlineare Gleichungen und Abbildungsgrade, Springer, Berlin (1974), Hochschultext

5 : A. DOLD : Lectures on algebraic topology , Springer, Berlin (1972)

6 : B.C. EAVES : A view of complementary pivot theory (or solving equations with homotopies), these proceedings

7 : G. EISENACK and C. FENSKE : Fixpunkttheorie , Bibliographisches Institut Mannheim (1978)

8 : B. KEARFOTT : An efficient degree-computation method for a generalized method of bisection, submitted to Numer. Math.

9 : C. MAUNDER : Algebraic topology , van Nostrand Reinhold Comp. (197o)

1o : H.O. PEITGEN and M. PRÜFER : The Leray-Schauder continuation method is a constructive element in the numerical study of nonlinear eigenvalue and bifurcation problems, these proceedings

11 : M. PRÜFER : Sperner Simplices and the Topological Fixed Point Index, University of Bonn, SFB 72, Preprint No. 134 (1977)

12 : J. SCHWARTZ : Nonlinear functional analysis, New York, Gordon and Breach (1969)

13 : H.W. SIEGBERG : Abbildungsgrade in Analysis und Topologie, Diplomarbeit, University of Bonn (1977)

14 : H.W. SIEGBERG : Brouwer degree : history and numerical computation, to appear

15 : F. STENGER : Computing the topological degree of a mapping in $\mathbb{R}^n$, Numer. Math. 25 (1975), 23 - 38

16 : M. STYNES : An algorithm for the numerical calculation of the degree of a mapping , Thesis, Oregon State University (1977)

17 : H. WHITNEY : Geometric integration theory, Princeton University Press (1964)

Michael Prüfer
Fachbereich 4 der Universität
Kufsteiner Str.
28oo Bremen
Fed. Rep. Germany

Hans Willi Siegberg
Institut für Angewandte Mathematik
der Universität
Wegelerstr.
53oo Bonn, Fed. Rep. Germany

PERTURBATIONS IN FIXED POINT ALGORITHMS

R. Saigal
Y. S. Shin
Department of Industrial Engineering and Management Sciences
Northwestern University
Evanston, Illinois 60203

ABSTRACT

If the Jacobian of a differentiable function is singular at a zero of the function, any piecewise linear approximation to it may not have a zero, even when the function has one. In this paper we present a technique for perturbing such functions so that the recent fixed point algorithms that trace zeros of piecewise linear homotopies will succeed in finding a zero of such functions. We also show how to unperturb in case the Jacobian is nonsingular at the solution, and thus not impede the super linear convergence attained by these algorithms.

This work is supported by the grant MCS77-03472 from the National Science Foundation.

1. Introduction: Let R^n be the n dimensional Euclidean space and let $\|x\|$ be the Euclidean norm of a vector $x = (x_1,\ldots,x_n)$ in R^n. Given a function $\ell \equiv (\ell_1,\ell_2,\ldots,\ell_n)$ from R^n into R^n, in this paper we consider the problem of computing an x in R^n such that $\ell(x) = 0$. In particular, we consider the recent fixed point algorithms, pioneered by Scarf [12], for computing such an x. These algorithms, namely the continuous deformation method of Eaves and Saigal [2] and the restart method of Merrill [6] and Kuhn and MacKinnon [5] have the ability to compute fixed points (or equivalently, zeros) of mappings under a variety of conditions on ℓ, which include the conditions of the Brouwer fixed point theorem [1] and the Leray-Schauder theorem [7, 6·3·3]. Appropriate generalizations of these fixed point theorems to point-to-set mappings exist, and these algorithms are effective in computing fixed points of such mappings as well; see for example [2, Theorem 4.2].

Fixed point algorithms are effective in computing a zero under weaker conditions on ℓ than the traditional methods. These conditions are usually "global" in nature. There is a shortcoming, however. As these algorithms create a piecewise linear homotopy L between a sufficiently good piecewise linear approximation to ℓ and some simple (usually one-to-one linear) mapping and follow its path of zeros, for a certain class of mappings, including $\ell(x) = x^2$, no piecewise linear approximations have zeros. For these functions this strategy will fail, though the traditional methods may succeed. For such a mapping ℓ, the derivative $D\ell$ is singular at the zero. It is the aim of this study to present one possible way to overcome this difficulty.

Our approach is to perturb the function ℓ such that sufficiently good piecewise linear approximations to the perturbed function have zeros. Since the perturbation has a detrimental effect on the accuracy of the solution, we then devise a scheme to decrease the perturbation as the algorithm moves closer to the solution. Further, in case the perturbation is not necessary, which would be the case if the derivative $D\ell$ is nonsingular at the solution, we devise a procedure to drop the perturbation to zero. This then involves a hybridization of the fixed point algorithm with a generalized Katzenelson algorithm [3].

In section 2 we present a brief overview of the continuous deformation method [2]. In section 3 we present a perturbed version of the algorithm in section 2, and in section 4 we present a hybridized algorithm for dropping the perturbation if it is not required.

2. The Continuous Deformation Method:

In this section we describe the continuous deformation method of Eaves and Saigal [2]. In this method, an initial one-to-one affine mapping r is chosen. Let its unique zero be x^1. Starting with x^1, a piecewise linear path x^t is traced such that if such a path converges, it converges to a zero of ℓ. This scheme shares many similarities with the recent "globalization schemes" for Newton's method proposed by Kellogg, Li and Yorke [4] and Smale [13], where a similar, but differentiable, path is traced via a solution to a differential equation.

The usual implementation of this algorithm is on the triangulation J_3 of $R^n\times(0,1]$, Todd [14]. One creates a piecewise linear homotopy L on $R^n\times[0,1]$ such that L restricted to $R^n\times\{1\}$ is r, and restricted to $R^n\times\{0\}$ is ℓ. The algorithm then traces a component of $L^{-1}(0)$, which contains $(x^1,1)$.

The triangulation J_3 is a collection of $(n+1)$ simplexes which partition $R^n\times(0,1]$, and whose vertices lie in $R^n\times\{2^{-k}\}$ for $k = 0,1,\cdots$. The homotopy L is then created by defining

$$L(v,2^{-k}) = \begin{cases} \ell(v) & \text{if } k > 0 \\ r(v) & \text{if } k = 0 \end{cases}$$

where $(v,2^{-k})$ is a vertex of the triangulation. The mapping L is then extended linearly on each simplex σ in J_3. In case ℓ is continuous, and we set $L(x,0) = \ell(x)$, then $L:R^n\times[0,1] \to R^n$ is a piecewise linear homotopy from r to ℓ. The method then traces the unique component of $L^{-1}(0)$ which contains $(x^1,1)$ in the following manner. Starting with the usually unique simplex σ_0 of J_3 containing $(x^1,1)$, it traces the line $L_{\sigma_0}^{-1}(0)$ where L_σ is the restriction of L to σ. This line crosses some other facet of σ_0 (assumed unique by a nondegeneracy assumption) at some point (x',t'). This procedure is then continued by finding the other simplex σ' of J_3 which contains this point, and tracing the line $L_{\sigma'}^{-1}(0)$. Under certain general conditions on ℓ and r this piecewise linear path converges to a zero of ℓ; see for example [2, Theorem 4.2].

3. A perturbed algorithm:

Throughout this and the remaining sections we will assume that ℓ is continuously differentiable with its derivative $D\ell$ satisfying the Hölder condition

$$\|D\ell(x) - D\ell(y)\| \leq K\|x - y\|^\alpha \tag{3.1}$$

for some positive constants K and α. For an $n\times n$ matrix, $\|A\| = \max_{\|x\|=1} \|Ax\|$.

Let $p \neq 0$ and $\alpha > 0$ be as above. Consider the piecewise linear homotopy L' from $R^n\times(0,2\varepsilon]$ into R^n defined by the triangulation J_3, where, for a vertex (x,t)

we set

$$L'(x,t) = \begin{cases} r(x) & t = 2\varepsilon \\ \ell(x) - t^{1+\alpha}\, p & t = \varepsilon \cdot 2^{-k},\ k = 0,1,\ldots \end{cases} \tag{3.2}$$

where, as before, r is a one-to-one affine mapping whose zero is x^1. We will now give conditions under which the component of $L'^{-1}(0)$ containing $(x^1, 2\varepsilon)$ converges to a zero of ℓ. We first state a condition on the choice of p.

Condition 3.2: There exists an open bounded set C containing x^1, and an open convex set $G \subset \ell(C)$ such that 0 lies in the closure $\overline{G}$ of G, p is in G and $G \cap \ell(\partial C) = \emptyset$, where ∂C is the boundary of the set C.

We now state the main theorem which specifies the conditions under which the fixed point algorithm implementing (3.2) will converge to a zero of ℓ.

Theorem 3.3: Let C and G satisfy Condition 3.2. Also, let $\ell(x) - p + \rho r(x) \neq 0$ for all $\rho \geq 0$ and x in the boundary of C. Then, for some sufficiently small initial grid size, the algorithm implementing the homotopy L' will generate a sequence of points that converge to a zero of ℓ.

We now give an outline of the proof of the theorem. The full details can be found in Saigal and Shin [11]. Define, for some sufficiently small $\varepsilon > 0$,

$$\varepsilon_k = \varepsilon \cdot 2^{-k} \qquad k = 0,1,\ldots \tag{3.3}$$

and for k = 1,2,..., the sets

$$G_k = \{y \in G : \|y - y'\| > \frac{K}{2}\varepsilon_k^{1+\alpha} \quad \text{for all } y' \in \partial G\}. \tag{3.4}$$

Lemma 3.4: There exists an $\varepsilon' > 0$ such that for all ε_k in $(0, \varepsilon')$, $\beta p \in G_k$ for $(\frac{1}{2})^{k(1+\alpha)} \leq \beta \leq 1$, for k = 1,2,....

Proof: Can be readily confirmed using (3.3) and (3.4).

For a given t, let $\ell_t(x) = L'(x,t)$. Then, it can be readily shown that:

Lemma 3.5: Let $t \leq \varepsilon \cdot 2^{-k}$. Then $\|\ell(x) - \ell_t(x)\| \leq \frac{1}{2} K \varepsilon_k^{1+\alpha}$.

Proof: Is identical to the proof of [Theorem 3.1, 10], using the fact that the grid of the induced triangulation by J_3 of $R^n \times \{t\}$ is bounded by $\varepsilon \cdot 2^{-k}$ for $t \leq \varepsilon \cdot 2^{-k}$.

For the proof of Theorem 3.3 it can be shown, using Lemmas 3.4 and 3.5, that, for sufficiently small initial grid size $\varepsilon > 0$, the path will never cross the boundary of C. This is done in a manner identical to the proof of [Thm. 1, 9].

In practical problems, selecting a vector p such that it satisfies the condition of Theorem 3.3 may be difficult. One possible way is to set $p = \ell(x^1)$ where x^1 is an initial estimate of the solution. Then if the algorithm fails, that is, it diverges after a certain level k, we may obtain some information about the function which may help to modify the selection of p.

4. A Hybrid Algorithm:

As is now well known, if the Jacobian is nonsingular at a zero of the function to which the algorithm is converging, these algorithms can be made to converge quadratically, Saigal [8]. The algorithm presented in section 3 loses this convergence property even when the Jacobian is nonsingular at the solution found. Thus, to improve the efficiency of the algorithm, we need to drop the perturbation from further applications in case it is not needed. To do this, we introduce the generalized Katzenelson algorithm, which is then used to eliminate the perturbation, when not needed. The hybrid algorithm we thus consider in this paper is the continuous deformation method that has been perturbed, along with the generalized Katzenelson algorithm. In this hybridization, we use the latter algorithm on an induced subdivision by J_3 of $R^n \times \{t\}$ for some sufficiently small t. We now present an overview of the latter algorithm with the convergence conditions. Detailed analysis of the algorithm and the proofs of the theorems can be found in Saigal and Shin [11].

Let $\bar{\ell}: R^n \to R^n$ be a piecewise linear mapping, and let $\bar{\ell}_\sigma(x) = A_\sigma x - a_\sigma$, where $\bar{\ell}_\sigma$ is the restriction of the mapping $\bar{\ell}$ to the piece σ of linearity of $\bar{\ell}$, A_σ is an $n \times n$ matrix and a_σ is an n vector. Define $\bar{L}: R^{n+1} \to R^n$ by

$$\bar{L}(x,\theta) = \bar{\ell}(x) - \theta p. \qquad (4.1)$$

We will usually pick $p = \bar{\ell}(x^1)$ for some vector x^1. We now assume that $p = \bar{\ell}(x^1)$. Then, $\bar{L}(x^1,1) = 0$. We now seek a solution $\bar{L}(x,0) = \bar{\ell}(x) = 0$. We obtain the Katzenelson algorithm, [3], by initiating the algorithm at point $(x^1,1)$, and in the direction for which θ decreases. The algorithm then traces the component of $\bar{L}^{-1}(0)$ which contains $(x^1,1)$, and which possibly meets $R^n \times \{0\}$.

We now state a condition for the success of this algorithm. Let

$$D = \{x: \bar{\ell}(x) = \bar{\ell}(x^1)\}$$

and we assume that the set D is finite. Also, let the point $x^k \in D$ lie in the piece of linearity σ_k of $\bar{\ell}$, for each $k = 1,\ldots,m$ (for some m).

<u>Theorem 4.1</u>: Let $\det A_{\sigma_k} \cdot A_{\sigma_1} \geq 0$ for all $k = 1,\ldots,m$. Then, the path generated by the above algorithm will converge to a zero of $\bar{\ell}$.

Proof: It can be readily shown that the algorithm will either compute a zero of $\bar{\ell}$, or will find another solution x^k, $k \neq 1$. In the latter case, it can be shown that $\det A_{\sigma_k} A_{\sigma_1} < 0$, and so we have our result.

As a corollary to this theorem we prove the following result: let ℓ be a C^1 mapping as before, and let ℓ_T be a piecewise linear approximation to ℓ on some subdivision T of R^n. Let the grid size of T be $\varepsilon > 0$.

Corollary 4.2: Let $D = \{x: \ell(x) = \ell_T(x')\}$ for some x^1. If $\operatorname{Det} D\ell(x)D\ell(x^1) > 0$ for all x in D, and the grid size ε of T is sufficiently small, then the generalized Katzenelson algorithm will compute a zero of ℓ_T.

Proof: Using [Theorem 3.2, 10], the result follows readily from Theorem 4.1.

There is now an obvious hybrid algorithm combining the one in section 3 with the one above. Given the homotopy L' of (3.2), define the piecewise linear mapping

$$\begin{aligned} \ell'_k(x) &= L'(x,t_k) \\ &= \bar{\ell}_k(x) - (\tfrac{1}{2})^{k(1+\alpha)} \cdot p \cdot \varepsilon^{1+p} \qquad t_k = \varepsilon \cdot 2^{-k} \end{aligned}$$

for $k = 0,1,2,\ldots$, where $\bar{\ell}_k(x)$ is the piecewise linear approximation of ℓ generated by the induced triangulation by J_3 of $R^n \times \{\varepsilon \cdot 2^{-k}\}$, with the grid size $\varepsilon \cdot 2^{-k}$. Now, whenever the perturbed algorithm generates an approximate solution x^k such that $L'(x^k,t_k) = 0$ for the first time at level k, the generalized Katzenelson algorithm can be applied to decrease the perturbation. If the conditions of Theorem 4.1 are satisfied, which would be so if the algorithm is converging to a solution where the Jacobian is nonsingular (see corollary 4.2), the latter algorithm will reduce the perturbation to zero.

We now state a condition under which the hybrid algorithm will converge. For this we introduce

Condition 4.3: There exists an open bounded set C containing x^k. Let x^ℓ in C, $\ell = 1,\ldots,m_k$ be points in C such that $\bar{\ell}_k(x^\ell) = \bar{\ell}_k(x^k)$. Also, let σ_k contain x^k and σ_ℓ contain x^ℓ, $\ell = 1,\ldots,m_k$, respectively. Then $\det A_{\sigma_k} A_{\sigma_\ell} > 0$ for each $\ell = 1,\ldots,m_k$ and $\beta p \notin \bar{\ell}_k(\partial C)$ for all β in $[0,1]$.

We are now ready to state our theorem:

Theorem 4.4: Let C and ℓ satisfy the condition 4.3. Then, the hybrid algorithm will converge to a zero of ℓ. Also, asymptotically the algorithm will drop the perturbation.

Proof: Since ℓ satisfies condition 4.3, if x^t is the path traced by the Katzenelson algorithm, then as $\bar{\ell}_k(x^t) = \theta p$ for some $0 \leq \theta \leq 1$, or as $\beta p \notin \bar{\ell}_k(\partial C)$ for all β in [0,1], the path x_t lies inside C. The result now follows by an argument similar to that of the proof of Theorem 4.1.

From the computational point of view, if applying the second algorithm repeatedly at each new level k does not reduce the perturbation, this may be due to the nature of the mapping ℓ. In this case it would be advisable not to apply the latter algorithm after a certain level k has been reached.

References

[1] L. E. J. Brouwer, "Über Abbildung von Mannigfaltigkeiten," Math. Ann., 71 (1912), pp. 97-115.

[2] B. C. Eaves and R. Saigal, "Homotopies for computing fixed points on unbounded regions," Math. Programming, 3 (1972), pp. 225-237.

[3] J. Katzenelson, "An Algorithm for solving the Nonlinear Resistor Networks," Bell Telephone Tech. J., 44 (1965), pp. 1605-1620.

[4] R. B. Kellogg, T. Y. Li and J. Yorke, "A constructive proof of the Brouwer fixed point theorem and computational results," SIAM J. Numer. Analysis, 13 (1976), pp. 473-483.

[5] H. W. Kuhn and J. G. MacKinnon, "Sandwich Method for Finding Fixed Points," J.O.T.A., 17 (1975), pp. 189-204.

[6] O. H. Merrill, "Applications and Extensions of an Algorithm that Computes Fixed Points of Certain Upper Semi-continuous Point-to-set Mappings," Ph.D. Dissertation, Univ. of Michigan, Ann Arbor, Michigan, 1972.

[7] J. M. Ortega and W. C. Rheinboldt, Iterative Solutions of Nonlinear Equations in Several Variables, Academic Press, New York, 1970.

[8] R. Saigal, "On the convergence rate of algorithms for solving equations that are based on methods of complementary pivoting," Math. of Operations Res., 2 (1977), pp. 108-124.

[9] R. Saigal, "Fixed Point Computing Methods," Encyclopedia of Computer Science and Technology, Marcel Dekker, New York, 1977.

[10] R. Saigal, "On piecewise linear approximations to smooth mappings," to appear in Math. of Operations Res.

[11] R. Saigal and Y. S. Shin, "Perturbations in fixed point algorithms," in preparation.

[12] H. E. Scarf, "The approximation of fixed points of a continuous mapping," SIAM J. Appl. Math., 15 (1967), pp. 1328-1343.

[13] S. Smale, "A convergent process of price adjustment and global Newton methods," J. Math. Econ., 3 (1976), pp. 107-120.

[14] M. J. Todd, The Computation of Fixed Points and Applications, Springer-Verlag, New York, 1976.

Bifurcation of a stationary solution of a dynamical system into n-dimensional tori of quasiperiodic solutions

Jürgen Scheurle

1. Introduction

Let the autonomous system

$$(1.1) \qquad \dot{\underline{x}} = \underline{X}(\underline{\lambda},\underline{x})$$

be given, where $\underline{x}$ denotes a function of $t \in \mathbb{R}$ with values in $\mathbb{R}^m$, $\underline{\lambda} \in \mathbb{R}^\ell$ is a constant vector and the mapping $\underline{X} : \mathbb{R}^\ell \times \mathbb{R}^m \to \mathbb{R}^m$ is smooth. We shall consider the components λ_k of $\underline{\lambda}$ as free parameters and we assume that (1.1) has the trivial solution $\underline{x} = \underline{0}$ for all values of $\underline{\lambda}$. Then we ask the question,whether there exist points $(\underline{\lambda}_o,\underline{0})$ on the basic solution each neighborhood of which (in a suitable topology of the product space) contains a quasiperiodic solution. Such points are said to be <u>bifurcation points</u> of (1.1).

The function $\underline{x}$ is called <u>quasiperiodic</u> with frequencies $\omega_1, \ldots ,\omega_n$, if these numbers are rationally independent and if $\underline{x}$ has the form

$$\underline{x}(t) = \underline{f}(\omega_1 t, \ldots ,\omega_n t) ,$$

where $\underline{f}(u_1, \ldots ,u_n)$ is continuous and periodic in $u_1, \ldots ,u_n$ with period 2π.

So we are studying oscillations with a finite number of rationally independent basic frequencies. Here we always assume that $n \leq m$ and that $\underline{f} : [0,2\pi]^n \to \mathbb{R}^m$ is a diffeomorphism of class C^1 with rank $\frac{\partial \underline{f}}{\partial \underline{u}}(\underline{u}) \geq n$ for all $\underline{u} = (u_1, \ldots ,u_n)$. Then the equation

$$\underline{v} = \underline{f}(u_1, \ldots u_n)$$

evidently describes an embedding of an n-dimensional torus in the m-dimensional $\underline{v}$-space. $u_1, \ldots ,u_n$ are coordinates on the torus if they are identified mod 2π (angle variables). $u_k = \omega_k t$ is a straight line in the $\underline{u}$-space which, identified mod 2π, covers the torus densely

In the usual way we try to answer the above question by linearizing the system (1.1) in the basic solution:

$$(1.2) \qquad \dot{\underline{x}} = \frac{\partial \underline{X}}{\partial \underline{x}}(\lambda,\underline{0})\underline{x}$$

It is well known that, for a fixed $\underline{\lambda} = \underline{\lambda}_o$, there is an n-parameter family of invariant tori

$$(1.3) \qquad \underline{v} = \sum_{k=1}^{n} c_k \mathrm{Re}(\underline{v}_k e^{iu_k}) \ , \quad c_k \in \mathbb{R}$$

of (1.2), provided that the Jacobian $\frac{\partial \underline{X}}{\partial \underline{x}}(\underline{\lambda}_o,\underline{0})$ has at least $n \leq \frac{m}{2}$ pairs of purely imaginary eigenvalues $\pm i\omega_k$, $1 \leq k \leq n$. Here $\underline{v}_k$ is an eigenvector corresponding to the eigenvalue $i\omega_k$. The solutions on the tori are described by

$$u_k = \omega_k t + d_k \ , \quad d_k \in \mathbb{R} \ .$$

Thus we have quasiperiodic solutions. $(\underline{\lambda}_o,\underline{0})$ is a bifurcation point of (1.2), where the bifurcation happens in the vertical direction with respect to $\underline{\lambda}$, i.e. the value $\underline{\lambda}_o$ remains fixed, and all the bifurcating solutions have the same basic frequencies.

Let us assume now, that $\frac{\partial \underline{X}}{\partial \underline{x}}(\underline{\lambda}_o,\underline{0})$ has exactly n pairs of purely imaginary eigenvalues $\pm i\omega_k$, which are simple. Then it can be shown by the same methods as in [5], that there is at most a 2n-parameter family of solutions of the non-linear system

$$(1.4) \qquad \dot{\underline{x}} = \underline{X}(\underline{\lambda}_o,\underline{x}) \ ,$$

which are small in the sup norm over the whole real line. Hence, because of the invariance under translations of the independent variable t, (1.4) has at most an n-parameter family of invariant tori near the origin. This is a uniqueness result in some sense, which in particular implies, that $(\underline{\lambda}_o,\underline{0})$ cannot be a bifurcation point of (1.1), if the real parts of all eigenvalues of $\frac{\partial \underline{X}}{\partial \underline{x}}(\underline{\lambda},\underline{0})$

are different from zero for $\underline{\lambda}$ near $\underline{\lambda}_o$.

However, equation (1.4) does not have so many "small" solutions in general. Indeed, in the case $2n = m$, the existence of a 2n-parameter family of uniformly bounded small solutions would be equivalent to the fact that the stationary solution $\underline{x} = \underline{0}$ is stable with respect to the past and to the future. But already for $n = 1$, a perturbed stable equilibrium becomes generically a repellor either for $t \to +\infty$ or $t \to -\infty$ (see [6]). Systems describing a non-dissipative motion such as Hamiltonian systems or reversible systems are exceptional cases. If (1.4) belongs to one of those classes of differential equations, then the perturbation theory for quasiperiodic solutions developed by A.N. Kolmogorov, V.I. Arnol'd, and J. Moser yields at least a weakened stability result, provided that the numbers ω_k are rationally independent and that the third order terms of a certain normal form satisfy a non-degeneracy condition (see [8]): For any $\varepsilon > 0$ there exists a neighborhood $U = U(\varepsilon)$ of the origin containing a set Q of quasiperiodic solutions such that $\mu(U-Q) < \varepsilon\mu(U)$, where μ denotes the m-dimensional Lebesgue measure. Thus the majority of the solutions in U are quasiperiodic. Examples show, that the solutions of the excluded set may even be unbounded.

In the present paper we shall consider the general case $m \geqq 2n$. (1.1) is assumed belonging to a class of the above mentioned conservative systems for all values of the parameters λ_k. It is our purpose to show the local persistence of the quasiperiodic solutions of the linearized equation without any further restriction for the non-linear terms and without any assumption on those eigenvalues of $\frac{\partial \underline{X}}{\partial \underline{x}}(\underline{\lambda}_o,\underline{0})$ lying off the imaginary axis.

An important contribution in this direction is given by the Hopf bifurcation ([2]). A single parameter $\lambda \in \mathbb{R}$ is sufficient for this kind of bifurcation phenomenon. It is usually assumed that, at $\lambda = \lambda_o$, a pair of simple, non-real, conjugated eigenvalues $\nu(\lambda)$, $\overline{\nu}(\lambda)$ of the matrices $\frac{\partial \underline{X}}{\partial \underline{x}}(\lambda,\underline{0})$ crosses the imaginary axis with non-vanishing velocity, say at $\pm i\omega_1$. That is $\nu(\lambda_o) = i\omega_1$ and $\frac{d\nu}{d\lambda}(\lambda_o) \neq 0$. Further, for $\lambda = \lambda_o$, there should be no other eigenvalue which is an integer multiple of $i\omega_1$. Then, for a fixed phase, a 1-parameter branch of periodic solutions of (1.1) emanates from the basic solution. Their periods change continuously near $2\pi/\omega_1$. There are also results concerning the global existence of the branching solutions and generalizations for multiple eigenvalues ([1], [3]). Furthermore let us mention G. Joos who, in his lecture notes [4], considers the bifurcation under a Hopf-type condition of a periodic solution into 2-dimensional invariant tori.

However, conservative dynamical systems have the property, that a pair of simple eigenvalues ν, $\bar{\nu}$, which are continuous functions of $\underline{\lambda}$, remains on the imaginary axis provided it is there for $\underline{\lambda} = \underline{\lambda}_o$. That is to say, if ν is an eigenvalue of $\frac{\partial \underline{X}}{\partial \underline{x}}(\underline{\lambda},\underline{0})$, so is $-\nu$. Hence, as long as we are dealing only with real systems, the pair ν, $\bar{\nu}$ cannot leave the imaginary axis. Otherwise, the pair would have to split up into 4 eigenvalues.

Therefore it is convenient to suppose that, for all $\underline{\lambda}$, the Jacobian $\frac{\partial \underline{X}}{\partial \underline{x}}(\underline{\lambda},\underline{0})$ possesses exactly $n \leqq \frac{m}{2}$ pairs of purely imaginary eigenvalues. In the following these are assumed to be simple and to move along the imaginary axis with non-vanishing velocities. With these assumptions it turns out, that $(\underline{\lambda}_o,\underline{0})$ is a bifurcation point of (1.1). Quasi-periodic branching solutions can arise with any number of frequencies from 1, ... ,n. In order to produce these solutions we require that the number of available parameters λ_k should not be less than n-1. The parameters are needed to control the frequencies along each solution branch. To shorten the description of this work, we shall only sketch the proofs here and content ourselves mainly with stating the results. We shall restrict ourselves to reversible systems, but analogous results may of course be formulated in the case of Hamiltonian systems.

2. Results

We shall now consider a set of reversible systems of the form (1.1) near the equilibrium $\underline{x} = \underline{0}$.

Definition 2.1 (J. Moser [8]):

The system (1.1) is said to be reversible, if there exists a linear reflection R of $\mathbb{R}^m$ with

(2.1) $$R^2 = \text{id} ,$$

such that

(2.2) $$\underline{X}(\underline{\lambda}, R\underline{x}) = -R\underline{X}(\underline{\lambda}, \underline{x})$$

holds for all $\underline{\lambda} \in \mathbb{R}^\ell$, $\underline{x} \in \mathbb{R}^m$. That is to say, if $\underline{x}(t)$ is a solution, so is $R\underline{x}(-t)$.

Because of (2.1), the reflection R has only the eigenvalues ± 1. Therefore, R may be represented as a diagonal matrix with entries ± 1 respectively to some basis of $\mathbb{R}^m$. So reversibility implies a certain kind of conservative character of the system. For example, the system

(2.3) $$\ddot{q}_k + \lambda_k q_k = Q_k(\underline{q}, \underline{p}) \ , \ 1 \leq k \leq n$$

$Q_k(\underline{q}, \underline{p}) = o(\|q\| + \|p\|)$, $\lambda_k > 0$, of weakly coupled oscillations corresponds to a reversible system of the form

(2.4) $$\begin{aligned} \dot{q}_k &= p_k \\ \dot{p}_k &= -\lambda_k q_k + Q_k(\underline{q}, \underline{p}) \quad , \end{aligned}$$

if either

(2.5) $$Q_k(\underline{q}, -\underline{p}) = Q_k(\underline{q}, \underline{p})$$

or

(2.6) $$\begin{aligned} Q_k(\underline{q}, -\underline{p}) &= -Q_k(\underline{q}, \underline{p}) \\ Q_k(-\underline{q}, -\underline{p}) &= Q_k(\underline{q}, \underline{p}) \end{aligned}$$

holds. Here the terms λ_k may be interpreted as free parameters, and

$\|\cdot\|$ denotes any norm in $\mathbb{R}^n$. In the case of (2.5) the correspon-ding reflection of $\mathbb{R}^{2n}$ is given by

$$R\underline{x} = \begin{pmatrix} \underline{q} \\ -\underline{p} \end{pmatrix} , \quad \underline{x} = \begin{pmatrix} \underline{q} \\ \underline{p} \end{pmatrix}$$

and in the case of (2.6) by

$$R\underline{x} = \begin{pmatrix} -\underline{q} \\ \underline{p} \end{pmatrix} , \quad \underline{x} = \begin{pmatrix} \underline{q} \\ \underline{p} \end{pmatrix} .$$

Another example for a reversible system is the N-body problem.

We start with the following hypothesis:

(H) The parameter-vector $\underline{\lambda}$ has n components and the Jacobian $\frac{\partial \underline{X}}{\partial \underline{x}}(\underline{\lambda},\underline{0})$ has exactly n pairs of simple eigenvalues $\pm i\omega_k(\underline{\lambda})$ for all $\underline{\lambda}$. The real parts of the other eigenvalues never vanish and

$$\det \frac{\partial \underline{\omega}}{\partial \underline{\lambda}}(\underline{\lambda}_o) \neq 0 , \tag{2.7}$$

where $\underline{\omega} = (\omega_1, \ldots ,\omega_n)$.

If the mapping $\underline{X} : \mathbb{R}^n \times \mathbb{R}^m \to \mathbb{R}^m$ is sufficiently smooth, it follows immediately under (H), that $(\underline{\lambda}_o,\underline{0})$ is a bifurcation point of equation (1.1) provided that the values $\omega_k = \omega_k(\underline{\lambda}_o)$ are rationally independent. That is, for all k, there is a branch of periodic solutions emanating from the basic solution at $(\underline{\lambda}_o,\underline{0})$ with periods $2\pi/\omega_k$. Just one of the parameters λ_k is enough to control the period along any one solution branch, while the remaining n-1 components of $\underline{\lambda}$ remain fixed. By choosing appropriate spaces of periodic functions, the proof of this result reduces at once to the classical bifurcation theory in a simple eigenvalue (c.f. [5]). Using the results of [9] one is even able to make assertions concerning the global existence of the non-trivial solution branches.

However, simple examples show that in general the set of those frequencies ω_k, $1 \leq k \leq n$, for which (1.1) possesses a quasiperiodic solution, is of the first category in the sense of Baire. It is usually assumed in the theory of quasiperiodic solutions that the frequencies ω_k satisfy the following condition for all integers j_k with $\|\underline{j}\| = \Sigma|j_k| \geq 1$ and some positive constants γ, τ:

$$(2.8) \qquad \left|\sum_{k=1}^{n} j_k\omega_k\right| \geq \gamma\|\underline{j}\|^{-\tau}$$

Note, that the set of $\underline{\lambda}$'s near $\underline{\lambda}_o$, for which the values $\omega_k(\underline{\lambda})$ in (H) do not fulfil (2.8), have Lebesgue measure zero.

Theorem 2.2:

Let the system (1.1) be reversible. Let the mapping $\underline{X} : \mathbb{R}^n \times \mathbb{R}^m \to \mathbb{R}^m$ be real analytic when $|\underline{\lambda} - \underline{\lambda}_o|$ and $\|\underline{x}\|$ are small, and suppose that $\underline{X}(\underline{\lambda},\underline{0}) = \underline{0}$. Moreover assume that (H) is satisfied, and suppose that the frequencies $\omega_k = \omega_k(\underline{\lambda}_o)$ fulfil the condition (2.8). Then, for every invariant torus (1.3) of the linearization at $\underline{\lambda} = \underline{\lambda}_o$ with $\|\underline{c}\| = 1$, $\underline{c} = (c_1, \ldots, c_n)$, there exists a 1-parameter family of invariant tori

$$\underline{v}(\varepsilon,\underline{u}) = \varepsilon \sum_{k=1}^{n} c_k \mathrm{Re}(\underline{v}_k e^{iu_k}) + \varepsilon\underline{\Psi}(\varepsilon,\underline{u})$$

of the system (1.1), where $\underline{\Psi} = O(\varepsilon)$ is periodic in $u_1, \ldots, u_n$ with period 2π, $|\varepsilon| < \varepsilon_o(\underline{c})$ and $\underline{\lambda} = \underline{\lambda}(\varepsilon) = \underline{\lambda}_o + O(\varepsilon)$. The dimension of the tori depends on the bifurcation direction and is identical to the number of non-vanishing components of $\underline{c}$, i.e. $\underline{v}$ does not depend on u_k provided that $c_k = 0$. The functions $\underline{\Psi} : \mathbb{R} \times \mathbb{R}^n \to \mathbb{R}^m$ and $\underline{\lambda} : \mathbb{R} \to \mathbb{R}^n$ are analytic in $|\varepsilon| < \varepsilon_o$. The solutions on the tori are quasiperiodic and described by

$$u_k = \omega_k t + d_k \,, \quad d_k \in \mathbb{R}.$$

In special cases one can even show, that the range of the parameter ε is bounded independently of $\underline{c}$. Then one obtains an n-dimensional analytic manifold of invariant tori near $\underline{x} = \underline{0}$, where the modified parameter $\underline{\lambda} = \underline{\lambda}(\varepsilon,\underline{c})$ is just so analytic in ε and $\underline{c}$. The system (2.3) of weakly coupled oscillations is an example of this, provided $Q_k|_{q_k=p_k=0} \equiv 0$ holds for every k.

We remark, that in fact just n-1 free parameters λ_k suffice to guarantee the existence of quasiperiodic solutions near the point $(\underline{\lambda}_o,\underline{0})$:

Theorem 2.3:

Suppose that the assumptions of Theorem 2.2 are satisfied for the system (1.1), but with $\underline{\lambda} \in \mathbb{R}^{n-1}$. Furthermore, suppose

$$(2.9) \qquad \det \left(\frac{\partial \underline{\omega}}{\partial \underline{\lambda}}(\underline{\lambda}_o) \;\; \underline{\omega}(\underline{\lambda}_o) \right) \neq 0$$

instead of (2.7) in (H). Then the conclusions of Theorem 2.2 hold, except that the solutions on the invariant tori are now of the form

$$u_k = \alpha\omega_k t + d_k \,, \quad d_k \in \mathbb{R},$$

where $\alpha = \alpha(\varepsilon)$ is an analytic function in $|\varepsilon| < \varepsilon_o$ with $\alpha(0) = 1$.

Thus,the frequencies of the quasiperiodic solutions change along each solution branch; but their pairwise ratios remain constant.

To the proof of Theorem 2.2 and Theorem 2.3 :

After some similarity transformations in the phase space and in the parameter space and after the introduction of so-called normal coordinates, in both cases we may assume that the system (1.1) is of the form

$$(2.10)_1 \qquad \left.\begin{aligned} \dot{\theta}_k &= \omega_k + \mu_k + g_k/\rho_k \\ \dot{\rho}_k &= h_k \end{aligned}\right\} \quad 1 \leq k \leq \bar{n}-1$$

$$(2.10)_2 \qquad \left.\begin{aligned} \dot{\theta}_k &= \omega_k \\ \dot{\eta}_k &= -(\mu_k/\omega_k)\cos\theta_k \sin\theta_k\, \eta_k - (\mu_k/\omega_k)\cos^2\theta_k\, \xi_k + h_k \\ \dot{\xi}_k &= (\mu_k/\omega_k)\sin^2\theta_k\, \eta_k - (\mu_k/\omega_k)\cos\theta_k \sin\theta_k\, \xi_k + g_k \end{aligned}\right\} \quad k \in [\bar{n},n]$$

$$\dot{\underline{y}}^1 = A_1\underline{y}^2 + \underline{h}$$
$$\dot{\underline{y}}^2 = A_2\underline{y}^1 + \underline{g} \quad .$$

Here μ_k, $k = 1, \ldots, n$, are n free parameters. The functions g_k, h_k, $\underline{g}$ and $\underline{h}$ are analytic in the variables $\theta_j \in \mathbb{R}$ as well as in μ_j, ρ_j, η_j, ξ_j resp. $\underline{y}^1$, $\underline{y}^2$ near the origin in $\mathbb{R}$ resp. $\mathbb{R}^\ell$, $\ell = \frac{m}{2} - n$. Moreover, they are 2π-periodic in θ_j and

$$g_k, h_k, \underline{g}, \underline{h} = O(\ |\rho_j|^2 + |\eta_j|^2 + |\xi_j|^2 + \|\underline{y}^1\|^2 + \|\underline{y}^2\|^2\ )$$

holds independently of θ_j and μ_j. The block matrix

$$\begin{pmatrix} 0 & A_1 \\ A_2 & 0 \end{pmatrix}$$

depends analytically on μ_j, and its spectrum is strictly bounded away from the imaginary axis for small $|\mu_j|$. The reflection R is now described by

$$\begin{aligned}
\rho_j &\mapsto \rho_j \\
\theta_j &\mapsto -\theta_j \\
\eta_j &\mapsto \eta_j \\
\xi_j &\mapsto -\xi_j \\
\underline{y}^1 &\mapsto \underline{y}^1 \\
\underline{y}^2 &\mapsto -\underline{y}^2
\end{aligned}$$

i.e., g_k and $\underline{g}$ are even in θ_j, ξ_j, $\underline{y}^2$, whereas h_k and $\underline{h}$ are odd.

Thus, we are able to construct the branch of $(\bar{n}-1)$-dimensional invariant tori of (1.1) emanating from the basic solution at $\underline{\lambda} = \underline{\lambda}_0$ in the direction of $\underline{c} = (c_1, \ldots, c_{\bar{n}-1}, 0, \ldots, 0)$, $c_k \neq 0$. Setting $\rho_k = \varepsilon c_k + \hat{\rho}_k$ we consider θ_j, $\hat{\rho}_j$, η_j, ξ_j, $\underline{y}^1$ and $\underline{y}^2$ as functions of $z_k = \omega_k t$ and introduce the differential operator

$$L = \sum_{k=1}^{n} \omega_k \partial_{z_k} \quad .$$

Then $(2.10)_1$, $(2.10)_2$ becomes a system of partial differential equations, where the first derivatives with respect to t are replaced by the corresponding values of L. Evidently, $\theta_k = z_k$, $\hat{\rho}_k = \eta_k = \xi_k = 0$, $\underline{y}^1 = \underline{y}^2 = \underline{0}$ solves these equations for $\varepsilon = 0$. Now, by choosing appropriate Banach spaces X_n, Y_n, $0 \leq n \leq \infty$, of real analytic functions (c.f. [7]) and by constructing the required operator sequences, the following iteration procedure and resulting hard implicit function theorem apply to solve $(2.10)_2$ for small $|\varepsilon| \neq 0$ inspite of the arising difficulties with small divisors. $(2.10)_1$ has to be considered separately. The parameters μ_k are needed to guarantee the solvability in the used spaces.

Let $i_{n,m} : X_n \to X_m$ resp. $j_{n,m} : Y_n \to Y_m$, $n \leq m \leq \infty$, be continuous linear operators such that $\|i_{n,m}\|$, $\|j_{n,m}\| \leq 1$ and

$$i_{n,m} \circ i_{m,k} = i_{n,k} \quad \text{resp.} \quad j_{n,m} \circ j_{m,k} = j_{n,k}$$

A mapping $F_m : X_m \to Y_m$ is called a <u>continuation</u> of $F_n : X_n \to Y_n$, $n \leq m$, if $F_m(i_{n,m}x) = j_{n,m}F_n(x)$ holds for every $x \in X_n$. If E is another Banach space, for given elements $\varepsilon_o \in E$, $x_o \in X_o$ and positive numbers s,r, the set $\{(\varepsilon,x) \in E \times X_n \ / \ \|\varepsilon - \varepsilon_o\| < s$ and $\|x - i_{o,n}x_o\| < r\}$ is denoted by $U^n_{s,r}(\varepsilon_o,x_o)$. $\mathcal{L}(X_n,Y_m)$ denotes the space of bounded linear operators between X_n and Y_m equipped with the induced norm.

Let us consider a family of analytic mappings $G_n : U_n \to Y_n$, $0 \leq n \leq \infty$. Moreover, let operators $\tilde{L}_n \in \mathcal{L}(Y_n,X_{n+1})$, $L_n \in \mathcal{L}(X_{n+1},Y_{n+1})$ be given and define $S_n \in \mathcal{L}(Y_n, Y_{n+1})$ and $\tilde{S}_{n+1} \in \mathcal{L}(X_{n+2}, Y_{n+2})$ by the following iteration schema, n = 0,1,2, ... :

$$(2.11)_{n+1} \quad \begin{aligned} L_n\tilde{L}_n &= j_{n,n+1} + S_n \\ x_{n+1} &= i_{n,n+1}x_n - K_nG(\varepsilon,x_n), \quad K_o = \tilde{L}_o \\ \Lambda_n &= D_xG_{n+1}(\varepsilon,x_{n+1})\Gamma_n - L_n, \quad \Gamma_o = i_{1,1} \\ \tilde{S}_{n+1} &= L_{n+1} - \overline{L}_nM_n - \overline{L}_n - \overline{\Lambda}_n \\ \Gamma_{n+1} &= \overline{\Gamma}_n(i_{n+2,n+2} + \underline{M}_n) \\ K_{n+1} &= \Gamma_{n+1}\tilde{L}_{n+1} \end{aligned}$$

Here $\bar{L}_n$, $\bar{\Lambda}_n \in \mathcal{L}(X_{n+2}, Y_{n+2})$ and $\bar{\Gamma}_n \in \mathcal{L}(X_{n+2})$ are continuations of L_n, Λ_n and Γ_n such that

$$\bar{\Lambda}_n = D_x G_{n+2}(\varepsilon, i_{n+1,n+2} x_{n+1}) \bar{\Gamma}_n - \bar{L}_n$$

$$\|\bar{\Lambda}_n\| \leq \|\Lambda_n\|$$

$$\|\bar{\Gamma}_n\| \leq \|\Gamma_n\|$$

and $M_n \in \mathcal{L}(X_{n+2})$.

Theorem 2.4:

Let constants $\delta_o > 0$, $\kappa \in [\sqrt{2}, 2]$ and $c_1, c_2, c_3, c_4, c_5 \geq 1$ be given such that in $\|\varepsilon - \varepsilon_o\| < s$

$$(2.12)_n \qquad \|D_x G_n(\varepsilon, x) - D_x G_n(\varepsilon, y)\| \leq c_1^{\kappa^n} \|x - y\|$$

$$(2.13)_n \qquad \|M_n\| \leq c_2^{\kappa^n} \|\Lambda_n\|$$

$$(2.14)_n \qquad \|\tilde{L}_n\| \leq c_3^{\kappa^n}$$

and with

$$C = c_2(1 + c_5 + c_3 c_1^{\kappa}(c_1 + c_3 + c_4)(1 + \delta_o)^2)$$

$$q = \delta_o C^{\frac{\kappa}{2-\kappa}}$$

$$\delta_n = q^{2^n} C^{\frac{\kappa}{\kappa - 2}\kappa^n}$$

$$(2.15)_n \qquad \sum_k \delta_k < r$$

$$(2.16)_n \qquad \|S_n\| \leq c_4^{\kappa^n} \delta_n$$

$$(2.17)_n \qquad \|\tilde{S}_{n+1}\| \leq c_5^{\kappa^{n+1}} \delta_n^2$$

hold. Furthermore, assume that the operators L_o, $i_{n,m}$, M_n, $\tilde{L}_n$ and $\bar{\Gamma}_n$ depend analytically on ε, let $G_m(\varepsilon, \cdot)$ be a continuation of $G_n(\varepsilon, \cdot)$, $n \leq m \leq \infty$, and suppose that

$$(2.18)\qquad \begin{aligned} G_o(\varepsilon_o,x_o) &= 0 \\ D_x G_1(\varepsilon_o,x_o) &= L_o(\varepsilon_o) \end{aligned}$$

hold. Then, there exists a positive s_o, such that the equation

$$(2.19)\qquad G_\infty(\varepsilon,x) = 0$$

has a solution $x = g(\varepsilon)$ with $(\varepsilon,g(\varepsilon)) \in U^\infty_{s_o,r}(\varepsilon_o,x_o)$ for all $\|\varepsilon - \varepsilon_o\| \leq s_o$. The mapping $\varepsilon \mapsto g(\varepsilon)$ is analytic. The sequence (x_{n+1}) defined by $(2.11)_{n+1}$ converges towards $g(\varepsilon)$ and one has the estimate

$$(2.20)\qquad \|x_{n+1} - g(\varepsilon)\| \leq q^{2^{n+1}} C^{\frac{\kappa}{\kappa-2}} \sum_{k=0}^{\infty} q^{2^k-1} .$$

For a proof of this theorem see [10]. Finally we remark that the reduced system

$$\begin{aligned} \dot\theta_k &= \omega_k + \mu_k + g_k/\rho_k \\ \dot{\hat\rho}_k &= h_k \end{aligned} \quad , \quad \rho_k = \varepsilon c_k + \hat\rho_k, \quad c_k \neq 0$$

is an example of J. Moser in [7], where g_k and h_k do not depend on the μ_j. He has presented there a perturbation theory for quasi-periodic solutions. Certain transformation techniques for the differential equations are used to overcome the problems with the small divisors. However, it seems to be impossible to prove our results in full generality by the method of [7].

3. References

[1] J.C. Alexander and J.A. York, Global bifurcation of periodic orbits, Amer. J. Math., to appear

[2] E. Hopf, Abzweigung einer periodischen Lösung von einer stationären Lösung eines Differentialgleichungssystems, Ber. Verl. Sächs. Akad. Wiss. Leipzig, Math.-nat. Kl. 95, no. 1, 1943, 3 - 22

[3] G. Ize, Bifurcatión Global de Órbitas Periodicas, Preprint, CIMAS, Mexico City 1974

[4] G. Joos, Topics in Bifurcation of Maps and Applications, Lecture notes, University of Minnesota 1978

[5] K. Kirchgässner and J. Scheurle, On the bounded solutions of a semilinear elliptic equation in a strip, J. Diff. Equations, to appear

[6] I.G. Malkin, Theorie der Stabilität einer Bewegung, Verlag R. Oldenburg, München 1959

[7] J. Moser, Convergent Series Expansions for Quasi-Periodic Motions, Math. Annalen 169, 1967, 136 - 176

[8] J. Moser, Stable and random motions in dynamical systems, Annales of Math. Studies, Princeton University Press 1973

[9] P. Rabinowitz, Some global results for non-linear eigenvalue problems, J. Funct. Anal. 7, 1971, 487 - 573

[10] J. Scheurle, Newton iterations without inverting the derivative, Preprint 1978, to appear

Mathematisches Institut A
der Universität Stuttgart
Pfaffenwaldring 57

7 Stuttgart 80

PERIODIC SOLUTIONS OF DELAY-DIFFERENTIAL EQUATIONS

by

Klaus Schmitt

1. INTRODUCTION

In this paper we consider the functional differential equation

$$x' = f(t,x_t), \tag{1.1}$$

where $f: R \times C([-\tau,0], R^n) \to R^n$ is continuous and T-periodic with respect to t and provide some existence theorems for T-periodic solutions of this equation. While we provide results which are quite general in nature we shall be particularly interested in the case where f has the form

$$f(t,x_t) = g(x_t) + e(t), \tag{1.2}$$

i.e. it consists of an autonomous part perturbed by a T-periodic input, and the question of the existence of T-periodic solutions becomes the question whether a T-periodic input will cause an output of the same period. This question is again of particular interest in case the autonomous equation

$$x' = g(x_t) \tag{1.3}$$

has nonconstant periodic solutions or in case the stable states of

(1.3) are non-periodic and one desires a periodic output. Such situations have recently been found in modelling certain physiological phenomena. Both Mackey and Glass [12] and Lasota and Ważewska [11] have obtained autonomous nonlinear equations (1.3) as models describing the production of red blood cells, the nonlinearities in their respective situations being:

$$g(x_t) = -\gamma x(t) + \frac{\beta x(t-\tau)}{1 + x^{2n}(t-\tau)} \tag{1.4}$$

$$g(x_t) = -\sigma x(t) + (cx(t-\tau))^s \, e^{-x(t-\tau)}, \tag{1.5}$$

where all constants appearing are positive and the time delay τ may vary over a certain parameter range. In both of these models it turns out that for certain values of the parameters the equation has, aside from the trivial state, another critical point which is stable for small values of τ and from which nontrivial periodic solutions bifurcate as τ increases through a certain critical value (Hopf bifurcation takes place). Computer studies show that as τ is increased a stable periodic solution exists up to another critical value of τ after which the stable state seems no longer periodic but exhibits a highly irregular behavior (chaotic behavior). For such equations it is therefore of interest to see whether a regular (periodic) output (which is stable) can be maintained through external forcing.

Another possible motivation for the study of such problems is analytic in nature. Namely one may consider the difference $x' - g(x_t)$ as a nonlinear operator on a space of periodic functions (of a fixed period), it is then natural to ask for a characterization of the range of this operator, or less generally, at least identify subsets of its range.

In studying the existence of periodic solutions of period T one may use initial value problem methods, i.e. study the (Poincaré) operator which associates with each initial function ϕ (defined on $[-\tau,0]$) the function $P\phi = x_T(\phi)$, i.e. the segment of the solution defined on $[T-\tau, T]$, and then consider the fixed point problem $\phi = P\phi$ (if initial value problems are uniquely solvable) or $\phi \in P\phi$ otherwise. Using this approach one has to overcome several (nontrivial) problems, namely extendability of solutions to $[0,T]$ and compactness of the solution operator. In the presence of asymptotic stability of the unforced problem these difficulties have been successfully overcome by many investigators, e.g. Hale [7], Stephan [9] and Chow [1] among others. Secondly one may formulate equivalent integral equations and study these in the appropriate space of T-periodic functions or view (1.1) as a nonlinear equation where the domain and range are different function spaces (the approach used in this discussion). Both approaches will depend upon tools from nonlinear functional analysis, and both have their advantages and disadvantages.

2. EXISTENCE RESULTS

Let $f\colon [0,T] \times C([-\tau,0], R^n) \to R^n$ be continuous, where $\tau < T$. Our first general result is the following theorem which we shall verify using a continuation theorem of Mawhin [3, 14]. The result may also be derived from a result in [5] (see also [20]).

Theorem 2.1. Let there exist a bounded open neighborhood G of $0 \in R^n$ such that the following conditions hold:

(a) $\forall\ \varepsilon \in (0,1)$ every T-periodic solution x of

$$x' = \varepsilon f(t,x_t) \tag{2.1}$$

such that $x\colon [0,T] \to \overline{G}$ has the property that $x\colon [0,T] \to G$.

(b) The Brouwer degree $d(h,G,0)$ is defined and unequal to 0, where h is the mapping from R^n to R^n given by

$$a \longmapsto \frac{1}{T}\int_0^T f(x,a)ds. \tag{2.2}$$

Then equation (2.1) has a T-periodic solution x such that $x\colon [0,T] \to \overline{G}$, for every $\varepsilon \in [0,1]$.

PROOF: We let E denote the Banach space of continuous T-periodic functions and let $\mathcal{O} = \{x \in E\colon\ x\colon [0,T] \to G\}$. Then $\mathcal{O}$ is open in E and for no $x \in \partial\mathcal{O}$, $0 < \varepsilon < 1$, is it true that $Lx = \varepsilon Nx$, where $L\colon \operatorname{dom} L \subset E \to E$ ($\operatorname{dom} L = \{x \in E\colon\ x$ is continuously differentiable$\}$) is defined by $x \longmapsto x'$ and $N\colon E \to E$ is given by $x(t) \longmapsto f(t,x_t)$. L is a Fredholm operator of index zero and one may easily check that its right inverse is a compact linear operator. Using hypothesis (b) and the mean value operator as a projection onto the kernel of L one sees that all hypotheses of the continuation theorem of Mawhin [3] are satisfied.

In order to be able to apply this result, one, of course, needs to construct a set $G \subset R^n$ which verifies the conditions of this theorem. The following situations permit the construction of such a set G.

Corollary 2.2. Let $f(t,x_t)$ be of the form

$$f(t,x_t) = h(t,x(t),x_t)$$

where h is continuous and T-periodic. Assume G is a bounded open convex neighborhood of $0 \in R^n$ and assume at every $x \in \partial G$ there exists an outer normal vector $n(x)$ to G such that

$$n(x) \cdot h(t,x,y) > 0,\ y \in \{z \in C[-\tau,0]: z(t) \in \bar{G}\},\ 0 \leq t \leq T. \tag{2.3}$$

Then eq. (2.1) has a T-periodic solution $x: [0,T] \to \bar{G}$ for every $\varepsilon \in [0,1]$.

REMARK. Requirement (2.3) may be replaced by

$$n(x) \cdot h(t,x,y) < 0. \tag{2.4}$$

Furthermore, using approximation arguments one may show that the strict inequalities in (2.3) and (2.4) may be replaced by weak ones.

To specialize further one obtains

Corollary 2.3. Let there exist $r > 0$ such that

$$x \cdot h(t,x,y) \geq 0\ \ (\leq 0),\ |x| = r,\ |y| \leq r,\ t \in [0,T], \tag{2.5}$$

then eq. (2.1) has a T-periodic solution x with $|x(t)| \leq r$, for every $\varepsilon \in [0,1]$ (here "$\cdot$" denotes the scalar product of R^n and $|\cdot|$ the Euclidean norm).

Corollaries 2.2 and 2.3 may be found in [6]. Also similar results exist for second order functional differential equations [5], [16], [17].

If one specializes h further one may show that periodic solutions of the functional differential equation may actually be obtained by successive approximations as a limit of a sequence of ordinary differential equations. To illustrate we consider a special case.

Corollary 2.4. Let the equation be scalar and assume h is increasing with respect to y (using the natural ordering in $C[-\tau,0]$) and assume that (2.5) holds with inequalities strict, further assume h satisfies a local Lipschitz condition with respect to x, then (2.1) has a periodic solution which is the limit of the sequence of T-periodic functions $\{x_n\}$ which are solutions of

$$x_n'(t) = h(t, x_n, x_{n-1\ t}) \quad n = 1, 2, \ldots$$

$$x_0(t) \equiv r.$$

The above corollary may be found in [6], it is based on an earlier result of Mikolajska [15].

We now give some applications of these results. It is apparent (a simple change of variables) that the hypothesis that $0 \in G$ may be deleted; this remark is useful in some of the considerations to follow.

Consider the equation

$$(2.6) \qquad x'(t) = -\sigma x(t) + |x(t-\tau)|^s \exp(-|x(t-\tau)|) + e(t)$$

where σ and s are positive parameters and $e(t)$ is a T-periodic input.

We choose $G = (-r,r)$ where $r > 0$ is taken sufficiently large. Since the nonlinear term involving the delays is bounded it is easy to see that corollary 2.3 is applicable.

As mentioned in the introduction (2.6) serves as a model for certain physiological phenomena, it hence is of interest to consider the case where the input $e(t)$ is nonnegative and causes a nonnegative output. A result of this type follows by a use of variation of constants formula and application of the Schauder fixed point theorem. Since (2.6) has a T-periodic solution if and only if the integral equation

$$x(t) = \int_0^T G(t,\mu)(\{x(\mu-\tau)^s e^{-x(\mu-\tau)} + e(\mu)\}d\mu \qquad (2.7)$$

has such, where $G(t,\mu)$ is a positive piecewise continuous kernel. Thus if e is nonnegative the right hand side of (2.7) transforms the set of all continuous nonnegative T-periodic functions into a bounded subset of itself and hence (2.7) will have a solution by the Schauder fixed point theorem.

Another immediate consequence of Theorem 2.1 is the following result.

Corollary 2.5. Assume that there exists a constant $r > 0$ such that all possible T-periodic solutions $x(t)$ of (2.1) for $0 < \varepsilon < 1$ satisfy $|x(t)| < r$, $0 \leq t \leq T$ and let the map (2.2) have nonzero Brouwer degree at 0 relative to the ball in R^n of radius r.

Then (2.1) has a T-periodic solution $x(t)$ (for all $0 < \varepsilon \leqslant 1$), such that $|x(t)| < r$.

This result is particularly useful in case the equation at hand lends itself to a priori estimation of possible T-periodic solutions. To illustrate let us consider the following result.

Let A be a constant nonsingular $n \times n$ matrix and consider the linear equation

$$x'(t) = A\,x(t-\tau) + e(t), \tag{2.8}$$

where e is T-periodic and $\tau < T$. Then using L^2 estimates one obtains (see [5]) that whenever

$$T|A| < \sqrt{12} \tag{2.9}$$

(here $|A|$ is the operator norm induced by the Euclidean norm in R^n) T-periodic solutions of the association equation

$$x'(t) = \varepsilon(Ax(t-\tau) + e(t)),\ 0 < \varepsilon \leqslant 1,$$

satisfy

$$|x| \leqslant k|e|,$$

where k is a constant independent of ε. The associated map h has the form

$$A + \frac{1}{T}\int_0^T e(s)ds,$$

and since A is nonsingular it has nonzero Brouwer degree relative to large balls.

These arguments may also be used to consider equations of the type

$$x'(t) = Ax(t-\tau + \mu k(t,x(t)) + e(t), \tag{2.10}$$

where k is T-periodic with respect to t and μ is a small parameter and where A is constant or T-periodic. For example one can conclude that if (2.9) holds (2.10) has a T-periodic solution for all small μ and all T-periodic e.

In case the equation is a scalar equation and A is a negative constant similar results may be obtained using a stability analysis and asymptotic fixed point theory, whenever $a\tau < \pi/2$, here any period $T > \tau$ is permissable, see e.g. [1], [19]), however, on the range $\pi/2 \leqslant a\tau < \sqrt{12}$, the arguments in [19] do not apply.

Equations such as (2.10) have first been proposed by K. Cooke [2] and have also been studied in [1], where almost periodic inputs are treated as well. Chow [1], in fact shows for eq. (2.10), if $a\tau \neq \pi/2 + n\pi$, $n = 0, 1, 2, \ldots$, then (2.10) has a T-periodic solution for any T-periodic input $e(t)$.

As a further application let us consider the scalar equation

$$x'(t) = g(x(t))(x(t) - h(x(t-\tau)) + e(t), \tag{2.11}$$

where g, h and e are continuous functions and $e(t)$ is T-periodic, furthermore we require that h be bounded.

<u>Corollary 2.6.</u> Assume that g satisfies:

$$\text{(a)} \qquad \lim_{|x|\to\infty} \left|\frac{g(x)}{x}\right| = 0.$$

(b) There exists $M \geq 0$ and $\gamma > 0$ such that $|x| > M$ implies either $g(x) \geq \gamma$ or $g(x) < -\gamma$.

Then (2.11) has a T-periodic solution.

PROOF. We first show that for $0 < \varepsilon \leq 1$ the set of T-periodic solutions of

$$x'(t) = \varepsilon\{g(x(t))(x(t) - h(x(t-\tau))) + e(t)\}$$

is a priori bounded. Let t_0 be such that $|x(t_0)| = \max\{|x(t)|: t \in [0,T]\}$ then $x'(t_0) = 0$ and

$$|g(x(t_0))x(t_0)| \leq |g(x(t_0)|k + |e|,$$

where $k = \sup_{|y|<\infty} |h(y)|$. Dividing by $|x(t_0)|$ and using (a) and (b) we obtain that $|x(t_0)| \leq N$ where $N \geq M$ is chosen such that $|x| > N$ implies $\left|\frac{g(x)}{x}\right| k + \left|\frac{e}{x}\right| < \gamma$.

Finally we need to consider the scalar map

$$g(a)(a - h(a)) + \frac{1}{T}\int_0^T e(s)ds,$$

for $|a| > N$. This map will have opposite signs for $a > N$ and $a < -N$, hence has nonzero Brouwer degree

Using similar arguments one may establish the following

Corollary 2.7. Let g satisfy:

(a) $$\liminf_{|x|\to\infty} |xg(x)| > |e|.$$

(b) There exists $M \geq 0$ such that $|x| \geq M$ implies either $g(x) \geq 0$ or $g(x) \leq 0$.

Then (2.11) has a T-periodic solution.

In case e has mean value 0 the conditions on g may be considerably weakened.

Theorem 2.8. Let e have mean value 0 and assume there exists a constant $M > 0$ such that either (i) $g(x) \geq 0$ if $|x| \geq M$ or (ii) $g(x) \leq 0$ if $|x| \geq M$ and $g(x)$ is bounded. Then (2.11) has a T-periodic solution.

In proving Theorem 2.8 we again rely on Theorem 2.1. The necessary a priori bounds on possible T-periodic solutions follow from the one-signedness of g for large $|x|$ and the boundedness of h, and standard integral inequalities. Using some more detailed estimates one may replace the second assumption in (ii) by the weaker assumption that

$$\limsup_{|x|\to\infty} \left|\frac{g(x)}{x}\right| = B, \text{ where } BK < \frac{2\pi}{2\pi + 1}, \text{ and } K = \sup_{|y|<\infty} |h(y)|.$$

The detailed verification may be found in [13].

REMARK. We note that both of the interesting nonlinearities (1.4) and (1.5) are covered by the above results, however not much information can be gathered concerning the location of the periodic response. For example if the unperturbed equation has the properties described in the introduction then the bifurcating periodic solutions and the "chaotic" solutions tend to oscillate around the nonzero critical point of the equation and thus have a mean value "close" to the numerical value of the critical point. Thus when forcing the equation with a T-periodic $e(t)$ it is of interest to see whether periodic responses can be obtained whose mean value equals or is close to the value of the critical point. A positive answer to this question is contained in the following result (see [13]).

Theorem 2.9. Assume that $\int_0^T e(t)dt \neq 0$ and let the assumptions of corollary 2.7 hold. Then for every $r \neq 0$ there exists $\mu \neq 0$ such that the equation

$$x'(t) = g(x(t)(x(t) - h(x(t-\tau))) + \mu e(t) \tag{2.12}$$

has a T-periodic solution $x(t)$ with $\int_0^T x(t)dt = r$.

3. CONCLUDING REMARKS - OPEN PROBLEMS

Instead of considering inputs which solely depend upon t one equally well could (and should) consider cases where e also depends upon the state of the system at time t or some previous time. The tools used in establishing these results and the nature of theorem 2.1 and corollary 2.5 will yield such results. Also the nonlinear variation of constants formulae of Hastings [10], Shanholt [18] and Gustafson [4] should prove useful in such a study when combined with the tools of nonlinear functional analysis (see [4] for some results in this direction). A shortcoming of the results of section 2 is that nothing can be asserted about the stability properties of the solutions obtained. In fact the stability analysis of periodic solutions of (2.11) (even in the special case of nonlinearities like (1.4) and (1.5)) has totally eluded us until now. However in computer studies we have observed the following in the case where $g(x)$ is a negative constant. Choosing $e(t)$ to be a T-periodic function with nonzero mean value and varying the parameter μ we found that for small values of μ the unforced system dominates (as is to be expected) and thus in the case μ is such that the autonomous equation has a stable "chaotic" solution such "chaotic" solutions will be seen, but as μ is increased entrainement will occur, i.e. the system will have a stable T-periodic output.

It would be of interest to obtain results such as discussed in section 2 for other nonlinear functional differential equations. For example if $e(t)$ is T-periodic, do there exist T-periodic solutions of

$$x'(t) = -\alpha x(t-1)(1+x(t)) + e(t) \tag{3.1}$$

or

$$x'(t) = -\alpha x(t-1)(1-x^2(t)) + e(t)? \tag{3.2}$$

In case $e(t)$ is of the form $e(t) = \varepsilon h(t)$, then for small values of ε the answer is yes as shown by Gustafson [4], otherwise, however, the problem is open.

In addition Hale and Mawhin [9] have obtained existence theorems for neutral equations; however, very few other papers on the existence of periodic solutions of forced neutral equations exist.

REFERENCES

[1] S.-N. Chow, On a conjecture of K. Cooke, J.D.E., 14(1973), 307-325.

[2] K.L. Cooke, Functional differential systems: some models and perturbation problems, Internat. Symp. Dyn. Systems, Acad. Press, 1965.

[3] R. E. Gaines and J. L. Mawhin, Coincidence Degree and Nonlinear Differential Equations, Springer Lecture Notes in Math., vol. 568, 1977.

[4] G. B. Gustafson, An integral equation for perturbed nonlinear functional differential equations, with applications to periodic solutions and nonlinear boundary value problems, Arch. Reine Angew. Mathematik, 271(1974), 35-62.

[5] G. B. Gustafson and K. Schmitt, Periodic solutions of hereditary differential systems, J.D.E., 13(1973), 567-587.

[6] G. B. Gustafson and K. Schmitt, A note on periodic solutions for delay-differential systems, P.A.M.S., 42(1973), 161-166.

[7] J. Hale, Periodic and almost periodic solutions of functional-differential equations, Arch. Rat. Mech. Anal., 15(1964), 291-304.

[8] J. Hale, Functional Differential Equations, Springer-Verlag, New York, 1976.

[9] J. K. Hale and J. L. Mawhin, Coincidence degree and periodic solutions of neutral equations, J.D.E., 15(1974), 295-307.

[10] S. P. Hastings, Variation of parameters for nonlinear differential difference equations, P.A.M.S., (1968), 1211-1216.

[11] A. Lasota and M. Ważewska, Mathematical models of the red cell system, Mat. Stosowana, 6(1976), 25-40.

[12] M. Mackey and L. Glass, Oscillation and chaos in physiological control systems, Science, 197(1977), 287-289.

[13] M. Martelli, K. Schmitt and H. Smith, Periodic solutions of some nonlinear delay differential equations, Univ. Utah preprint, 1978.

[14] J. Mawhin, Periodic solutions of some vector retarded functional differential equations, J. Math. Anal. Appl., 45(1974), 455-467.

[15] Z. Mikolajska, Sur l'éxistence d'une solution périodique d'une équation differentielle du premier ordre avec le parametre retardé, Ann. Polon. Math., 23(1970), 25-36.

[16] K. Schmitt, Periodic solutions of linear second order differential equations with deviating argument, P.A.M.S., 26(1970), 282-285.

[17] K. Schmitt, Intermediate value theorems for periodic functional differential equations, Equations différentielles et fonctionelles nonlineaires, Hermann, Paris, 1973, 65-78.

[18] G. A. Shanholt, A nonlinear variation of constants formula for functional differential equations, Math. Systems Theory, 6(1973), 343-352.

[19] B. H. Stephan, On the existence of periodic solutions of $z'(t) = -az(t-r + \mu k(t,z(t))) + F(t)$, J.D.E., 6(1969), 408-419.

[20] V. V. Strygin, A certain theorem on the existence of periodic solutions of systems of differential equations, Mat. Zamet., 8(1970), 600-602.

Department of Mathematics
University of Utah
Salt Lake City, Utah 84112
USA

HAMILTONIAN TRIANGULATIONS OF R^n

by

Michael J. Todd*

School of Operations Research
and Industrial Engineering
College of Engineering
Cornell University
Ithaca, New York

ABSTRACT

We show that two well-known triangulations of R^n can be enumerated so that successive simplices are adjacent. As an application, we obtain efficient ways to search for initial completely labelled simplices in simplicial algorithms for nonlinear eigenvalue problems.

---

*This research was supported in part by National Science Foundation Grant ENG76-08749.

Allgower and Georg [1] conjectured that there exist sequential triangulations of R^n, i.e., triangulations of R^n whose associated graphs have (one-way infinite) hamiltonian paths. We show that two well-known triangulations have this property (which we call hamiltonian) for $n \geq 2$. Our proofs give recursive algorithms for constructing the hamiltonian paths. Indeed we prove a stronger result for $n \geq 4$. Let C be a union of unit cubes in R^n with vertices in the integer lattice Z^n. Then the induced triangulations of C are hamiltonian if there is a hamiltonian path in the graph of the southwest corners of the cubes in C.

M. Prufer and H.-O. Peitgen have developed a constructive approach to global bifurcation in nonlinear eigenvalue problems using simplicial algorithms--see for example [2]. These methods first search for a completely labelled simplex in some neighborhood of zero from which to initiate the algorithm. The hamiltonian paths we construct provide an efficient way to make these searches. Triangulations are also of practical importance in numerical integration [4,7], piecewise approximation of functions and their fixed points [2,6] and finite element methods [8].

Two Triangulations

Let $u^1,\ldots,u^n$ be the unit vectors in R^n. If v^0 lies in Z^n and π is a permutation of $N = \{1,2,\ldots,n\}$, then $k_1(v^0,\pi)$ denotes the simplex with vertices v^0, $v^i = v^{i-1} + u^{\pi(i)}$, $i \in N$. If, further, v_i^0 is odd for each i and $s \in R^n$ is a sign vector ($s_i = \pm 1$ for $i \in N$) then $j_1(v^0,\pi,s)$ denotes the simplex with vertices $\hat{v}^0 = v^0$, $\hat{v}^i = \hat{v}^{i-1} + s_{\pi(i)}u^{\pi(i)}$, $i \in N$. Let K_1 denote the collection of all such $k_1(v^0,\pi)$, and J_1 the collection of all such $j_1(v^0,\pi,s)$. Then K_1 and J_1 are triangulations of R^n; that is, they are locally finite collections of n-simplices which with all their faces partition R^n. For the history of these triangulations, see [5,6]. K_1 and J_1 are illustrated for $n = 2$ in Figure 1.

It is easy to see that if C is a union of unit cubes in R^n (with vertices in Z^n), then the simplices of K_1 or J_1 in C triangulate C.

The graph of a triangulation T has the simplices of T as vertices, two such being adjacent if they have a common (n-1)-face. T is hamiltonian if its graph admits a (one-way infinite) hamiltonian path. Figure 1 shows hamiltonian paths in K_1 and J_1 for $n = 2$.

Let $k_1(v^0,\pi)$ be a simplex in K_1. The adjacent simplices are: $k_1(v^0,\rho)$ where ρ is obtained from π by interchanging $\pi(i-1)$ and $\pi(i)$ for $i = 2,\ldots,n$; $k_1(v^0 + u^{\pi(1)}, (\pi(2),\ldots,\pi(n),\pi(1)))$; and $k_1(v^0 - u^{\pi(n)}, (\pi(n),\pi(1),\ldots,\pi(n-1)))$. Similarly, let $j_1(v^0,\pi,s)$ be a simplex in J_1. The adjacent simplices are: $j_1(v^0,\rho,s)$ with ρ as above; $j_1(v^0 - 2s_{\pi(1)}u^{\pi(1)}, \pi, s - 2s_{\pi(1)}u^{\pi(1)})$; and $j_1(v^0, \pi, s - 2s_{\pi(n)}u^{\pi(n)})$.

Permutations

Clearly the easiest way to construct a hamiltonian path in the graphs of K_1 and J_1 is to traverse all simplices within a given unit cube before passing to the next unit cube. Let C_1 and C_2 be adjacent unit cubes. If the common face of C_1 and C_2 has x_i constant, say $x_i = c$, then to enumerate the simplices of K_1 in C_1 and then pass to C_2 we would like to enumerate all permutations of N ending with one with i in first (if C_1 has $x_i \leq c$) or last (if C_1 has $x_i \geq c$) position. To remain within C_1 we can only interchange adjacent symbols in the permutation. Since we do not wish to limit the entry into C_1, we would like to find such a sequence for any prescribed starting permutation. Theorem 2 states that this is possible for $n \geq 4$.

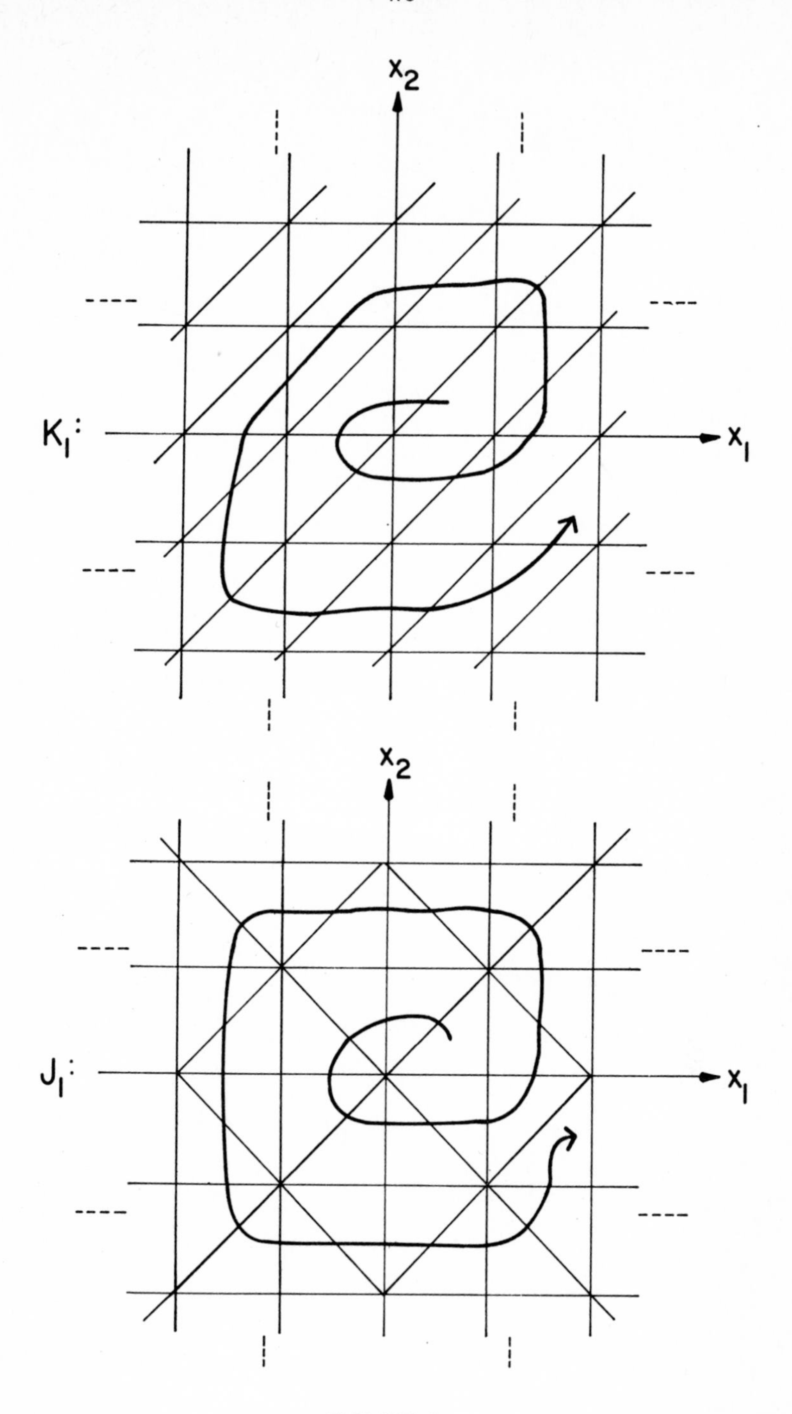

FIGURE 1

Definition. Let S_n be the set of permutations on N. Let G_n be the graph whose vertices are these permutations with π and ρ adjacent if ρ is obtained from π by interchanging $\pi(i-1)$ and $\pi(i)$ for $i = 2,\dots,n$. This edge, if traversed from π to ρ, will be said to be of type i.

The first theorem describes a standard way of generating all permutations of N. See for example the algorithms in [3].

Theorem 1. Suppose in G_{n-1} $(n \geq 3)$ there is a hamiltonian path from $(1,2,\dots,n-1)$ to $(\rho(1),\rho(2),\dots,\rho(n-1))$. Then for each $\pi \in S_n$ there is a hamiltonian path in G_n from π to $(\pi(\rho(1)),\dots,\pi(\rho(n-1)),\pi(n))$.

Proof. Suppose the path in G_{n-1} uses edges of types $i_1,i_2,\dots,i_p$ $(p = (n-1)!-1)$ in that order. Construct a path in G_n by using edges of types

$$n,n-1,\dots,2,i_1+1,2,\dots,n,i_2,n,\dots,2,\dots,i_p+1,2,\dots,n.$$

The first $n-1$ edges generate all permutations that have $\pi(1),\dots,\pi(n-1)$ in that order, with $\pi(n)$ moving from the last to the first position. The next edge leaves $\pi(n)$ in the first position, and the second permutation of $\{\pi(1),\dots,\pi(n-1)\}$ is generated; then $\pi(n)$ is moved back to the last position. Clearly all permutations are generated in this way. Also, since $n \geq 3$, p is odd and thus $\pi(n)$ finishes in the last position.

Corollary 1. There is a hamiltonian path in G_n for all $n \geq 2$.

We say that the hamiltonian path in G_n obtained as in the proof of theorem 1 is generated from the path in G_{n-1} by weaving in $\pi(n)$. By reversing all permutations we can similarly weave in $\pi(1)$.

Theorem 2. Suppose $\pi \in S_n$, $n \geq 4$, and $j \in N$. Then there is a hamiltonian path in G_n from π to some ρ with $\rho(1) = j$ and to some ρ' with $\rho'(n) = j$.

Proof. By reversing permutations, we need only prove the first assertion. We may also assume without loss of generality that $\pi = (1,2,\ldots,n)$. The proof is by induction.

For $n = 4$, see figure 2 where the required paths are shown together with G_4.

Now assume that $n \geq 5$ and the theorem is true for $n-1$. If $1 \leq j \leq n-1$, the inductive hypothesis gives a hamiltonian path in G_{n-1} from $(1,2,\ldots,n-1)$ to ρ with $\rho(1) = j$. The desired path in G_n is then obtained by weaving in n. Suppose now $j = n$. By the inductive hypothesis there is a hamiltonian path in G_{n-1} from $(1,2,\ldots,n-1)$ to ρ with $\rho(1) = n-1$. There is therefore a similar path P of permutations of $\{2,\ldots,n\}$ from $(2,\ldots,n)$ to ρ' with $\rho' = (n,\ldots)$. By weaving in 1, we get a hamiltonian path P' in G_n from $(1,2,\ldots,n)$ with final three vertices $(n,i,1,j,\ldots) \to (n,1,i,j,\ldots) \to (1,n,i,j,\ldots)$, with the same final sequence in each case. Somewhere in P' the edge $(1,n,j,i,\ldots) \to (n,1,j,i,\ldots)$ occurs, with the same final sequence. (Parity arguments show that the edge is traversed in this direction.) We then replace this edge with the three edges $(1,n,j,i,\ldots) \to (1,n,i,j,\ldots) \to (n,1,i,j,\ldots) \to (n,1,j,i,\ldots)$ and omit the two final vertices of P'. The result is a hamiltonian path in G_n from $(1,2,\ldots,n)$ to $(n,i,1,j,\ldots)$ as desired. The induction step is thus complete.

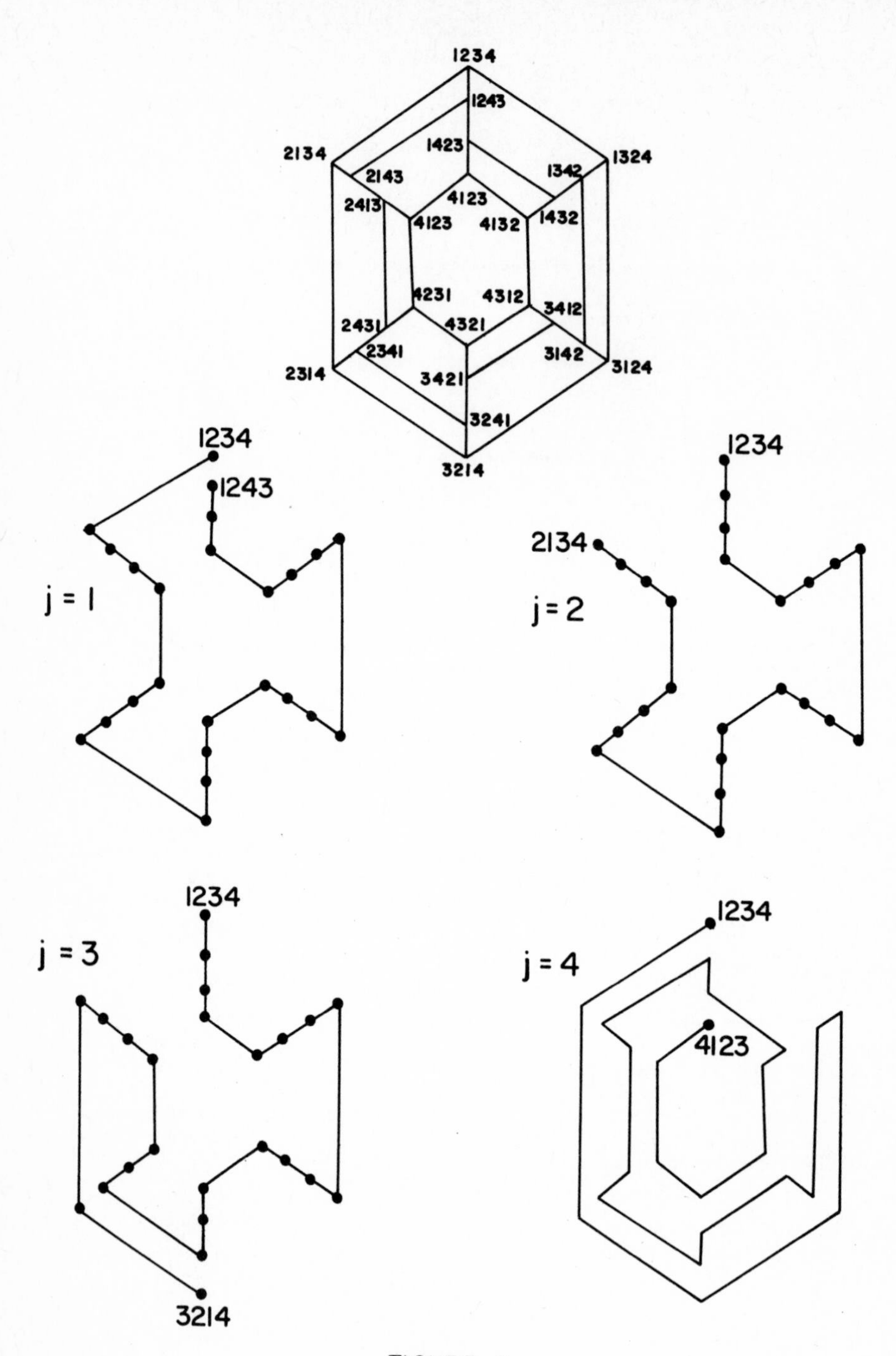

1234
1243
1423
2134
1324
2143
1342
4123
2413
4123
4132
1432
4231
4312
3412
2431
4321
2341
3142
2314
3421
3124
3241
3214
1234
1243
j = 1
1234
2134
j = 2
1234
j = 3
3214
1234
j = 4
4123

FIGURE 2

The Main Results

Let C be a union of unit cubes in R^n. Then $C' \subseteq Z^n$ denotes the set of southwest corners of these cubes. $G(C')$ is the corresponding graph, with $x,y \in C'$ adjacent if $\sum_N |x_i - y_i| = 1$.

Theorem 3. For $n \geq 4$, the restriction of K_1 or J_1 to C is hamiltonian if $G(C')$ is hamiltonian.

Proof. Let $x^1 \to x^2 \to \ldots$ be a hamiltonian path in $G(C')$, and let C_i be the unit cube with southwest corner x^i. We show that there is a hamiltonian path in the graph of the triangulations that traverses all the simplices in C_1, then all those in C_2, etc.

For K_1, suppose that a path including all simplices of $C_1,\ldots,C_{i-1}$ and one of C_i is known. Let $k_1(x^i,\pi)$ be the final simplex in this path. Let the common face of C_i and C_{i+1} be $x_j = c$. Then any simplex $k_1(x^i,\rho)$ with j in the first (if C_i has $x_j \leq c$) or last (if C_i has $x_j \geq c$) position in ρ has an $(n-1)$-face in C_{i+1}. By theorem 2 the simplices in C_i can be traversed ending with $k_1(x^i,\rho)$. There is then an adjacent simplex in C_{i+1}. Continuing, we obtain the desired hamiltonian path.

For J_1 we use a similar argument. If c is odd we generate all simplices of the form $j_1(y^i,\rho,s^i)$ ending with one with $\rho(n) = j$. If c is even we end with $\rho(1) = j$. Here y^i is the vertex of C_i with all coordinates odd and s^i is the appropriate sign vector. Again one more step gives us a simplex in C_{i+1} and the process continues.

It is well-known that $G(Z^n)$ has a hamiltonian path. Indeed this follows from lemma 1 below. We therefore have

Corollary 2. For $n \geq 4$, K_1 and J_1 are hamiltonian.

Indeed, lemma 1 shows that we can construct hamiltonian paths in K_1 and J_1 of the following form. First all simplices lying in the cube $[0,1]^n$ are enumerated. Then the remaining simplices in $[-1,2]^n$ are enumerated, and so on. Such paths provide efficient searches for completely labelled simplices in a neighborhood of the origin.

We may wish to enumerate the simplices lying in a neighborhood of the boundary of some compact convex set, for example to approximate the degree of a mapping defined on this set. If the set is cubical, this can be done. Let C^n_m denote $\{x \in Z^n | -m \leq x_i \leq m+1 \text{ for } i \in N\}$ for $m \geq 0$ (and integer) and let $C^n_{-1} = \emptyset$. We use ∂C^n_m to denote $C^n_m \backslash C^n_{m-1}$. From lemma 1 below and theorem 3 we obtain

Corollary 3. For $m \geq 0$ and $n \geq 4$, the restriction of K_1 or J_1 to ∂C^n_m is hamiltonian.

Lemma 1. (a) For every $m \geq 0$, there is a hamiltonian cycle in $G(\partial C^2_m)$. For every $m \geq 0$ and $n \geq 3$ there is a hamiltonian path P^n_m in $G(\partial C^n_m)$ from $(0,\ldots,0,-m)$ to $(0,\ldots,0,m+1)$.

(b) For every $m \geq 0$ and $n \geq 2$, there is a hamiltonian path Q^n_m in $G(C^n_m)$ from 0 to a vertex in ∂C^n_m; for $n \geq 3$ the final vertex is $(0,\ldots,0,m+1)$ if m is even and $(0,\ldots,0,-m)$ for m odd. For $0 \leq k < m$, Q^n_k is an initial segment of Q^n_m.

Proof. For $n = 2$ (a) and (b) are trivial. We proceed by induction; however, $n = 3$ must be treated separately.

(a) Write ∂C^3_m as $(C^2_m \times \{-m\}) \cup (\partial C^2_m \times \{-m+1,\ldots,m\}) \cup C^2_m \times \{m+1\}$. We now construct P^3_m. We start with the path $Q^2_m \times \{-m\}$; that is, a third coordinate $-m$ is added to each vertex on Q^2_m. If the final vertex is

$(\bar{x}_1,\bar{x}_2,-m)$, we then move to $(\bar{x}_1,\bar{x}_2,-m+1)$. If $m = 0$, we then add $(Q_m^2)^{-1} \times \{1\}$ to obtain P_0^3. If $m > 0$, let P_m^2 be a hamiltonian path in ∂C_m^2 with initial vertex $(\bar{x}_1,\bar{x}_2)$ (obtained from the hamiltonian cycle). Since $|\{-m+1,\ldots,m\}|$ is even, we obtain P_m^3 by joining in the natural way $Q_m^2 \times \{-m\}$, $P_m^2 \times \{-m+1\}$, $(P_m^2)^{-1} \times \{-m+2\},\ldots,(P_m^2)^{-1} \times \{m\}$ and $(Q_m^2)^{-1} \times \{m+1\}$. This same construction works also for $n > 3$, where now P_m^{n-1} is given by the induction hypothesis; if m is even we use $(P_m^2)^{-1} \times \{-m+1\}$, $P_m^2 \times \{-m+2\},\ldots,P_m^2 \times \{m\}$.

Part (b) follows readily from (a). Q_0^n is the same as P_0^n. Given Q_{m-1}^n we obtain Q_m^n by adding the edge $(0,\ldots,0,m) \to (0,\ldots,0,m+1)$ and $(P_m^n)^{-1}$ if m is odd; and by adding $(0,\ldots,0,-m+1) \to (0,\ldots,0,-m)$ and P_m^n if m is even. This completes the induction step.

The Cases n = 2 and n = 3

The set of adjacent transpositions is sufficiently rich to allow the method above to work only for $n \geq 4$. We show here that K_1 and J_1 are hamiltonian for $n = 2$ and $n = 3$.

For $n = 2$ we merely indicate the associated paths in figure 1. Allgower and Georg have previously shown that K_1 is hamiltonian for $n = 2$ and have given an algorithm to generate the path [1]. Figure 1 clearly shows how such a path in J_1 can be constructed.

For $n = 3$ we use the general method except that we use cubes of side two rather than unit cubes, with vertices having even coordinates. We will always pass between such double cubes in such a way that the cube with larger coordinates is entered via a simplex in its unit cube with smallest coordinates. It turns out that the graph of the restriction of K_1 or J_1 to such a double cube admits hamiltonian paths of sufficient variety to enable the double cubes to be traversed in any order. Clearly the double cubes (with the natural adjacencies) admit a hamiltonian path, and the hamiltonian path in K_1 or J_1 follows.

The graphs of the restrictions of K_1 and J_1 to double cubes are shown in Figures 3 and 4; we do not show all the required hamiltonian paths, but two are indicated. In both graphs each small hexagon corresponds to the six tetrahedra within each unit cube; by indicating the vertex of the hexagon corresponding to the permutation (1,2,3) and that corresponding to (1,3,2) all vertices are determined.

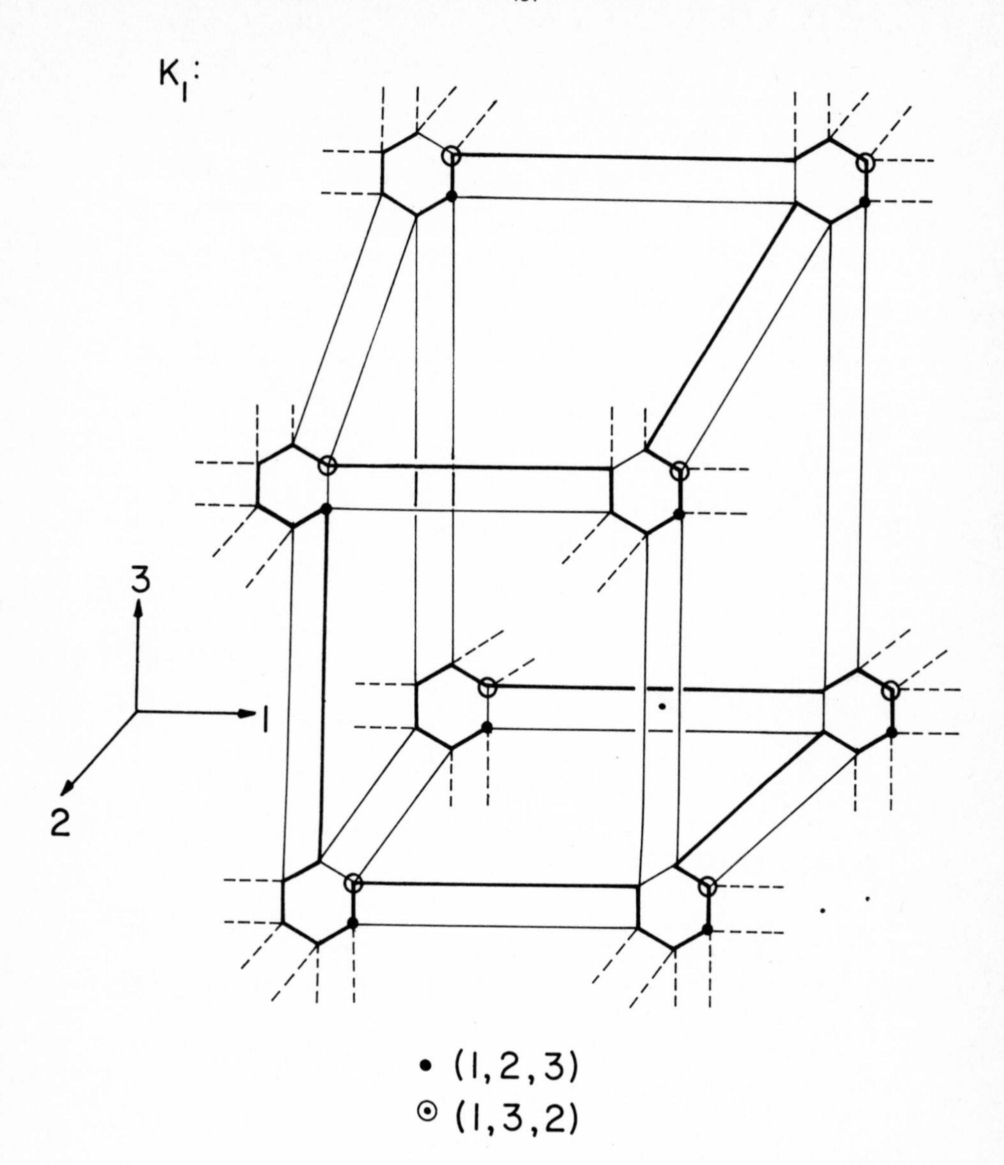

FIGURE 3

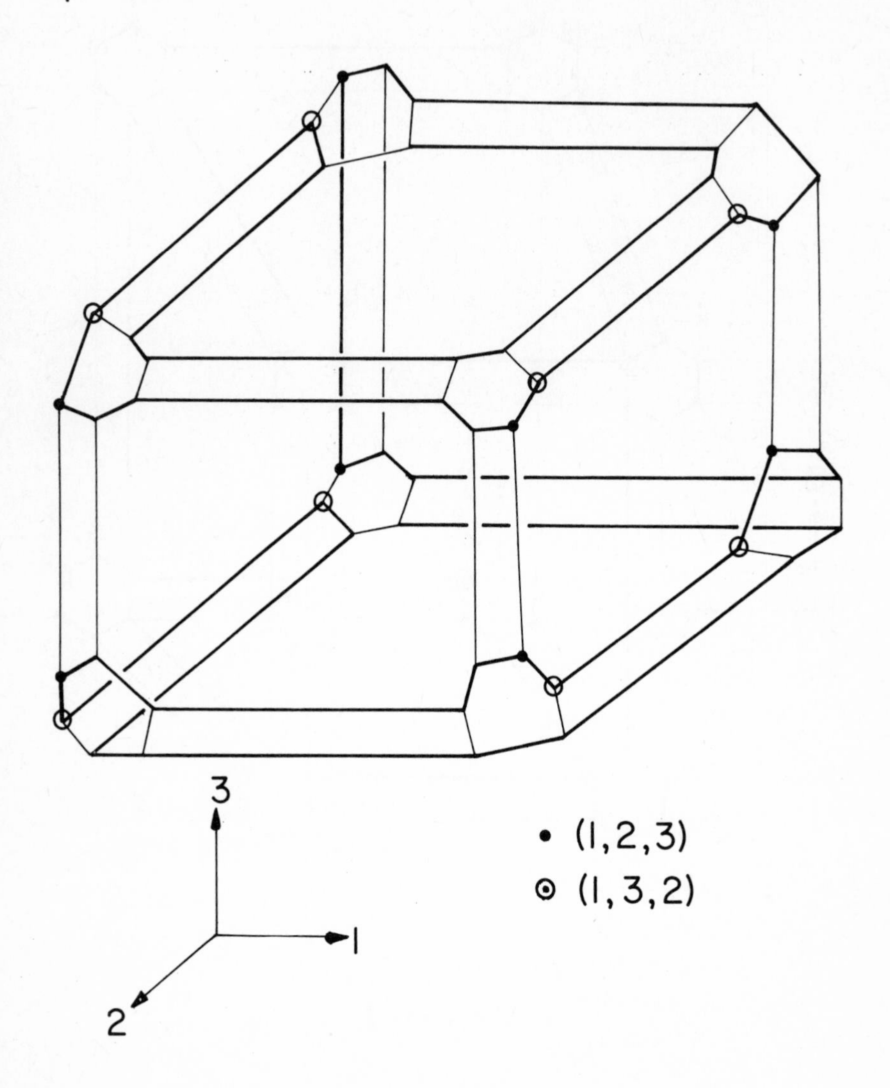

FIGURE 4

REFERENCES

1. Allgower, E., and K. Georg, "Triangulations by Reflections with Applications to Approximation," Proceedings of the Oberwolfach Conference on Numerische Methoden der Approximationstheorie, 1978

2. Allgower, E., and K. Georg, "Simplicial and Continuation Methods for Approximating Fixed Points and Solutions to Systems of Equations," to appear in SIAM Review

3. Reingold, E.M., J. Nievergelt and N. Deo, _Combinatorial Algorithms: Theory and Practice_, Prentice Hall, Englewood Cliffs, N.J.,1977.

4. Stroud, A.H., _Approximate Calculation of Multiple Integrals_, Prentice-Hall, Englewood Cliffs, N.J., 1971.

5. Todd, M.J., "Union Jack Triangulations," in _Fixed Points: Algorithms and Applications_, S, Karamardian (ed.), Academic Press, New York, 1977

6. Todd, M.J., _The Computation of Fixed Points and Applications_, Springer Lecture Notes in Economics and Mathematical Systems No. 124, Springer-Verlag, Berlin, 1976.

7. Whitney,H., _Geometric Integration Theory_, Princeton University Press, Princeton, N.J., 1957.

8. Zienkiewicz, O.C., _The Finite Element Method in Engineering Science_, McGraw-Hill, London, 1971.

The Beer Barrel Theorem

A new proof of the asymptotic conjecture in fixed point theory

by A. J. Tromba

In a seminar in Bonn during the summer of 1977, H. Peitgen challenged the author to give an easy proof of the differentiable case of the classic Schauder conjecture; namely (simply stated) let $T : E \circlearrowleft$ be a C^1 map with T^2 compact then T has a fixed point. The prize was a barrel of beer. This theorem had already been proved by Nussbaum [2] but involved rather long, technical and sophisticated methods in fixed point theory. The purpose of this note is to present a rather simple proof of the Nussbaum result which uses only the mod 2 degree of Smale introduced in [3], and a transversality result [5].

§ 1 The Mod 2 degree of Smale

Let $W \subset E$ be an open subset of a Banach space. Let $f:\bar{W} \to F$ be a continuous proper map (inverse image of compact sets is compact) such that $f:W \to F$ is Fredholm of index zero; i.e. $df(x):E \to F$ is linear Fredholm of index zero. Assume further that $0 \notin f(\partial W)$. Let O be a component of $F-f(\partial W)$ containing 0. By Smale's version of the Sard theorem there is a regular value $y \in O$ for $f|W$. Then the number of points in $f^{-1}(y)$ will be finite and this number mod 2 is called the mod 2 degree of f relative to W and O, and is denoted by $\deg_2(f,\bar{W},O)$. Smale shows, using cobordism arguments going back to Pontryagin that this number is independent of the choice of regular value $y \in O$ and is invariant under proper homotopies f_t such that for all t $0 \notin f_t(\partial W)$.

§ 2 Fredholm families and a transversality theorem

Let $f:A\times\bar{W} \to \mathbb{F}$ be C^1 where A is a smooth Banach manifold. We say that f is a <u>Fredholm family of index zero</u> if for each $a \in A$, $f_a = f(a,\cdot)$

is Fredholm of index zero on W. We say that f is a <u>zero proper</u> family if whenever $a_n \to a$ and $f_{a_n}(x_n) \to 0$, then x_n has a convergent subsequence.

<u>Definition</u>. If $g: W \to F$ is Fredholm of index zero, a zero x_0 of g ($g(x_0)=0$) is said to be <u>non-degenerate</u> if $dg(x_0)$ is an isomorphism.

A proof of the following transversality result in case A is Hilbert can be found in [5].

<u>Theorem 1</u>. Suppose $f: A \times \bar{W} \to F$ is a zero proper Fredholm family of index zero and such that for each $x \in f^{-1}(0)$, $x=(a,e)$, the total derivative $df(x): T_a A \times E \to F$ is surjecture. Then there exists an open and dense set $\hat{A} \subset A$ such that whenever $a \in \hat{A}$ <u>all the zeros of f_a</u> are non-degenerate.

§ 3 <u>Proof of the Schauder conjecture</u>

<u>Theorem 2</u>. Let $\Omega \subset E$ be open and convex and let $T: \bar{\Omega} \to \Omega$ be C^1 with T^2 compact. Then T has a fixed point.

<u>Remark</u>. By the classic Schauder theorem, T^2 has a compact set of fixed points $\mathscr{F}$. Nussbaum only had to assume that T is C^1 on some neighbourhood of $\mathscr{F}$. The same will be true for our proof, however, for pedagodical reasons, we shall make no reduction in our hypothesis.

The basic idea of the proof is quite simple. Suppose the fixed point set $\mathscr{F}$ of T^2 is non-degenerate; i.e. $id-d(T^2)(x)$ is an isomorphism for each $x \in \mathscr{F}$. Since T^2 is compact, the map $g(x)=x-T^2x$ is proper Fredholm of index zero and therefore has a finite number of zeros (which are of course the fixed points of T). Using the mod 2 degree with W=E and considering the homotopy $(\lambda x) \to x-\lambda T^2 x$ for $\lambda \in [0,1]$, we see that the mod 2 degree of g is 1 and consequently T^2 must have an odd number of fixed points. But T acts as a permutation on this finite set, and each element is of order 2 under T. Since there are an odd number of fixed points, one of them must have order 1. This is a fixed point of T, and the Schauder conjecture is proved in this case. In the remainder of the paper, we will show how to apply theorem 1 to perturb T so that the fixed points of the perturbed mapping are non-degenerate.

We shall now have the standing assumption that T has no fixed points; also without loss of generality we can assume that $0 \in \Omega$.

Let $\mathscr{A}$ be the Banach space of C^1 compact maps on E with bounded differentials; i.e. $k \in \mathscr{A}$ if $\sup_{x \in E} \|dk(x)\| < \infty$ and let $\mathscr{G}$ be the set of C^1 diffeomorphisms $G: E \circlearrowleft$ of the form $G = id + k$, $k \in \mathscr{A}$ and $\sup_{x \in E} \|dk(x)\| < 1$. Then $\mathscr{G}$ is clearly a C^∞ Banach manifold modelled on $\mathscr{A}$ with the global chart $G \to k$. Moreover each $G \in \mathscr{G}$ is properly homotopic to the identity though diffeomorphisms via the homotopy $t \to id + tk$.

As before, let $\mathscr{F}$ denote the compact fixed point set of T^2 and hence the zero set of $f_I(x) = x - T^2x$. Since $\mathscr{F} \subset \Omega$ is compact, there exist neighbourhoods V^* of $\mathscr{F}$, and U^* of $I \equiv id$ in $\mathscr{G}$ such that both $V^* \subset \Omega$ and $(GT)(V^*) \subset \Omega$ for all $G \in U^*$. Thus the double composition $(GT)^2$ is defined on V^*. Consider the family of mappings $f: U^* \times V^* \to E$ defined by $f(G,x) = x - (GT)^2x$. f need not necessarily be a Fredholm family but $f_I = f(I,\cdot)$ is Fredholm of index zero. It therefore follows from the implicit function theorem that for each $x \in \mathscr{F}$ there exists a splitting of $E = E_1(x) \oplus E_2(x)$, $\dim E_2(x) < \infty$, and bounded neighbourhoods $U(x)$ of $I \in U^*$ and $V(x) \subset V^*$ of x such that for $G \in U(x)$ there is a local diffeomorphism Ψ_G of a neighbourhood of $0 \in E$ to $V(x)$ with

(1) $\quad f_G \circ \Psi_G(X_1, X_2) = (X_1, \Theta_G(X_1, X_2))$

where $(G, X_1, X_2) \to \Psi_G(X_1, X_2)$ and $(G, X_1, X_2) \to \Theta_G(X_1, X_2)$ are C^1.

Remark. This normal form has been used by several authors (see e.g. [1], and [3]).

Since $\mathscr{F}$ is compact, we can find a neighbourhood U of $I \in U^*$ and a finite number $V_1, \ldots, V_l$ of neighbourhoods covering $\mathscr{F}$ so that for $G \in U$ and $x \in V_j \subset V^*$ representation (1) holds.

Recall that we are assuming that T has no fixed points from which it follows that $\|T^2x - Tx\| \geq \delta > 0$ for all $x \in \mathscr{F}$. Therefore we can find neighbourhoods $\tilde{U}$ of I in $\mathscr{G}$ and $\tilde{V}$ of $\mathscr{F}$ such that

(2) $\quad \|TGTx - Tx\| > \delta/2$

for all $G \in \tilde{U}$ and $x \in \tilde{V}$.

Consequently, we can assume that neighbourhoods $U \subset U^*$ of I and $V = \cup V_j \subset V^*$ have been chosen so that with respect to these neighbourhoods representation (1) and inequality (2) holds. Let W be a neighbourhood of $\mathscr{F}$ with $\bar{W}\subset V$. From representation (1) it follows that there is a neighbourhood $A\subset U$ of I so that for all $G\epsilon A$, $0\notin f_G(\partial W)$.

By the normal form (1), $f:A\times\bar{W}\to E$ is a zero proper Fredholm family of index zero. Moreover for each $G\epsilon A$, f_G is proper and hence has a mod 2 degree. Since, by construction, each $G\epsilon A$ is properly homotopic to the identity, one can immediately conclude that $\deg_2(f_G,\bar{W},0) = \deg_2(f_I,\bar{W},0)$. But f_I is proper Fredholm of index zero on all of Ω and the zero set of f_I is contained in W. Therefore $\deg_2(f_I,\bar{W},0) = \deg_2(f_I,\bar{\Omega},0)$. Since Ω is convex, the homotopy $(\lambda,x) \to x-\lambda T^2x$ is a proper homotopy of f_I to the identity introducing no zeros on $\partial\Omega$. Thus $\deg_2(f_I,\bar{\Omega},0) = 1 = \deg_2(f_I,\bar{W},0)$.

The next lemma is the main step in the proof of the Schauder conjecture.

Lemma. The family $f:A\times\bar{W}\to E$ has the property that whenever $f_G(x)=0$ the total derivative $df(G,x)$ is surjecture

Proof. By direct computation

$$df(G,x)[H,h] = h - DG_{TGT(x)} \circ DT_{GT(x)} \circ DG_{T(x)} \circ DT_x[h]$$

$$- H(TGTx) - DG_{TGT(x)} \circ DT\circ H(Tx)$$

We know that $x\epsilon V$ and that on V

$TGTx \neq Tx$ for $G\epsilon A$.

So in order to show surjectivity given $w\epsilon E$, we must produce an $[H,h]$ so that $df(G,x)[H,h] = w$. Choose $h=w$ and H so that $H(Tx)=0$ and $H(TGTx) = -DG\circ DT\circ DG\circ DT_x(w)$.

This concludes the lemma.

By theorem 1, we can conclude that there exists a sequence $G_n\epsilon A$, $G_n \to I$ such that $(G_n\circ T)^2$ has non-degenerate fixed points in W. By degree there must be an odd number and by parity G_nT has a fixed point x_n. However $G_nTx_n=x_n$ implies that $f_{G_n}(x_n)=0$. Since the family f is zero proper, we can conclude that x_n has a subsequence converging to $z\epsilon W$. Clearly $Tz=z$ which concludes the proof of the

Schauder conjecture.

Remarks. Nussbaum's original proof worked for the case T^n compact, $n>2$. A slightly more technical argument than that presented here should also work for this case. It is also clear that these methods yield a clean proof of the Steinlein mod (p) result [4].

References

[1] K. D. Elworthy and A. J. Tromba; Differential structures and Fredholm maps on Banach manifolds; Prac. Symp. pure math., AMS vol XV, 45-94

[2] R. D. Nussbaum; Some asymptotic fixed point theorems, Trans. Amer. Math. Soc. vol 171 (1972), 349-375

[3] S. Smale; An infinite dimensional version of Sard's theorem, Amer. J. Math., 87 (1965), 861-866

[4] H. Steinlein; A new proof of the (mod p) theorem in asymptotic fixed point theory, Prac. Symp. on problems in nonlinear functional analysis, Universität Bonn, July 22-26, 1974. Ber. Ges. Math. Datenverarbeitung Bonn, No. 103, 29-42 (1975)

[5] A. J. Tromba; Fredholm vector fields and a transversality theorem, J. Funct. Anal. vol 23, No. 4 (1976)

A. J. Tromba
Mathematisches Institut
Universität Bonn, SFB 72
and
University of California
at
Santa Cruz

On instability, ω-limit sets and periodic solutions of nonlinear autonomous differential delay equations

Hans-Otto Walther

1. Introduction. This paper deals with slowly oscillating solutions of equation

$$\text{(f)} \qquad \dot{x}(t) = -f(x(t-1))$$

for continuous functions $f:R \to R$ which are differentiable at $\xi = 0$ and satisfy $\xi f(\xi) > 0$ for $\xi \neq 0$. We are interested in the behaviour of the trajectories $(x_t)_{t \geq 0}$ in the state space of continuous functions on the initial interval.

We prove that in case $f'(0) > \pi/2$ and f bounded below trajectories of slowly oscillating solutions are attracted by a set which may be viewed as a solid torus in function space (Theorem 2). - For a related result on $(x(t),\dot{x}(t))$-trajectories in the plane, see Kaplan and Yorke. - Theorem 2 implies the existence of solutions on the whole real line which are not necessarily periodic but regularly oscillating in some sense, with the set of zeros unbounded for $t > 0$ as well as for $t < 0$ (Corollary 2).

Slowly oscillating periodic solutions with minimal period correspond to fixed points of a Poincare operator on a cone of initial functions. This is Jones' well known idea to obtain periodic solutions. In case of equation (f), consider the operator $T:\phi \to x^{\phi}_{z_2+1}$ on the cone K of continuous increasing functions $\phi: [-1,0] \to R^{+}_{0}$ which are not identically zero.

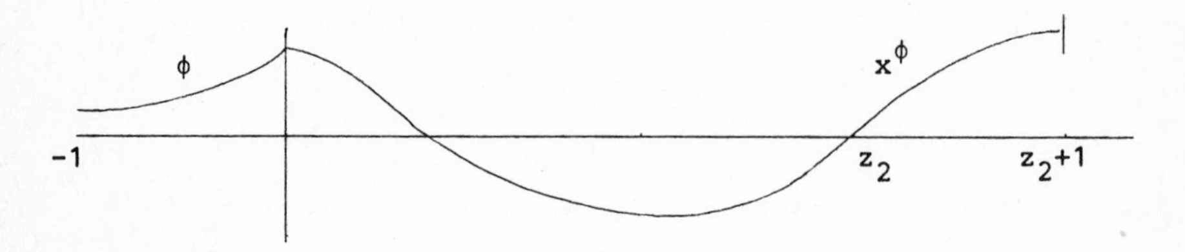

Our method shows that it is possible to get by with an application of the Schauder fixed point theorem if $f'(0) > \pi/2$ and, say, f bounded below. We construct a suitable closed subset $D \subset K$ with $T(D) \subset D$ (proof of Theorem 3). In particular D is bounded away from the critical point $\phi = 0$. Former proofs which use Jones' idea have in common that they use restrictions of the Poincare operator to sets with $\phi = 0$ in the closure. In order to get a nonzero fixed point they need fixed point theorems of expansion-compression type or the theorem of Browder on nonejective fixed points [G,N_1,N_2,C,A].

Our proofs make use of the Liapunov functional of Hale and Perello for unstable behaviour of functional differential equations. V is defined by a projection P onto an eigenspace of increasing exponential solutions of the linearized delay equation. A consequence of the result of Hale and Perello is our Theorem 1 saying that trajectories leave a neighbourhood of zero and, what is important, then keep away from it provided they lie in a set S' where an inequality

$$c\|\phi\| \le \|P\phi\| \quad \text{for all } \phi \in S' \tag{1}$$

with $c > 0$ is valid. The essential step is then to find sets which both satisfy an inequality of type (1) and contain trajectories of solutions with initial value in K. This is done in sections 5 - 6, beginning with Lemma 3 which exhibits the basic relation between characteristic values of the linear equation and slow oscillations. Due to Lemma 4, our method works without the "integral representation of the projection by the formal adjoint equation" [H, chapter 7].

It applies to equations with distributed delay too, see the proof of ejectivity in [Wa].

Finally, let us point out one difference to other proofs of instability properties which employ the functional of Hale and Perello. In [G,C,A] the integral representation just mentioned is used to deduce an inequality like (1) on the domain of definition of the Poincare operator. Note that these domains are invariant with respect to this operator but not for trajectories $(x_t)_{t\geq 0}$. In fact, the only segments x_t which they contain are those defined by the iterates of the Poincare operator.

2. <u>Notation</u>. X denotes the Banach space of continuous functions $\phi:[-\tau,0] \to R^n$. We assume $\tau > 0$ and $n \in N$ arbitrary in section 3, $\tau = 1 = n$ in sections 4 - 8. Y denotes the Banach space of continuous functions $\psi:[-\tau,0] \to C^n$.

Let $F:X \to R^n$ be given. A solution of the equation

$$\dot{x}(t) = F(x_t) \tag{F}$$

is either a continuous function $x:[-\tau,\infty) \to R^n$ which is differentiable for $t > 0$ and satisfies (F) for $t > 0$ or a differentiable function $x:R \to R^n$ which satisfies (F) for all $t \in R$. The segments $x_t \in X$ are defined by $x_t(a) := x(t+a)$, $a \in [-\tau,0]$. The trajectory of a solution x is the family $(x_t)_{t\geq 0}$, or $(x_t)_{t\in R}$ respectively.

3. Instability. We consider equation (F) for $F: X \to R^n$ continuous, $F(0) = 0$, F differentiable at $\phi = 0$. We assume that there is a non-empty set E of eigenvalues with positive real part of the infinitesimal generator of the semigroup in Y defined by the linear equation $(F'(0))$. Let P denote the eigenprojection which belongs to E. We state a variant of Theorem 3 [HP] as

Lemma 1: There exist a continuous quadratic functional $V: X \to R_o^+$ and positive constants c_1, c_2, c_3 with

(i) $c_1 \|P\phi\|^2 \leq V(\phi) \leq c_2 \|P\phi\|^2$ for all $\phi \in X$,

(ii) $\forall p > 0 \; \exists \delta_p > 0 \; \forall$ (x solution of (F)) $\forall\; t \geq 0$:

$$\|x_t\| \leq \delta_p \wedge p^2 \|x_t\|^2 \leq V(x_t) \Rightarrow c_3 V(x_t) \leq \dot{V}(x,t)$$

(with $\dot{V}(x,t) := \lim\limits_{\substack{0 \neq h \to 0 \\ 0 < t+h}} (V(x_{t+h}) - V(x_t))/h$ for $t \geq 0$).

Theorem 1: Let $S' \subset X \setminus \{0\}$ and $c > 0$ be given with

(1) $$c\|\phi\| \leq \|P\phi\| \text{ for all } \phi \in S'.$$

Let V be a functional as in Lemma 1. Conclusions:

(i) There exists $p > 0$ with

(V) $$p^2 \|\phi\|^2 \leq V(\phi) \text{ for all } \phi \in S'.$$

(ii) There exists $a > 0$ such that for every solution x of (F) with trajectory in S' there is a real s with

$$a \leq \|x_t\| \text{ for all } t \geq s.$$

(iii) Assume δ_p is chosen according to part (ii) of Lemma 1. Then we have

$$p^2 \delta_p^2 \leq V(x_t) \text{ for all } t \geq 0$$

for every solution x of (F) with trajectory in S' and with $p^2 \delta_p^2 \leq V(x_o)$.

Proof: $0 < p < c\sqrt{c_1}$ yields $p^2 \|\phi\|^2 \leq c^2 c_1 \|\phi\|^2 \leq c_1 \|P\phi\|^2 \leq V(\phi)$ for all $\phi \in S'$, by (1) and Lemma 1 (i). Choose δ_p as in part (ii) of Lemma 1. Consider a solution x of (F) with $x_t \in S'$ for all $t \geq 0$.

a) In case $\|x_o\| < \delta_p$ there is $u > 0$ with $\|x_u\| \geq \delta_p$. Proof: Assume $\|x_t\| < \delta_p$ for all $t \geq 0$. By Lemma 1 (ii), by (V) and $0 \notin S'$, $0 < c_3 V(x_t) \leq \dot{V}(x,t)$, hence $0 < V(x_o)e^{c_3 t} \leq V(x_t) \leq c_2\|Px_t\|^2 \leq$ $\leq c_2\|x_t\|^2 < c_2\delta_p^2$ for all $t \geq 0$, contradiction.

b) Suppose $\|x_u\| \geq \delta_p$, $u \geq 0$. For every $v > u$ with $\|x_v\| < \delta_p$ there exists $w \in [u,v)$ with $\|x_w\| = \delta_p$ and $\|x_t\| < \delta_p$ for all $t \in (w,v]$. By Lemma 1 (ii) and by (V), $V(x_t)$ increases on $[w,v]$. Hence $p^2\delta_p^2 = p^2\|x_w\|^2 \leq V(x_w) \leq V(x_v) \leq c_2\|Px_v\|^2 \leq c_2\|x_v\|^2$.

a) and b) imply the existence of $s \geq 0$ with

$$\min\{\delta_p, p\delta_p/\sqrt{c_2}\} \leq \|x_t\| \quad \text{for all } t \geq s.$$

c) Assume $p^2\delta_p^2 \leq V(x_o)$. Let $t > 0$. $\delta_p \leq \|x_t\|$ implies $p^2\delta_p^2 \leq$ $\leq p^2\|x_t\|^2 \leq V(x_t)$, by (V). $\|x_t\| < \delta_p$ and $\|x_s\| \leq \delta_p$ for all $s \in [0,t)$ imply $p^2\delta_p^2 \leq V(x_o) \leq V(x_t)$, by (V) and Lemma 1 (ii). In case $\|x_t\| < \delta_p < \|x_s\|$ with $0 \leq s < t$ there is $u \in (s,t)$ with $\|x_u\| = \delta_p$ and $\|x_v\| < \delta_p$ for all $v \in (u,t]$, hence $p^2\delta_p^2 = p^2\|x_u\|^2 \leq V(x_u) \leq$ $\leq V(x_t)$.

4. Slowly oscillating solutions. In the following we consider equation (f) for continuous functions $f: R \to R$ which are differentiable at $\xi = 0$ and satisfy $\xi f(\xi) > 0$ for $\xi \neq 0$. For every $\phi \in X$ there exists a unique solution $x: [-1,\infty) \to R$ of (f) with $x_o = \phi$. This solution will also be denoted by x^ϕ. On compact intervals solutions depend continuously on the initial values with respect to uniform convergence.

Lemma 2: Assume $f'(0) > 1$. Then we have:

(i) For every $\phi \in X \setminus \{0\}$ with $0 \leq \phi$ there is a sequence of non-negative zeros $z_j = z_j(\phi)$ of $x = x^\phi$ with $0 \leq x$ in $[-1, z_1]$, $x < 0$ in $(z_1, z_1+1]$, $\dot{x} \leq 0$ in $(0, z_1+1]$ and

(Z) $\begin{Bmatrix} 0 < \dot{x} \\ \dot{x} < 0 \end{Bmatrix}$ in $(z_j+1, z_{j+1}+1)$ for all $\begin{Bmatrix} \text{odd} \\ \text{even} \end{Bmatrix}$ j.

(ii) For every $r > 0$ there exists $d > 0$ with $z_1(\phi) < d$ for all $\phi \in X \setminus \{0\}$ with $0 \leq \phi \leq r$.

Assume in addition $\inf f > -\infty$. Set $r_1 := -\inf f$, $r_2 := \sup_{[0,r_1]} f$. Then we have

(iii) $-r_2 \leq x^\phi(t) \leq r_1$ for all $\begin{cases} t \geq z_2(\phi) \text{ if } \phi \in X \setminus \{0\},\ 0 \leq \phi. \\ t \geq -1 \text{ if } \phi \in X,\ 0 \leq \phi \leq r_1. \end{cases}$

Proof: See e. g. the proofs of Lemmas 2.3 and 2.2 in $[N_1]$.

Remark 1: Assertion (i) with signs of x and $\dot{x}$ reversed and assertion (ii) hold if $\phi \neq 0$ and $\phi \leq 0$, or $-r \leq \phi \leq 0$ respectively.

Definition: A solution x of (f) is called slowly oscillating iff there exists a sequence of zeros of x with property (Z).

Solutions which start in K are slowly oscillating, and their trajectories lie in the set of functions with at most one change of sign, that is in the set S of $\phi \in X \setminus \{0\}$ such that ϕ or $-\phi$ satisfies

$0 \leq \tilde{\phi}$ or

$\exists\, z \in [-1,0]: -1 \leq a < z < a' \leq 0 \Rightarrow \tilde{\phi}(a) \leq 0 \leq \tilde{\phi}(a')$.

S is a cone in the sense that we have $t\phi \in S$ whenever $t > 0$, $\phi \in S$. But it is not convex.

Proposition:

(i) For every slowly oscillating solution x of (f) there exists $s \in R$ with $x_t \in S$ for all $t \geq s$.

(ii) For every $\phi \in S$ there exists $t \in [0,1]$ with $0 \leq x_t^\phi$ or $x_t^\phi \leq 0$.

(iii) $x_t^\phi \in S$ for all $t \geq 0$ and all $\phi \in S$.

(iv) In case $f'(0) > 1$ x^ϕ is slowly oscillating for all $\phi \in S$.

(v) $\mathrm{cl}\, S = S \cup \{0\}$.

Proof of (ii): Let $\phi \leq 0$ in $[-1,z)$ and $0 \leq \phi$ in $(z,0]$. (f) implies $0 \leq \dot{x}^\phi$ in $(0,z+1]$, hence $0 \leq x_{z+1}^\phi$.

Proof of (iv): By (ii), $0 \leq x_t^\phi$ or $x_t^\phi \leq 0$ for some $t \geq 0$. We have $x_t^\phi \neq 0$. Apply Lemma 2 (i) and Remark 1 to the solution $-1 \leq u \to x^\phi(t+u)$ of (f).

5. Linear slow oscillations and inequality (1). Linearization of (f) near x = 0 yields equation

$$(\alpha) \qquad \dot{y}(t) = -\alpha y(t-1)$$

with $\alpha = f'(0)$, a special case of the type considered. Equation (α) defines a semigroup $(T_t)_{t \geq 0}$ of bounded linear operators in Y by

$$T_t\phi := (y^{\mathrm{Re}\, \phi})_t + i(y^{\mathrm{Im}\, \phi})_t.$$

The eigenvalues of the infinitesimal generator are the zeros of the entire function $\lambda \to \lambda + \alpha e^{-\lambda}$.

For $\alpha > 1/e$ they are all simple, and they form a sequence of pairs $\lambda_j, \overline{\lambda}_j$; $j \in N_0$, with

(2) $\quad u_{j+1} < u_j$ and $2\pi j < v_j < 2\pi j + \pi$ for all $j \in N_0$

($u_j := \mathrm{Re}\, \lambda_j$, $v_j := \mathrm{Im}\, \lambda_j$). We have: $0 < u_0 \Leftrightarrow \alpha > \pi/2$.

For a proof of these assertions, see [Wr].

We consider the projection P_0 which belongs to λ_0 and $\overline{\lambda}_0$. With regard to (1) let us first determine the maximal subset in X where P_0 does not vanish.

Lemma 3: Assume $\alpha > 1$. For all $\phi \in X$ we have

$$P_o\phi \neq 0 \iff (\text{Solution } y^\phi \text{ of } (\alpha) \text{ is slowly oscillating}).$$

Proof "<=": Suppose y^ϕ is a slowly oscillating solution of (α) with $P_o\phi = 0$. Let P_1 denote the projection which belongs to λ_1 and $\overline{\lambda}_1$.

a) Assume Re $P_1\phi \neq 0$. There are $\varepsilon > 0$, $c > 0$ with

(3) $\|T_t(\phi - (P_o + P_1)\phi)\| \leq ce^{(u_1-\varepsilon)t}\|\phi - (P_o + P_1)\phi\|$ for all $t \geq 0$

[H, ch. 7]. The solution with initial value Re $(P_o + P_1)\phi$ = Re $P_1\phi$ has the form

$$t \to e^{u_1 t}(a \cos v_1 t + b \sin v_1 t)$$

with a, b real, $|a| + |b| > 0$. By (3),

$$y^\phi(t)e^{-u_1 t} - a \cos v_1 t - b \sin v_1 t \to 0 \quad \text{as } t \to \infty.$$

$2\pi < v_1$ implies a contradiction to y^ϕ slowly oscillating.

b) Assume Re $P_1\phi = 0$. There is a neighbourhhod U of ϕ in X with y^ψ a slowly oscillating solution of (α) for every $\psi \in U$. (Proof: By $y_t^\phi > 0$ for some $t > 0$ and by continuous dependence, there is a neighbourhood U with $y_t^\psi > 0$ for all $\psi \in U$. The solution $-1 \leq s \to y^\psi(t+s)$ with initial value y_t^ψ is slowly oscillating, hence y^ψ too.)

We choose a real-valued $\chi \neq 0$ in $P_1 Y$ with $\psi := \phi + \chi \in U$. Then $P_o\psi = 0$, Re $P_1\psi = \chi \neq 0$. As in a), we derive a contradiction.

Proof "=>": The estimate $\|T_t(\phi - P_o\phi)\| \leq ce^{(u_0-\varepsilon)t}\|\phi - P_o\phi\|$ for $\phi \in X$ with $P_o\phi \neq 0$ implies Re $P_o\phi \neq 0$. From $0 < v_o < \pi$ and from $\|T_t\phi - T_t \text{ Re } P_o\phi\|e^{-u_0 t} \to 0$ as $t \to \infty$ we infer the existence of a segment $T_t\phi > 0$. It follows that ϕ defines a slowly oscillating solution.

The hypothesis $\alpha > 1$ in Lemma 3 and Lemma 4 below can be weakened to $\alpha > 0$ (with the definition of "slowly oscillating" changed to include monotone solutions which exist for $\alpha \leq 1/e$).

Corollary 1: We have $P_o\phi \neq 0$ for all $\phi \in S$.

Proof: Apply part (iv) of the Proposition to $f = \alpha$ id, use Lemma 3.

We look for subsets of S where estimates of type (1) hold.

Lemma 4: Assume $\alpha > 1$. For every non-empty cone $S' \subset S$, estimate

$(1)_o$ $\exists c > 0\ \forall \phi \in S': c\|\phi\| \leq \|P_o\phi\|$

is equivalent to

(4) $\exists k > 0\ \forall \phi \in S': k\|\phi\| \leq \|T_1\phi\|$.

Proof: Set $S_1' := \{\phi \in S': \|\phi\| = 1\}$.

"(4) => $(1)_o$": T_1 completely continuous implies cl T_1S_1' compact. This is a subset of $S \cup \{0\}$, by $T_1S \subset S$ (Proposition (iii)) and by cl $S = S \cup \{0\}$, and moreover a subset of S, by (4). Hence $P_o\psi \neq 0$ for all $\psi \in$ cl T_1S_1', and

$0 < \inf\{\|P_o\psi\|: \psi \in \text{cl } T_1S_1'\} \leq \inf\{\|P_oT_1\phi\|: \phi \in S_1'\} \leq$

$\leq \|T_1\| \inf\{\|P_o\phi\|: \phi \in S_1'\}$, by $P_oT_1 = T_1P_o$. This yields $(1)_o$.

"$(1)_o$ => (4)": $(1)_o$ gives $0 \notin$ cl P_oS_1'. This is a closed bounded subset of the finite-dimensional space P_oY. $T_1\psi \neq 0$ for all ψ in $P_oY \setminus \{0\}$ implies $0 < \inf\{\|T_1\psi\|: \psi \in \text{cl } P_oS_1'\}$, and the assertion follows as above.

Inequality (4) is easier to verify than $(1)_o$ since

$$T_1\phi(a) = \phi(0) - \alpha\int_{-1}^{a}\phi(t)dt \quad \text{for all } a \in [-1,0]$$

while integral formulas for P_o involve the functions $t \to e^{\lambda_0 t}$, see [H, ch. 7] and [Wa].

6. Verification of inequality (4).

Lemma 5: Assume $f'(0) > 1$. Let $r > 0$ be given. Set $\alpha := f'(0)$ and consider the operator T_1 of the semigroup defined by equation (α). Then there exists a constant $k = k(r,f) > 0$ such that

$$k\|x_t\| \le \|T_1 x_t\| \quad \text{for all } t \ge 0$$

for every solution x of (f) with $|x| \le r$ and with x_0 in the convex cone $K = \{\phi \in X\colon 0 \le \phi \text{ increasing and } 0 < \phi(0)\}$.

Proof: Choose $a > 0$, $b > 1$ such that $a|\xi| < |f(\xi)| < b|\xi|$ for $0 < |\xi| \le r$. Let x be a solution as in the assertion. We consider its local extrema $m_0 := 0$ and $m_j := z_j+1$, $j \in N$ (see Lemma 2). On the intervals $[m_j, m_j + 1/b]$ we have $|x| \ge |g_j|$ for the affine functions g_j with $g_j(m_j) = x(m_j)$ and $\dot{g}_j = -bx(m_j)$. This follows from $|\dot{x}(t)| = |f(x(t-1))| \le b|x(t-1)| \le b|x(m_j)|$ for $t \in (m_j, m_j+1]$ and from $b > 1$.

Let $t \ge 0$. For $t-1 \le u < v \le t$ we have

(5) $\quad \alpha|\int_u^v x(s)ds| \le 2\|T_1 x_t\|$.

Proof: $2\|T_1 x_t\| \ge |T_1 x_t(v-t) - T_1 x_t(u-t)| = \alpha|\int_{u-t}^{v-t} x_t(s')ds'| = \alpha|\int_u^v x(s)ds|$.

Case I: $t-1 \le m_j \le t$ for some $j \in N_0$.

Subcase 1: $\|x_t\| = |x(m_j)|$ and $t \le m_j + 1/2b$.

Then $\|T_1 x_t\| \ge |T_1 x_t(-1)| = |x(t)| \ge |g_j(t)| \ge |g_j(m_j + 1/2b)| = |x(m_j)|/2$.

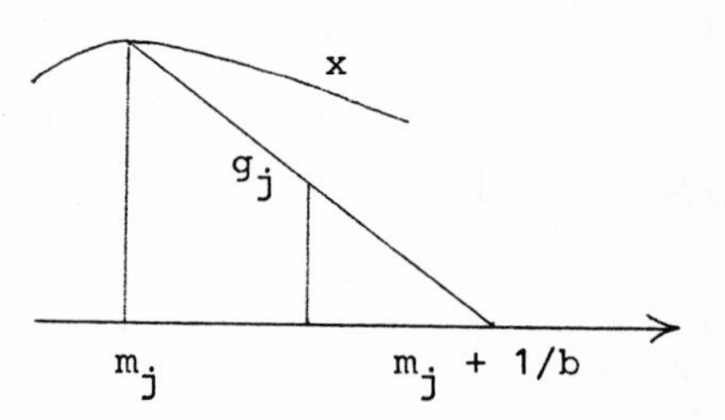

Subcase 2: $\|x_t\| = |x(m_j)|$ and $m_j + 1/2b < t$.

On $[m_j, m_j + 1/2b] \subset [t-1, t]$ we have $|x| \ge |g_j| \ge |x(m_j)|/2 = \|x_t\|/2 > 0$. (5) with $u = m_j$,

$v = m_j + 1/2b$ implies $2\|T_1x_t\| \geq \alpha\|x_t\|/4b$.

Subcase 3: $|x(m_j)| < \|x_t\|$.

$|x|$ increases on $[m_j-1,m_j]$, decreases on $[m_j,z_{j+1}]$ and increases on $[z_{j+1},z_{j+1}+1]$. This yields $z_{j+1} \leq t$ and $|x(t)| = \|x_t\|$, hence $\|T_1x_t\| \geq |x(t)| = \|x_t\|$.

Case II: $\forall j \in N_o$: $m_j \notin [t-1,t]$.

Then $-1 < t-2$, and x is monotone in $[t-1,t]$. Therefore $\|x_t\| = |x(t)|$ or $\|x_t\| = |x(t-1)|$.

Subcase 1: $\|x_t\| = |x(t)|$ implies $\|T_1x_t\| \geq \|x_t\|$.

Subcase 2: $\|x_t\| = |x(t-1)| > |x(t)|$ and $|x| \geq |x(t-1)|$ in $[t-1-1/2b, t-1]$.

For $s \in [t-1/2b, t]$ we infer $|\dot{x}(s)| = |f(x(s-1))| \geq a|x(s-1)| \geq$ $\geq a\|x_t\| > 0$, hence $|x(t) - x(t-1/2b)| \geq a\|x_t\|/2b$, therefore

(6) $\quad a\|x_t\|/4b \leq |x(t)| \leq \|T_1x_t\|$ or

(7) $\quad a\|x_t\|/4b \leq |x(t-1/2b)|$.

In case of (7) and $|x(t-1/2b)| \leq |x(t)|$ we obtain (6) once more.

In case of (7) and $|x(t-1/2b)| > |x(t)|$ the monotonicity of x in $[t-1,t]$ implies $|x| \geq |x(t-1/2b)|$ in $[t-1, t-1/2b]$. (7) and (5) yield $2\|T_1x_t\| \geq \alpha(1-1/2b)a\|x_t\|/4b$.

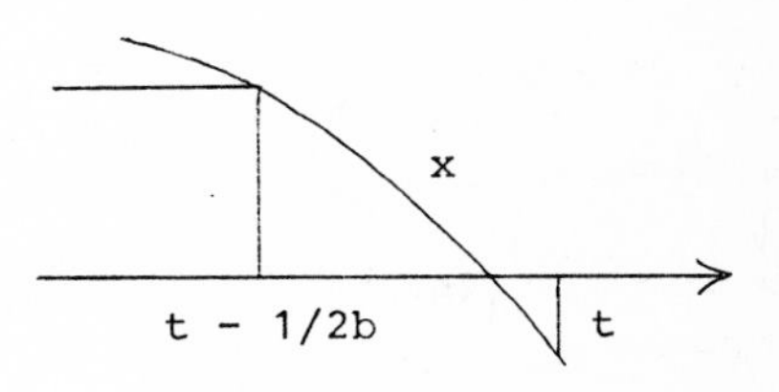

Subcase 3: $\|x_t\| = |x(t-1)| > |x(t)|$ and $|x(s)| < |x(t-1)|$ for some s in $[t-1-1/2b, t-1]$.

Lemma 2 (i) implies the existence of $j \in N_o$ with m_j in $[t-1-1/2b, t-1]$ and $|x(m_j)| \geq |x(t-1)|$. On $[t-1, t-1+1/4b] \subset [m_j, m_j+3/4b]$

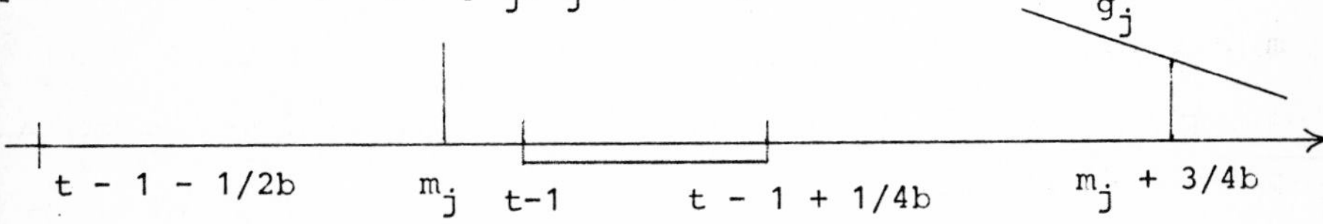

we obtain $|x| \geq |g_j| \geq |g_j(m_j + 3/4b)| = |x(m_j)|/4 \geq |x(t-1)|/4 =$ $= \|x_t\|/4$, and (5) yields $2\|T_1 x_t\| \geq \alpha\|x_t\|/16b$.

7. Attractor and ω-limit sets.

Theorem 2: Let a continuous function $f:R \to R$ be given which is differentiable at $\xi = 0$ and satisfies $\xi f(\xi) > 0$ for $\xi \neq 0$. Assume $f'(0) > \pi/2$ and $\inf f > -\infty$. Then there are constants $a > 0$, $r > 0$ such that for every slowly oscillating solution x of equation

(f) $$\dot{x}(t) = -f(x(t-1))$$

there is a number $s \geq 0$ with

$$x_t \in \{\phi \in S: a \leq \|\phi\| \leq r\} \quad \text{for all } t \geq s.$$

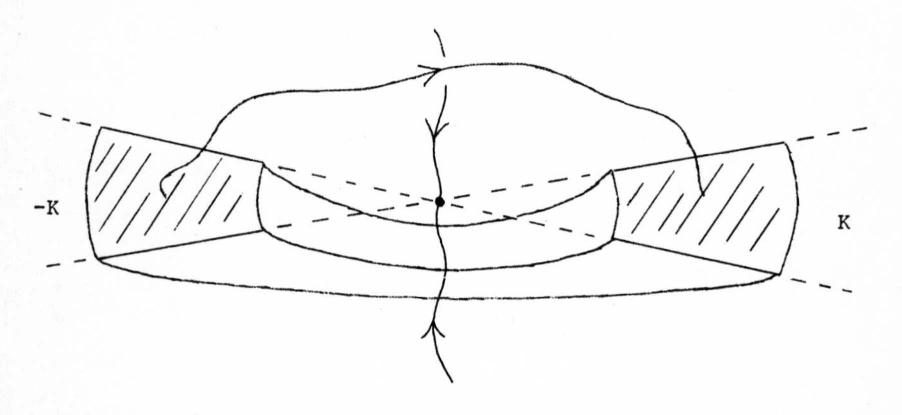

Proof: Set $r := \max\{r_1, r_2\}$. Define

$$S^f := \{\phi \in S: k\|\phi\| \leq \|T_1\phi\|\}$$

with $k = k(r,f)$ from Lemma 5. By Lemma 4 S^f satisfies $(1)_o$, and we obtain a constant $a > 0$ according to Theorem 1 (ii). Consider a solution x of (f) which has a sequence of zeros z_j with (Z). We have $x_{z_2+1} \in K$ and $|x(t)| \leq r$ for $t \geq z_4$, by Lemma 2 applied to the solution $-1 \leq t \to x(z_2 + 1 + t)$. Lemma 5 implies that the trajectory of the solution $\tilde{x}:-1 \leq t \to x(z_4 + 1 + t)$ is contained in S^f, and

Theorem 1 (ii) yields $a \leq \|\tilde{x}_t\| = \|x_{z_4 + 1 + t}\|$ in an unbounded interval in R^+.

For a bounded solution x of (f) the ω-limit set

$$\omega_x := \{\phi \in X: \exists (t_j)_{j\in N} \text{ in } R: \lim_{j\to\infty} t_j = \infty \text{ and } \lim_{j\to\infty} x_{t_j} = \phi\}$$

is non-empty. For every $\phi \in \omega_x$ there is a solution $y:R \to R$ of (f) with $y_o = \phi$ and $y_t \in \omega_x$ for all $t \in R$ [H, p. 82, Corollary 2.1].

Corollary 2:

(i) ω-limit sets of slowly oscillating solutions are contained in the set $\{\phi \in S: a \leq \|\phi\| \leq r\}$.

(ii) For every solution $y:R \to R$ with trajectory in the ω-limit set of a slowly oscillating solution the zeros of y form a family $(z_j)_{j\in Z}$ with property (Z). In particular: $\lim_{j\to -\infty} z_j = -\infty$.

Proof of (i): Use cl $S = S \cup \{0\}$ and Theorem 2.

Proof of (ii): It is enough to show the existence of $d' > 0$ such that for every $s \in R$ there is a sequence of zeros w_j, $j \in N$, of y in (s,∞) with $w_1 < s + d'$, $w_j+1 < w_{j+1}$, $0 < |y|$ in (w_j,w_{j+1}) for all j in N. Let $s \in R$ be given. $y_s \in S$ implies $0 \leq y_{s+t}$ or $y_{s+t} \leq 0$ for some $t \in [0,1]$ (Proposition (ii) applied to $-1 \leq v \to y(s+v)$).
By $0 < \|y_{s+t}\| \leq r$, by Lemma 2 (ii) and Remark 1 there is a sequence of zeros $w_j' \geq 0$, $j \in N$, of the solution $-1 \leq u \to y(s+t+u)$ with $w_1' < d$, $w_j' + 1 < w_{j+1}'$, $0 < |y(s+t+\cdot)|$ in (w_j',w_{j+1}') for all $j \in N$.
Set $d' := d + 1$.

Remark 2: Without boundedness assumptions on f we obtain: For every $r > 0$ there exists $a_r > 0$ such that for every slowly oscillating solution x bounded by r there is $s \in R$ with $a_r \leq \|x_t\|$ for all $t \geq s$.

8. Periodic solutions by Schauder's theorem.

Theorem 3 [N_1] : Let a continuous function $f:R \to R$ be given which is differentiable at $\xi = 0$ and satisfies $\xi f(\xi) > 0$ for $\xi \neq 0$. Assume $f'(0) > \pi/2$ and $\inf f > -\infty$. Then there exists a slowly oscillating periodic solution of

$$\text{(f)} \qquad \dot{x}(t) = -f(x(t-1)).$$

Proof: Solutions $x:[-1,\infty) \to R$ with $x_0 \in K$ and $x_0 \leq r_1$ satisfy $-r_2 \leq x \leq r_1$ (Lemma 2), and their trajectories lie in the set S^f (see proof of Theorem 2) which together with the projection P_0 fulfills the hypotheses of Theorem 1. We choose $\delta_p < r_1$ according to parts (i) and (iii) of Theorem 1 and define

$$D := \{\phi \in K: p^2\delta_p^2 \leq V(\phi) \text{ and } \phi \leq r_1\}.$$

Theorem 1 (iii) and Lemma 2 (iii) show that the operator $T:\phi \to x^{\phi}_{z_2+1}$ maps D into D. T is completely continuous.

D is homeomorphic to a closed bounded convex set: First, we have $D \neq \emptyset$ ($\phi \in K$ and $\delta_p \leq \|\phi\| \leq r_1$ imply $\phi \in S^f$, hence $p^2\delta_p^2 \leq p^2\|\phi\|^2 \leq$ $\leq V(\phi)$) and $V(\phi) > 0$ for all $\phi \in D$; in particular $0 \notin D$, see e. g. Lemma 1 (i). Set $\rho_\phi := \|\phi\| p\delta_p/\sqrt{V(\phi)}$ for $\phi \in D$. Then $\phi \in D$ iff $\phi \in K$ and $\rho_\phi \leq \|\phi\| \leq r_1$. We have $\rho_\phi < r_1$ (otherwise $r_1 \leq \rho_\phi =$ $= \|\phi\| p\delta_p/\sqrt{V(\phi)} < r_1\|\phi\| p/\sqrt{V(\phi)}$ by the choice $\delta_p < r_1$, contradiction to $\phi \in D \subset S^f$ and (V)). The map $\phi \to (1 + (\|\phi\| - \rho_\phi)/(r_1 - \rho_\phi))\phi/\|\phi\|$ is a homeomorphism of D onto the set $\{\phi \in K: 1 \leq \phi(0) \leq 2\}$.

The fixed point of T in D defines the periodic solution.

References:

A W. Alt: Some periodicity criteria for functional differential equations. Manuscripta math. 23, 295 - 318 (1978)

F.E. Browder: A further generalization of the Schauder fixed point theorem. Duke Math. J. 32, 575 - 578 (1965)

C S.N. Chow: Existence of periodic solutions of autonomous functional differential equations. J. Differential Equations 15, 350 - 378 (1974)

G R.B. Grafton: A periodicity theorem for autonomous functional differential equations. J. Differential Equations 6, 87 - 109 (1969)

H J.K. Hale: Theory of functional differential equations. Berlin-Heidelberg-New York: Springer 1977

HP J.K. Hale, C. Perello: The neighbourhood of a singular point for functional differential equations. Contributions to Differential Equations 3, 351 - 375 (1964)

G.S. Jones: The existence of periodic solutions of $f'(x) = -\alpha f(x-1)\{1 + f(x)\}$. J. Math. Anal. Appl. 5, 435 - 450 (1962)

J.C. Kaplan, J.A. Yorke: On the stability of a periodic solution of a differential delay equation. SIAM J. Math. Anal. 6, 268 - 282 (1975)

J.C. Kaplan, J.A. Yorke: On the nonlinear differential delay equation $x'(t) = -f(x(t),x(t-1))$. J. Differential Equations 23, 293 - 314 (1977)

N_1 R.D. Nussbaum: Periodic solutions of some nonlinear autonomous functional differential equations. Ann. Mat. Pura Appl. 101, 263 - 306 (1974)

N_2 R.D. Nussbaum: Periodic solutions of some nonlinear autonomous functional differential equations II. J. Differential Equations 14, 360 - 394 (1973)

Wa H.O. Walther: Über Ejektivität und periodische Lösungen bei autonomen Funktionaldifferentialgleichungen mit verteilter Verzögerung. Habilitationsschrift München 1977

Wr E.M. Wright: A non-linear differential-difference equation. J. Reine Angew. Math. 194, 66 - 87 (1955)

Address: H.O. Walther, Mathematisches Institut der Universität, Theresienstr. 39, D 8000 München 2, Federal Republic of Germany

Vol. 580: C. Castaing and M. Valadier, Convex Analysis and Measurable Multifunctions. VIII, 278 pages. 1977.

Vol. 581: Séminaire de Probabilités XI, Université de Strasbourg. Proceedings 1975/1976. Edité par C. Dellacherie, P. A. Meyer et M. Weil. VI, 574 pages. 1977.

Vol. 582: J. M. G. Fell, Induced Representations and Banach *-Algebraic Bundles. IV, 349 pages. 1977.

Vol. 583: W. Hirsch, C. C. Pugh and M. Shub, Invariant Manifolds. IV, 149 pages. 1977.

Vol. 584: C. Brezinski, Accélération de la Convergence en Analyse Numérique. IV, 313 pages. 1977.

Vol. 585: T. A. Springer, Invariant Theory. VI, 112 pages. 1977.

Vol. 586: Séminaire d'Algèbre Paul Dubreil, Paris 1975–1976 (29ème Année). Edited by M. P. Malliavin. VI, 188 pages. 1977.

Vol. 587: Non-Commutative Harmonic Analysis. Proceedings 1976. Edited by J. Carmona and M. Vergne. IV, 240 pages. 1977.

Vol. 588: P. Molino, Théorie des G-Structures: Le Problème d'Equivalence. VI, 163 pages. 1977.

Vol. 589: Cohomologie l-adique et Fonctions L. Séminaire de Géométrie Algébrique du Bois-Marie 1965–66, SGA 5. Edité par L. Illusie. XII, 484 pages. 1977.

Vol. 590: H. Matsumoto, Analyse Harmonique dans les Systèmes de Tits Bornologiques de Type Affine. IV, 219 pages. 1977.

Vol. 591: G. A. Anderson, Surgery with Coefficients. VIII, 157 pages. 1977.

Vol. 592: D. Voigt, Induzierte Darstellungen in der Theorie der endlichen, algebraischen Gruppen. V, 413 Seiten. 1977.

Vol. 593: K. Barbey and H. König, Abstract Analytic Function Theory and Hardy Algebras. VIII, 260 pages. 1977.

Vol. 594: Singular Perturbations and Boundary Layer Theory, Lyon 1976. Edited by C. M. Brauner, B. Gay, and J. Mathieu. VIII, 539 pages. 1977.

Vol. 595: W. Hazod, Stetige Faltungshalbgruppen von Wahrscheinlichkeitsmaßen und erzeugende Distributionen. XIII, 157 Seiten. 1977.

Vol. 596: K. Deimling, Ordinary Differential Equations in Banach Spaces. VI, 137 pages. 1977.

Vol. 597: Geometry and Topology, Rio de Janeiro, July 1976. Proceedings. Edited by J. Palis and M. do Carmo. VI, 866 pages. 1977.

Vol. 598: J. Hoffmann-Jørgensen, T. M. Liggett et J. Neveu, Ecole d'Eté de Probabilités de Saint-Flour VI – 1976. Edité par P.-L. Hennequin. XII, 447 pages. 1977.

Vol. 599: Complex Analysis, Kentucky 1976. Proceedings. Edited by J. D. Buckholtz and T. J. Suffridge. X, 159 pages. 1977.

Vol. 600: W. Stoll, Value Distribution on Parabolic Spaces. VIII, 216 pages. 1977.

Vol. 601: Modular Functions of one Variable V, Bonn 1976. Proceedings. Edited by J.-P. Serre and D. B. Zagier. VI, 294 pages. 1977.

Vol. 602: J. P. Brezin, Harmonic Analysis on Compact Solvmanifolds. VIII, 179 pages. 1977.

Vol. 603: B. Moishezon, Complex Surfaces and Connected Sums of Complex Projective Planes. IV, 234 pages. 1977.

Vol. 604: Banach Spaces of Analytic Functions, Kent, Ohio 1976. Proceedings. Edited by J. Baker, C. Cleaver and Joseph Diestel. VI, 141 pages. 1977.

Vol. 605: Sario et al., Classification Theory of Riemannian Manifolds. XX, 498 pages. 1977.

Vol. 606: Mathematical Aspects of Finite Element Methods. Proceedings 1975. Edited by I. Galligani and E. Magenes. VI, 362 pages. 1977.

Vol. 607: M. Métivier, Reelle und Vektorwertige Quasimartingale und die Theorie der Stochastischen Integration. X, 310 Seiten. 1977.

Vol. 608: Bigard et al., Groupes et Anneaux Réticulés. XIV, 334 pages. 1977.

Vol. 609: General Topology and Its Relations to Modern Analysis and Algebra IV. Proceedings 1976. Edited by J. Novák. XVIII, 225 pages. 1977.

Vol. 610: G. Jensen, Higher Order Contact of Submanifolds of Homogeneous Spaces. XII, 154 pages. 1977.

Vol. 611: M. Makkai and G. E. Reyes, First Order Categorical Logic. VIII, 301 pages. 1977.

Vol. 612: E. M. Kleinberg, Infinitary Combinatorics and the Axiom of Determinateness. VIII, 150 pages. 1977.

Vol. 613: E. Behrends et al., L^p-Structure in Real Banach Spaces. X, 108 pages. 1977.

Vol. 614: H. Yanagihara, Theory of Hopf Algebras Attached to Group Schemes. VIII, 308 pages. 1977.

Vol. 615: Turbulence Seminar, Proceedings 1976/77. Edited by P. Bernard and T. Ratiu. VI, 155 pages. 1977.

Vol. 616: Abelian Group Theory, 2nd New Mexico State University Conference, 1976. Proceedings. Edited by D. Arnold, R. Hunter and E. Walker. X, 423 pages. 1977.

Vol. 617: K. J. Devlin, The Axiom of Constructibility: A Guide for the Mathematician. VIII, 96 pages. 1977.

Vol. 618: I. I. Hirschman, Jr. and D. E. Hughes, Extreme Eigen Values of Toeplitz Operators. VI, 145 pages. 1977.

Vol. 619: Set Theory and Hierarchy Theory V, Bierutowice 1976. Edited by A. Lachlan, M. Srebrny, and A. Zarach. VIII, 358 pages. 1977.

Vol. 620: H. Popp, Moduli Theory and Classification Theory of Algebraic Varieties. VIII, 189 pages. 1977.

Vol. 621: Kauffman et al., The Deficiency Index Problem. VI, 112 pages. 1977.

Vol. 622: Combinatorial Mathematics V, Melbourne 1976. Proceedings. Edited by C. Little. VIII, 213 pages. 1977.

Vol. 623: I. Erdelyi and R. Lange, Spectral Decompositions on Banach Spaces. VIII, 122 pages. 1977.

Vol. 624: Y. Guivarc'h et al., Marches Aléatoires sur les Groupes de Lie. VIII, 292 pages. 1977.

Vol. 625: J. P. Alexander et al., Odd Order Group Actions and Witt Classification of Innerproducts. IV, 202 pages. 1977.

Vol. 626: Number Theory Day, New York 1976. Proceedings. Edited by M. B. Nathanson. VI, 241 pages. 1977.

Vol. 627: Modular Functions of One Variable VI, Bonn 1976. Proceedings. Edited by J.-P. Serre and D. B. Zagier. VI, 339 pages. 1977.

Vol. 628: H. J. Baues, Obstruction Theory on the Homotopy Classification of Maps. XII, 387 pages. 1977.

Vol. 629: W. A. Coppel, Dichotomies in Stability Theory. VI, 98 pages. 1978.

Vol. 630: Numerical Analysis, Proceedings, Biennial Conference, Dundee 1977. Edited by G. A. Watson. XII, 199 pages. 1978.

Vol. 631: Numerical Treatment of Differential Equations. Proceedings 1976. Edited by R. Bulirsch, R. D. Grigorieff, and J. Schröder. X, 219 pages. 1978.

Vol. 632: J.-F. Boutot, Schéma de Picard Local. X, 165 pages. 1978.

Vol. 633: N. R. Coleff and M. E. Herrera, Les Courants Résiduels Associés à une Forme Méromorphe. X, 211 pages. 1978.

Vol. 634: H. Kurke et al., Die Approximationseigenschaft lokaler Ringe. IV, 204 Seiten. 1978.

Vol. 635: T. Y. Lam, Serre's Conjecture. XVI, 227 pages. 1978.

Vol. 636: Journées de Statistique des Processus Stochastiques, Grenoble 1977, Proceedings. Edité par Didier Dacunha-Castelle et Bernard Van Cutsem. VII, 202 pages. 1978.

Vol. 637: W. B. Jurkat, Meromorphe Differentialgleichungen. VII, 194 Seiten. 1978.

Vol. 638: P. Shanahan, The Atiyah-Singer Index Theorem, An Introduction. V, 224 pages. 1978.

Vol. 639: N. Adasch et al., Topological Vector Spaces. V, 125 pages. 1978.

Vol. 640: J. L. Dupont, Curvature and Characteristic Classes. X, 175 pages. 1978.

Vol. 641: Séminaire d'Algèbre Paul Dubreil, Proceedings Paris 1976–1977. Edité par M. P. Malliavin. IV, 367 pages. 1978.

Vol. 642: Theory and Applications of Graphs, Proceedings, Michigan 1976. Edited by Y. Alavi and D. R. Lick. XIV, 635 pages. 1978.

Vol. 643: M. Davis, Multiaxial Actions on Manifolds. VI, 141 pages. 1978.

Vol. 644: Vector Space Measures and Applications I, Proceedings 1977. Edited by R. M. Aron and S. Dineen. VIII, 451 pages. 1978.

Vol. 645: Vector Space Measures and Applications II, Proceedings 1977. Edited by R. M. Aron and S. Dineen. VIII, 218 pages. 1978.

Vol. 646: O. Tammi, Extremum Problems for Bounded Univalent Functions. VIII, 313 pages. 1978.

Vol. 647: L. J. Ratliff, Jr., Chain Conjectures in Ring Theory. VIII, 133 pages. 1978.

Vol. 648: Nonlinear Partial Differential Equations and Applications, Proceedings, Indiana 1976–1977. Edited by J. M. Chadam. VI, 206 pages. 1978.

Vol. 649: Séminaire de Probabilités XII, Proceedings, Strasbourg, 1976–1977. Edité par C. Dellacherie, P. A. Meyer et M. Weil. VIII, 805 pages. 1978.

Vol. 650: C*-Algebras and Applications to Physics. Proceedings 1977. Edited by H. Araki and R. V. Kadison. V, 192 pages. 1978.

Vol. 651: P. W. Michor, Functors and Categories of Banach Spaces. VI, 99 pages. 1978.

Vol. 652: Differential Topology, Foliations and Gelfand-Fuks-Cohomology, Proceedings 1976. Edited by P. A. Schweitzer. XIV, 252 pages. 1978.

Vol. 653: Locally Interacting Systems and Their Application in Biology. Proceedings, 1976. Edited by R. L. Dobrushin, V. I. Kryukov and A. L. Toom. XI, 202 pages. 1978.

Vol. 654: J. P. Buhler, Icosahedral Golois Representations. III, 143 pages. 1978.

Vol. 655: R. Baeza, Quadratic Forms Over Semilocal Rings. VI, 199 pages. 1978.

Vol. 656: Probability Theory on Vector Spaces. Proceedings, 1977. Edited by A. Weron. VIII, 274 pages. 1978.

Vol. 657: Geometric Applications of Homotopy Theory I, Proceedings 1977. Edited by M. G. Barratt and M. E. Mahowald. VIII, 459 pages. 1978.

Vol. 658: Geometric Applications of Homotopy Theory II, Proceedings 1977. Edited by M. G. Barratt and M. E. Mahowald. VIII, 487 pages. 1978.

Vol. 659: Bruckner, Differentiation of Real Functions. X, 247 pages. 1978.

Vol. 660: Equations aux Dérivée Partielles. Proceedings, 1977. Edité par Pham The Lai. VI, 216 pages. 1978.

Vol. 661: P. T. Johnstone, R. Paré, R. D. Rosebrugh, D. Schumacher, R. J. Wood, and G. C. Wraith, Indexed Categories and Their Applications. VII, 260 pages. 1978.

Vol. 662: Akin, The Metric Theory of Banach Manifolds. XIX, 306 pages. 1978.

Vol. 663: J. F. Berglund, H. D. Junghenn, P. Milnes, Compact Right Topological Semigroups and Generalizations of Almost Periodicity. X, 243 pages. 1978.

Vol. 664: Algebraic and Geometric Topology, Proceedings, 1977. Edited by K. C. Millett. XI, 240 pages. 1978.

Vol. 665: Journées d'Analyse Non Linéaire. Proceedings, 1977. Edité par P. Bénilan et J. Robert. VIII, 256 pages. 1978.

Vol. 666: B. Beauzamy, Espaces d'Interpolation Réels: Topologie et Géometrie. X, 104 pages. 1978.

Vol. 667: J. Gilewicz, Approximants de Padé. XIV, 511 pages. 1978.

Vol. 668: The Structure of Attractors in Dynamical Systems. Proceedings, 1977. Edited by J. C. Martin, N. G. Markley and W. Perrizo. VI, 264 pages. 1978.

Vol. 669: Higher Set Theory. Proceedings, 1977. Edited by G. H. Müller and D. S. Scott. XII, 476 pages. 1978.

Vol. 670: Fonctions de Plusieurs Variables Complexes III, Proceedings, 1977. Edité par F. Norguet. XII, 394 pages. 1978.

Vol. 671: R. T. Smythe and J. C. Wierman, First-Passage Perculation on the Square Lattice. VIII, 196 pages. 1978.

Vol. 672: R. L. Taylor, Stochastic Convergence of Weighted Sums of Random Elements in Linear Spaces. VII, 216 pages. 1978.

Vol. 673: Algebraic Topology, Proceedings 1977. Edited by P. Hoffman, R. Piccinini and D. Sjerve. VI, 278 pages. 1978.

Vol. 674: Z. Fiedorowicz and S. Priddy, Homology of Classical Groups Over Finite Fields and Their Associated Infinite Loop Spaces. VI, 434 pages. 1978.

Vol. 675: J. Galambos and S. Kotz, Characterizations of Probability Distributions. VIII, 169 pages. 1978.

Vol. 676: Differential Geometrical Methods in Mathematical Physics II, Proceedings, 1977. Edited by K. Bleuler, H. R. Petry and A. Reetz. VI, 626 pages. 1978.

Vol. 677: Séminaire Bourbaki, vol. 1976/77, Exposés 489–506. IV, 264 pages. 1978.

Vol. 678: D. Dacunha-Castelle, H. Heyer et B. Roynette. Ecole d'Eté de Probabilités de Saint-Flour. VII-1977. Edité par P. L. Hennequin. IX, 379 pages. 1978.

Vol. 679: Numerical Treatment of Differential Equations in Applications, Proceedings, 1977. Edited by R. Ansorge and W. Törnig. IX, 163 pages. 1978.

Vol. 680: Mathematical Control Theory, Proceedings, 1977. Edited by W. A. Coppel. IX, 257 pages. 1978.

Vol. 681: Séminaire de Théorie du Potentiel Paris, No. 3, Directeurs: M. Brelot, G. Choquet et J. Deny. Rédacteurs: F. Hirsch et G. Mokobodzki. VII, 294 pages. 1978.

Vol. 682: G. D. James, The Representation Theory of the Symmetric Groups. V, 156 pages. 1978.

Vol. 683: Variétés Analytiques Compactes, Proceedings, 1977. Edité par Y. Hervier et A. Hirschowitz. V, 248 pages. 1978.

Vol. 684: E. E. Rosinger, Distributions and Nonlinear Partial Differential Equations. XI, 146 pages. 1978.

Vol. 685: Knot Theory, Proceedings, 1977. Edited by J. C. Hausmann. VII, 311 pages. 1978.

Vol. 686: Combinatorial Mathematics, Proceedings, 1977. Edited by D. A. Holton and J. Seberry. IX, 353 pages. 1978.

Vol. 687: Algebraic Geometry, Proceedings, 1977. Edited by L. D. Olson. V, 244 pages. 1978.

Vol. 688: J. Dydak and J. Segal, Shape Theory. VI, 150 pages. 1978.

Vol. 689: Cabal Seminar 76–77, Proceedings, 1976–77. Edited by A.S. Kechris and Y. N. Moschovakis. V, 282 pages. 1978.

Vol. 690: W. J. J. Rey, Robust Statistical Methods. VI, 128 pages. 1978.

Vol. 691: G. Viennot, Algèbres de Lie Libres et Monoïdes Libres. III, 124 pages. 1978.

Vol. 692: T. Husain and S. M. Khaleelulla, Barrelledness in Topological and Ordered Vector Spaces. IX, 258 pages. 1978.

Vol. 693: Hilbert Space Operators, Proceedings, 1977. Edited by J. M. Bachar Jr. and D. W. Hadwin. VIII, 184 pages. 1978.

Vol. 694: Séminaire Pierre Lelong – Henri Skoda (Analyse) Année 1976/77. VII, 334 pages. 1978.

Vol. 695: Measure Theory Applications to Stochastic Analysis, Proceedings, 1977. Edited by G. Kallianpur and D. Kölzow. XII, 261 pages. 1978.

Vol. 696: P. J. Feinsilver, Special Functions, Probability Semigroups, and Hamiltonian Flows. VI, 112 pages. 1978.

Vol. 697: Topics in Algebra, Proceedings, 1978. Edited by M. F. Newman. XI, 229 pages. 1978.

Vol. 698: E. Grosswald, Bessel Polynomials. XIV, 182 pages. 1978.

Vol. 699: R. E. Greene and H.-H. Wu, Function Theory on Manifolds Which Possess a Pole. III, 215 pages. 1979.

Vol. 700: Module Theory, Proceedings, 1977. Edited by C. Faith and S. Wiegand. X, 239 pages. 1979.

Vol. 701: Functional Analysis Methods in Numerical Analysis, Proceedings, 1977. Edited by M. Zuhair Nashed. VII, 333 pages. 1979.

Vol. 702: Yuri N. Bibikov, Local Theory of Nonlinear Analytic Ordinary Differential Equations. IX, 147 pages. 1979.

Vol. 703: Equadiff IV, Proceedings, 1977. Edited by J. Fábera. XIX, 441 pages. 1979.

Vol. 704: Computing Methods in Applied Sciences and Engineering, 1977, I. Proceedings, 1977. Edited by R. Glowinski and J. L. Lions. VI, 391 pages. 1979.

Vol. 705: O. Forster und K. Knorr, Konstruktion verseller Familien kompakter komplexer Räume. VII, 141 Seiten. 1979.

Vol. 706: Probability Measures on Groups, Proceedings, 1978. Edited by H. Heyer. XIII, 348 pages. 1979.

Vol. 707: R. Zielke, Discontinuous Čebyšev Systems. VI, 111 pages. 1979.

Vol. 708: J. P. Jouanolou, Equations de Pfaff algébriques. V, 255 pages. 1979.

Vol. 709: Probability in Banach Spaces II. Proceedings, 1978. Edited by A. Beck. V, 205 pages. 1979.

Vol. 710: Séminaire Bourbaki vol. 1977/78, Exposés 507-524. IV, 328 pages. 1979.

Vol. 711: Asymptotic Analysis. Edited by F. Verhulst. V, 240 pages. 1979.

Vol. 712: Equations Différentielles et Systèmes de Pfaff dans le Champ Complexe. Edité par R. Gérard et J.-P. Ramis. V, 364 pages. 1979.

Vol. 713: Séminaire de Théorie du Potentiel, Paris No. 4. Edité par F. Hirsch et G. Mokobodzki. VII, 281 pages. 1979.

Vol. 714: J. Jacod, Calcul Stochastique et Problèmes de Martingales. X, 539 pages. 1979.

Vol. 715: Inder Bir S. Passi, Group Rings and Their Augmentation Ideals. VI, 137 pages. 1979.

Vol. 716: M. A. Scheunert, The Theory of Lie Superalgebras. X, 271 pages. 1979.

Vol. 717: Grosser, Bidualräume und Vervollständigungen von Banachmoduln. III, 209 pages. 1979.

Vol. 718: J. Ferrante and C. W. Rackoff, The Computational Complexity of Logical Theories. X, 243 pages. 1979.

Vol. 719: Categorial Topology, Proceedings, 1978. Edited by H. Herrlich and G. Preuß. XII, 420 pages. 1979.

Vol. 720: E. Dubinsky, The Structure of Nuclear Fréchet Spaces. V, 187 pages. 1979.

Vol. 721: Séminaire de Probabilités XIII. Proceedings, Strasbourg, 1977/78. Edité par C. Dellacherie, P. A. Meyer et M. Weil. VII, 647 pages. 1979.

Vol. 722: Topology of Low-Dimensional Manifolds. Proceedings, 1977. Edited by R. Fenn. VI, 154 pages. 1979.

Vol. 723: W. Brandal, Commutative Rings whose Finitely Generated Modules Decompose. II, 116 pages. 1979.

Vol. 724: D. Griffeath, Additive and Cancellative Interacting Particle Systems. V, 108 pages. 1979.

Vol. 725: Algèbres d'Opérateurs. Proceedings, 1978. Edité par P. de la Harpe. VII, 309 pages. 1979.

Vol. 726: Y.-C. Wong, Schwartz Spaces, Nuclear Spaces and Tensor Products. VI, 418 pages. 1979.

Vol. 727: Y. Saito, Spectral Representations for Schrödinger Operators With Long-Range Potentials. V, 149 pages. 1979.

Vol. 728: Non-Commutative Harmonic Analysis. Proceedings, 1978. Edited by J. Carmona and M. Vergne. V, 244 pages. 1979.

Vol. 729: Ergodic Theory. Proceedings, 1978. Edited by M. Denker and K. Jacobs. XII, 209 pages. 1979.

Vol. 730: Functional Differential Equations and Approximation of Fixed Points. Proceedings, 1978. Edited by H.-O. Peitgen and H.-O. Walther. XV, 503 pages. 1979.

This series reports new developments in mathematical research and tea
ing – quickly, informally and at a high level. The type of material conside
for publication includes:

1. Preliminary drafts of original papers and monographs
2. Lectures on a new field or presentations of a new angle in a classical f
3. Seminar work-outs
4. Reports of meetings, provided they are
   a) of exceptional interest and
   b) devoted to a single topic.

Texts which are out of print but still in demand may also be conside
if they fall within these categories.

The timeliness of a manuscript is more important than its form, wh
may be unfinished or tentative. Thus, in some instances, proofs may
merely outlined and results presented which have been or will later
published elsewhere. If possible, a subject index should be incluc
Publication of Lecture Notes is intended as a service to the internatic
mathematical community, in that a commercial publisher, Springer-V
lag, can offer a wide distribution of documents which would otherw
have a restricted readership. Once published and copyrighted, they
be documented in the scientific literature.

Manuscripts

Manuscripts should be no less than 100 and preferably no more than 500 pages in length.
They are reproduced by a photographic process and therefore must be typed with extreme care. Syn
not on the typewriter should be inserted by hand in indelible black ink. Corrections to the type
should be made by pasting in the new text or painting out errors with white correction fluid. Authors re
75 free copies and are free to use the material in other publications. The typescript is reduced sligh
size during reproduction; best results will not be obtained unless the text on any one page is kept v
the overall limit of 18 x 26.5 cm (7 x 10½ inches). On request, the publisher will supply special paper
the typing area outlined.

Manuscripts should be sent to Prof. A. Dold, Mathematisches Institut der Universität Heidelberg, Im Ne
heimer Feld 288, 6900 Heidelberg/Germany, Prof. B. Eckmann, Eidgenössische Technische Hochsc
CH-8092 Zürich/Switzerland, or directly to Springer-Verlag Heidelberg.

Springer-Verlag, Heidelberger Platz 3, D-1000 Berlin 33
Springer-Verlag, Neuenheimer Landstraße 28–30, D-6900 Heidelberg 1
Springer-Verlag, 175 Fifth Avenue, New York, NY 10010/USA

ISBN 3-540-**09518**-7
ISBN 0-387-**09518**-7